ANDERSONIAN LIBRARY
WITHDRAWN
FROM
LIBRARY
STOCK
UNIVERSITY OF STRATHCLYDE

# MECHANICAL RELIABILITY

**Macmillan title of related interest**

David J. Smith, *Reliability and Maintainability in Perspective (Second Edition)*

# MECHANICAL RELIABILITY

A. D. S. CARTER

Second Edition

First edition 1972
Second edition 1986

Published by
MACMILLAN EDUCATION LTD
Houndmills, Basingstoke, Hampshire RG21 2XS
and London
Companies and representatives
throughout the world

Typeset by TecSet Ltd,
Sutton, Surrey
Printed in Hong Kong

British Library Cataloguing in Publication Data
Carter, A. D. S.
Mechanical reliability.–2nd ed.
1. Reliability (Engineering)
I. Title
620′.00452 TS173

ISBN 0–333–40586–2
ISBN 0–333–40587–0 Pbk

*TO*
*HAZEL CARTER*
*IN MEMORIAM*

# Contents

# Preface to the first edition

My attention was first drawn to problems of mechanical reliability in the early sixties by my wife's utter disgust at the behaviour of items of everyday use which a simple engineering assessment soon showed to be woefully inadequate. This led to a more professional appraisal of the situation and of some of the problems involved. Although there was a great deal of information on the reliability of electrical and electronic equipment and the accompanying mathematical theory, very little had been written from the point of view of the mechanical engineer. The essential difference between mechanical and electrical behaviour and the much wider issues involved in mechanical reliability seemed to be ignored in standard texts, though a number of specialist studies were available. The need for a balanced integrated account of the subject was obvious even at that stage. From everyday items my interest naturally turned to Service equipment. As a result of these studies an attempt was made to introduce some of the concepts of mechanical reliability, albeit on a small scale, into some of my specialist courses as long ago as 1965. In 1970 I was able to introduce a course of lectures into the first degree at RMCS. In general I have found students very keen to come to grips with the subject, which they find interesting, stimulating, and challenging, and also of real practical relevance and importance. This book has been based on my studies of the subject, and the experience gained in teaching over the last few years.

My knowledge and approach to the subject has been influenced by many friends and colleagues both in industry and the Services, and in the teaching world. In addition to the source material quoted in the references I must have unwittingly drawn on the experience of many who I cannot acknowledge in detail. In particular I am especially grateful for all the helpful comments I received on my original notes. It would be invidious to name a few out of the many, but special mention must be made of Frank Nixon for many helpful discussions and encouragement to publish this work. Finally, I would acknowledge detail assistance I have received from many colleagues including Mr A. J. Laurie in preparing many of the figures and Mr D. D. Appleford in acting as a willing guinea pig for most of the exercises.

A.D.S.C.

(*Shrivenham, 1972*)

# Preface to the second edition

A lot of water has flowed under the bridge since the first edition of this book was published, although to those of us directly concerned progress has often been disappointingly slow. However, ideas have developed significantly, and the time does seem just about right for a full re-appraisal of the subject matter presented in the earlier edition. In many ways this new edition is a complete rewrite of the earlier version, though the approach is exactly the same. The general format has been retained, and the exercises are still regarded as an integral part of the treatment, so the author's preferred solution (which, as typical of practical engineering problems, should never be regarded as the only one) is once again always given in full.

Throughout the book a number of detail changes have been made to improve clarity on the basis of past teaching and lecturing experience, but the significant changes can be summarised as follows. Apart from rather more emphasis on definitions, which teaching experience has shown to be necessary, the opening chapters remain essentially unchanged. The first substantive difference occurs in the treatment of random failures, which formed a large part of the first edition. This has been extended to include loading roughness. It is an important factor that had not been properly evaluated at the time the first edition was written, and no modern treatment of mechanical reliability would be complete without it. This leads to an appraisal of intrinsic reliability and similar concepts. The most significant difference between the two editions would, however, be the treatment of wear-out failures. Treatment in the first edition was superficial, to say the least, and since this accounts for the vast majority of mechanical failures I have given it a much greater coverage. In particular fatigue, which is probably the most important mechanical failure mechanism having no counterpart in the electronic world, is studied in detail. The treatment of maintenance has had to be revised to line up with the new approach to wear-out failures, and I believe is thus much improved. The chapter on design required corresponding revision, though the general flavour remains much the same. In this chapter I have devoted rather more space to estimating factors of safety and to design for fatigue, as well as giving more specific examples of FMEA and FTA techniques than in the first edition. The chapter on development follows much more along the lines of the early edition, but with the addition of reliability growth evaluation techniques, which were in their infancy when the first edition was written and not then included. The chapter on management has been extended, and rounds off the treatment of the subject as before.

In one respect this edition remains exactly the same as the first: it is a book written about mechanical reliability for mechanical engineers; and although mathematical rigour has been called upon when necessary, an essentially practical engineering approach again has been followed throughout.

A.D.S.C.
(*Oxford, 1985*)

# Notation

| | |
|---|---|
| $A$ | availability, acquisition cost, constant used in the SIL equation |
| $a$ | distance or length |
| $B$ | constants used in cumulative costs and the SIL equations |
| $B_x$ | life to $x$ per cent cumulative failures |
| $b$ | distance or length |
| $C$ | cost, constant in SIL equation |
| $C_c$ | cumulative cost |
| $\overline{C}_c$ | mean cumulative cost |
| $C_M$ | maintenance cost |
| $C_m$ | mean cumulative maintenance cost |
| $C_0$ | basic running costs |
| $C_R$ | cost of equipment to achieve reliability $R$ |
| $C_s$ | scheduled maintenance cost |
| $C_u$ | unscheduled maintenance cost |
| $c$ | constant defined in various contexts |
| $D$ | durability |
| $d$ | diameter, damage due to fatigue |
| $E$ | Young's modulus |
| $E(s)$ | probability density function of endurance (fatigue limit) |
| $e$ | exponential function |
| $F$ | probability of failure, that is, cumulative failures |
| $F(x)$ | cumulative failures in terms of variable $x$ |
| $f$ | frequency of vibration |
| $f(x)$ | probability density function of failure in terms of variable $x$ |
| $g$ | acceleration due to gravity |
| $h$ | small time interval |
| $h_L$ | constant value of load probability density function (rectangular distribution) |
| $h_s$ | constant value of strength probability density function (rectangular distribution) |
| $I$ | initial cost |
| $i$ | order number of successive events |

| | |
|---|---|
| $j$ | order number of successive events |
| $K$ | cost ratio of unscheduled to scheduled maintenance |
| $K_{1,2,3}$ | constants defined in text |
| $k$ | constants defined in text |
| $L$ | load |
| $\overline{L}$ | mean load or sometimes nominal load |
| $L_s$ | load causing failure in single application after fatigue damage |
| $L_i$ | successive loads in fatigue load cycle |
| $L(s)$ | probability density function of occurrence of load |
| $M$ | maintainability |
| $M_x$ | maintainability for specified value of variate $x$ |
| MTBF | mean time between failures |
| MTBO | mean time between overhauls |
| MTR | mean time to repair |
| MTTF | mean time to failure |
| $m$ | mean time between failures, slope of linear curves |
| $N, n$ | number (of components, cycles, etc.) |
| $N_i$ | number of fatigue cycles to failure at loads $L_i$ |
| $N_s$ | number of fatigue cycles to failure at load $L_s$ |
| $N(s)$ | function giving fatigue $s$–$N$ curve |
| $n_i$ | number of cycles at load $L_i$ |
| $P, p$ | probability |
| $q$ | probability |
| $R$ | probability of success = reliability |
| $R(t)$ | reliability expressed as a function of time |
| $r(t)$ | repair limit cost |
| $S$ | strength |
| $\overline{S}$ | mean strength |
| $S(s)$ | probability density function of occurrence of strength |
| $s$ | stress |
| $s_f$ | fatigue stress of an individual item |
| $T$ | time for, or between, specific events |
| $t$ | time as a general variable |
| $t_c$ | locating constant in modified Weibull distribution |
| $t_0$ | locating constant in Weibull and other distributions |
| $V$ | volume |
| $X, Y$ | Cartesian coordinates |
| $x$ | arbitrary variable |
| $\alpha$ | log-normal dispersion, Duane growth parameter |
| $\beta$ | Weibull shaping parameter |
| $\Gamma$ | Gamma function |
| $\Delta x$ | small increment in any variable $x$ |
| $\eta$ | Weibull scaling parameter (characteristic life) |
| $\lambda$ | failure rate |

| | |
|---|---|
| $\lambda(t)$ | failure rate expressed as a function of time |
| $\rho$ | density |
| $\sigma$ | standard deviation |
| $\tau$ | time |

*Suffixes*

| | |
|---|---|
| 1, 2, 3 . . . | successive values of series of events, items, components, etc. |
| L | refers to load |
| 0 | a particular value of |
| S | refers to strength |
| s | scheduled |
| u | unscheduled |
| $n$ | a combination of $n$ items |

# Glossary

*Availability*. The probability that an item, at any instant in time, will be available.

*Available*. The state of an item such that it can perform its function under stated conditions of use and maintenance in the required location.

*B-percentile Life*. The length of time at which a stated proportion (*B* per cent) of a sample of items has failed.

*Conditional Failure Rate*. A more precise description for failure rate – see Failure Rate.

*Cumulative Failure Rate*. Total failures divided by total equipment operating time.

*Defect*. Any non-conformance of an item with its specifications and drawings.

*Durability*. The ability to resist the adverse effects of environment, use, and maintenance with the progress of time.

*Early Life Failures*. Failures which occur in the weaker members of a population during the early life period.

*Failure*. Any defect which results in the inability of the item to perform its specified functions.

*Failure Rate*. The rate at which failures arise in a given non-maintained population expressed as a proportion of the survivors. It is defined mathematically by equation (2.5).

*Failure-inducing Load*. The particular load, possibly of many, that causes failures.

*FMEA*. Failure Modes and Effects Analysis.

*FTA*. Fault Tree Analysis.

*Hazard* or *Hazard Rate*. An alternative phraseology for failure rate.

*Inherent Availability*. Availability based on active repair time.

*Instantaneous Failure Rate*. A more precise phraseology for failure rate – see Failure Rate.

*Maintainability*. The ability to retain an item in, or restore it to, specification under the stated conditions of maintenance. It may be measured as the probability to achieve the objective within a stated time period, or as the time to achieve the objective for a given proportion of the population.

*MART*. Mean Active Repair Time.

*MDT*. Mean Down Time.

*MTBD*. Mean Time Between Defects.

*MTBF*. Mean Time Between Failures. [*Note* that some variants, such as MDBF (Mean Distance Between Failures) etc. are often used.]

*MTTF*. Mean Time To Failure.

*MTTR*. Mean Time To Repair.

*Occurrence of Failures*. The rate at which failures arise in a changing (for example, maintained) population.

*On Condition Maintenance*. Maintenance carried out before an item fails, but only when its condition, established by continuous monitoring, indicates that failure is imminent.

*Preventive* or *Scheduled Maintenance*. Maintenance carried out to keep an equipment in satisfactory operational conditions by providing systematic replacement of components before they are expected to fail.

*Random Failures*. Failures which arise on a random basis, characterised by a constant failure rate.

*Reliability*. The ability of an item to perform a required function under stated conditions of use and maintenance for a stated period of time. It may be measured as a probability of success at any time, or the time for which a level of success can be assured.

*Reliability Growth*. The increase of reliability which occurs during development.

*Repair Limit*. The condition in which it is no longer economic to continue maintaining an item.

*Repair Maintenance*. Maintenance carried out on a non-scheduled basis to restore an item to a satisfactory condition by providing correction of a failure after it has occurred.

*Scheduled Maintenance*. See Preventive or Scheduled Maintenance.

*Stochastic*. Pertaining to a conjecture (OED), that is, lacking precise definition; involving a probability.

*Stress*. Force/area.

*Wear-out Failures*. Failures which are due to loss of strength with time or use.

*Whole* (or *Total*) *Life Costs*. The total cost of possessing and using an equipment – comprising initial, maintenance, and operational costs such as fuel, staff etc., interest payable on capital, and so on – over its life span.

# 1

# Introduction

Everyone knows what they mean when they use the word 'reliability', be they the housewife buying a small item in the local supermarket or the Minister of Defence finalising the contract for the most sophisticated guided weapon system costing millions of pounds. Colloquially, they expect what they have bought to do the job they want when they want. Equally, it is becoming clear that the general public is becoming exasperated with products that do ***not*** do the job they want when they want, as evidenced by numerous complaints in the general press and the growth of consumer associations. On a more sophisticated level the managements of many large concerns are becoming more aware of the cost of repair and especially the cost of lost revenue when their equipment fails. Unreliability costs money: thus reliability has become a commercial and saleable commodity to both the large-scale and individual user. On the other hand, most – though not all – do appreciate that they cannot have 'Rolls Royce reliability at Mini prices'. How then do we make a value judgement?

We start in the next chapter of this book by defining in precise terms, which can be quantified, exactly what it is we mean by reliability. Our definitions may be tentative at first, but they allow us to isolate and assess those factors which control the reliability of mechanical equipment. In the subsequent half dozen or so chapters of this book we then try to build up a body of technical knowledge of the subject, on the basis of which we can turn to the more practical steps to be taken if our objective is to be achieved. However, one of the difficulties in presenting a logical step by step development of the thesis is the cross-correlation of the many facets which go to make up the subject. This is illustrated in figure 1.1 which shows how each chronological and physically identifiable step in the life history of an item is closely linked to those which precede and succeed it. It is essential to keep this aspect in mind throughout our study of reliability. In the end, however, reliability and any associated maintenance has to be costed; what is the price we have to pay to achieve reliability? Hence in the latter chapters of this book the cost of any and every activity assumes greater attention than is customary in a technical book.

We shall see as our theme develops that the source of the occasional failure which leads to unreliability is the variability in both the strength of the item with

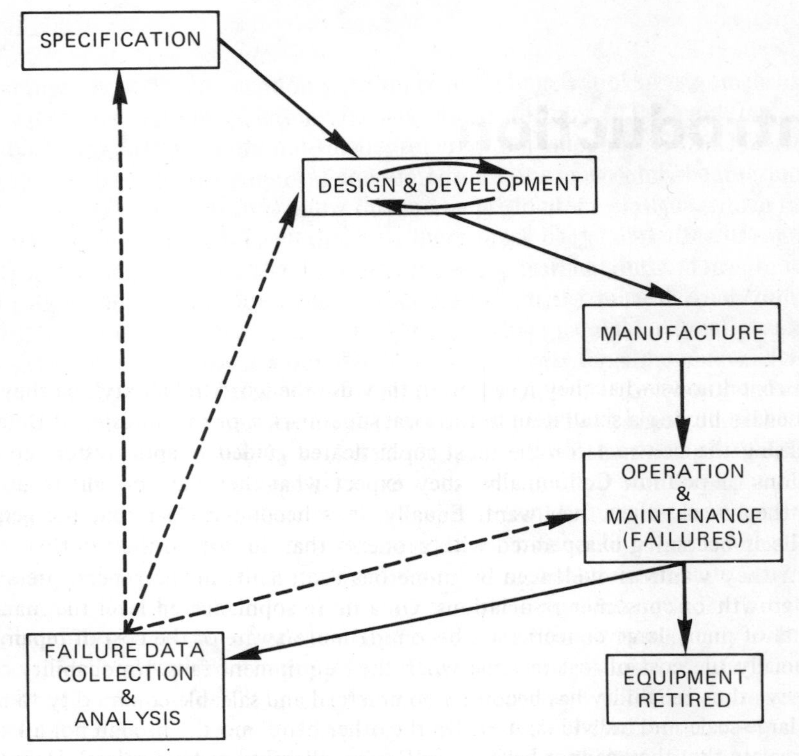


**Figure 1.1** Diagrammatic representation of flow of information during life span of a piece of equipment. → Forward flow of information and action. ---→ Feedback of information

which we are concerned and in the load to which it is usually subjected. The study of variability involves statistics. Statistics is an essential element of reliability studies, but it would be a mistake to identify reliability with statistics. Perfectly satisfactory and highly reliable equipment was designed long before statistics was introduced into the subject. Such designs were based on sound engineering judgement and practice, and these still form the basis of our approach. Statistics only enable us to take some of the empiricism out of these procedures. Thus to me there is no such person as a reliability engineer – *per se*. Those concerned with reliability must first and foremost be specialists in the product with which they are concerned, and it is to this prime activity that they bring some of the tools and techniques with which we are concerned in this book. By far the greater part of reliability practice is good sound fundamental engineering backed by plain (engineering) common sense and an ability to anticipate consequences. A feeling for work on reliability can only be obtained by practice. An attempt to simulate this practice is provided by a number of simple exercises, which are really an integral part of the text. So far as possible they have been centred around subjects of common knowledge (hence the prevalence of the motor car in these exercises), so avoiding the need for specialised knowledge often necessary for solving many actual problems. The thought processes involved are much more important than the actual answers. So although some readers may treat the exercises and solutions (which are always given in full in chapter 15) as part of the text, it is hoped that most readers will work out their own solutions, in greater or lesser detail, before reading the author's preferred solution. Exercises which have been included as illustrations of mathematical techniques have been marked with an asterisk, as have some sections of the text which constitute a mathematical justification for physical principles. Sections so marked may be omitted without loss of continuity. This book is essentially a physical approach to the phenomena involved in reliability intended to provide a physical understanding of mechanical reliability, but it is also backed by sound technical and statistical justification.

# 2 Definitions

An essential requirement of a logical approach to any subject is an agreed vocabulary and accepted definitions. For a scientific approach the definitions must be precise. Now, we must accept that the description of reliability given in the introduction is inadequate for technical purposes. However, more formal definitions do not always agree. Thus we find the following definitions of reliability from highly respectable bodies

European Organisation for Quality Control (1965)
The measure of the ability of a product to function successfully when required for the period required in the specified environment. It is measured as a probability.

U.S. Military Handbook (217B, 1970)
The probability that an item will perform its function under stated conditions of use and maintenance for a stated measured of the variate (time, distance etc.).

British Standards Institution (1971)
The ability of an item to perform a required function under stated conditions for a stated period of time.

U.K. Army Department Engineering Terminology Committee (1976)
As U.S. definition above.

U.K. Ministry of Defence, Def. Stan. 00-5 (Part I) Issue 3 (1979)
The ability of an item to perform, or to be capable of performing, a required function without failure under stated conditions for a stated period of time or unit of operation. It may be expressed as a probability.

U.K. Ministry of Defence, Def. Stan. 00-5 (Part I) Issue 5 (1982)
As B.S. above.

Although there are many similarities there are also marked differences, in particular it does not seem to be agreed in the U.K. whether reliability is to be measured as a probability or not, though the U.S. have consistently done so.

Defining reliability as a probability is very attractive because it enables us to quantify reliability in a manner that is generally understood, and enables us to call on the immense understanding of probability contained in the subject of statistics. It remains to be seen, however, whether such a definition is appropriate or whether defining reliability as an 'ability' is less restrictive and allows a more general approach. The specific inclusion or otherwise of maintenance in the various definitions is another source of difference. Some believe 'conditions' should be interpreted as 'conditions of use and maintenance'. Others do not. We shall not commit ourselves to any definition at present, but will accept probability as a first order measure of reliability. We do this primarily to quantify reliability. It has often been stated that unless one can quantify a concept one cannot understand it. It may then be observed that there are four quantities that appear in these and other definitions of reliability. They are

(1) a specified performance is expected
(2) it is expected only under specified conditions of use
(3) it is expected over a specified period of time
(4) reliability over that time is expressed as a statistical probability.

The features as listed above appear in an apparent order of increasing tractability, yet each is open to a wide range of interpretations. Further differences in the overall concept of reliability stem from a differing emphasis on these features. Our first objective is therefore to obtain as clear a picture as possible of the nature of reliability and of the factors involved by studying each of the above features. We shall then be in a better position to formulate an all-embracing definition.

Before placing too great a store on a definition of reliability, particularly as measured by probability, it would be as well to note that other measures of success or failure do exist. For example, in most practical problems we are concerned with the survivors only, that is the situation as it exists rather than its historical aspects. Furthermore, it is often practically more convenient to think in terms of failures than in terms of probability of success. Hence we define (following substantially the ideas introduced by R. R. Carhart (ref. 1))

$R(t)$ = cumulative probability function of occurrence of survival (that is, probability of success, or reliability)
$F(t)$ = cumulative probability function of occurrence of failure
so that

$$F(t) = 1 - R(t) \text{ by definition} \tag{2.1}$$

We can now define a failure rate. In its most elementary form it is the rate at which failures occur in time, or other suitable variate. It may be represented mathematically as

$$\frac{\mathrm{d}F(t)}{\mathrm{d}t} \tag{2.2}$$

However, it is customary practice to express the failure rate as a proportion of the population existing at the instant to which the rate refers. This would seem a common sense approach, because the absolute magnitude of the failure is of little significance unless related to the population, and in real life the original population may well not be known. Thus mathematically

$$\text{failure rate} = \lambda(t) = \left\{\frac{\mathrm{d}F(t)}{\mathrm{d}t}\right\} R(t) \tag{2.3}$$

in which $R(t)$ is the proportion of the original population surviving at time $t$. Since from (2.1)

$$\frac{\mathrm{d}F(t)}{\mathrm{d}t} = -\frac{\mathrm{d}R(t)}{\mathrm{d}t} \tag{2.4}$$

$$\text{failure rate} = \lambda(t) = -\left\{\frac{\mathrm{d}R(t)}{\mathrm{d}t}\right\} \Big/ R(t) \tag{2.5}$$

It should be noted that $\lambda(t)$, like $R(t)$ and $F(t)$, is in general a quantity which varies with time. The symbol $(t)$ emphasises this fact by making each of the quantities a function of time, but will be omitted in special cases when the values are essentially invariant with respect to time.

Some purists, mainly statisticians, object to this definition of failure rate on the score that it is not a rate as mathematically defined (that is, of the form $\mathrm{d}y/\mathrm{d}x$). Consequently it is given a variety of other names, such as 'force of mortality' from actuarial practice, or 'mortality intensity', 'hazard', or 'hazard function' or even 'hazard rate', 'age-specific failure rate', 'instantaneous failure rate', and so on. All these mean the same thing! For simplicity we shall ignore these pedanticisms, and use the phrase 'failure rate' for the quantity appearing in equation (2.5) in conformity with general engineering practice, relying on such precedents as growth rate, inflation rate, etc. to provide respectability. To avoid any confusion, however, it should be made clear that $\lambda(t)$ is not a probability density function. The probability density function of failure is given by

$$f(t) = \frac{\mathrm{d}F(t)}{\mathrm{d}t} \tag{2.6}$$

and is, of course, a rate, as mathematically defined. The expected total proportion of failures, that is, the cumulative failures, in the interval $T_1$ to $T_2$ will be given by

$$\int_{T_1}^{T_2} \{\mathrm{d}F(t)/\mathrm{d}t\}\,\mathrm{d}t \quad \text{or} \quad \int_{T_1}^{T_2} f(t)\,\mathrm{d}t$$

Such matters are important in the application of statistics to reliability, and the reader should make himself familiar with the exact meaning behind all the terminology.

The practical importance of the failure rate, $\lambda(t)$, is perhaps best demonstrated by the product $\lambda(t)\,\delta t$ where $\delta t$ is a small interval of time. Writing equation (2.3) as

$$\delta F(t)/R(t) = \lambda(t)\,\delta t \tag{2.7}$$

it is seen that this product represents either

(1) the proportion of survivors at time $t$ from a large initial population that fail in the next interval $\delta t$; or
(2) the expected (or average) proportion of survivors from a sample that fail during the next interval $\delta t$; or
(3) the probability that an individual item that has survived to time $t$ will fail during the next interval $\delta t$.

The above interpretations are obviously of practical importance: perhaps more practically relevant in many circumstances than the definition in (2.5). Some authorities, of which the British Standards is one, consequently use interpretation (3) as the definition of failure rate. U.S. authorities usually use the definition given by equation (2.5). The author has also found that the definition (2.5) is more readily understood by those approaching the subject for the first time, *provided* it is appreciated that the probabilities $R(t)$ and $F(t)$ refer to the original and not the current population. The above three interpretations follow as consequences. Nevertheless, with the B.S. definition the failure rate is undoubtedly a 'rate'. Statistical terminological exactitude would be achieved by calling $\lambda(t)$ the 'conditional failure rate', which seems the most logical alternative if simple 'failure rate' is unacceptable. The adjective emphasises the condition that the failure rate applies to the survivors, not the original population. The phrase is, however, a bit of a mouthful for everyday use, and would not appeal to the average mechanical engineer who would normally think of the population as current items anyway (in the same way that the population of a town does not include the deceased, that is, failed individuals). It will not therefore be used, though the statistical condition is always implied when the words 'failure rate' are used without any other qualification.

As a further aid to the physical significance of this measure of unreliability, the various quantities appearing in the above analysis have been represented schematically in figure 2.1. The distribution of failures has been assumed roughly normal, though of course any distribution is possible, and the failure rate may increase, as shown, remain constant, or decrease with time, or indeed be any combination of these. The total proportion failured at $t_1$ will be the area under the failure probability density function, $f(t)$, up to time $t_1$; and the area beyond, shown shaded, represents the not-failed proportion of the population, that is, the reliability. Then

$$\text{failure rate} = \lambda(t) = \frac{\mathrm{d}F(t)/\mathrm{d}t}{R(t)} = \frac{f(t)}{R(t)} \tag{2.8}$$

to give the wholly statistical definition and the graphical representation shown on figure 2.1.

Before leaving this subject it would be as well to note that the many other methods of defining failure rate imply other definitions; for example, during development work and early service it is sometimes the practice to total all the failures and divide by the total operating time, even though the figures relate to

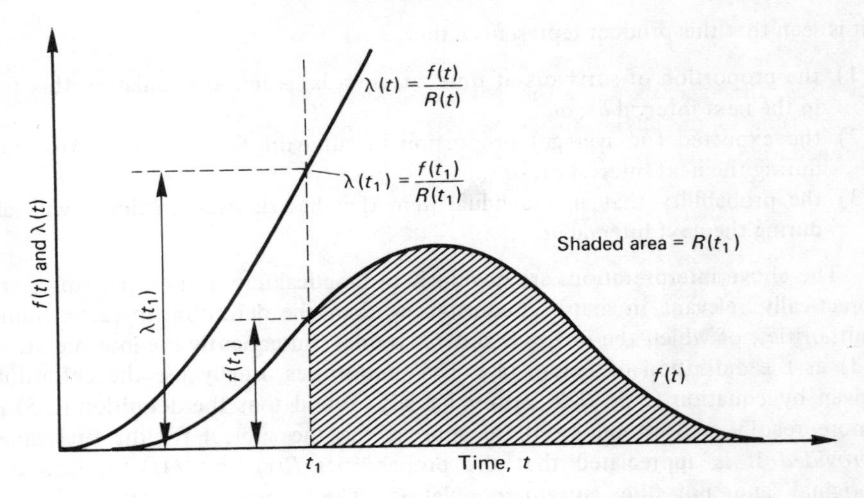


**Figure 2.1** Graphical representation of failure rate

different standards of build and possibly modified units. This quantity is sometimes called the 'cumulative failure rate'. It is valuable for demonstrating the improvement obtained as a result of experience and development and is discussed in chapter 10. It is inapplicable to forward planning, of course, because it contains data on obsolete items. Other workers simply divide the total number of failures occurring in a given time interval by that interval to obtain a failure rate. If divided by the total number of items in the population it may appear to be a practical approximation to (2.5), but if the population consists of items of different ages and different build standards, which are perhaps subject to different maintenance, the correlation can be very remote. Some care has therefore to be taken to ensure that the failure rate is properly defined in any particular circumstance.

It would be as well to recognise at the outset that high reliability (or low failure rate) of itself is rarely the prime objective. To a greater or lesser extent a low reliability can be offset by easy cheap maintenance. It remains to see how far this is possible, but other criteria of quality can be devised to represent this view. One of these criteria is 'availability', which is defined as follows

Available: The state of an item such that it can perform its function under stated conditions of use and maintenance in the required location

Availability: The probability that an item, at any instant in time, will be available.

Following from these definitions it is possible to evaluate availability as

$$\text{availability} = \frac{\text{up time}}{\text{up time + down time}} \tag{2.9}$$

where 'up time' is the time that the item is available as defined above and 'down time' is the total repair time including any time spent on administrative procedures involved, time spent waiting for spares, in transit and so on, as well as the active repair time. Sometimes a further quantity, 'inherent availability' is quoted. It is defined as

$$\text{inherent availability} = \frac{\text{up time}}{\text{up time} + \text{active repair time}} \tag{2.10}$$

In this definition 'active repair time' is taken to be that actually spent on the job, and excludes all administrative, transport, waiting time etc. This is clearly more relevant to an engineering appraisal. The various forms of availability are used extensively by the armed services of all countries, since any realistic costing for emergency use is not possible. It is also used by large process industries in cases where large capital expenditure has already been committed, and availability is then a measure of the ability of the concern to meet a demand. The ultimate criterion must, of course, be cost.

It should be self-evident that very high reliability has to be paid for, both in design and production. On the other hand the smaller number of failures in service will require less maintenance action and necessitate fewer spares. The total upkeep costs will therefore reduce with increasing reliability. This has been illustrated in figure 2.2, which shows the individual procurement and operating costs plotted against reliability. Also shown on this curve is the sum of the individual items, which represents the 'whole life cost' of the equipment. The phrases 'total cost' or 'cost of ownership' are sometimes used in place of 'whole life cost'. This curve shows a characteristic minimum at a point where the costs of supply and maintenance are properly balanced. Also plotted on figure 2.2 is a curve taking into account the loss of revenue due to lack of reliability, conservatively estimated from an assumed revenue of 15 per cent of the whole life cost at the optimum point, and neglecting goodwill. Both whole life cost curves show the characteristic minimum, but the value of the reliability at which the minimum occurs is different. Once again we need to be sure in any application of the precise manner in which the whole life or total cost is evaluated.

Unfortunately little practical data is available to construct real graphs corresponding to figure 2.2. However, we do see that we should be aiming at an optimum combination of reliability and maintenance rather than the highest possible reliability. As our theme develops we shall therefore refer to whole life cost as defined above and to optimum reliability as that reliability which results in the minimum whole life costs, it being assumed that loss of revenue is included.

Before we can proceed with a study of reliability there are some other concepts that need to be defined. Some texts on reliability use the word 'stress' to imply load. This is confusing. In this book we shall retain the conventional mechanical engineering definition of stress, namely force/area. In this way it may be a measure of strength, as in yield stress, or it may be a measure of load. The words 'load' and 'strength' will then be used in the connotation normally adopted by mechanical

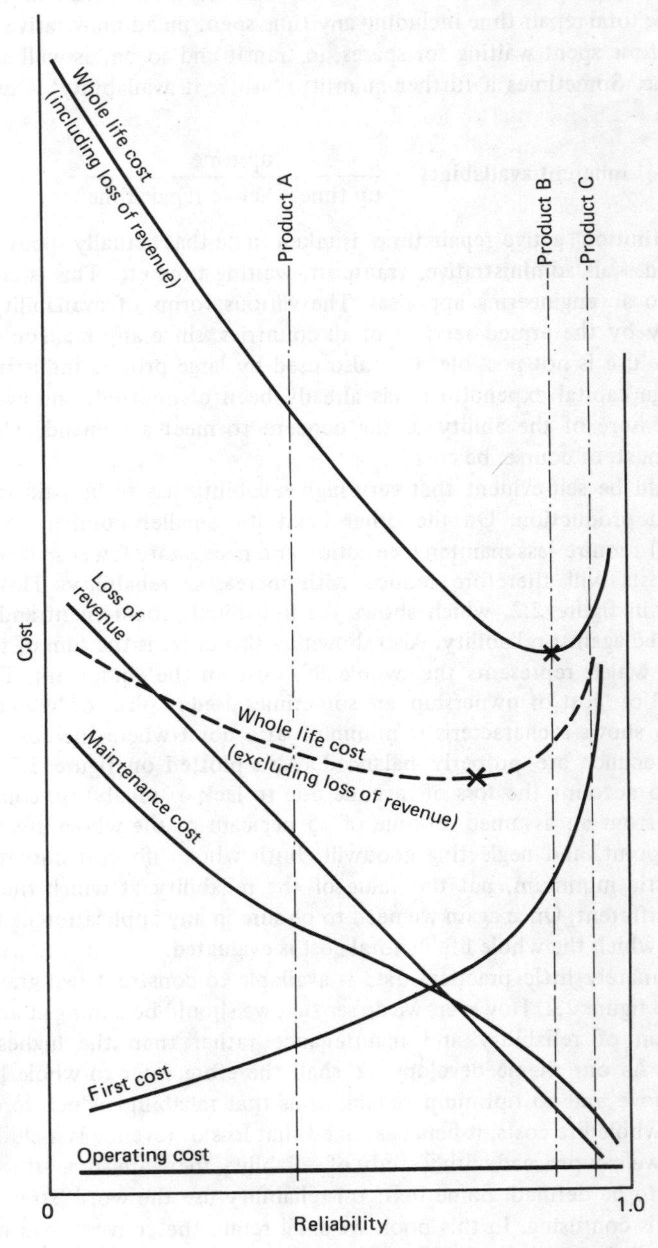


**Figure 2.2** Effect of reliability on cost

engineers and their meaning is self-evident. In this way a consistent terminology with other branches of mechanical engineering is maintained, even though it involves some differences with specialised reliability texts, which are usually not directed towards mechanical engineering.

Likewise it is essential to know what we mean by the terms 'component' and 'equipment' or 'system', since the same physical item can be either, depending on the viewpoint. As an illustration, an automotive engineer may regard the engine as a component of the vehicle with the gear box, rear axle, and so on as other components of the vehicle, which he regards as the item of equipment. To the engine manufacturer, however, the engine is the piece of equipment having crankshaft, pistons, and so on as components. Thus the same item – the engine in this case – can be regarded as a component *or* a piece of equipment, depending on the context. One man's system is another man's component. To be absolutely clear we shall reserve the word 'component' for a non-maintained integral item. In any context, if a component fails it is removed and discarded from the population under consideration. It may or may not be replaced by a new component. It should be noted that this definition does not exclude a failed and discarded component being salvaged and repaired to original standards and returned to service, but it then constitutes a new component. The words 'equipment' or 'system' are used interchangeably to denote an assembly of components and the term 'complex' is sometimes added when a very large number of components is involved. Both these terms may be preceded by the words 'non-maintained' or 'maintained' whenever there is any doubt, though these adjectives may be omitted if the context makes it clear which applies. Sometimes the words 'item' or 'product' are used to denote that the text applies to either a component or a piece of equipment.

Finally 'time': for we shall use this word in preference to the all-embracing, but abstract statistical word 'variate', because the word 'time' conveys the physical concept we wish to employ. It must be understood, however, that it can be measured in any of the usual temporal units, or in other more convenient units such as distance in the case of, say, a vehicle, or number of operations in the case of an item which is operated intermittently, or in any other appropriate units. Consistent with the word 'time' we shall use the word 'age' to denote the time (in appropriate units) that has elapsed since the completion of manufacture to the instant under discussion. The word 'life' will be used to denote the age of a component at which failure occurred or may be expected to occur, and after which it ceases to exist. The life of a maintained equipment will obviously be greater than that of its components, indeed with unrestricted maintenance it would be infinite, being only limited in practice by cost, obsolescence, or fashion. We shall use the expressions 'economic life', now often replaced by the word 'durability' (see section 9.7), to denote the age at which a piece of equipment is discarded on these grounds.

Attention has been given to the definitions of some of the basic words, because in a developing subject covering a wide range of disciplines the same word can be used in different senses. In my experience considerable confusion can arise from such simple matters. Further definitions will be given at suitable points in the text as the general theme is developed. However, a complete glossary and notation has been included for ready reference.

As a matter of usage rather than definition the stress *s* due to any load is taken to be the effective failure-inducing stress. An item may be subject simultaneously to many stresses such as tension, bending, torsion, etc., but all can be reduced to three principal stresses, and these can be equated to a single failure-inducing stress by one of the Maximum Principal Stress (Tresca, Von Mises or Mises–Hencky) theories. This is not a textbook on stress analysis or materials, and so it will be assumed that evaluation of the single failure-inducing stress due to any load has already been carried out by one of these standard techniques. We shall then be concerned with the ability of an item to withstand this stress when both load and strength are distributed.

# 3

# Probability as a measure of reliability

## 3.1 Physical approach

Let us suppose that reliability is defined simply as the probability that an item can perform a required function under the stated conditions for a specified period of time. Reliability can then be expressed by some number lying between zero and unity, where zero represents certain failure and unity certain success in the accepted manner. In this concept an item either succeeds or fails – there is no halfway house. (This matter is further discussed in section 7.1.) We will denote reliability, as the probability of success, by the symbol $R$, and the probability of failure by the symbol $F$. Then

$$R + F = 1 \tag{3.1}$$

Reliability so defined may refer to a complete system, or to individual components of a system. Suppose now we have a system, consisting of a number of components in series, for which

(1) the failure of any one component causes failure of the complete system, and
(2) failure of any component is entirely independent of the failure of any other component.

The reliability of the complete system when no failed component is repaired or replaced may be expected to be related to the reliability of the individual components by the Product Rule. Thus, if $R_1, R_2, R_3, \ldots R_n$ are the reliabilities of $n$ individual components, the reliability $R$ of the system is given by

$$R = R_1 \times R_2 \times R_3 \times \ldots \times R_n \tag{3.2}$$

If the component reliabilities $R_1$ to $R_n$ are of roughly the same magnitude we can write

$$R = \overline{R}^n \tag{3.3}$$

where $\overline{R}$ is the mean component reliability.

If we have a system consisting of a number of components in parallel which also satisfy condition (2) above, the unreliability of the complete system is related to the unreliability of the individual components by the Product Rule. In this case the unreliability of the parallel system is taken as the probability that all components fail. Then

$$F = F_1 \times F_2 \times F_3 \times \ldots \times F_n \tag{3.4}$$

or

$$(1 - R) = (1 - R_1)(1 - R_2)(1 - R_3) \ldots (1 - R_n) \tag{3.5}$$

Again, if the reliabilities $R_1$ to $R_n$ are of roughly the same magnitude we can write

$$(1 - R) = (1 - \bar{R})^n \tag{3.6}$$

We are thus able to calculate the reliability of a complex system consisting of any number of components in series or in parallel from equations (3.2) and (3.4) or by combining these equations for any series/parallel system.

This approach leads to a demand for very high component reliability in any series system consisting of many parts, as illustrated by figure 3.1, which has been taken from ref. 2 and is often quoted in reliability texts. We see, for example, that to obtain 80 per cent reliability in a series system of 400 components requires a mean component reliability of nearly 99.95 per cent. To achieve a reliability of 99 per cent would require a component reliability of 99.9975 per cent in the same system. On the other hand the same reliability, 99 per cent, could be obtained from three parallel components each having a reliability of only 53.6 per cent. Thus very high reliabilities can be achieved from parallel systems employing redundant elements, but if redundancy is not possible very high component reliabilities are essential to achieve acceptable reliabilities in complex systems.

Simple everyday experience of automotive transport, to take just one everyday illustration, suggests this approach is over-simplified. At peak hour traffic conditions, 20 or 30 vehicles or more may be held up at a traffic light. Each vehicle has at least 100 components in series in its transmission system, giving some 2000 components in series at each traffic light. Yet how often does the queue fail to move because of mechanical failure when the traffic lights change to green? Chaddock has carried out a more scientific investigation of the supposed correlation between reliability and number of components, studying a number of weapons for which accurate data existed. He concludes there is no such correlation.

The truth is that success is achieved when the weakest or least adequate individual component of a system is capable of coping with the most severe loading or environment to be encountered, that is "the strength of the chain is that of its weakest link." This has been emphasised by both Lusser (ref. 3) and Nixon (refs 4 and 5) who at the same time "recognise the fact of variability, or scatter, both in the capability or strength of the product, and in the duty it will have to face," that is the load which will be imposed on it.

The statements in the above paragraph are considered fundamental to any concept of mechanical reliability and need some amplification and justification before proceeding.

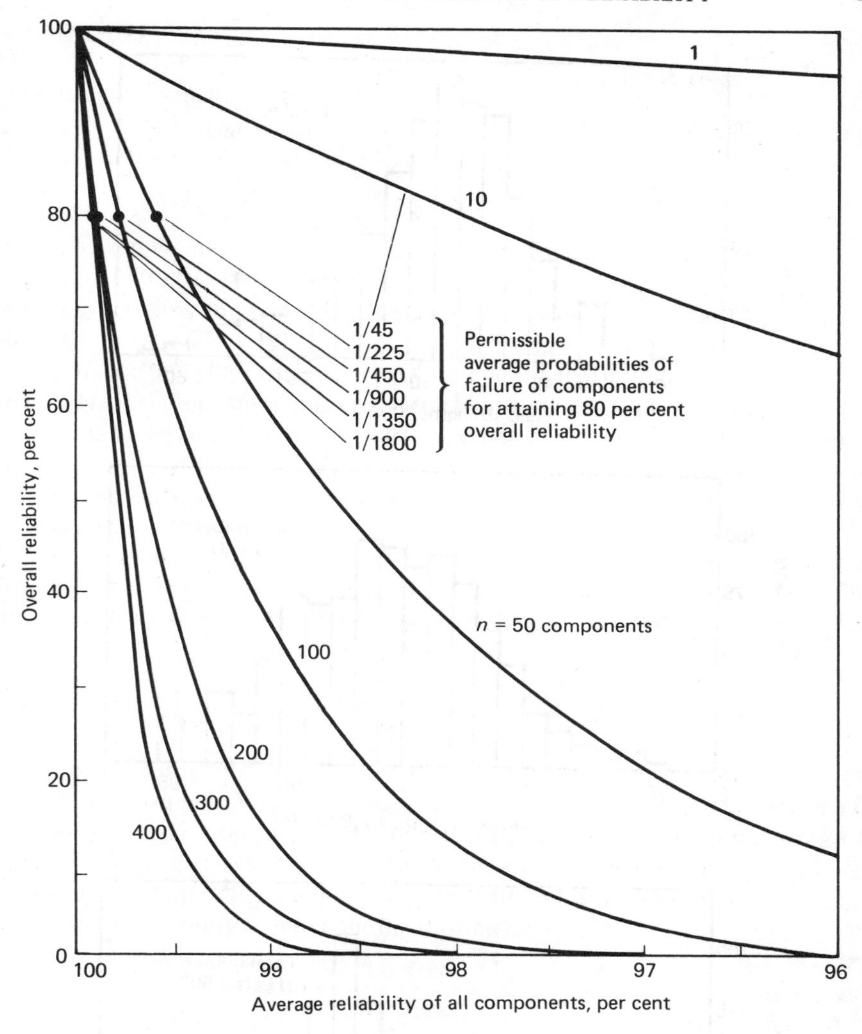


**Figure 3.1** Overall reliability of a complex series system

Let us start with the basic material from which our components are all made. It has to be recognised that material cannot be produced economically without any variation in properties. As illustration, figure 3.2 shows the distribution of yield strength, ultimate (tensile) strength, and elongation of a low-carbon hot-rolled steel to specification SAE 1035 (the data is from ref. 6). It shows that we can expect, for example, a yield strength of (40 to 60) $\times 10^3$ lbf/in$^2$ or (275 to 415) MN/m$^2$. This is a wide range indeed. Further examination of similar data will show that this is by no means an exceptional case; the variation in some special and high-duty materials can be greater still. If we turn to the more important dynamic properties such as fatigue, fracture toughness, notch sensitivity, creep,

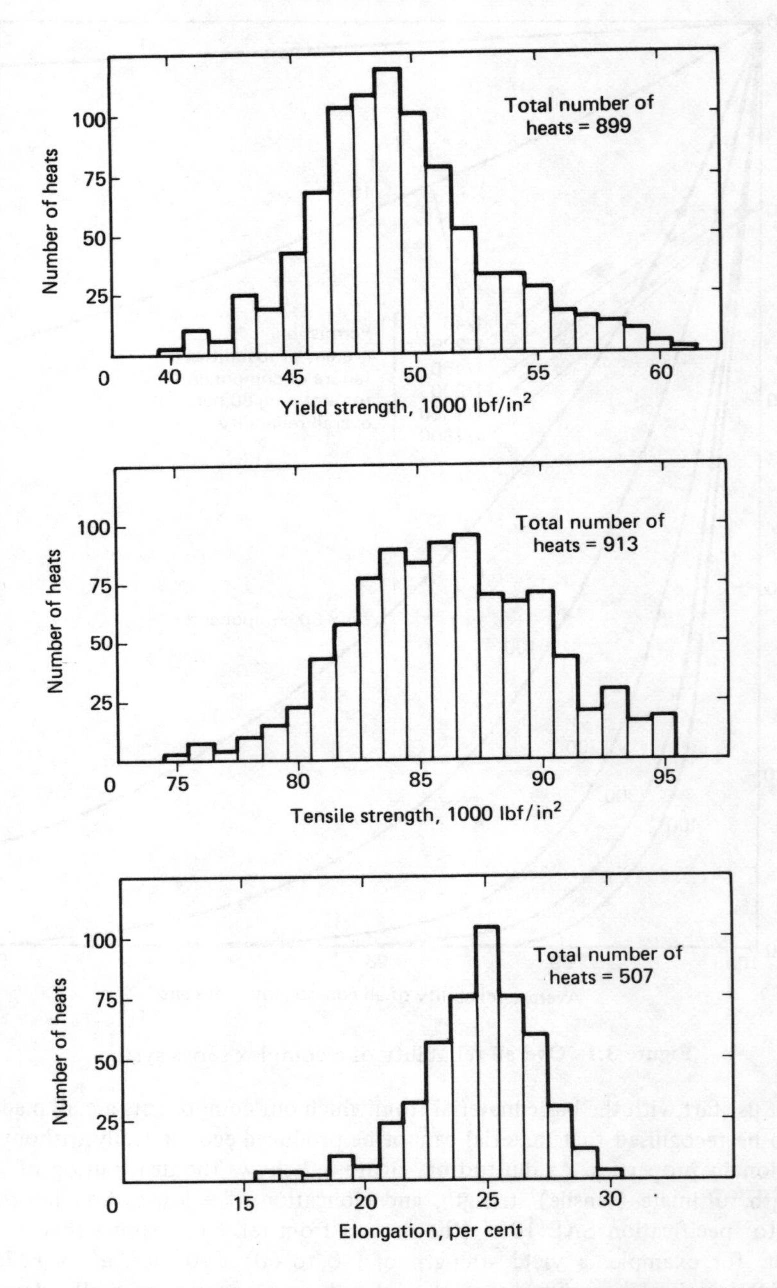


**Figure 3.2** Distribution of static tensile properties for SAE 1035 hot finished carbon steel (as hot-rolled 1–9 inch diameter round bars). Adapted from ref. 6

abrasive and erosive properties, corrosion resistance and so on the variation can be even wider. As illustration, figure 3.3 adapted from ref. 4, shows the endurance in fatigue bending tests of Nimonic alloy specimens taken from aero engine turbine blade forgings. Even in this important example, involving safety, we have a long tail of low-strength specimens. Although steps can be taken to reduce the scatter in properties – electro-slag refining is an example – it must be recognised that a variation does exist in the properties of the materials from which all our products are made, and that generally the variation in natural materials is much higher than in man-made materials.

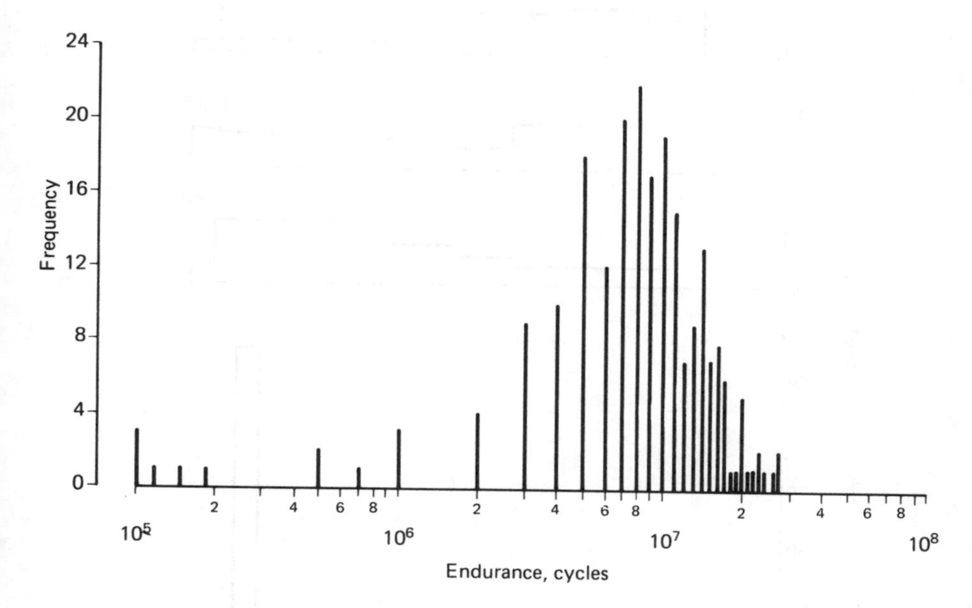


**Figure 3.3** Distribution of strength, illustrating the importance of the tail of the distribution. Results refer to rotating bending fatigue tests on Nimonic alloy specimens taken from blade forgings. All tests conducted at constant stress at 800°C (from ref. 4)

In addition to the variation in the basic material properties there will be further variations due to the manufacturing process itself. These may arise from working of the material during the manufacturing process, from deliberate or unavoidable heat treatment, from the surface finish left on the material, or from a variety of other causes. The manufacturing process will also introduce variations of an entirely different nature in that absolute dimensional accuracy cannot be attained and variations within specified tolerances are a necessary feature of all manufactured products. Thus any general theory of reliability must take into account substantial variation in the strength of all items.

The same applies to load, but more so; for one instinctively feels that most

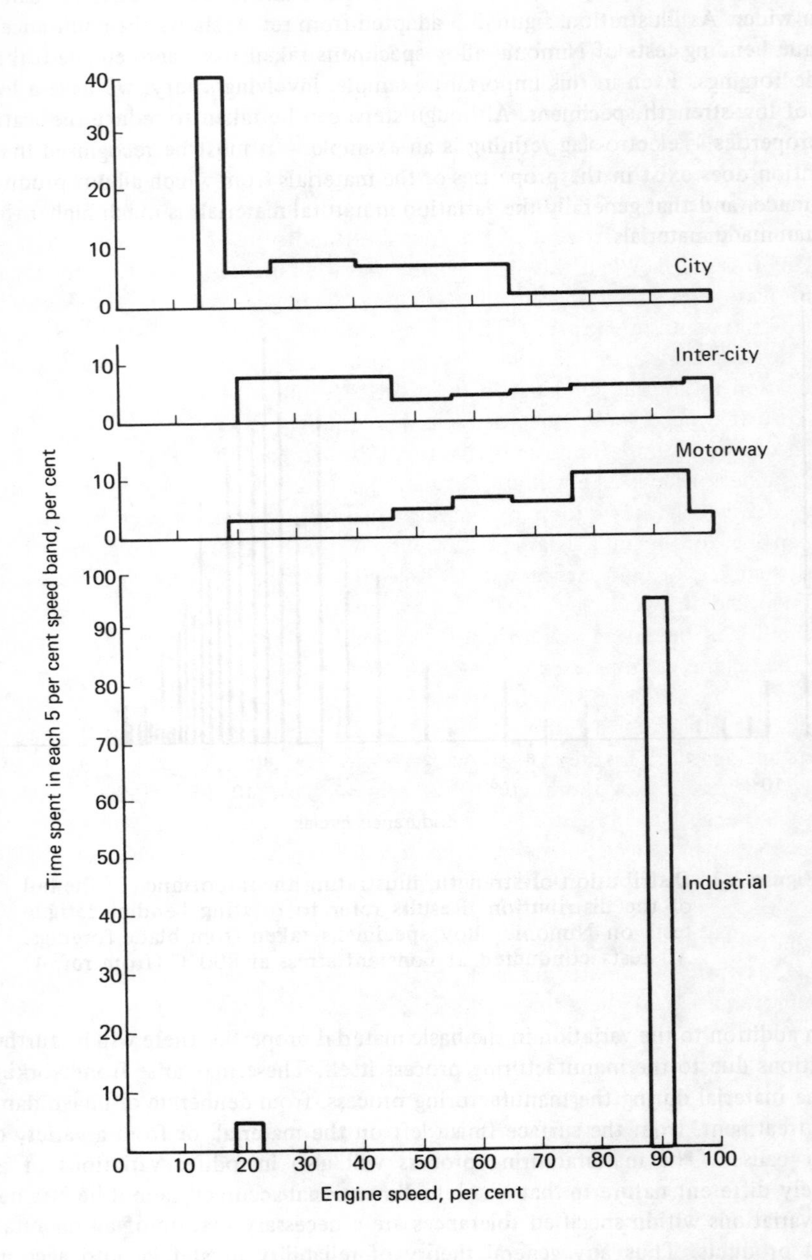


Figure 3.4 Speed–time histogram for a 11.3 litre Diesel engine in different applications. Adapted from ref. 7

mechanical equipment will be subjected to loads whose variation is many times the strength variation. Additionally the variation can differ in different applications. As illustration, figure 3.4 shows the time spent at various speed by an 11.3 litre Diesel engine which has been used in a number of applications. The upper three histograms illustrate the distribution in nominally the same appplication – a bus. However, the distribution shows significant differences when the bus was used on city, intercity, and motorway routes. As may be expected, the city route shows a great deal of time (nearly 40 per cent) spent idling, while the motorway route shows 40 per cent of the time spent uniformly between 75 and 95 per cent of full speed. When this same engine was used industrially for power generation it was operated for 95 per cent of its time at constant speed (chosen as 90 per cent of maximum rated speed). There is nothing surprising about these curves; but they do emphasise that the load is by no means constant, that it varies from one application to another, and that it bears little resemblance to the normal distribution. Furthermore, only one load has been quoted. There will be other loads which can be all important, and which may be linked. For example, in the above case engine torque is an equally if not more important load. This is brought out in figure 3.5 which shows joint probability distribution contours for engine speed and engine torque in a heavy vehicle application. These are joint probability density curves, or contours, in which the total volume under the contours is equal to unity and in which the volume under any speed and torque ranges gives the probability of operating under those conditions. We may note that in this application the engine was operated most of the time in special combinations of torque and speed on which design could concentrate. However, there are still other factors. Thus the rate of application of torque (including any shock) is a very important feature. Figure 3.6 shows a record of torque, among other factors, during the running of the above heavy vehicle. The very high transient torques during some gear changes should be noted, for such spurious loads often play the deciding role in determining the ultimate reliability of mechanical systems. The maximum torque (albeit transient) recorded in this instance is 2.5 times the maximum rated torque! Additionally we can expect variations in levels of vibration, in operating temperature and other environmental conditions, in quality of fuels and lubricants, in standards of installation and maintenance etc. The statistical representation of these loads, and in particular their simultaneous occurrence, soon becomes a prohibitive task. Furthermore, it must be recognised that many of these loads further react on the strength distribution, through fatigue, wear, creep etc., which has then to be appropriately modified. It is readily appreciated that a comprehensive mathematical representation of the real situation is impossible and that considerable simplification is necessary.

To achieve this simplification we shall confine our investigation in the first instance to a model in which all forms of loading can be isolated and therefore treated separately, that is, we shall consider the interaction between a single distributed load and its resisting distributed strength.

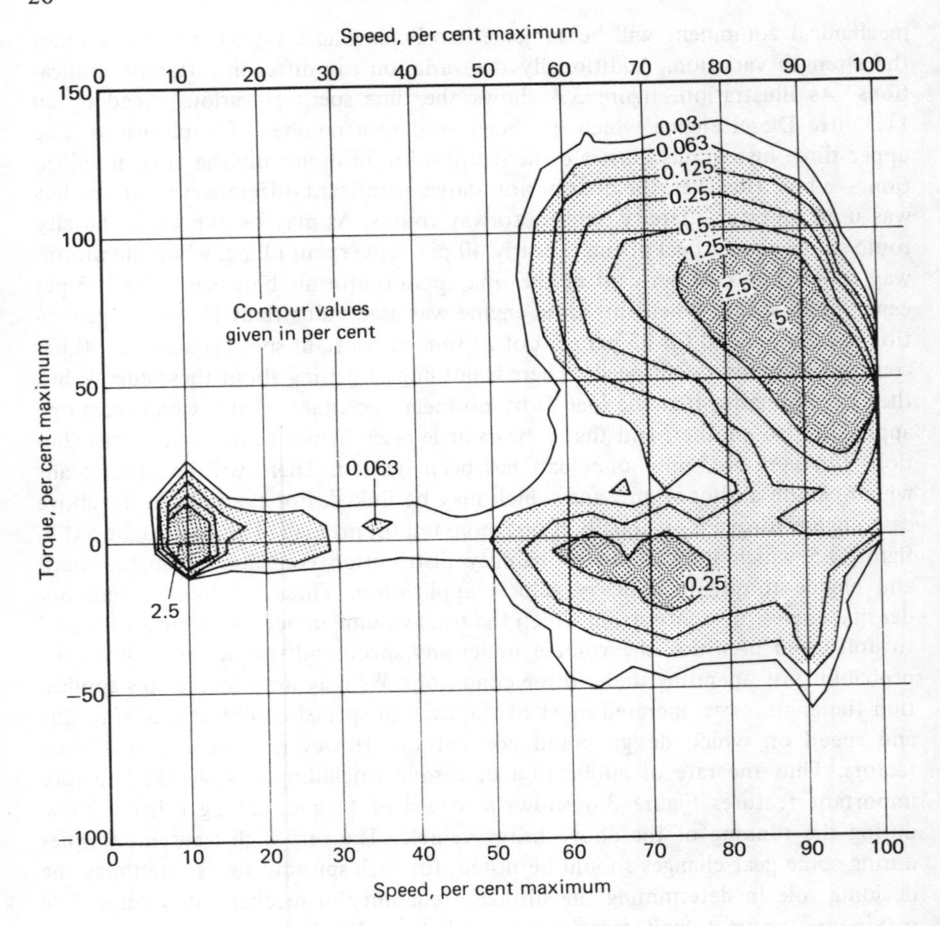


**Figure 3.5** Joint speed–torque probability density contours for an engine in a heavy vehicle application (from ref. 8)

Let $\overline{S}$ = the mean or nominal strength

$S(s)$ = the probability density function of occurrence of any strength in a family of products

$\overline{L}$ = the mean or nominal load applied

$L(s)$ = the probability density function of occurrence of any load.

Then the model we propose to investigate can be illustrated diagrammatically as in figure 3.7. It must be emphasised once again that although both load and strength have been represented by roughly normal distributions, in reality the distributions will often be far from normal.

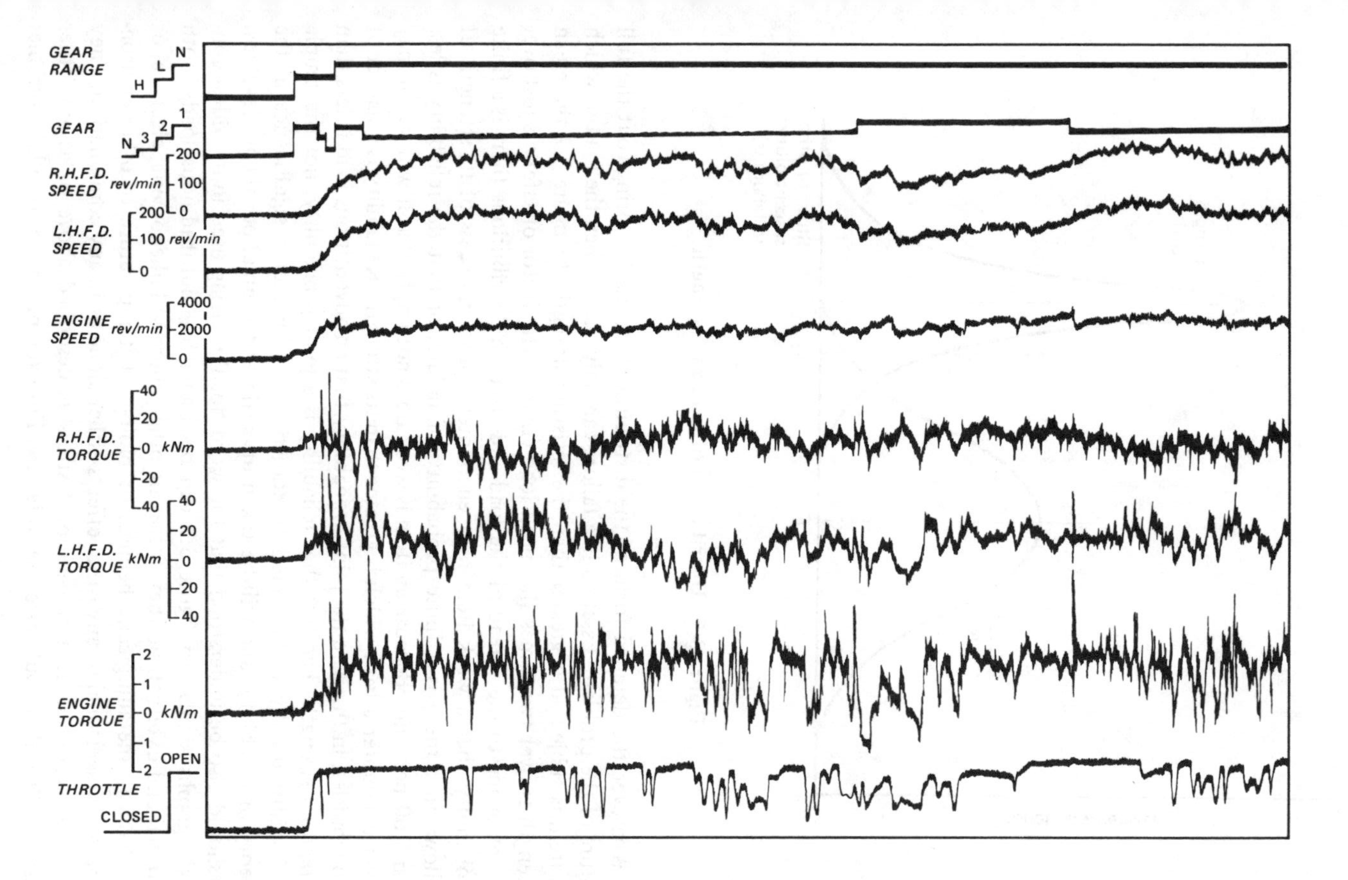


**Figure 3.6** Examples of transient operating conditions for engine whose mean load distribution is given in figure 3.5 (from ref. 9)

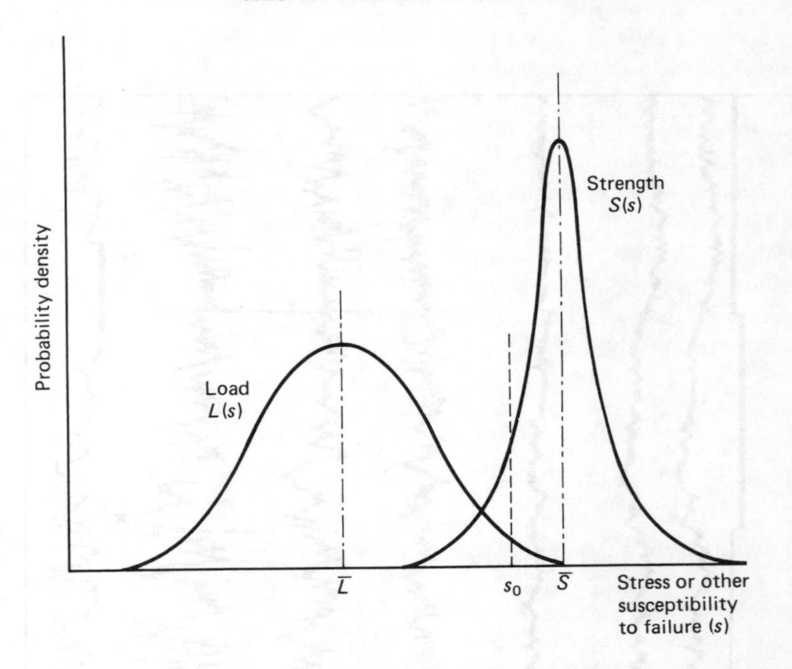


**Figure 3.7** Distribution of load and strength

Whatever the detailed form of the distributions, figure 3.7 brings out the full nature of the problem. Essentially a failure can only occur when the load to which an item is subjected is greater than the resisting strength. In figure 3.7 the mean strength is over two times the mean load (that is, the factor of safety is just over 2) yet owing to the spread of the load and strength distributions there is a finite probability that a weak item may encounter a load in excess of its strength. It follows that there is a finite probability of failure, or that the reliability is less than 100 per cent. Because an item is weak does not imply that it will fail: it may never encounter a load which is greater than its strength. Reliability is a matter of probabilities, in this case the joint probability that an adventitious load is less than a random strength. The fact that reliability is a joint probability, not the simple probability usually encountered in statistics, is of immense significance in the theory of reliability, and will be taken up again and again as our theme develops. It should also be understood that the word 'load' is being used in its widest context: graphs similar to figure 3.7 can be drawn for multitudinous loads, both internal and external, be they forces or other impressed phenomena, electrical or mechanical, metallurgical, chemical or biological, temperature or other environmental, dimensional or any other effects either alone or in any combination. They may be expressed as a conventional stress as described above or as any other susceptibility that could give rise to failure. For example, if we consider a turbine:

for one mode of failure the strength may be measured as the stator diameter and the load as the rotor diameter (over the blade tips), both being distributed in a manner similar to figure 3.7. Failure will occur, as before, when the load (that is the rotor diameter) exceeds the strength (that is, the stator diameter); ***en passant***, we note that the theory of tolerancing is also a particular example of the theory of reliability. Other examples spring to mind in which the susceptibility to failure (the abscissa of figure 3.7) can have a wide range of units. All possibilities must be considered. Nevertheless it is readily seen from figure 3.7 that the spreads of both distributions play a vital role in determining the reliability, and that it is essential to take both into account when developing a basic understanding of the processes which control reliability.

*Exercise 3.1*

Make rough sketches showing the probability density functions you would expect for the major load and withstanding stress in the following instances

(a) the strings of a tennis or squash racquet
(b) a lead pencil
(c) the rungs of a wooden ladder
(d) the escapement mechanism of a grandfather clock
(e) the braking system of an ordinary passenger car.

(The author's suggested solution and comments on this and all other exercises are given in full in Chapter 15.)

*Exercise 3.2*

It has been shown that reliability depends on the intersection of the load and strength distributions, that is, on the tails of the distributions. Discuss how far these extreme tails may be represented by any of the conventional statistical distributions.

Although the graphs of figure 3.7 and those you have drawn as part of exercise 3.1 define the situation completely, it is desirable to have at hand some simple parameters which can be used to provide a first order representation of those curves. In older engineering terminology the factor of safety, already referred to above, was used universally. It is equal to, or can be expressed in terms of, the ratio $\overline{S}/\overline{L}$. However, this is clearly inadequate. If both distributions were moved bodily along the $s$-axis, without changing their relative position, the model we are investigating would remain unchanged and hence result in the same reliability for all positions of the load and strength distributions, but the ratio $\overline{S}/\overline{L}$ would change. The factor of safety is not therefore an adequate parameter in that we cannot associate a factor of safety with a given reliability. In spite of its long established position in engineering design it will not stand up to the very simple appraisal of its rationale given above. It must therefore be discarded from any logical approach and replaced by a more valid parameter. A quantity that would remain constant as the distributions were moved is $(\bar{S} - \bar{L})$, which measures the separation of the means of the two distributions. The extent to which they should be separated will depend on the combined spread of the individual distributions. The relevant combination is the difference between all stresses $(S - L)$ which will be distributed with a standard deviation equal to the square root of the sum of the squares of the individual distributions. Hence it is possible to formulate a non-dimensional quantity, known as the safety margin, which is defined as

$$\text{safety margin} = \frac{\bar{S} - \bar{L}}{\sqrt{(\sigma_S^2 + \sigma_L^2)}} \tag{3.7}$$

where $\sigma_S$ = standard deviation of strength
and $\sigma_L$ = standard deviation of load
with which to represent the separation of the two distributions. However, we still need some parameter to define the relative shapes of the distributions. In this case the prime quantity we wish to represent is the spread of the load distribution, for this measures the various loads with which an item must cope. It is quantified by $\sigma_L$, the standard deviation of the load, and can also be expressed non-dimensionally as a fraction of the spread of the load minus strength distribution. This parameter is known as the loading roughness, and is defined as

$$\text{loading roughness} = \frac{\sigma_L}{\sqrt{(\sigma_S^2 + \sigma_L^2)}} \tag{3.8}$$

We thus have two parameters, the safety margin and the loading roughness, which together with the shape of the individual distributions define the model we have to investigate. The significance of safety margin is self-evident: engineers will see it as a logical refinement and replacement of the factor of safety, but the loading roughness may repay a closer examination.

The shapes of the loading and strength distribution for some representative values of the loading roughness have been plotted in figure 3.8, assuming all distributions are normal and a safety margin of 4.5 applies in all cases. For low values of the loading roughness, that is for smooth loading, the distribution of load is confined to a small range, but that of the strength is much wider. As the loading roughness is increased we pass through medium roughness where the distributions are comparable, to rough loading where the distribution of loading is much wider than that of strength. Practical examples cover the whole spectrum of loading roughness. At the lower end we can quote electronic equipment. Here the electrical load that is applied has necessarily to be controlled within very close limits if the performance is to be achieved, so that the loading roughness is not only low but is confined to a narrow range. The mechanical roughness to which electronic equipment is subjected may differ, and arises from environmental conditions. It can be much higher but is often maintained at low values by special environ-

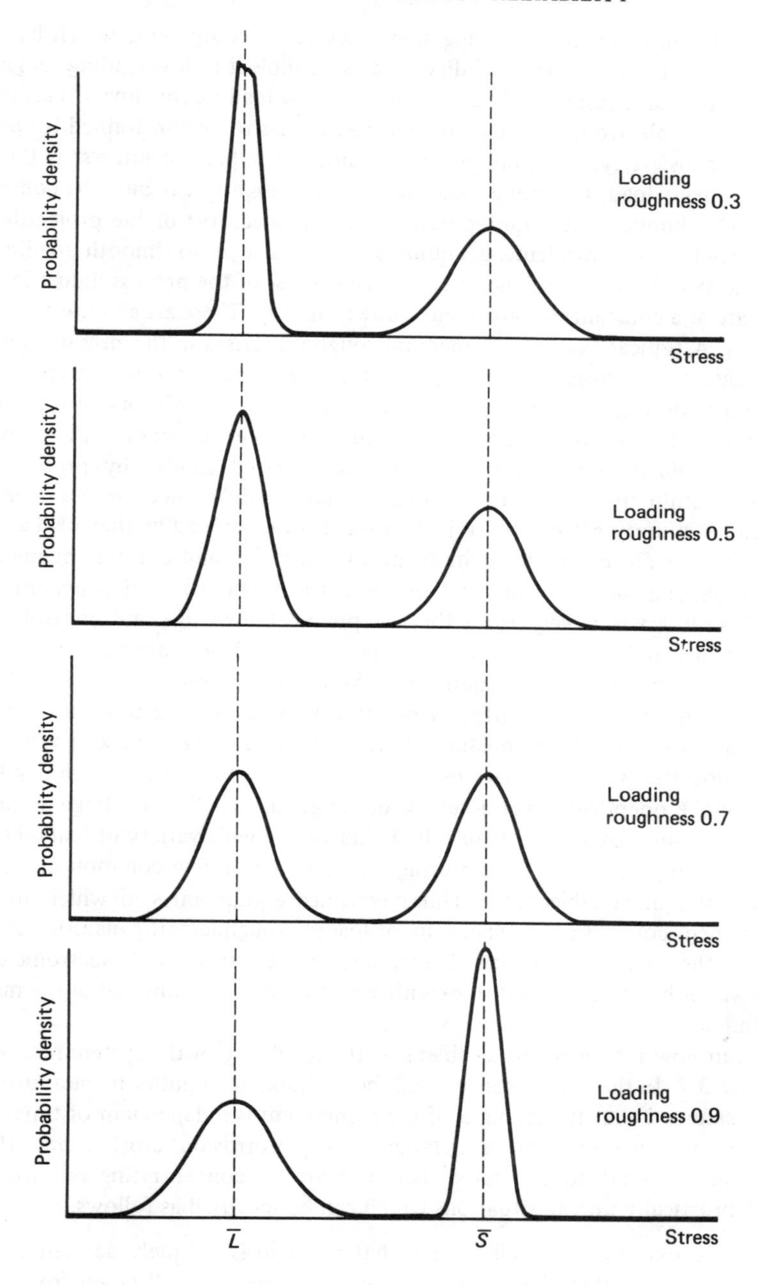


**Figure 3.8** Distributions of load and strength curves for various loading roughnesses at a safety margin of 4.5

mental control. Generally speaking then, electronic equipment, which has been the subject of considerable reliability studies, is subject to low loading roughness. Some mechanical equipment is also subject to low loading roughness. This occurs when, as with electronic equipment, the loading has to be constrained for performance reasons. A typical example in the author's sphere of interest is the gun. Here the major load, the firing load, may be very high but it must be controlled within close limits if the required performance (trajectory) of the projectile is to be achieved. Some mechanical equipment is confined to smooth loading for economic as well as performance reasons. Thus most of the process industries tend to operate at a constant output or close down entirely. These are all smooth loading examples. A typical example of medium roughness arises in the aircraft industry. In this case the environment, chiefly the atmosphere, and otherwise prepared runways and taxiways, is reasonably the same everywhere, and although variations do occur the particular environment that an aircraft will encounter is reasonably well established. Furthermore, the equipment is operated chiefly by professionals. Even so, within this technological area substantial differences are apparent; for example, a military aircraft is subject to much rougher loading than civil aircraft, arising from a more diverse flight plan, unpredictable and extreme manoeuvres, etc. The general run of mechanical equipment is subject to much rougher loading than this. It arises largely from the less precise knowledge and control of the environment, and also from the wide range of different applications of most mechanical components and equipment. As an illustration, a motor vehicle may be driven on any type of surface from the 'billiard table' smoothness of some motorways to, say, the impossible conditions of a jungle track. Even within normal operations wide variations are encountered. Furthermore, the skill and ability of the operators cover a very wide range indeed! Watch a large number of people start their cars on a steep hill. Think of the wide variety of loadings a nut and bolt, a pipe union, a ball bearing, to quote just a few common mechanical components, can be subjected to. Thus mechanical equipment, with which this book is concerned, covers the full spectrum of loading roughness, the majority of items falling at the rough loading end. It compares and contrasts with electronic equipment, which has been studied so extensively, and which is confined in the main to smooth loading.

We can now return to the problem of the reliability of the system represented by figure 3.7. In the first instance it will be desirable to simplify it even further. It will be assumed that the strength of the component is independent of time. By so doing we neglect such features as fatigue, creep, corrosion, erosion, etc., though these will be re-introduced later. Let us start by concentrating on two ideal, though practically unreal, situations which can be specified as follows

(1) As one extreme we shall assume that the load is uniquely defined by some value, $\bar{L}$, but that the components of the system are all taken from a large population of distributed strength $S(s)$. This is ideally smooth loading.
(2) As the other extreme we shall assume that the load is taken at random from

a distribution $L(s)$ but that the components of the system all have a unique value of the strength $\bar{S}$. This would be infinitely rough loading.

The simplified versions of figure 3.7 corresponding to these ideal situations are given in figure 3.9.

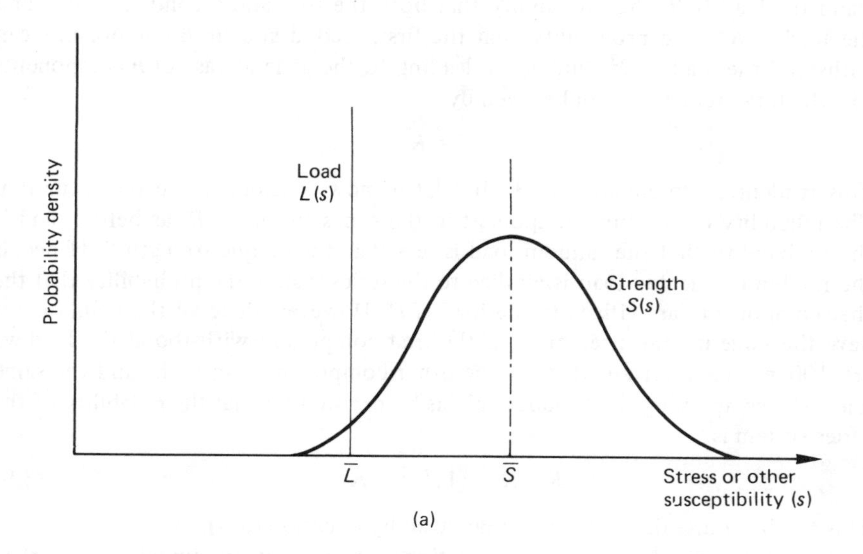


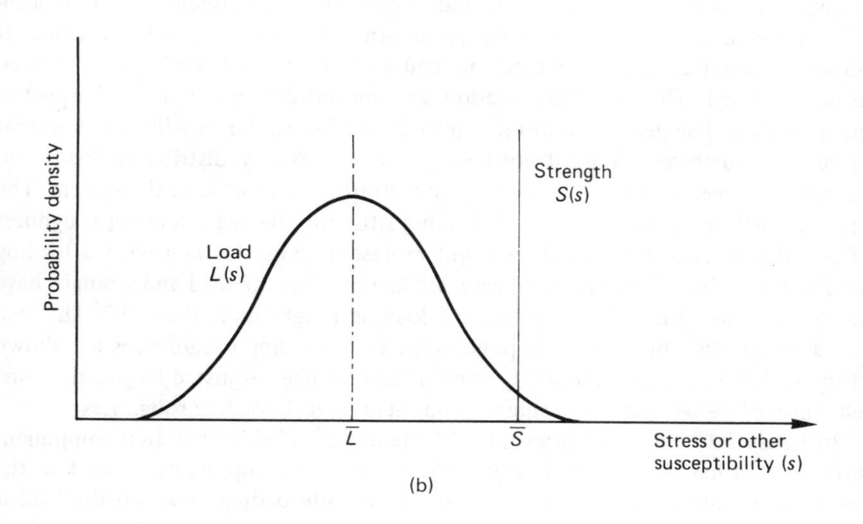


**Figure 3.9** Extreme loading conditions: (a) ideally smooth loading, (b) infinitely rough loading

Concentrating in the first place on the series systems we have already discussed, in situation (1) the reliability of the first component in the system is $\bar{R}$ which is the probability that its strength is greater than the unique load $\bar{L}$. Similarly, the reliability of each of the other components is $\bar{R}$. Hence, if the unique load $\bar{L}$ is applied to the series system the probability that the first component can withstand the load is $\bar{R}$; the probability that both the first and second can withstand the load is $\bar{R}^2$; the probability that the first, second and third components can withstand the load is $\bar{R}^3$; and so on leading to the general case of $n$ components for which the reliability will be given by

$$R = \bar{R}^n$$

This is identical to equation (3.3), but let us now consider the second situation. The reliability of the first component in the series system is $\bar{R}$, as before, and is the probability that the random load is less than the unique strength $\bar{S}$. Hence, if the randomly selected load is applied to the series system the probability that the first component can withstand the load is $\bar{R}$. However, since all the components have the same unique strength $\bar{S}$, if the first component withstoood the load we are 100 per cent certain that all the other components can withstand the same load. Hence applying the product rule as before we find that the reliability of the series system is

$$R = \bar{R} \times (1)^{n-1} = \bar{R} \tag{3.9}$$

This result is quite different from that given by equation (3.3).

As already stated, the extreme situations taken above are unrealistic, but they do illustrate how different premises lead to quite different conclusions. In section 3.2 a more comprehensive mathematical treatment of the subject is presented. It allows the practical cases in which the load and the strength are both distributed to be examined. The rest of this section will concentrate on the physical aspect in this situation. The general solution is given in section 3.2 but it will be convenient to restrict ourselves for illustrative purposes to normally distributed loads and strengths. Three cases will be used to demonstrate the response of the system. The first case will represent the smooth loading situation, having a loading roughness of about 0.3, and the second case will represent rough loading with a loading roughness of about 0.9. The third case will assume that the load and strength have much the same distributions so that the loading roughness is about 0.7. The load and strength distributions corresponding to these loading roughnesses are shown in figure 3.8 except that the safety margin has now been adjusted to give the same reliability of 99 per cent for a single component at each loading roughness.

In figure 3.10 (adapted from ref. 10) the overall reliability of an $n$ component series system has been plotted against the number of components for each of the cases quoted above. It will be seen that for smooth loading in which the load is well defined, the overall reliability drops rapidly with an increase in the number of components, and follows very closely the relationship (3.3) which has also been plotted on figure 3.9. However, it will be seen that for rough loading, in which the

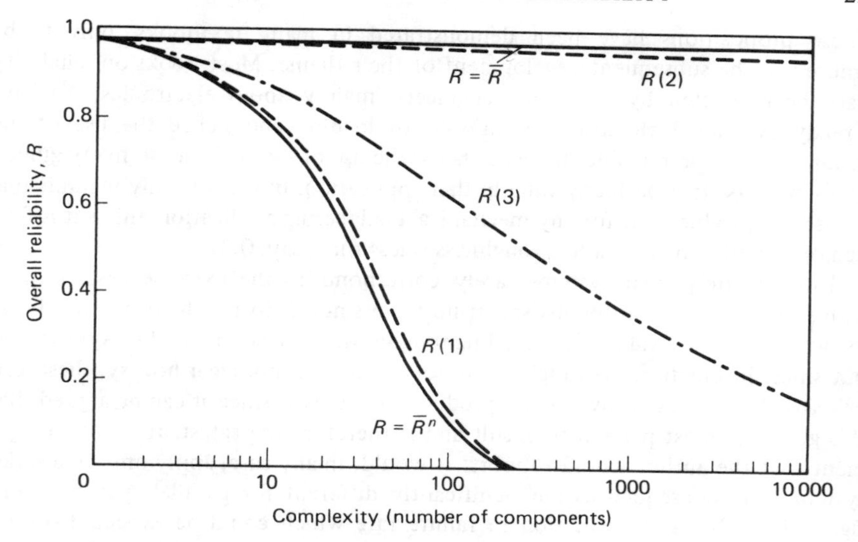


**Figure 3.10** Overall reliability of a complex system as a function of the mean component reliability (distributed loads and strengths). (1) Smooth loading, that is, the distribution of the load is restricted to a narrow range ($\sigma_L \ll \sigma_s$). (2) Rough loading, that is, the distribution of strength is restriced to a narrow range ($\sigma_s \ll \sigma_L$). (3) Distributions of load and strength roughly equal ($\sigma_L \approx \sigma_L$)

scatter in load is great in comparison with strength, the curve departs considerably from this relationship and is indeed much closer to the approximation $R$ = constant = $\bar{R}$, also plotted on the figure. This agrees with the simple physical treatment of the idealised cases. These are the extreme examples, and medium loading lying between (1) and (2) shows we can get intermediate results. The indiscriminate use of the relationship

$$R = \bar{R}^n \tag{3.3}$$

would therefore appear completely unjustified. The truth is that although reliability can be evaluated as a probability, it is not the simple probability of everyday statistics: as we have already seen, it is a joint probability. Consequently, the theorems of simple probability do not apply, and in particular as we have demonstrated above, the product rule is not always valid. If the loading is smooth it would appear that the product rule is applicable, but as the loading roughness increases it becomes more and more misleading until it is widely incorrect for rough loading. A more rigorous mathematical justification for this assertion has been given in section 3.2 in which it is shown that if a series system of components of varying reliability is subject to infinitely rough loading, the reliability of the system is equal to that of the weakest component (the weakest link of the chain).

These propositions have been demonstrated in many textbooks, only to be ignored in the subsequent development of their theme. Most books on reliability have been written by electronic engineers, mainly about electronics. We have already associated electronics with smooth loading and hence the use of the product rule is permissible. It accounts for the use of figure 3.1 in so many guided weapon texts. It is perfectly valid in that application. But it is totally inadmissible to use the product rule for any mechanical engineering application unless it is first demonstrated that the loading roughness is less than, say, 0.2.

Unfortunately, real systems rarely correspond to the extreme cases quoted above, and form a continuous spectrum from smooth to rough. In theory, if we knew the numerical value of the loading roughness, the system could be synthesised but since, in practice, the roughness is not known it is not clear how synthesis can be carried out. Nearly always the product rule is used since it can be argued that this gives the most pessimistic result and is therefore the safest. It is not an argument that the author would support: it would, in any case, apply only to a series system. The consequences are significantly different for parallel systems. Thus figure 3.11 shows the variation in failure rate which could be expected from a number of identical channels in parallel, subject to smooth and rough loading. With smooth loading the advantages of parallel systems are considerable, but with rough loading the advantage to be gained is nil. It should be noted that these conclusions apply to similar channels. If entirely different channels are used the

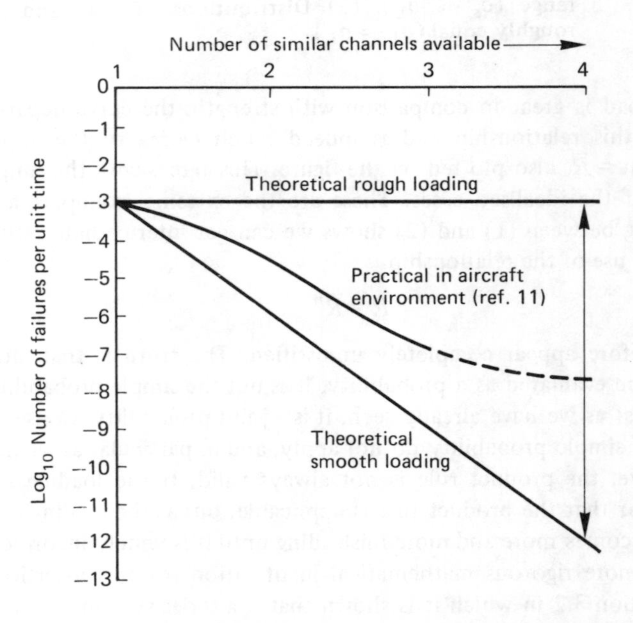


**Figure 3.11** Failure rate for varying number of identical channels in parallel

system is decoupled from a reliability standpoint and hence the product rule could apply. However, it is extremely difficult to achieve in practice. In the last resort it would require a totally different maintenance team for each channel! Still the product rule is used, in spite of the fact we know it is not valid and have to explain away discrepancies by such terms as 'common mode failures', a phrase which may aptly describe the phenomenon but does little else. The truth is that more often than not, the calculations have not been carried out correctly from basic concepts. This matter is expanded in chapter 11.

It is particularly important that the physical basis underlying the conclusions reached in this section should be appreciated, essentially so if any modelling is to be undertaken. The invalidity of the product rule does not conflict with any aspect of standard statistics, which recognises that the product rule applies only to independent events. The fact is that in a normal mechanical system the components are *not* independent but are subject to a common load. The whole object of a mechanism is to transmit a load or force, and this is common to the system, albeit modified in magnitude and failure-inducing effects from component to component. Thus the prime load seen by each component is a function of the operational load. If the prime load is the failure-inducing load then the product rule is invalid, since it implies dependence, except in the special case of ideally smooth loading when the load is not distributed and its stochastic effects are eliminated. If the prime load is not the failure-inducing load for some components then the product rule is applicable to those components, or groups of components, whose failure characteristics are controlled by independent loads. To give an example, suppose I wish to calculate the reliability of my getting to work by car each morning. The car drive line consists of a large number of components in series, the transmission load being common and generally the failure-inducing load. The reliability of the drive line is therefore the reliability of the least reliable component. However, before I can drive off I must fasten my safety belt. Now the loads applied to the safety belt are quite independent of those applied to the drive line. It would be reasonable therefore to take the reliability of starting off as the product of the drive line reliability and the safety belt reliability. Before getting to work other parts must function, of course. Thus braking reliability must be included, amongst others. This may be thought independent of the drive line, but is in fact connected by driving style and so forth. On a more sophisticated level the power transmitted by a shaft is proportional to the product of speed and torque. Torque may be the failure-inducing load for the shaft itself, for any splines and so on, but speed could be the failure-inducing load for the bearings. Thus in this case the shaft and bearings are independent in that they have independent failure-inducing loads. However, the speed and torque may not be truly independent in all cases – see the example on figure 3.5 for example. The subject is not pursued here but left for your own contemplation. System reliability will be considered in chapter 11, when component reliability, wear-out, and maintenance have been covered.

In the case of parallel components the operational or functional load is essentially a common load, and generally invalidates the product rule. It could be that some other load induces failure of some of the parallel channels: in that case the product rule applies to those channels that are truly independent, and a gain is to be achieved from parallel redundancy. One must always beware, however, of common loads giving rise to common mode failures in parallel systems. In many ways it is a pity that the description 'common mode' has been introduced: 'common load' would be a truer description of the phenomenon, and indeed of the term in the exact mathematical equations which represent it. The author notes that the term 'common cause' is currently being used more often and would support its use as a better physical description of the phenomenon.

An exact representation of any system is, of course, much more complicated. Ideally smooth or infinitely rough loading is never achieved, and hence the product rule or the lowest component reliability rule (highest for parallel systems) is also never valid, and the truth lies somewhere in between. It can only be determined if the exact load and strength distributions are known and inserted into a full mathematical analysis. Different models may be necessary for different failure mechanisms. Correct modelling of mechanical systems is a formidable undertaking, and if not carried out correctly can be very misleading indeed. It is an area for honest specialists.

In the current chapter we started by accepting probability as a measure of reliability, but have shown that because it is a joint probability the normal rules of statistics do not necessarily apply and are particularly invalid with rough loading. In developing our understanding of reliability we must be careful therefore always to start from first principles. It should also be recognised at this stage that the real problem is very much more complicated than the appraisal given here, which has been a study of only one load and its resisting strength, compared with the multiple loads that are imposed on any real item. Some of these loads may be statistically independent, some related, and the appropriate model must be used. Nevertheless, the investigation does bring out some fundamental factors and sets the background for subsequent studies. Before going on to an examination of reliability as a function of time we must, however, give a detailed mathematical justification for some of the assertions that have been made. Meanwhile the general reader may care to tackle one or perhaps both of the exercises below which are intended to explore the response of systems with different load and strength distributions. The suggested solution to these problems is set out in chapter 15.

*Exercise 3.3*

(a) A manufacturer of railway goods truck coupling claims he achieves a reliability of 99.9 per cent at 100 000 miles. What would be the reliability of goods trains having an average of 30 trucks over that distance so far as failure of the couplings alone is concerned?

(b) If the reliability of a steam turbine stage is assessed at 99.99 per cent at 50 000 hours, what would be the reliability of a ten-stage turbine driving a base load generator at 50 000 hours.

*Exercise 3.4*

At the *Conference on Safety and Failure of Components* (Institute of Mechanical Engineers, Sussex, 1969) W. T. Truscott wrote

> "One school of thought suggests that most links in any system are essentially failure free and that just a few elements or failure modes give rise to unreliability. The mathematical model which fits this philosophy is indicated by an example, in which a system is made up of 100 elements, 97 of which are essentially failure free and the other three have reliabilities of 90, 92 and 93 per cent respectively. System reliability is given by
>
> $$1.0^{97} \times 0.90 \times 0.92 \times 0.93 = 1.0 \times 0.77$$
> $$= 77 \text{ per cent.}$$
>
> This representation of the model leads one to concentrate on the small number of dominant failure modes; reliability by exception.
>
> Another body of opinion suggests that little is gained by concentrating on weak links. The plausible model which supports this philosophy is indicated by an example of a 100 link element system, 97 of which have reliabilities of 99 per cent and the remaining three reliabilities of 92, 94, and 93 per cent. System reliability is given by
>
> $$0.99^{97} \times 0.92 \times 0.94 \times 0.93 = 0.38 \times 0.80$$
> $$= 30 \text{ per cent.}$$
>
> If one eliminates all three weak links, the system reliability improves to only 38 per cent. Thus one is led here to considering all the elements in the system."

Discuss the contradiction in the above two paragraphs and suggest any criterion which could be used generally to determine the correct course of action.

## *3.2 Detailed mathematical analysis

The mathematical model we shall examine will conform to the general notation adopted in this book, and in particular to the distributed strengths and loads illustrated in figure 3.7. We start by concentrating on a component of strength $s_0$, as illustrated in that figure. The probability that the component will have a strength lying between $s_0$ and $(s_0 + \mathrm{d}s)$ is $S(s_0)\mathrm{d}s$. Additionally, the probability that the stress due to the load is less than the stress $s_0$ is

$$\int_0^{s_0} L(s)\mathrm{d}s$$

Since these events are essentially independent we can use the product rule to evaluate the probability of obtaining a component of strength $s_0$ and a load which will not cause it to fail as

$$S(s_0)\mathrm{d}s \int_0^{s_0} L(s)\mathrm{d}s$$

If we now remove the restriction that the strength of the component is $s_0$ and allow this to take any value in accordance with the distribution $S(s)$, we obtain a value of the overall reliability as

$$R = \int_0^{\infty} \left[ S(s) \int_0^{s} L(s)\mathrm{d}s \right] \mathrm{d}s \tag{3.10}$$

This is a general mathematical expression giving the reliability of a family of products of distributed strength when subject to a load which is also distributed. It is readily seen that this is not a simple probability, but involves two distributions, $L(s)$ and $S(s)$.

This general expression can now be applied to the series system introduced in section 3.1. For simplicity we shall assume that the system is made up of $n$ nominally identical components, each of which is taken at random from a distribution $S(s)$. Now $S(s)$ is the probability density function for the strength of each component, so that the probability of obtaining a component of strength greater than $s_0$ is

$$P(s > s_0) = \int_{s_0}^{\infty} S(s)\mathrm{d}s \tag{3.11}$$

The probability of obtaining $n$ components of strength greater than $s_0$ is

$$P_n(s > s_0) = \left[ \int_{s_0}^{\infty} S(s)\mathrm{d}s \right]^n \tag{3.12}$$

In order to make use of the general expression (3.10) for reliability, we need to know the probability density function corresponding to the cumulative distribution (3.12), that is the probability density function of the series system considered as a single item. This is given by

$$S_n(s) = -\frac{\mathrm{d}}{\mathrm{d}s}\left[ P_n(s) \right] \tag{3.13}$$

That is

$$S_n(s) = n\left[\int_s^{\infty} S(s)\,\mathrm{d}s\right]^{n-1} S(s) \tag{3.14}$$

Hence substituting for $S_n(s)$ in place of $S(s)$ in equation (3.10) gives the reliability as

$$R = \int_0^{\infty}\left\{n\left[\int_0^{\infty} S(s)\,\mathrm{d}s\right]^{n-1} S(s)\left[\int_0^{s} L(s)\,\mathrm{d}s\right]\right\}\mathrm{d}s \tag{3.15}$$

Although cumbersome, this expression can be evaluated once the functions $S(s)$ and $L(s)$ are known. It must again be emphasised that these are unlikely to be normal distributions, though in the absence of adequate evidence on actual distributions, the normal is probably the best assumption. It is most likely that the trends will be the same whatever the detail distribution, and hence the normal distribution was used to derive the behaviour illustrated in figure 3.10.

*Exercise 3.5*

Show that as an alternative to equation (3.10) the reliability of a system with distributed loads $L(s)$ and strength $S(s)$ may be written as

$$R = \int_0^{\infty}\left[L(s)\int_s^{\infty} S(s)\,\mathrm{d}s\right]\mathrm{d}s$$

where $R$ is the reliability and $s$ the strength.

In section 3.1 we examined the validity of the product rule through a graphical interpretation, as in figure 3.10. However, a more rigorous examination of this aspect can be made in a purely mathematical approach, as follows.

If we have a series system of $n$ dissimilar components, equation (3.12) can be written as

$$P_n(s > S_0) = \prod_{i=1}^{i=n}\left[\int_{S_0}^{\infty} S_i(s)\,\mathrm{d}s\right] \tag{3.16}$$

Hence the probability density function of the strength for the series system becomes

$$S_n(s) = -\frac{\mathrm{d}}{\mathrm{d}s}\left\{\prod_{i=1}^{i=n}\left[\int_s^{\infty} S_i(s)\,\mathrm{d}s\right]\right\} \tag{3.17}$$

Substituting this in the equation deduced in exercise 3.5 gives an expression for the reliability of the system as

$$R_n = \int_0^\infty L(s) \left\{ \prod_{i=1}^{i=n} \left[ \int_s^\infty S_i(s)\,ds \right] \right\} ds \tag{3.18}$$

It is readily seen that the overall reliability $R_n$ is not equal to the product of the reliabilities of the individual components, which is given by the expression in curly brackets, so long as $L(s)$ is allowed to take any arbitrary form.

If, however, we are dealing with infinitely smooth loading, the density function $L(s)$ has the value zero for all values of $s$ except those lying between $\overline{L}$ and $(\overline{L} + ds)$, in which range it has the value of $1/ds$. Evaluation of the integral then gives the reliability as

$$R_n = \prod_{i=1}^{i=n} \left[ \int_{\overline{L}}^\infty S_i(s)\,ds \right] \tag{3.19}$$

that is, in the special case of infinitely smooth loading the overall reliability is equal to the product of the individual component reliabilities, which was the conclusion reached physically and graphically in section 3.1. In no other circumstances is it true.

If the loading is infinitely rough so that the strength density function is zero for all values of $s$ except those lying between $\overline{S}$ and $(\overline{S} + ds)$ in which range it has the value $1/ds$, then

$$\int_s^\infty S_i(s)\,ds = \begin{cases} 1 \text{ for } s < \overline{S}_i \\ 0 \text{ for } s > \overline{S}_i \end{cases} \tag{3.20}$$

Whence it follows that

$$\prod_{i=1}^{i=n} \int_s^\infty S_i(s)\,ds = \begin{cases} 1 \text{ for } s < (\overline{S}_i)_{\min} \\ 0 \text{ for } s > (\overline{S}_i)_{\min} \end{cases} \tag{3.21}$$

where $(\overline{S}_i)_{\min}$ is the minimum value of the unique strengths, that is, the strength of the weakest component. Therefore substituting in (3.18) we obtain the reliability of the series system as

$$R_n = \int_0^{(\overline{S}_i)_{\min}} L(s)\,ds \tag{3.22}$$

$$= \text{reliability of weakest component}$$

Hence with infinitely rough loading the reliability is equal to the reliability of the weakest component, which agrees with the conclusion reached physically and graphically in section 3.1.

We can thus conclude that for a series system the overall reliability will be given by the product of the component reliabilities (product rule) if the loading is infinitely smooth and by the reliability of the weakest component if it is infinitely rough. For intermediate values of the loading roughness the reliability will lie between these extremes.

*Exercise 3.6*

It is reasonable to expect that several types of failures may occur in a component. If these are all independent so that there is no interaction of one failure process on another prior to actual failure, and if the occurrence of a single failure mode implies the total failure of the item, show that the overall failure rate is the arithmetic sum of all the individual failure rates.

*Exercise 3.7*

A mass-produced part is 100 per cent inspected to ensure two critical dimensions meet the specification. The inspection is carried out by semi-skilled personnel using purpose-designed gauges, each measurement being made independently. Each inspection process may fail in two ways

(1) A part which fails to meet the specification may be gauged as acceptable
(2) A part which meets the specification may be rejected by the gauge.

The inspection process as a whole fails if an unacceptable part is not rejected or if an acceptable part is rejected.

If $P_1$ is the probability of each gauge accepting a part that fails to meet the specification and $P_2$ is the probability of each gauge rejecting a part which does meet the specification, show that the reliability of the inspection process is given by

$$R = 1 - \{2\,\alpha P_1 + (1 - 2\alpha) P_2\,(2 - P_2)\}$$

where $\alpha$ is the probability of failure to meet each critical dimension.

*Exercise 3.8*

An aeroplane is propelled by four engines but is capable of flying on any three of them. If each of the engines is totally independent and each has a reliability of $R$ at a specified number of hours of operation, show that the reliability of the propulsive system, at the specified number of hours is given by

$$R^4 + 4R^3\,(1 - R)$$

Comment on the validity of this expression in the real situation, and derive an alternative expression for the reliability, taking into account any relevant factors not included in the above expression if you think it necessary.

**Exercise 3.9*

Show that for a parallel system in which all the channels are always active the probability of system failure is equal to the product of the probability of failure of the individual channels if the loading is ideally smooth, but that if the loading is infinitely rough the reliability of the system is equal to the reliability of the most reliable channel.

# 4 The variation of reliability with time

## 4.1 General physical approach

The second factor appearing in the list of essential features of reliability is time. To investigate the effects of this factor we shall again assume that we have a product whose strength is obtained from the distribution $S(s)$, and a load which has a distribution $L(s)$. The load is repeated many times from this distribution, the number of repetitions being some function of time. In this way time is introduced into the analysis.

We shall approach the problem of repeated loads in the same way as we tackled multiple components, by first examining physically the extreme cases of infinitely smooth and infinitely rough loading. As already stated, we shall assume in the first place that the strength of the product does not vary with time (that is, we shall be neglecting such features as fatigue, erosion, wear, corrosion, creep, etc.). Dealing then with infinitely smooth loading, if the load could be given a unique value $\bar{L}$ then it is obvious that all components whose strengths were less than the load $\bar{L}$ would fail on the application of the first load, while all components whose strength is greater than $\bar{L}$ would never fail, no matter how many times the load were repeated. The reliability would therefore remain constant with time. The variation of reliability and failure rate with time will thus take the form shown in figure 4.1 in which the failure rate will be greater than zero on the first application of the load, immediately dropping to zero and retaining this value throughout the remaining life, since $\mathrm{d}R/\mathrm{d}t$ is always zero.

Now let us suppose we are dealing with infinitely rough loading in which the items have an unique strength $\bar{S}$ and the load is taken at random from a distribution of probability density function $L(s)$. The reliability is the probability that no load taken at random from the distribution at any specified number of attempts exceeds the strength $\bar{S}$; or put in statistical phraseology it is the probability of the zero occurrence of an event in which the event is that the load exceeds the strength $\bar{S}$. The occurrences of events in a random situation is represented by the Poisson distribution

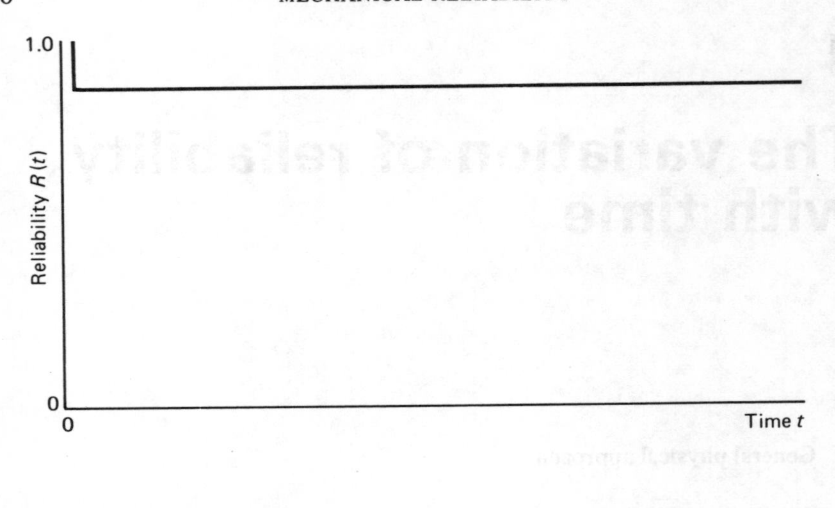


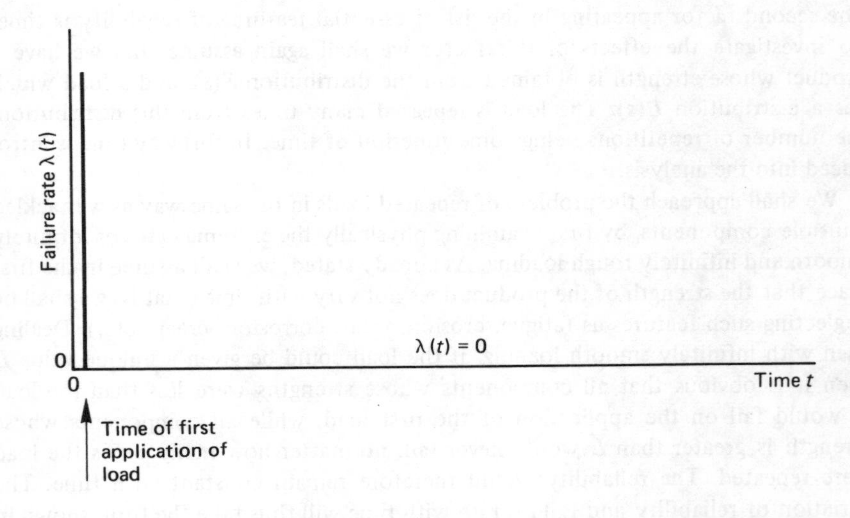


**Figure 4.1** Variation of reliability and failure rate with time for ideally smooth loading

$$e^{-z}\left(1 + z + \frac{z^2}{2!} + \frac{z^3}{3!} + \ldots\right) \tag{4.1}$$

in which the successive terms represent the probabilities of zero, one, two and so on, occurrences of a truly random event. Hence the zero occurrence is given by the first term. Identifying the variate $z$ with $\lambda t$ enables us to write

$$R(t) = e^{-\lambda t} \tag{4.2}$$

where λ is some arbitrary constant and $t$ is time. The failure rate, $\lambda(t)$, is obtained from equation (2.5) as

$$\lambda(t) = -\frac{\mathrm{d}R(t)}{\mathrm{d}t}/R(t) = \lambda \quad (4.3)$$

For infinitely rough loading, then, the reliability will always decrease with time according to a negative exponential law in which the constant (λ) is equal to the failure rate. The failure rate will be constant, and of finite value equal to λ, for all values of $t$, as illustrated in figure 4.2. As our theme develops we shall encounter the constant failure rate depicted above on many occasions. It arises directly from

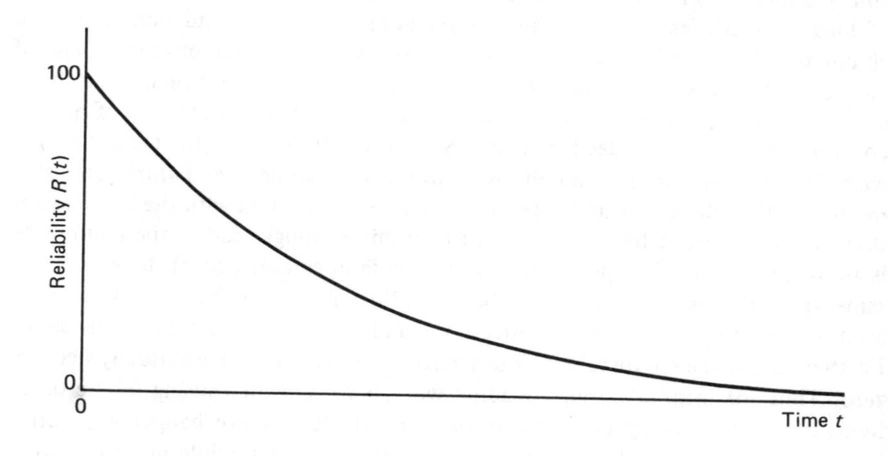


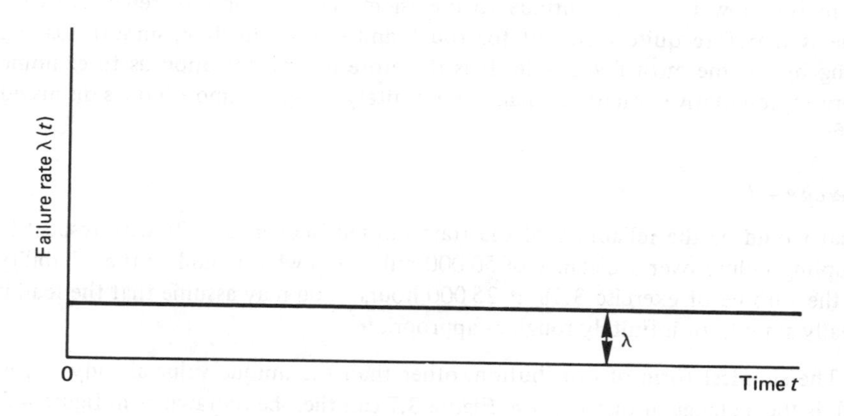


**Figure 4.2** Variation of reliability and failure rate with time for infinitely rough loading

the random occurrence of a load from an arbitrary distribution. Such a regime, characterised by a constant failure rate, is consequently referred to as the 'random phase' and the failure rate as the 'random failure rate'.

We immediately see that, as in the case of the series and parallel systems examined in the previous chapter, there are important differences between the rough and smooth loading conditions. Although we have only looked at the infinitely rough and smooth conditions, it is of advantage to isolate these differences before going on to more general cases. Of greatest significance is the failure rate after the initial application of load. If the loading is infinitely smooth the failure rate is zero, irrespective of the standard of design or the load involved! This may not be such an attractive position as it appears at first sight, since the reliability could be very low, all the failures having occurred on the first application of load. Nevertheless, if the manufacturer applies the first load himself he can eliminate all the failed items so that the user sees a perfect product. This, of course, is the basis of 'burn-in' or 'proofing' applied to real electronic or mechanical products. It is a very powerful technique – effectively a method of quality control – so effective indeed that in this ideal situation even if the mean strength were *less* than the mean load the user would encounter zero failure rate, as a moment's thought would indicate (*note*: redraw figure 3.9a with the load greater than the mean strength). By contrast, for infinitely rough loading the failure rate is finite (except in the special case when no failures ever occur). It retains this same value throughout the life of the item. Burn-in or proofing would be of no avail at all under these circumstances: the failure rate would remain the same. Furthermore if the failure rate, $\lambda$, is not zero, the reliability will eventually become zero. Thus for infinitely rough loading we can be sure of causing failure of all items if we continue operating them long enough (it may not happen in practice because operation could be terminated for other reasons), while in an infinitely smooth situation some items may be expected to survive (hopefully most of them) no matter how long we continue to use them. The response of reliability with time is therefore quite different for rough and smooth loading, smooth loading being by far the most favourable. It is therefore incumbent upon us to examine more representative situations than the infinitely rough or smooth ones discussed so far.

*Exercise 4.1*

What would be the reliability of the train quoted in exercise 3.3a with respect to coupling failure over a distance of 50 000 miles, and what would be the reliability of the turbine of exercise 3.3b at 25 000 hours? You may assume that the load is ideally smooth or infinitely rough as appropriate.

The simplest form of distribution, other than the unique value already discussed, is the rectangular distribution. Figure 3.7 can then be redrawn as in figure 4.3. This has been drawn so that the load distribution intersects the strength distribution in order to replicate the original model. If the difference between the mean

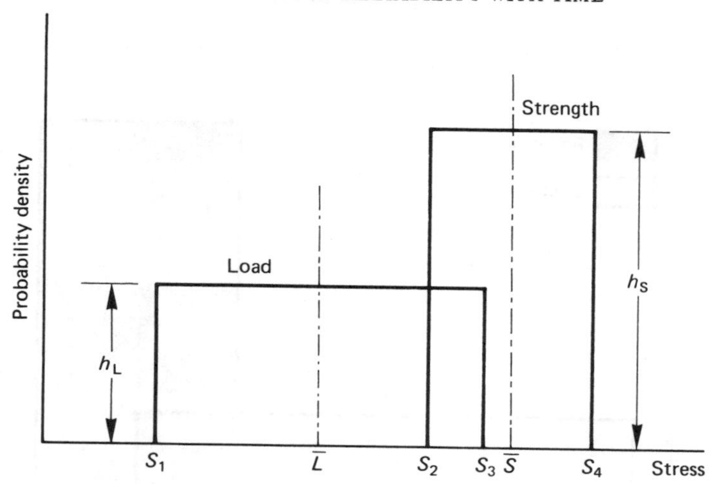


**Figure 4.3** Rectangular distributions

load and the mean strength were so great that no intersection occurred there could, of course, be no interaction and the reliability would always be 100 per cent and the failure rate zero. If $s_1$, $s_2$, $s_3$, and $s_4$ are the limiting stresses of the rectangular distributions as indicated on figure 4.3, then it is immediately obvious that no item whose strength is greater than $s_3$, the maximum load, can ever fail – no matter how many times the load is repeated at random from its distribution. All items whose strength lies between $s_3$ and $s_4$ will always survive and the ultimate reliability when all failures have taken place will be $h_S(s_4 - s_3)$, the areas shown lightly shaded in figure 4.4. Since in that ultimate regime the reliability is constant the failure rate will be zero. Hence the life pattern will consist of an initial fairly high failure rate, falling as the potentially vulnerable items (those represented by the heavily shaded areas in figure 4.4) are gradually eliminated, until the ultimate reliability is reached, when the failure rate becomes zero. A falling failure rate with age which occurs when an item is first brought into service is a characteristic feature of a number of components and systems, and such failures are called 'early life failures'. The period over which they occur is designated 'early life'. The early life period may be followed by any failure pattern. During the early life the weak components are gradually eliminated to leave only those that can survive the loads applied. The early life is therefore one of 'survival of the fittest' or 'natural selection' or 'natural quality control'.

**Exercise 4.2*

Prove mathematically that when the load and strength distributions are intersecting rectangles the whole life failure rate-age characteristic is of early life form and that the ultimate reliability is given by

$$R(\infty) = h_S(s_4 - s_3)$$

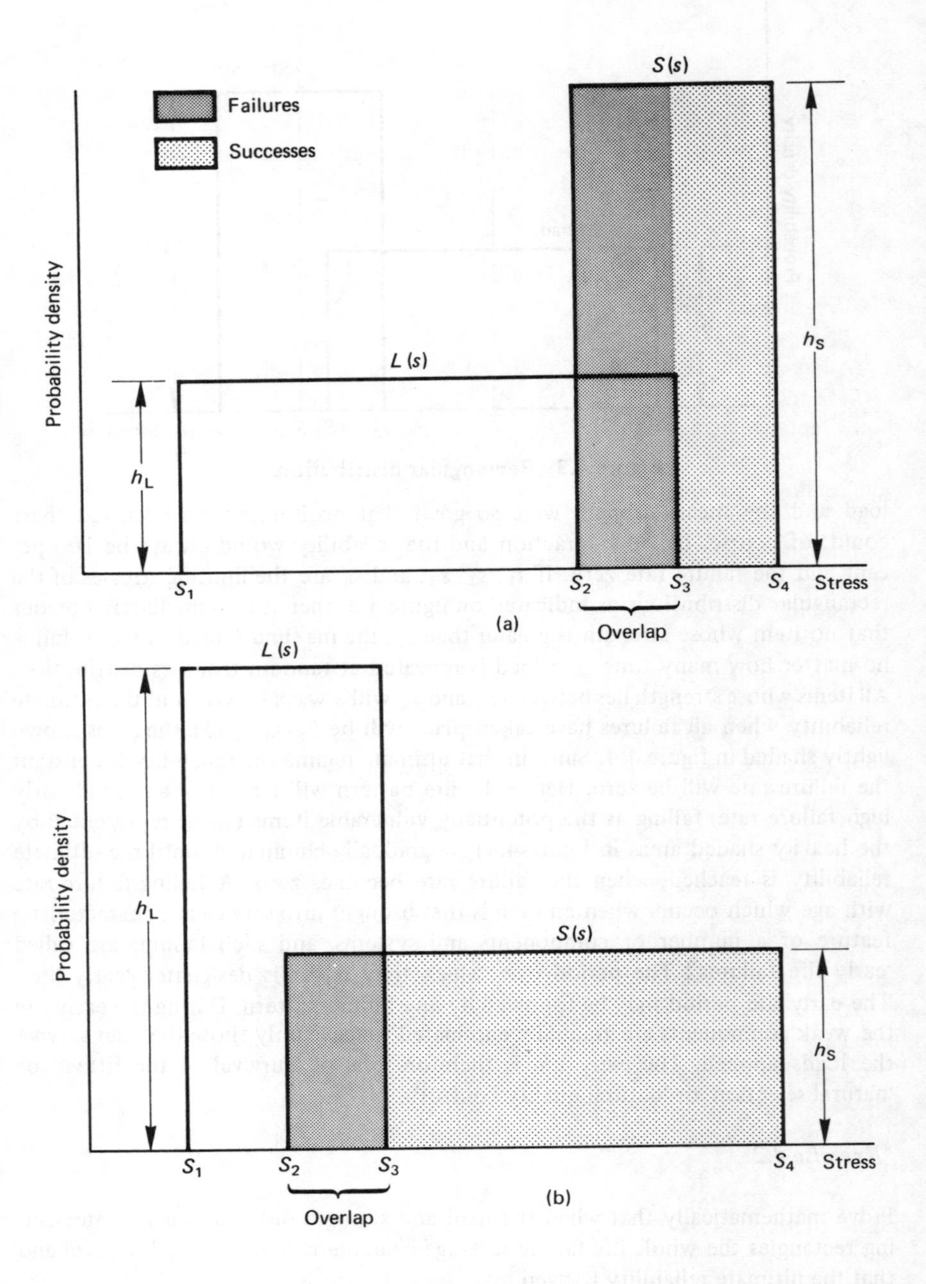


**Figure 4.4** Comparison of distributions for rough (a) and smooth (b) loading

This treatment of simple rectangular distributions brings out two important points. In the first place we see just why loading roughness is so important. Failures can only occur in the overlap region and by virtue of the fact that loads are repeated – in practice a very large number of times, if not the infinite number of the theoretical calculations – all items whose strength lies in the overlap region must eventually fail. A narrow distribution of strength associated with rough loading implies a high number of items fall in this region and that the number of failures is high. Conversely smooth loading implies that the strength is well distributed, that fewer items have a strength falling in the overlap region, and hence the number of failures is much smaller. This is well brought out by comparing figures 4.4a and 4.4b. The distributions in figure 4.4b have been obtained by interchanging those of figure 4.4a. This moves the model from the rough end of the loading roughness spectrum to the smooth end. The total failures, represented by the dark shaded areas, are considerably reduced. In the second place these wholly rectangular distributions are representative of the bulk of practical populations. Hence when the main bulk of the populations react we always have a falling failure rate–age curve.

It may well be argued that the rectangular distributions just discussed are not very relevant in as much as no real item would show such a dramatic cut-off at the limiting stresses $s_1$, $s_2$, $s_3$ and $s_4$, but would exhibit at least some rudimentary tail (but see ref. 12, giving the strength distribution for aircraft wings). In order to achieve a slightly more realistic model, therefore, small rectangular tails have been added to the main rectangular distributions as shown on figure 4.5. In this case the tail of the relevant distribution has been made to interact with the main body of the other distribution, interaction of the main bodies themselves being covered by our previous discussion. Once again the limiting stresses are denoted by $s_1$, $s_2$, $s_3$ and $s_4$.

We may proceed in exactly the same way as for wholly rectangular distributions by stating that no item whose strength is in excess of $s_3$, that is, all those whose strength lies between $s_3$ and $s_4$, can ever fail. Provided the load is repeated a sufficient number of times, all those whose strength is less than $s_3$, that is, those whose strength lies between $s_2$ and $s_3$, must fail. In figure 4.6 the successes and failures have been indicated by light and dark shading as before. In this figure we have again made a comparison by interchanging the load and strength distributions. Here the effect is more dramatic than with wholly rectangular distributions. If the strength distribution has the tail then it is only those weak items in the tail itself, whose strength is less than the maximum load, which fail – a small number. All the others succeed, so that the ultimate reliability is fairly high, the failure rate–age curve being one which falls quickly in an early life type of characteristic to this value. In the case of a tail to the load distribution, however, all those items whose strength is less than the maximum load must eventually fail; and this proportion can be quite high as shown by the dark shaded area on figure 4.6a. The proportion ultimately failing will depend solely on the extent of the load tail; if the tail is long enough it could be all items, that is, the ultimate reliability is 0 in

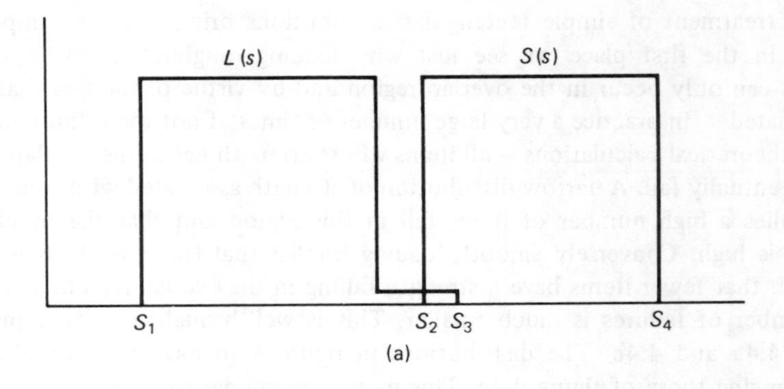


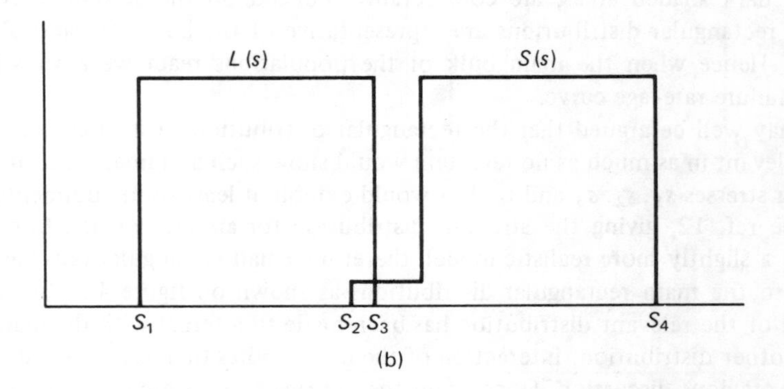


**Figure 4.5** Rectangular distributions with rectangular tails: (a) load distribution with tail, (b) strength distribution with tail

these circumstances. This raises an important point: the ultimate reliability of zero obtained from the negative exponential distribution ($e^{-\lambda t}$ for $t \to \infty$) is a consequence of the infinite tail of the normal distribution. One is forced therefore to reconsider how far these statistical distributions are representative of the actual situation, and are valid in reliability studies. The reader may at this stage like to re-appraise his solution to exercise 3.2 and/or the author's observations on this exercise. The conclusions reached there are of prime significance, since the length of the tail of actual distributions cannot be easily measured, but can have a profound effect on the total number of failures. We shall need, therefore, to be sure we are modelling the real situation, not some imaginary world of our own creation, before making any practical deductions.

There is one redeeming feature in the role played by the tail. When the strength tail intersects the main body of the load distribution the probability density of the load is high so that all values of the load may be expected to occur frequently. We can then expect consequential failure at an early stage with an early life character-

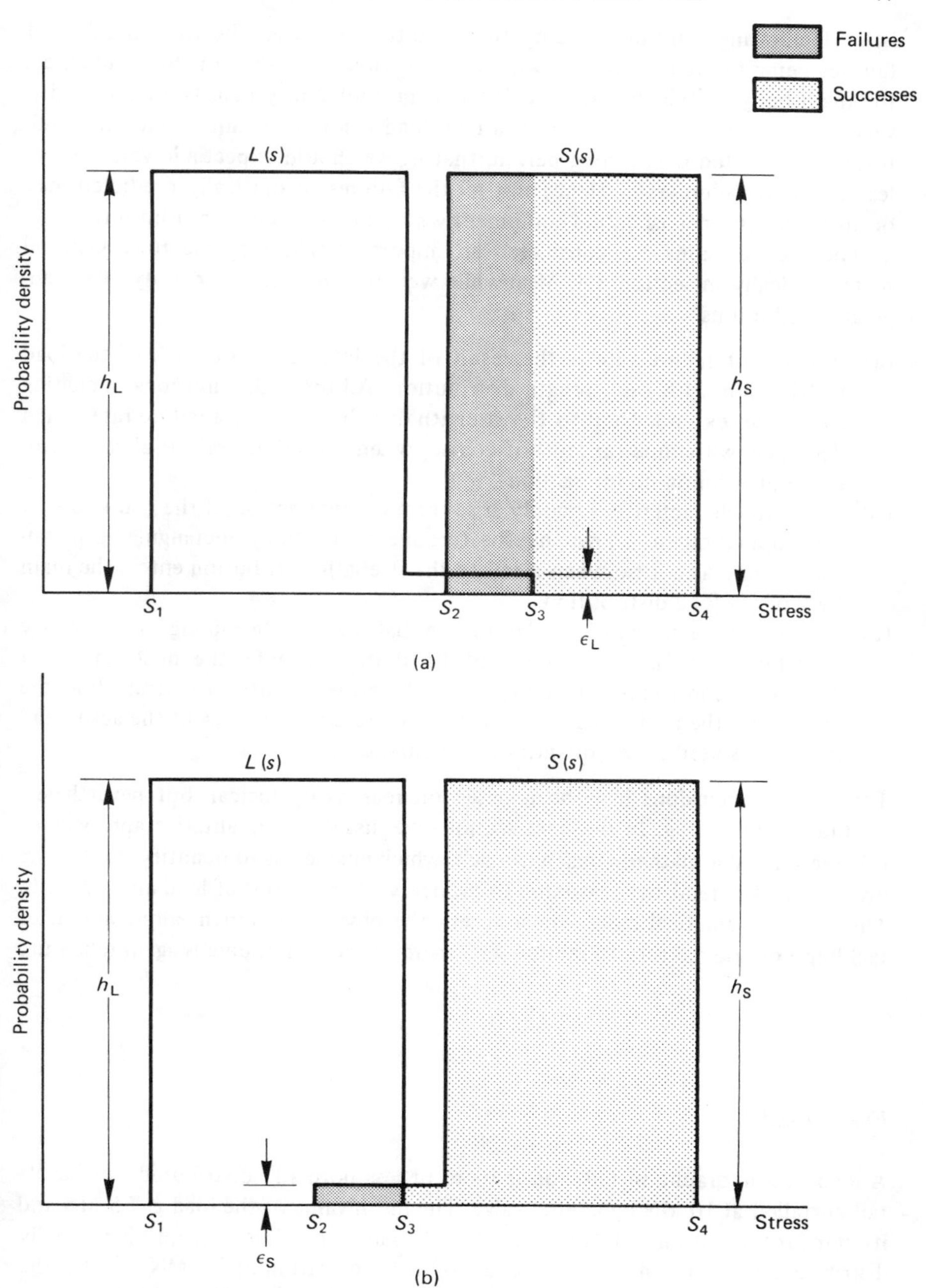


**Figure 4.6** Distributions with tails: (a) load distribution with tail, (b) strength distribution with tail

istic (facilitating burn-in), leading to the ultimate stage. The total number of failures being limited to the tail will be small. However, when the load tail enters the main body of the strength distribution the probability density of the load is very low, so that the occurrence of a high load is rare and failures are more uniformly distributed over a long period, that is, we should expect a lower, more or less constant failure rate. Hence not all the failures theoretically predicted may occur in the life span of actual equipment whose life is certainly not infinite.

These conclusions are considered so important that they are fully justified mathematically in section 4.4. Meanwhile we can summarise our study of rectangular distributions.

(a) A constant failure rate is the result of the interaction of a tail of the load distribution with the strength distribution. All other distributions, including that of an extended tail to the strength distribution, give a failure rate which decreases with time (except, of course, when the failure rate is always zero, consequent upon no interaction).
(b) Early life characteristics mainly arise from the interaction of the main bulk of the populations as shown by the treatment of wholly rectangular distributions. They also arise when a tail to the strength distribution enters the main bulk of the load distribution.
(c) The tails of the distributions play an essential role in determining the reliability or failure rate. The tail of the load distribution is by far the most significant parameter controlling reliability. It is therefore essential to ensure that the tail of any theoretical classical distribution does truly represent the actual tail before it is used in any quantitative calculations.

The above conclusions have been based on reasonably logical, but nevertheless qualitative, reasoning. In the next sections we justify the qualitative approach of this section by a mathematical approach, which enables us to quantify the factors involved and extend the reasoning to arbitrary distributions of load and strength. The non-mathematical reader must accept the physical argument alone and after tackling exercise 4.3 turn to section 4.4 where the general theme is again taken up.

*Exercise 4.3*

A load in a hydraulic installation is more or less normally distributed but has its tail curtailed at 10 MN by a relief valve. The mean value of the load is 7.5 MN and its standard deviation is 1 MN. It is applied to an item whose strength is normally distributed with a mean of 15 MN and a standard deviation of 2.5 MN. What is the ultimate reliability of this product? (Note the cut-off at 10 MN would also be distributed in actual practice but for the purpose of this exercise you can take it as absolutely defined.)

## *4.2 Mathematical analysis

To create our mathematical model we retain the model described by figure 3.7 and the assumption that the strength does not deteriorate with time. Assume we have a product whose strength is obtained from the distribution $S(s)$. Now let a load taken each time from a distribution $L(s)$ be applied $n$ times. For a single application of load the probability that it has a stress less than $s_0$ is given by

$$P(s < s_0) = \int_0^{s_0} L(s)\,\mathrm{d}s \tag{4.4}$$

The probability that the stress is less than $s_0$ for $n$ applications of load is then

$$P_n(s < s_0) = \left[\int_0^{s_0} L(s)\,\mathrm{d}s\right]^n \quad \text{for } n > 1 \tag{4.5}$$

This is a cumulative probability and cannot be used directly in equation (3.10). The corresponding probability density function is

$$\begin{aligned} L_n(s) &= \frac{\mathrm{d}}{\mathrm{d}s}\left[P_n(s < s_0)\right] \\ &= \frac{\mathrm{d}}{\mathrm{d}s}\left[\int_0^{s} L(s)\,\mathrm{d}s\right]^n \end{aligned} \tag{4.6}$$

Hence substituting in the general expression (3.10) for reliability with distributed load and strength

$$R(n) = \int_0^{\infty} \left\{S(s)\left[\int_0^{s} L(s)\,\mathrm{d}s\right]^n\right\} \mathrm{d}s \tag{4.7}$$

Putting $n = \psi(t)$

$$R(t) = \int_0^{\infty} \left\{S(s)\left[\int_0^{s} L(s)\,\mathrm{d}s\right]^{\psi(t)}\right\} \mathrm{d}s \tag{4.8}$$

where the symbol $R(t)$ denotes that the reliability is now a function of time.

In general one can expect a fairly complicated form of the function $\psi(t)$, but for many purposes it will be quite adequate to take a simple linear relation, and so

$$R(t) = \int_0^{\infty} \left\{S(s)\left[\int_0^{s} L(s)\,\mathrm{d}s\right]^{kt}\right\} \mathrm{d}s \tag{4.9}$$

The failure rate corresponding to equation (4.9) is given by the analytical expression

$$\lambda(t) = -\frac{\mathrm{d}R(t)}{\mathrm{d}t} \Big/ R(t) \tag{4.10}$$

and theoretically the failure rate corresponding to (4.7) can be obtained by replacing $t$ by $n$ in (4.10). However $n$ is essentially an integral quantity, and in practical terms the exact failure rate per application has to be expressed by the 'approximation'

$$\lambda(n) = \frac{R(n-1) - R(n)}{R(n-1)} \tag{4.11}$$

where

$$R(0) = 1 \tag{4.12}$$

since neither (4.7) nor (4.9) is valid for $n$ less than one, and the reliability before any load has been applied must be 1 (that is, 100 per cent) in the absence of any strength deterioration.

Equations (4.7) and (4.9) are probably the most important in the theory of reliability. They enable us to calculate how reliability varies with time for any load and strength distributions. Furthermore they can easily be extended to cover the case where the strength deteriorates, or grows as in development, that is, where the strength distribution is a function of time. Ideally they could be used to calculate the reliability of specific items. The difficulty, if we try to do this, is assigning values to the distributions $L(s)$ and $S(s)$, and also to $k$ if we wish to express the reliability as a function of time. Very often the functions are unknown. This difficulty is compounded by the fact that the reliability is, as we shall see, very sensitive to the numerical values assigned to these functions. Nevertheless we can learn a great deal about reliability by studying its response to change of different variables when $L(s)$ and $S(s)$ are represented by specific distributions. We shall find that we can isolate those factors to which the reliability is so sensitive and hence formulate a very practical approach to the subject.

**Exercise 4.4*

All the usual statistical distributions have extended, if not infinite, tails, whereas real strength distributions may have a cut-off at some low value of the strength introduced by quality control action, as discussed in the solution to exercise 3.2. Derive alternative forms for equations (4.7) or (4.9) which allow this factor to be introduced into any calculations involving manufactured items to which quality control has been applied.

**Exercise 4.5*

During the repeated application of a load a number of items may fail, so that on the application of the $n$th load the distribution of strength of the remaining items could differ from the original distribution. Show that the probability density function of the strength distribution at the $n$th application of load will be given by

$$S'_n(s) = S(s)\left\{\int_0^s L(s)\,ds\right\}^{n-1}$$

**Exercise 4.6*

During the application of load to an item a certain amount of damage may be done so that its strength deteriorates. The simplest assumption that can be made is that the mean strength is reduced by an amount $\Delta s_n$ on the application of the $n$th load but the standard deviation of the strength remains unchanged. Show that the reliability of such an item will be given by

$$R(n) = \int_0^\infty \left\{\prod_{i=1}^{i=n} \int_0^{s-\Delta s_{i-1}} L(s)\,ds\right\} ds$$

where $S(s)$ = initial probability density function of strength
$L(s)$ = initial probability density function of load
$i$ = applications of load
$\Delta s_i$ = loss of mean strength on $i$th application of load
$n$ = total number of load applications.

## *4.3 Mathematical treatment of rectangular distributions with tails

The general expression for the reliability of the systems depicted in figure 3.7 is given by equation (4.7) as

$$R(n) = \int_0^\infty S(s)\left\{\int_0^s L(s)\,ds\right\}^n ds \tag{4.13}$$

which can be written as

$$R(n) = \int_0^\infty \phi\,ds \tag{4.14}$$

where

$$\phi = S(s)\left\{\int_0^s L(s)\,ds\right\}^n ds$$

Then we can write

$$R(n) = \int_0^{s_1} \phi\,ds + \int_{s_1}^{s_2} \phi\,ds + \int_{s_2}^{s_3} \phi\,ds + \int_{s_3}^{s_4} \phi\,ds + \int_{s_4}^{\infty} \phi\,ds \tag{4.15}$$

Now let the probability density of the main load and strength distributions be $h_L$ and $h_S$ respectively and of the tails be $\epsilon_L$ and $\epsilon_S$ respectively, both of which are small. Considering first the situation in which the load has a tail, as in figure 4.10a, the individual integrals in equation (4.15) can be evaluated

$$\int_0^{s_1} \phi \, ds = 0 \quad \text{since both } L(s) \text{ and } S(s) \text{ are zero in this range} \tag{4.16}$$

$$\int_{s_1}^{s_2} \phi \, ds = 0 \quad \text{since } S(s) \text{ is zero in this range} \tag{4.17}$$

$$\int_{s_2}^{s_3} \phi \, ds = \int_{s_2}^{s_3} h_s \left\{ \int_{s_1}^{s} L(s) ds \right\}^n ds \tag{4.18}$$

But we can write

$$\left\{ \int_{s_1}^{s} L(s) ds \right\}^n = \left\{ 1 - \epsilon_L (s_3 - s) \right\}^n \tag{4.19}$$

obtained by taking the total integral of $L(s)$ as unity and subtracting the small rectangle bounded by the stresses $s_3$ and $s$. We may then consider two possibilities. If $n$ is very large indeed, since the expression in curly brackets is less than unity it follows that

$$\left\{ \int_{s_1}^{s} L(s) ds \right\}^n = 0 \tag{4.20}$$

Alternatively we may assume that $n$ is small or only moderately large, so that the product $n \, \epsilon_L$ is reasonably small. In that case

$$\left\{ \int_{s_1}^{s} L(s) ds \right\}^n = \left\{ 1 - \epsilon_L (s_3 - s) \right\}^n$$

$$= \{1 - n \, \epsilon_L (s_3 - s)\} \tag{4.21}$$

Then

$$\int_{s_2}^{s_3} \phi \, ds = \int_{s_2}^{s_3} h_S \{1 - n \, \epsilon_L (s_3 - s)\} \, ds$$

$$= \left[ h_S s + \frac{h_S n \, \epsilon_L (s_3 - s)^2}{2} \right]_{s_2}^{s_3}$$

$$= h_S \left\{ (s_3 - s_2) - \frac{n \, \epsilon_L (s_3 - s_2)^2}{2} \right\} \tag{4.22}$$

Additionally

$$\int_{s_3}^{s_4} \phi \, ds = \int_{s_3}^{s_4} h_S \{1\}^n \, ds$$

$$= h_S (s_4 - s_3) \tag{4.23}$$

and

$$\int_{s_4}^{\infty} \phi \, \mathrm{d}s = 0 \quad \text{since both } L(s) \text{ and } S(s) \text{ are zero in this range} \tag{4.24}$$

Hence we obtain two values of the reliability depending on the value of $n$. If $n$ is very large

$$R(n) = h_S(s_4 - s_2) \tag{4.25}$$

If $n$ is moderately large

$$R(n) = h_S(s_3 - s_2) \left\{1 - \frac{n\, \epsilon_L(s_3 - s_2)}{2}\right\} + h_S(s_4 - s_3) \tag{4.26}$$

The failure rate is obtained by differentiating with respect to $n$. Hence in the former case $\mathrm{d}R(t)/\mathrm{d}n = 0$ and the failure rate is zero. In the latter case

$$\frac{\mathrm{d}R(n)}{\mathrm{d}n} = -\frac{h_S\, \epsilon_L(s_3 - s_2)^2}{2} \tag{4.27}$$

and the failure rate is given by

$$\lambda = -\frac{\mathrm{d}R(n)}{\mathrm{d}n}/R$$

$$= \frac{\frac{1}{2} h_S\, \epsilon_L(s_3 - s_2)^2}{h_S(s_3 - s_2)\left\{1 - \frac{n\, \epsilon_L(s_3 - s_2)}{2}\right\} + h_S(s_4 - s_3)} \tag{4.28}$$

If $n\, \epsilon_L$ is small in relation to unity we may neglect the term $\frac{1}{2} n\, \epsilon_L(s_3 - s_2)$ so that the denominator of equation (4.28) becomes $h_S(s_4 - s_2)$ which is equal to unity. We can thus write

$$\lambda = \tfrac{1}{2} h_S\, \epsilon_L(s_3 - s_2)^2 \tag{4.29}$$

Hence if the load has a tail there is no early life characteristic, and the failure rate is constant for small or moderate values of $n$, decreasing to zero for very large values of $n$, when the ultimate reliability given by equation (4.25) will be achieved. It should be noted that the finite failure rate is directly proportional to the probability density of the load tail, but to the square of its penetration into the strength distribution.

We turn now to the case in which the strength distribution has a tail. In that case.

$$\int_0^{s_1} \phi \, \mathrm{d}s = 0 \tag{4.30}$$

$$\int_{s_1}^{s_2} \phi \, \mathrm{d}s = 0 \tag{4.31}$$

$$\int_{s_2}^{s_3} \phi \, \mathrm{d}s = \int_{s_2}^{s_3} \epsilon_S \left\{ \int_{s_1}^{s} h_L \, \mathrm{d}s \right\}^n \mathrm{d}s$$

$$= \int_{s_2}^{s_3} \epsilon_S \{h_L s - h_L s_1\}^n \, \mathrm{d}s$$

$$= \frac{\epsilon_S}{n+1} \frac{[\{h_L s_3 - h_L s_1\}^{n+1} - \{h_L s_2 - h_L s_1\}^{n+1}]}{h_L}$$

$$= \frac{\epsilon_S}{n+1} \frac{\{1 - (h_L s_2 - h_L s_1)^{n+1}\}}{h_L} \tag{4.32}$$

$$\int_{s_3}^{s_4} \phi \, \mathrm{d}s = \int_{s_3}^{s_4} S(s) \{1\}^n \, \mathrm{d}s$$

$$= \int_{s_3}^{s_4} S(s) \mathrm{d}s$$

$$= 1 - \epsilon_S(s_2 - s_3) \tag{4.33}$$

$$\int_{s_4}^{\infty} \phi \, \mathrm{d}s = 0 \tag{4.34}$$

whence the expression for reliability becomes

$$R(n) = \frac{\epsilon_S}{n+1} \frac{\{1 - (h_L s_2 - h_L s_1)^{n+1}\}}{h_L} + 1 - \epsilon_S(s_2 - s_3) \tag{4.35}$$

The failure rate will be given by

$$\lambda = -\frac{\mathrm{d}R(n)}{\mathrm{d}n}$$

$$= \frac{\dfrac{\epsilon_S}{(n+1)^2} \dfrac{\{1 - (h_L s_2 - h_L s_1)^{n+1}\}}{h_L} + \dfrac{\epsilon_S}{n+1} \dfrac{(h_L s_2 - h_L s_1)^{n+1} \log_e(h_L s_2 - h_L s_1}{h_L}}{\dfrac{\epsilon_S}{n+1} \dfrac{\{1 - (h_L s_2 - h_L s_1)^{n+1}\}}{h_L} + 1 - \epsilon_S(s_2 - s_3)} \tag{4.36}$$

It is difficult to see directly from equation (4.36) how the failure rate will vary with age (that is, with $n$), but if $n$ is large the second term in the numerator is always considerably less than the first, while the denominator approximates to $\{1 - \epsilon_S(s_2 - s_3)\}$. Hence for large values of $n$

$$\lambda \approx \frac{\epsilon_S}{(n+1)^2 \, h_L \, \{1 - \epsilon_S(s_2 - s_3)\}} \approx \frac{\epsilon_S}{(n+1)^2 \, h_L} \tag{4.37}$$

For this form of load and strength distributions we therefore have a falling failure rate over the whole life, that is, an early life type of characteristic, the failure rate becoming zero for large values of $n$, when, from equation (4.35) we see that the reliability is

$$R(n) = 1 - \epsilon_S(s_2 - s_3) \tag{4.38}$$

If the density of the strength tail, $\epsilon_S$, is small, the ultimate reliability will be close to 100 per cent. Consequently the early life characteristic will be limited.

The mathematical analysis has enabled us to quantify, and hence verify, the qualitative conclusions reached previously. As illustration, the ultimate reliabilities given by equations (4.25) and (4.38) represent the areas lightly shaded in figure 4.9. If the strength distribution has a tail the failure rate decreases rapidly by virtue of the term in $1/(n + 1)^2$, whereas if the load distribution has a tail the failure rate remains constant, though maybe at a low value depending on the density of the tail and its penetration into the bulk of the strength distribution. The net effect is that the ultimate reliability of a situation in which the load has a tail is much lower than that in which the strength has a tail. The mathematical reason for this is apparent from the basic equation (4.13): since the term in $L(s)$ in this equation is raised to the power $n$, small changes in $L(s)$ are considerably amplified when $n$ is large, whereas small changes in $S(s)$ are only reflected by a proportional effect. Thus the load is much more significant than the strength, as already concluded at (c) in section 4.1.

## 4.4 Variation of reliability with time for normally distributed loads and strengths

As soon as we try to examine the behaviour of real items we have to discard the idealised distributions we have used so far and employ more representative ones. The normal distribution is the one most favoured since it represents the variation encountered in the presence of the random effects we may expect in connection with the load and strength distributions, $L(s)$ and $S(s)$. In these circumstances no physical or even analytical treatment is possible and so we shall have to resort to a numerical evaluation of equation (4.7). To facilitate this evaluation, and for reasons we shall discuss later, the normal distribution has been replaced by the Weibull distribution when obtaining all the numerical results quoted in this book. The Weibull distribution is fully discussed in section 6.4, and all we need note now is that it is a very flexible distribution which can be made to approximate very closely to the normal distribution.

Employing this technique, the reliability and hence the failure rate was calculated for a range of safety margins and loading roughnesses. The values obtained have been plotted in figures 4.7, 4.8, 4.9, 4.10 and 4.11, as $\log_{10}$(failure rate) against number of load applications. The above figures refer to loading roughnesses of 0.1, 0.3, 0.5, 0.7 and 0.9 respectively, and each figure contains curves for

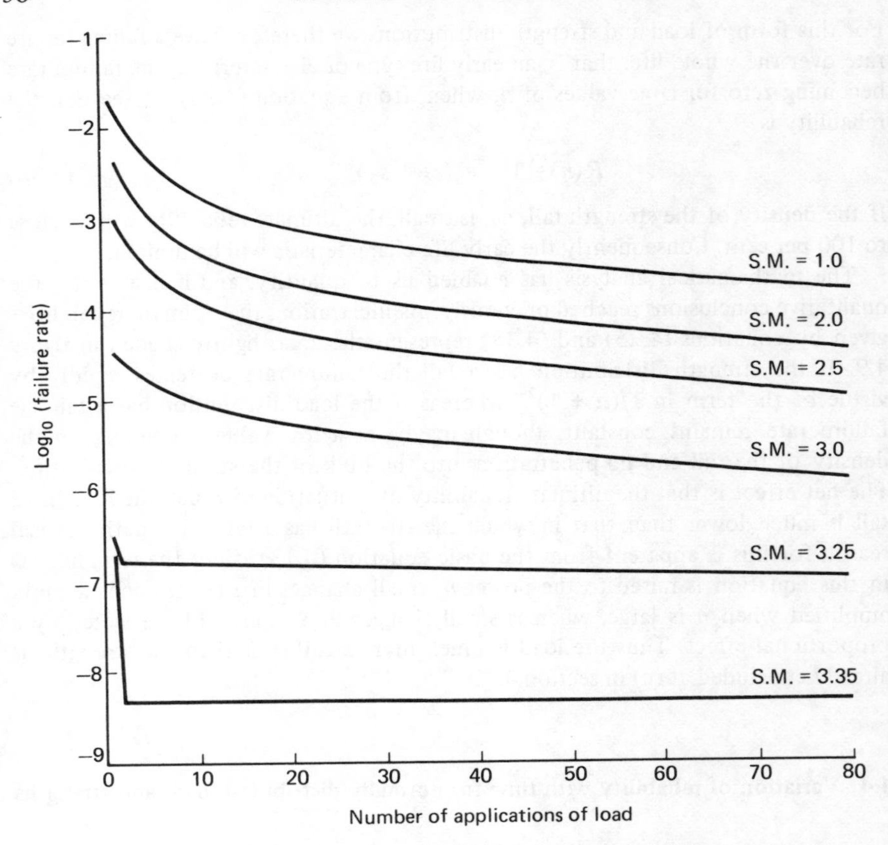


**Figure 4.7** Failure rate against number of applications of load for various safety margins (S.M.) at a constant loading roughness of 0.1. Strength distributions invariant. Near normal distributions

safety margins appropriate to the loading roughness. The logarithmic scale for failure rate was adopted to cover the wide range of values, so some distortion of the shape of individual curves is involved and must be allowed for in any detailed appraisal. Even with the logarithmic plotting it has not been possible to represent the curves for very high safety margins, which take the form of a single finite value for the first application followed by zero failure rate for all subsequent applications, within the accuracy of the calculations (1 in $10^{-12}$).

The general nature of these curves follows the pattern we should expect from our study of infinite rough and smooth distributions and rectangular distributions, with or without tails. Thus at low safety margins the curves show a failure rate which falls with age for all values of loading roughness. This is just what would be expected from our study of plain rectangular distributions, for at low safety margins the bulk of the normally distributed populations will be interacting, and

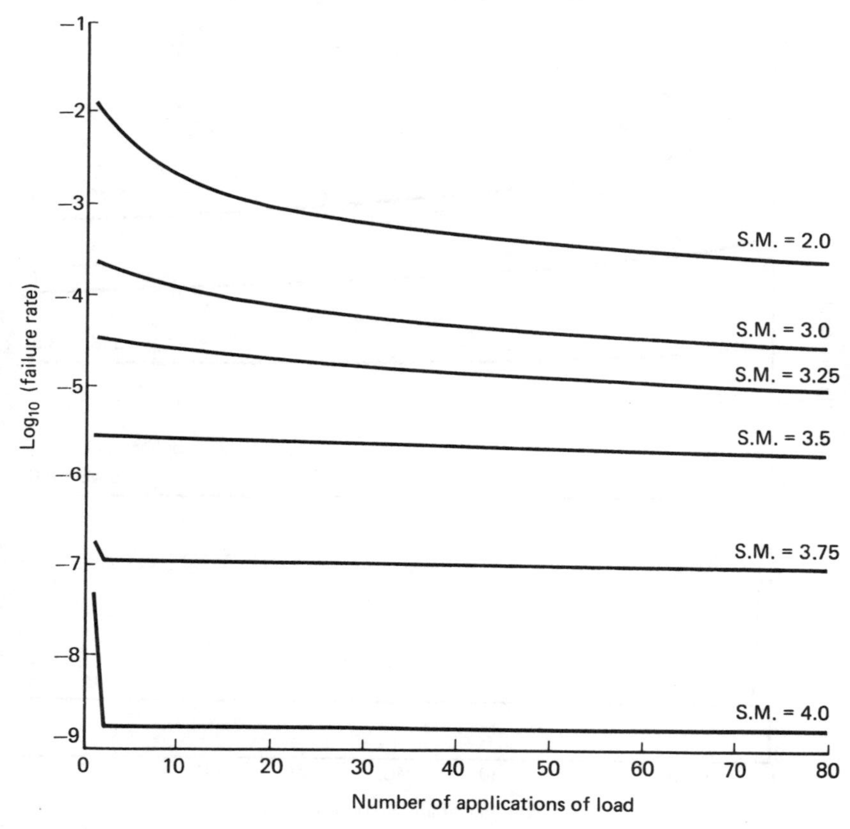


**Figure 4.8** Failure rate against number of applications of load for various safety margins (S.M.) at a constant loading roughness of 0.3. Strength distributions invariant. Near normal distributions

as we have seen, this always gives rise to a falling characteristic. Also, as we have seen from rectangular distributions, this is more pronounced at low loading roughness but less evident at high roughness. At the higher safety margins there is often a pronounced drop in the failure rate following the first application of load. It was anticipated in our study of concentrated loads when the loading is infinitely smooth, but persists for normally distributed loads and strengths to much higher values of the loading roughness than at first might be expected. It arises from the lower tail of the strength distribution. The extent to which it will occur in real instances consequently depends on that distribution.

Neglecting the first application of load, the shape of the curves at higher safety margins – the ones which would be employed in practice – can nearly always be defined by a constant failure rate. At high safety margins only the tails can be interacting, and from our study of rectangular distributions with tails we can

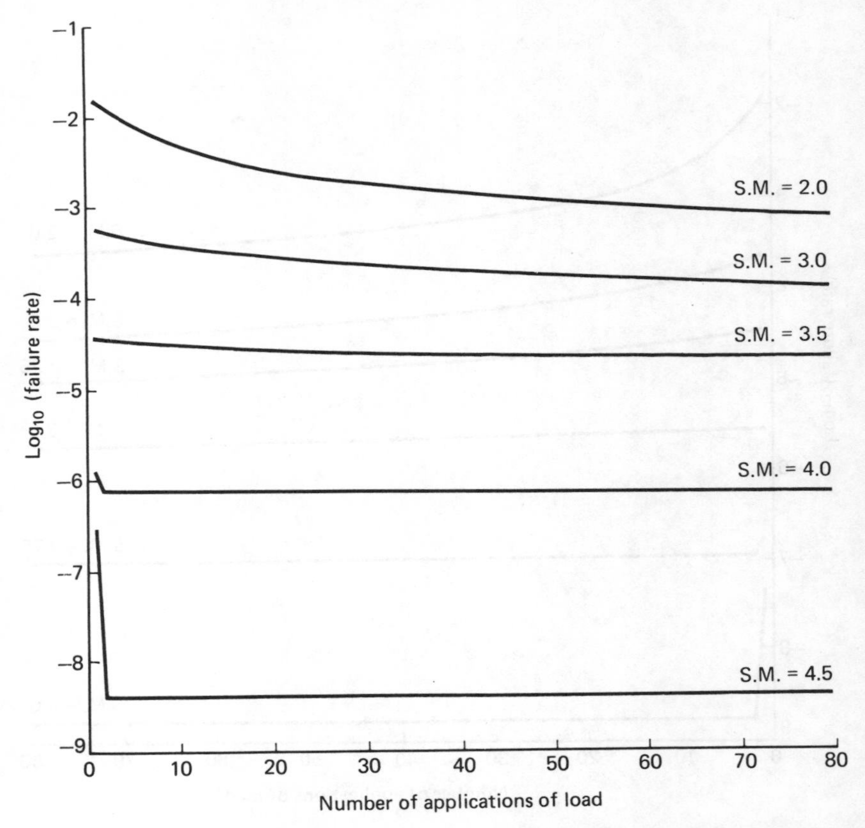


**Figure 4.9** Failure rate against number of applications of load for various safety margins (S.M.) at a constant loading roughness of 0.5. Strength distributions invariant. Near normal distributions

attribute constant failure rate to the tail of the load distribution. In these circumstances, therefore, it is the tail of the load distribution which is dictating the shape of the curves, and, of course, the numerical value of the failure rates. Since this is a very important practical area we should note the dominance of the tail of the load distribution. Control of the load, certainly its tail, would appear to be the most effective method of controlling reliability, and this conclusion must influence very considerably our approach to practical reliability. We shall have occasion to refer again to this feature as our thesis develops. It follows that the phrase 'under stated conditions (of use)' or its equivalent is an essential part of any definition of reliability. Any quoted values of reliability only apply to the exact loading conditions for which the data was collected, or the calculations carried out, and any reading across from one application to another is totally unjustified.

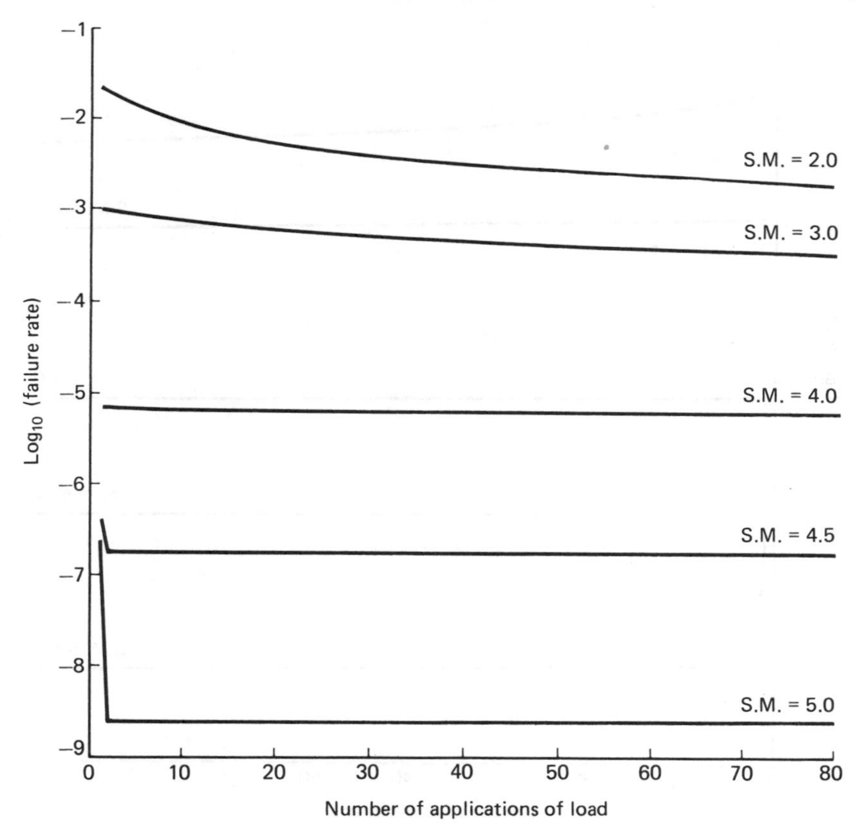


**Figure 4.10** Failure rate against number of applications of load for various safety margins (S.M.) at a constant loading roughness of 0.7. Strength distributions invariant. Near normal distributions

## 4.5 Intrinsic reliability

It has been demonstrated in the two previous sections that the tails of the distributions are of paramount importance in determining the reliability, and furthermore that the tail of the load distribution is of more importance than that of the strength over most of the effective life span. We have been able to make a number of other observations on the reliability and the variation to be expected from some load/strength distributions. Even so, the analysis we have carried out is only sufficient to give us an elementary understanding of the failure process. Further development could only be justified by a closer representation of the real situation. We then immediately run into the difficulties discussed in the solution to

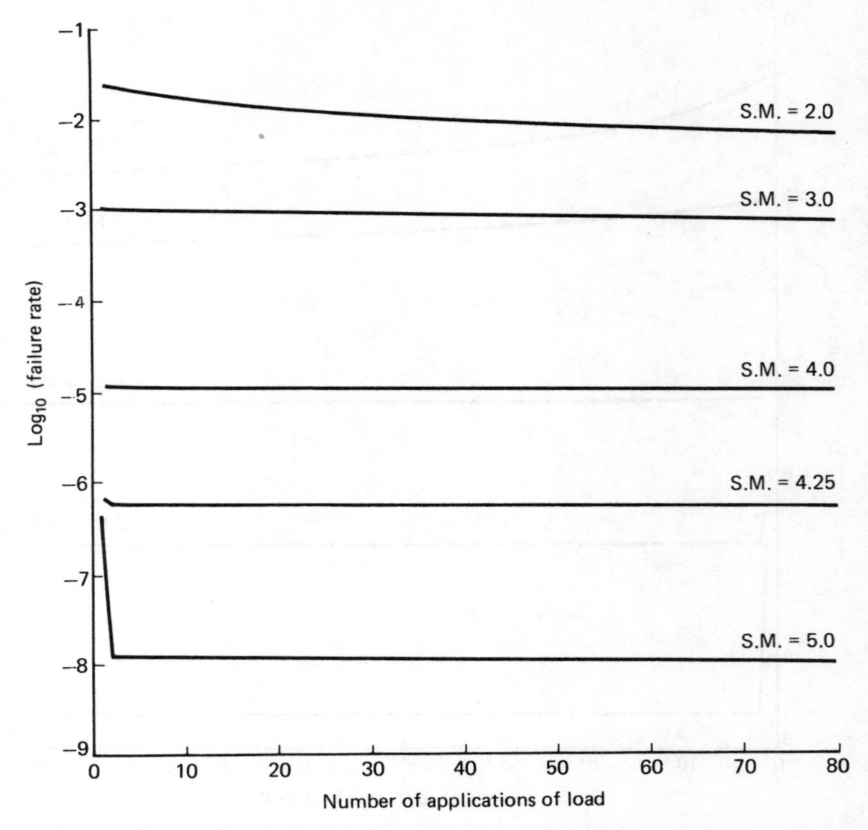


**Figure 4.11** Failure rate against number of applications of load for various safety margins (S.M.) at a constant loading roughness of 0.9. Strength distributions invariant. Near normal distributions

exercise 3.2: the information required to build up representative models is not available and is prohibitively expensive to obtain. It is unlikely ever to be available. We must therefore develop a philosophy of reliability without this information, dealing with the problem on a macroscopic rather than a microscopic level.

Let us revert to the plain rectangular distributions shown on figure 4.3. If the distributions did not intersect as illustrated on figure 4.12a, that is, if $s_2 > s_3$, then no failures could ever occur and the reliability would be 100 per cent. This would represent an ideal situation, and provides a target for our activities. As a first step towards reality we might extend our model to include small rectangular tails, as in figure 4.12b. Even in this case the limit of the tail is defined, so if we arranged that the same condition is satisfied, that is, that $s_2 > s_3$, as in figure 4.12c, we shall achieve the same objective of 100 per cent reliability. We have of course to pay for the tails by way of a higher safety margin and hence a heavier

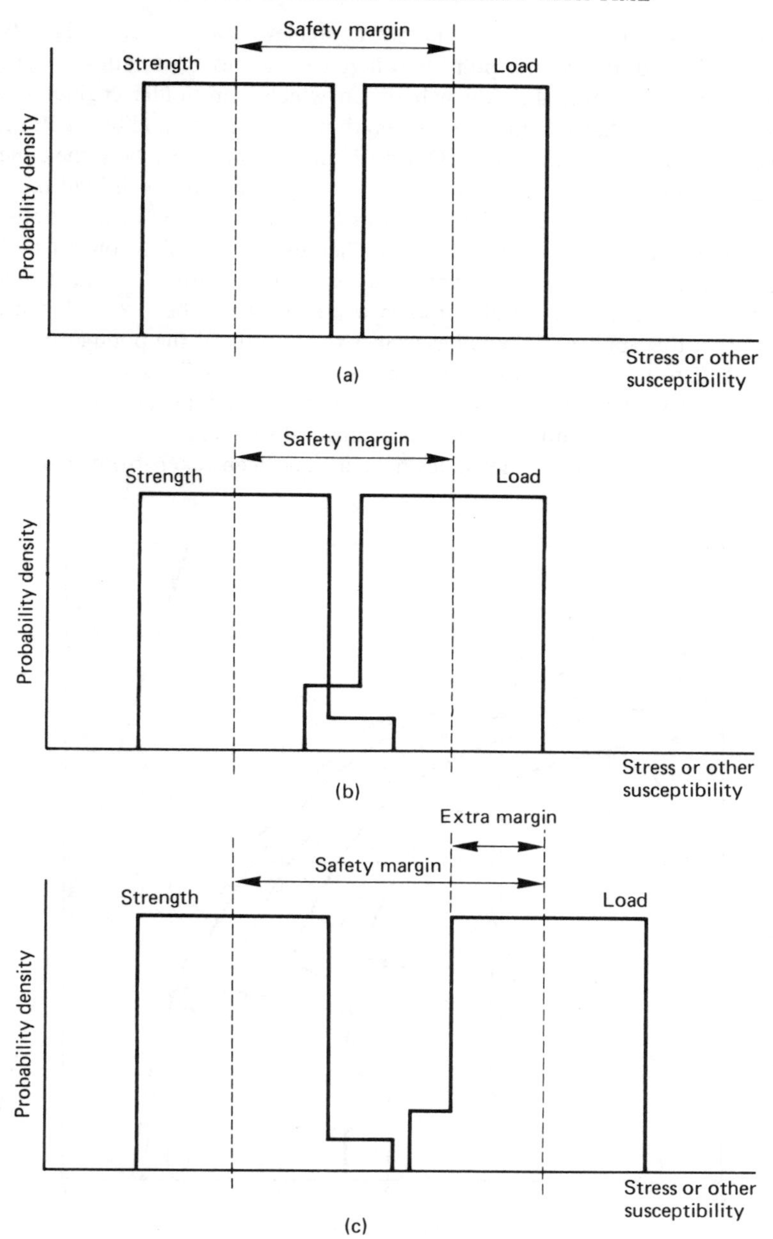


**Figure 4.12** The separation of rectangular distributions with and without tails to ensure zero failures, that is, intrinsic reliability. (a) No tails; no interaction – intrinsically reliable. (b) Tails interacting – intrinsically unreliable. (c) Tails not interacting – intrinsically reliable

and probably more expensive item, but we do achieve 100 per cent reliability. It may be worth recording at this point that here we are only postulating, through a quasi-statistical argument, the design basis on which many older engineers were brought up, namely that the minimum strength of any item should be greater than the maximum load it will encounter. Our difficulty is that statistically the concept of an identifiable maximum load is difficult to sustain. Any such load must be associated with a likelihood, which brings us back to statistical distributions, re-introducing the problem of tails. However, let us bypass this problem for the moment, and assume that real distributions can be represented by some standard distribution, which for analytical purposes we shall take to be a Weibull distribution, and that this applies to the tail as well as to the bulk of the population. Then we can calculate the failure rate from the basic equation (4.7) as in section 4.3 of this chapter. If we ignore the early life portion of the failure rate–age curves, we can plot the constant random failure rate against safety margin for various loading roughnesses in figure 4.13. (The values have in fact been taken from figures 4.7,

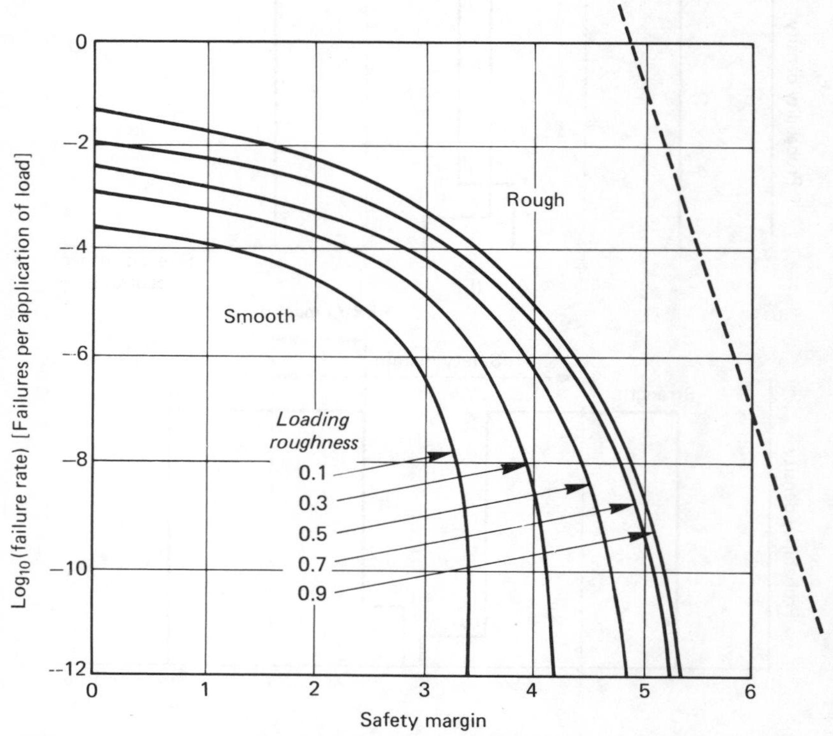


**Figure 4.13** Failure rate–safety margin curves for non-maintained items whose strength does not deteriorate with time. Both load and strength are approximately normally distributed (Weibull $\beta = 3.44$). — — — — indicates slope of a factor of 2 in failure rate for 0.1 increment/decrement in safety margin

4.8, 4.9, 4.10 and 4.11, with additional calculations to define the curves more accurately.) It will be recalled, and we re-emphasise here, that the values apply to items whose strength does not deteriorate with time and that the distributions of load and strength are near normal. These curves provide the clue to our handling of reliability. Let us concentrate on a fixed value of the loading roughness, giving the typical curve shown in figure 4.14. It will be seen that we can identify three regions, as marked on the figure. In the first region the failure rate is too high for practical use. The safety margin is much lower than would be considered in any practical instance, so on the basis of these calculations and on the basis of experience we can discard this region as being of no practical significance. The second region is much more interesting since it covers values of the safety margin that would be contemplated in actual designs. Nevertheless there are difficulties. First and foremost your attention is drawn to the very rapid change of failure rate with safety margin. In assessing this feature you should compare the slopes of the

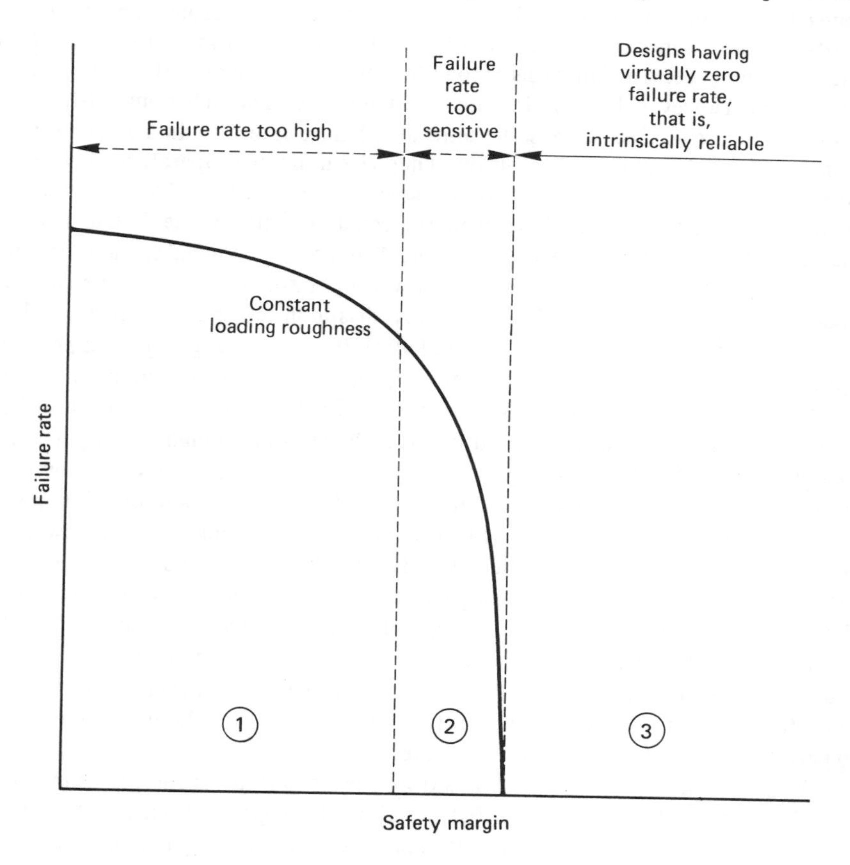


**Figure 4.14** Characteristic regions of a typical failure rate–safety margin curve

curves with that of the dotted line on figure 4.13. That line is drawn at a slope which represents a change by a factor of 2 in failure rate for an incremental change of 0.1 in safety margin. Think what this means. A factor of 2 in misestimating the failure rate would correspond to a 100 per cent error in the spares provisioning. It is totally unacceptable. If we aim at a more reasonable figure of about 10 per cent error in failure rate estimation we should need to know the safety margin to an accuracy of 0.005! The information to obtain such accuracy is just not available, and is not likely to be available in the foreseeable future, if ever. It would appear, therefore, that there is virtually no possibility of designing to a specified level of reliability in region 2 of figure 4.13. Even if by some miracle all the information could be provided we would still run into difficulties. We may rationally suppose that no two operators are equal in all respects, so that some operators will impose less and some more arduous conditions on the item than others. Since the design must be based on a common value, any item falling in region 2 of figure 4.13 is going to exhibit a very marked difference in failure rate from one operator to another when used under field conditions. The illustration in figure 4.15a shows the probability density functions of load for seven of many hypothetical operators of an item. However the manufacturer will base his design on a global distribution shown in figure 4.15b. Suppose he aims at a failure rate of 1 in $10^6$ per application based on this global distribution. Then the individual operators may expect to encounter the failure rates shown in the last column of the table accompanying figure 4.15. The failure rates have been expressed as a ratio of the design value as a measure of achievement. Clearly operators 4 and 5 will be delighted and operator 7 very pleased with this product. However operator 1 will be somewhat disappointed, operators 2 and 3 will be very disappointed, but operator 6 will be furious. (He will undoubtedly be featured by all the consumer groups and probably appear on TV to rail against his country's products.) His anger will be fuelled by the fact that his mean load is less than its rated value. We may well observe here that most judgements will be based on mean values even though the dispersion of load is of equal if not more importance – it is the high standard deviation which is responsible for the high failure rate in this case. By contrast operator 5 is applying the highest mean load, but because he is controlling his operation to narrow limits (low standard deviation) his failure rate is effectively zero, and he is one of the satisfied customers. Even accepting that there will be other operators whose load density function approximates to the design, it is clear that the reputation of this product is not going to be too good. It must also be appreciated that in real life individual operators are likely to have only the vaguest idea of their own operating conditions, so that which operator is which will be largely unknown before experience is gained. Afterwards is too late.

The situation depicted above arises from the steep slope of the curves in region 2 of figure 4.14. It is postulated that design in this region is totally unacceptable – both because it implies a complete inability to predict equipment behaviour, with maybe disastrous results, and because such a disparity in operating experience results in the item acquiring a poor reputation: realistically the bad experience

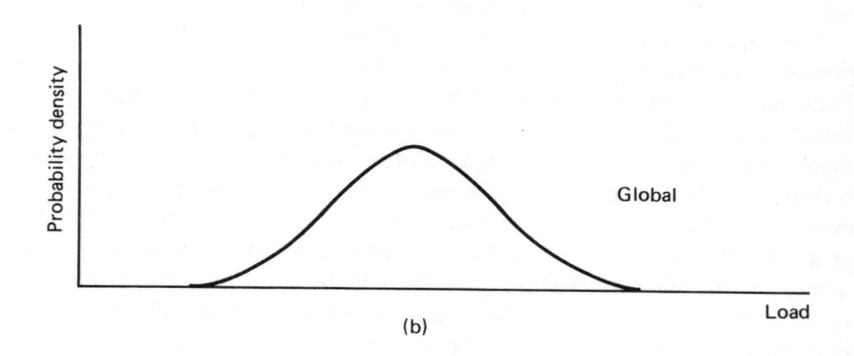


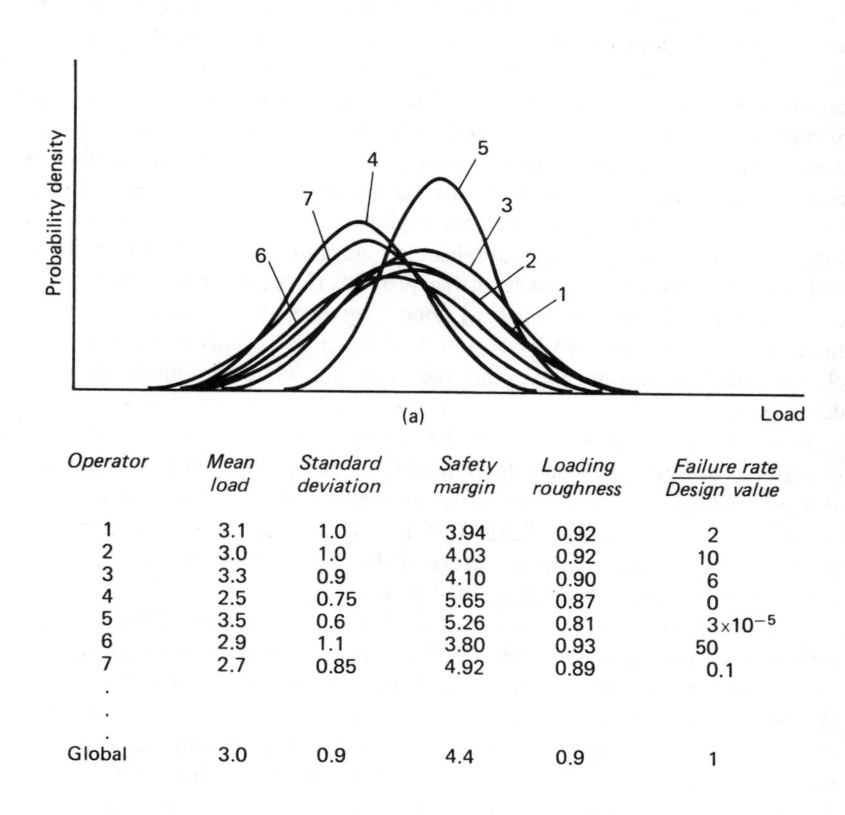


| *Operator* | *Mean load* | *Standard deviation* | *Safety margin* | *Loading roughness* | *Failure rate / Design value* |
|---|---|---|---|---|---|
| 1 | 3.1 | 1.0 | 3.94 | 0.92 | 2 |
| 2 | 3.0 | 1.0 | 4.03 | 0.92 | 10 |
| 3 | 3.3 | 0.9 | 4.10 | 0.90 | 6 |
| 4 | 2.5 | 0.75 | 5.65 | 0.87 | 0 |
| 5 | 3.5 | 0.6 | 5.26 | 0.81 | $3 \times 10^{-5}$ |
| 6 | 2.9 | 1.1 | 3.80 | 0.93 | 50 |
| 7 | 2.7 | 0.85 | 4.92 | 0.89 | 0.1 |
| . | | | | | |
| . | | | | | |
| . | | | | | |
| Global | 3.0 | 0.9 | 4.4 | 0.9 | 1 |

**Figure 4.15** Comparison of operator experience with a non-intrinsically reliable item

will drive out the good. Items falling in region 2 will therefore be of low quality and generally unacceptable. We conclude, therefore, that on both design and operating grounds region 2 is not practically available.

The third region is one of very low failure rate, so low that for all practical purposes the rate can be treated as zero. Theoretically, of course, the value is finite but from an examination of the shape of the curves on figure 4.13 it is readily seen that the values can have no practical significance. In legal jargon failure can be treated as 'an act of God'. It is an essential feature of the approach presented in this book that all good quality designs must lie in this region. However, within the region we are unable realistically to allocate either a failure rate or a reliability expressed as a probability; for all practical purposes the failure rate is zero and the reliability 100 per cent. Outside this region we can attribute figures to these properties but with no degree of confidence. Hence all we can claim is that an item is unreliable if situated in regions 1 and 2 of figure 4.14; or it is reliable, in the sense that it is so unlikely to fail that we may assume it will not fail, if situated in region 3. It is interesting to note that such a concept is more in accord with the approach of the common man than that of the statistician. Nevertheless it is submitted that there are good sound fundamental reasons for this proposition. Furthermore it is the outlook adopted by all the successful engineers of the past and today. What engineer designs his bridges with a chance they will fall down, his ships so that there is a possibility they will sink, his turbines with a chance they will distintegrate or his boilers with a chance they will explode? Engineers, and especially mechanical engineers, have always striven after certain success, always recognising that absolute certainty is unachievable. It is here maintained that even on true statistical grounds the objective of traditional engineering is the only tenable one. All good engineers should be seeking near zero failures such as are offered by region 3 of figure 4.14. This is the most basic conclusion reached in this book and the tenet on which its whole philosophy is developed.

In order to distinguish between the reliability possessed by an item belonging to region 3 of figure 4.14 and the reliability of other items which can be expressed as a probability, however uncertainly, we shall describe an item falling anywhere in region 3 as 'intriniscally reliable' (ref. 13). This phrase is used since the property we are describing is an intrinsic feature of the item and is independent of the loading roughness below a value prescribed by the designer.

It may be observed that we have achieved intrinsic reliability by separating conventional continuous distributions in the same way as we achieved absolute zero failure rate (or 100 per cent reliability) by separating rectangular distributions so that they did not intersect. We can do this even though the tails are of infinite extent because the probability densities are exceedingly low in that region, at least after the application of the first load: so low that the existence or otherwise of the tail is then irrelevant. Thus by dealing only with intrinsically reliable items the nature of the tail ceases to be the issue it first assumed.

Intrinsic reliability is defined in relation to the random failure phase only, and is not necessarily applicable to the early life – it is also not applicable to wear out, as we shall see in chapter 5. Referring to the early life period, it is evident from figures 4.7, 4.8, 4.9, 4.10 and 4.11 that even when the failure rate after a fair number of load applications is substantially zero (and certainly cannot be evaluated) the failure rate at the first application can be still assessable. Intrinsic reliability is not therefore relevant to the first application of load. We could assume that any first load failures are acceptable – they would be very small anyway – but before doing so would usually wish to assess their significance. The problem of the tail now re-asserts itself, for it would appear that these first load failures arise from the extended tail of the strength distribution, which is effectively destroyed by the application of the first load. The failure rates marked on figures 4.7, 4.8, 4.9, 4.10 and 4.11 apply only to the precise tail for which the values were calculated. They may or may not represent real tails. It would appear therefore that we should either have to accept any first load failures as they arise, or eliminate them by application of the first load as part of the manufacturing process – a mild form of burn-in or proof running. As an alternative we can rely on quality control techniques to eliminate the extended tail of the strength distribution. Actually it does seem that good manufacture without any quality control may be all that is necessary for a cut-off at about the much quoted $3\sigma$ point could be all that is necessary. Still, the first load failure rate must remain a bit of an enigma with the present knowledge of strength tails. Tighter quality control is, of course, possible, enabling the safety margin for intrinsic reliability to be decreased, though at the expense of higher rejects. If the performance gains offset this cost, the procedure is worth while. The matter is taken up in chapter 10 under Design, since this procedure in no way affects the concept of intrinsic reliability with which we are at present concerned.

The reader is reminded that all the graphs presented in figures 4.7, 4.8, 4.9, 4.10 and 4.11 were calculated for near normal distributions (in fact Weibull with $\beta = 3.44$), but we noted that many other distributions are possible. In figures 4.16 and 4.17 a comparison is shown between the failure rates expected for these near normal distributions and skewed distributions. The comparison takes the form of a plot of the random phase failure rate against safety margin at a loading roughness of 0.9 in figure 4.16 and a loading roughness of 0.3 in figure 4.17. Significantly different curves (the full lines) are obtained for the skewing shown in the inserts. Now we have already seen in our treatment of rectangular distributions that the response was dominated by the intersecting tail. We may consequently suppose that a model which adequately represents the tails alone is sufficient. For the continuous actual distributions a near normal distribution can be fitted to the intersecting tails of the skewed distributions, so that the tails, though not the bulk of the population, can be represented by the near normal distribution. The curves of failure rate against safety margin calculated for the equivalent near normal distributions agree closely with the originals, as shown by a comparison of the full and dotted lines of figures 4.16 and 4.17. In the one case where agreement is not too good, it is shown that better agreement can be obtained by a closer fit to the

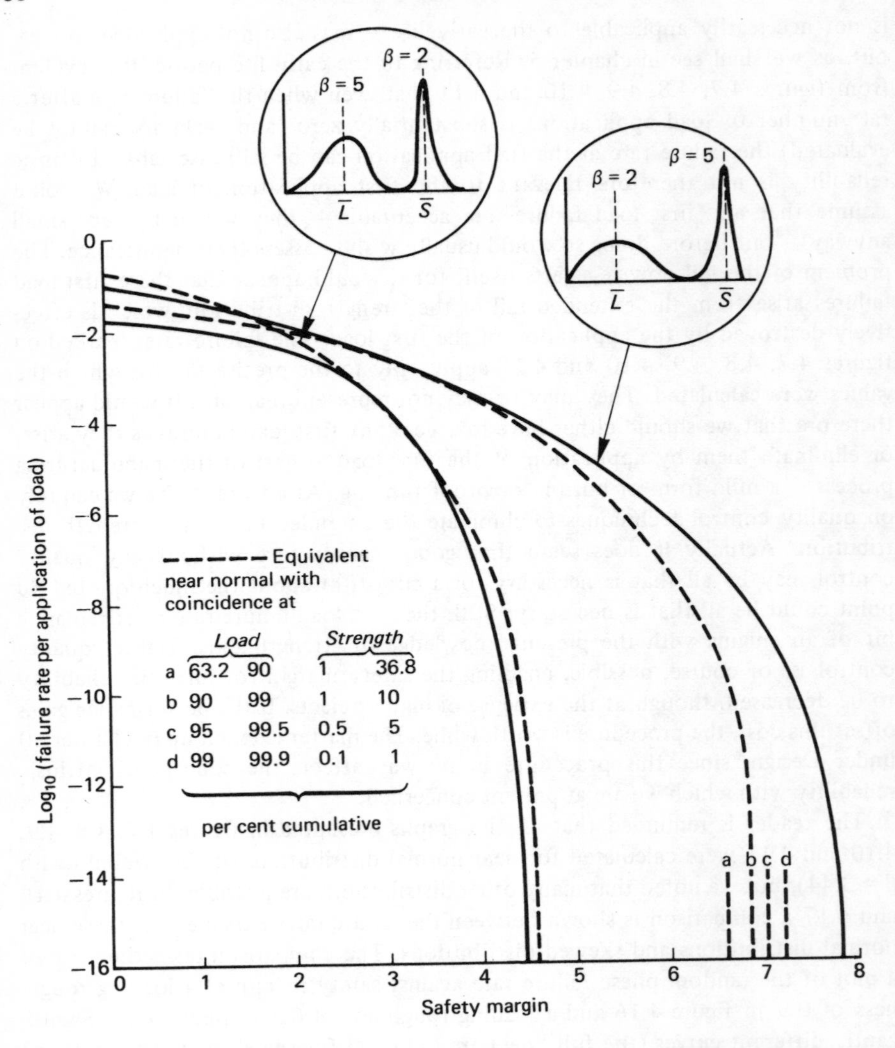


**Figure 4.16** Failure rate–safety margin curves for asymmetric distributions (loading roughness 0.9) and equivalent symmetric distributions

extremity of the skewed distribution. It would appear necessary in this instance because both the load and strength distributions have extended tails. It follows, therefore, that as with rectangular distributions, the tail dominates the response of the system, and that we can replace skewed distributions by equivalent near normal distributions. No generality is therefore lost in our treatment of intrinsic reliability by confining our attention to near normal distributions, provided the

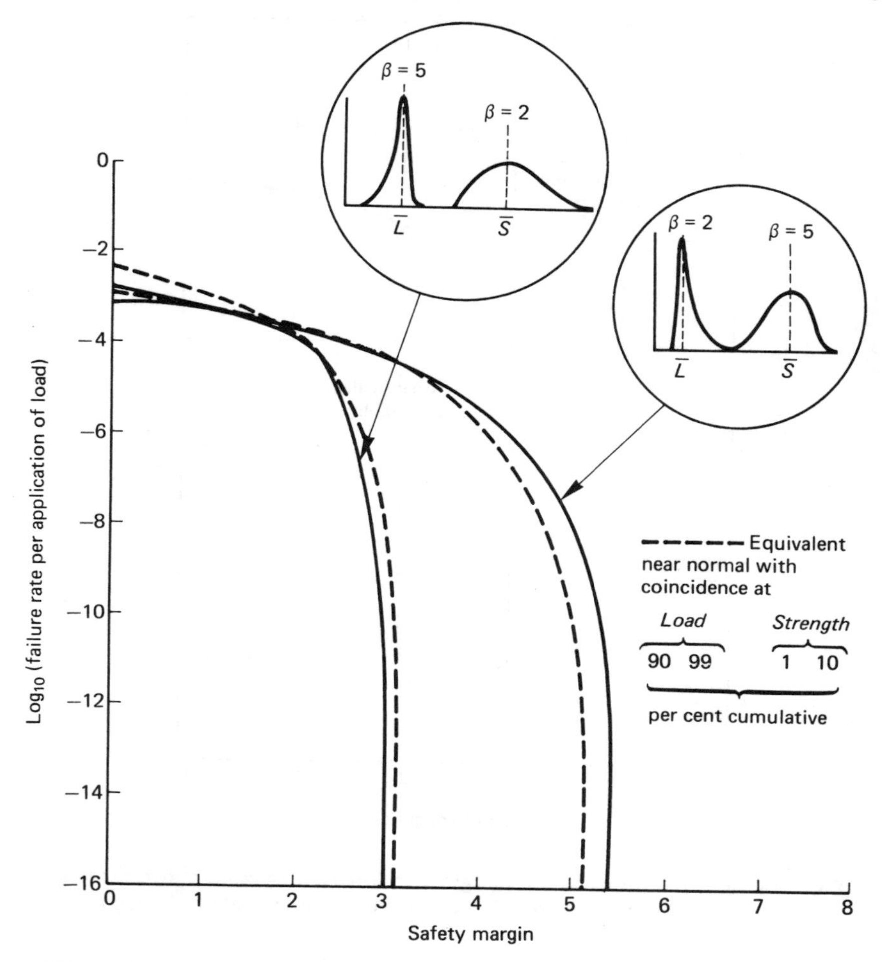


**Figure 4.17** Failure rate–safety margin curves for asymmetric distributions (loading roughness 0.3) and equivalent symmetric distributions

tails adequately represent the real load and strength tails. On this account the curves of figure 4.13 can be treated as general curves. Region 3 is defined by the 'asymptotic' values of the safety margin for each value of the loading roughness based on the equivalent near normal distributions. These values have been plotted against loading roughness in figure 4.18 to give the minimum value of the equivalent safety margin for intrinsic reliability. Clearly safety margins in excess of the values plotted will also provide intrinsic reliability. Figure 4.18 is therefore fundamental to our whole concept of reliability, because it enables us to design for, or to assess the ability to design for, intrinsic reliability, that is, 100 per cent

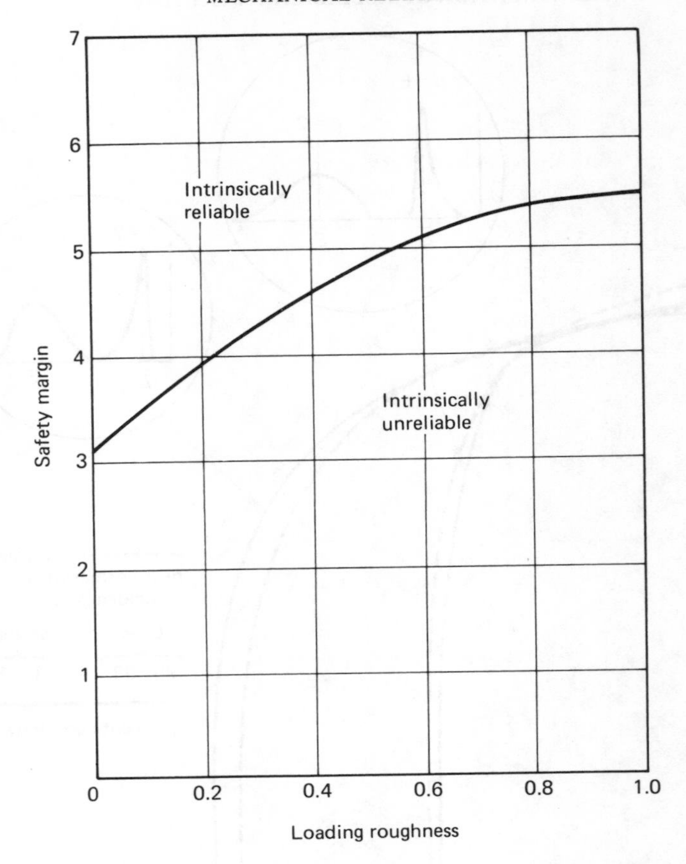


**Figure 4.18** Minimum safety margin for intrinsic reliability for substantially normal (that is, symmetrical) or equivalent distributions

reliability or zero failures. With its aid a prime step towards our goal has been taken. Ideally we can design mechanical components for 100 per cent reliability so long as the strength of the item does not deteriorate with time.

We must at this point advise some caution. Figure 4.18 only enables us to fashion existing concepts to give 100 per cent reliability. It does not provide the concepts themselves and the fashioning may prove uneconomic or impractical. As an illustration we may consider the domestic electricity supply to the author's house, which was by way of overhead cable. During high winds large branches (particularly from old pines) or whole trees were sometimes blown down, falling on the cable which broke under the load, to give a failure of the supply. We lost our supply roughly twice a year from this cause. Now the failure is a classical 'random' phenomenon, being the chance event of high wind. Using figure 4.18 the strength of the cable could be assessed to provide intrinsic reliability. However, without doing any sums we can see that the distribution of the relevant load has a very extended tail indeed. The size of the cable to provide intrinsic reliability would be totally impractical. The solution in this case is simple: bury the cable in the ground. But this is a different concept. Thus intrinsic reliability is not solely achieved by the application of the data on figure 4.18. Acceptable intrinsic reliability has to be achieved from concepts which enable the conditions of figure 4.18 to be met practically – often no easy task. Numerous other examples will readily spring to the mind of anyone who has worked in the reliability field. We shall not pursue it further here, but shall return to the topic under design. The example is quoted at this stage to warn against any belief that a single tenet, however important it may be, solves all reliability problems.

A second subject of caution is the relationship between component and equipment reliability. It may well appear that if all the components were intrinsically reliable then the equipment would be intrinsically reliable. This would be so whether the product rule or least reliable component were the criterion, that is, it could be true whatever the loading roughness. But a moment's thought would show this cannot be true. Consider a hammer consisting of a head, a shaft, and a wedge. All three components may be intrinsically reliable, but if the wedge were forgotten during assembly the hammer would be unreliable (not to say dangerous!). It seems customary practice to absorb the unreliability of assembly into the reliability of the component. This only blurs the logic, and is completely indefensible as the above example shows. It could well be, of course, that quantitative reliabilities are often so imperfectly known and often combined in such arbitrary manners that neglect of this factor is neither here nor there! However, if we are to aim at and achieve, as does seem possible, intrinsically reliable components then we must take into account the reliability of assembly when evaluating system reliability. It is essentially independent of the component reliability, but may well involve component design. For example, if it is possible to fit a non-return valve the wrong way round then one would think that any failure in this mode must be attributed to the valve design itself and not to the assembly process. Clearly there is no universal rule, except that all failure modes must be identified and clearly defined, and allocated to a source (see also section 7.1).

The imperfections of the assembly process may readily be represented by the model we have adopted for all our other calculations. Such imperfections must result in a reduction – probably a considerable reduction – in one of the strength attributes of some components or sub-assemblies. We may expect the number involved to be small if the assembly is reasonably efficient. One cannot imagine wedges being omitted from all hammers, to quote the example already introduced. We thus have a strength distribution with an extended tail. Often in these circumstances the tail is itself distributed, as in the distribution of figure 3.3. Such concentrations in the tail are sometimes referred to as 'freaks'. From the analysis presented in section 4.3, it follows that the failure rate will initially be high but fall (reasonably quickly one would hope in this case) to zero when the freaks have disappeared. The reliability would be equal to the proportion of successfully assembled items – those with wedges in the example of the hammer already quoted. The failure rate would subsequently be substantially zero as intrinsic reliability is achieved, and remain so until any wear-out phenomena set in.

Finally we must ask ourselves if intrinsic reliability is always a realistic proposition. It has been demonstrated that unreliability arises from the tails of the distributions and in particular from the tail of the load distribution. It should also be clearly recognised that we do pay a penalty to achieve intrinsic reliability, which could become excessive if no limit is placed on load distributions. Are the majority of users to be penalised for those who make up the extreme tail? Obviously not always. It does seem to the writer perfectly acceptable, therefore, to specify the loading condition for which the design is intrinsically reliable in recognition that more adverse loading will lead to random failures. It may well be argued that this is effectively designing for random failures. The author would disagree with this argument in as much as a defined reliability behaviour can be guaranteed up to a specified loading roughness but increases in roughness beyond that cannot be assessed, and must be made the responsibility of the user. There could be instances, therefore, in which random failures are to be expected, but these should be exceptional, and the responsibility for the failures lies essentially with the user, whether or not the control is fully in his hands. Numerous examples spring to mind. The packaging of goods is an excellent example. While it is essential to recognise that packages will be handled roughly it is clear that every eventuality cannot be catered for. It is therefore necessary to assess at what level rough handling becomes abuse. Realistically nothing is to be gained by setting the level so low that it does not encompass virtually all contemporary practice. Thus if I am sending a delicate article to my favourite aunt by post it would be as well to recognise that the parcel will be thrown (hurled?!) around and dropped on to concrete floors, but it would be unrealistic to assume it would be left there and run over by the local goods van. As in so much concerned with mechanical reliability it comes in the end to a balanced decision as to what is reasonable – or perhaps in this connection, as to what is quite unreasonable. The 'quite unreasonable' cannot be catered for in any way, but the 'mildly unreasonable' has to be. Even for the 'mildly unreasonable' intrinsic reliability is an essential target. Any ensuing random failures must then be treated as malpractice, though I would observe most strongly that the designer or supplier has a bounden duty to clearly inform the purchaser or user where that level has been set.

There is one set of circumstances when intrinsic reliability cannot be achieved. It occurs when both the load and strength distribution have extended tails or perhaps if the load alone has an extremely long tail. In these circumstances the failure rate–safety margin curve has not a sufficiently steep slope to define an 'asymptotic' value (see figure 4.16 for example). Intrinsic reliability does not exist. Designing for a specified reliability would seem possible, assuming the exact distributions are known, because values can be defined on the lower slope of such curves. A common example would be the windscreen of the normal motor car. The load due to impact from flying stones etc. is rare, except when road resurfacing is involved and deliberate risks are taken, and the loads are then so much greater than normal that the load tail is very extended. Even in these cases, however, it may be more realistic to ensure intrinsic reliability when the extreme loads are excluded and assess separately the effects of the exceptionally high load, than to design for the whole load spectrum. These circumstances should not arise often, because it does seem sensible to the writer to take measures to eliminate the exceptionally high load if at all possible. Extreme variation cannot be reconciled with customer satisfaction except in special cases. Quality control must apply equally, if not more so, to load as well as strength if high reliability is required.

*Exercise 4.7*

You are asked to calculate a satisfactory diameter for the circular rungs of the wooden ladder referred to in exercise 3.1 item (c). The only additional information you are given is that the rungs are made of ash for which the equivalent fibre stress at maximum load has a mean value of 80 N/mm$^2$ and a standard deviation of 15 N/mm$^2$. You must make appropriate assumptions for all other quantities as in normal design, and state these and any other assumptions you make to determine the diameter.

*Exercise 4.8*

Explain why the curves on figures 4.7 to 4.11 have a marked drop in failure rate after the application of the first load.

*Exercise 4.9*

Discuss the concept of intrinsic reliability in relation to the 'one shot' device in the light of your solution to exercise 4.8.

## 4.6 The non-intrinsically reliable item

In spite of the difficulties, the author would write impossibilities, of designing to a specified failure rate, attempts are continuously being made to do so because of some of the possible advantages, maybe illusory, which attract designers. For example, at a loading roughness of 0.9 it would be necessary to use a minimum safety margin of 5.4 to achieve intrinsic reliability. However, if a failure rate of 1 in $10^8$ applications is acceptable the safety margin can be reduced to 5. This implies that a simple tension member of fixed length could be made some 8 per cent lighter, with the corresponding increase in performance. Similar improvements can be achieved in all items. Since manufacturing cost is roughly proportional to weight for any given application, it implies a reduction of first cost even if performance is not at a premium. (Total cost may still be greater of course, but it is the customer who pays the repair bill – not the manufacturer!) There is consequently an incentive to aim at the non-intrinsically reliable item if performance is at a premium or if a means can be found to overcome the possibility of higher failure rates than planned.

One possibility arises from the shape of the failure rate curve at low loading roughnesses. Here the majority of the failures occur in the early life period. Hence the manufacturer himself can operate the item for the period of early life failures before passing it to the user. In this way the user does not see most failures and believes he has, as indeed he does have, a much more reliable item. The possibilities are brought out by figure 4.19 which shows the failure rate–age curves for a number of loading roughnesses, in each case the safety margin being chosen to give a random failure rate of about 1 in $10^6$ load applications. It will be seen that the necessary early life characteristic is only present when the loading roughness is below about 0.2. This technique is thus only available with smooth loading, but it does include the very important electronic field. It has become more and more the practice to run electronic equipment for a period before passing for operation. The technique is known as 'burn-in'. Several electronic engineers have told the author that without burn-in they could not have achieved the reliabilities now expected of some electronic equipment. One of the difficulties brought out by figure 4.19 is that even at very low loading roughnesses where the technique is so powerful, the early life period can be extensive, for example, the curve for a loading roughness of 0.1 appears to be continuously falling even after the concentrated failures of the very early life. This difficulty has been recorded in practice. Furthermore, the technique clearly cannot cater for instances where the loading roughness may be increased beyond the burn-in value. A full description of burn-in may be obtained from ref. 14.

An extension of this technique is the one of 'proving' or 'proof testing' in which a load in excess of the normal duty is applied during early life. It is the technique favoured for mechanical items because it does allow a range of loading roughness to be covered. Much burn-in does in fact also make use of this method. The idea is that a higher than normal load will eliminate all weak items and also allow a margin for possible load variation in actual operation. Figure 4.19b shows the failure rate–age curves that would result from the limited application of a load 20 per cent in excess of the rated load for each of the cases plotted in figure 4.19. It is readily seen that a dramatic improvement is obtained at low loading roughnesses: not only is the failure rate reduced by a factor of $10^3$ but the whole early life characteristic has disappeared at a loading roughness of 0.1. It would appear that an adequate margin is achieved to allow for reasonable changes in the loading conditions. Similar, though increasingly less dramatic, improvements are obtained at loading roughnesses of 0.2 and 0.3, but above a loading roughness of 0.5 no improvement at all is achieved. Thus although 'proof testing' is effective at higher loading roughnesses than simple 'burn-in', it also is essentially confined to smooth loading conditions. Proof testing was probably first introduced to establish

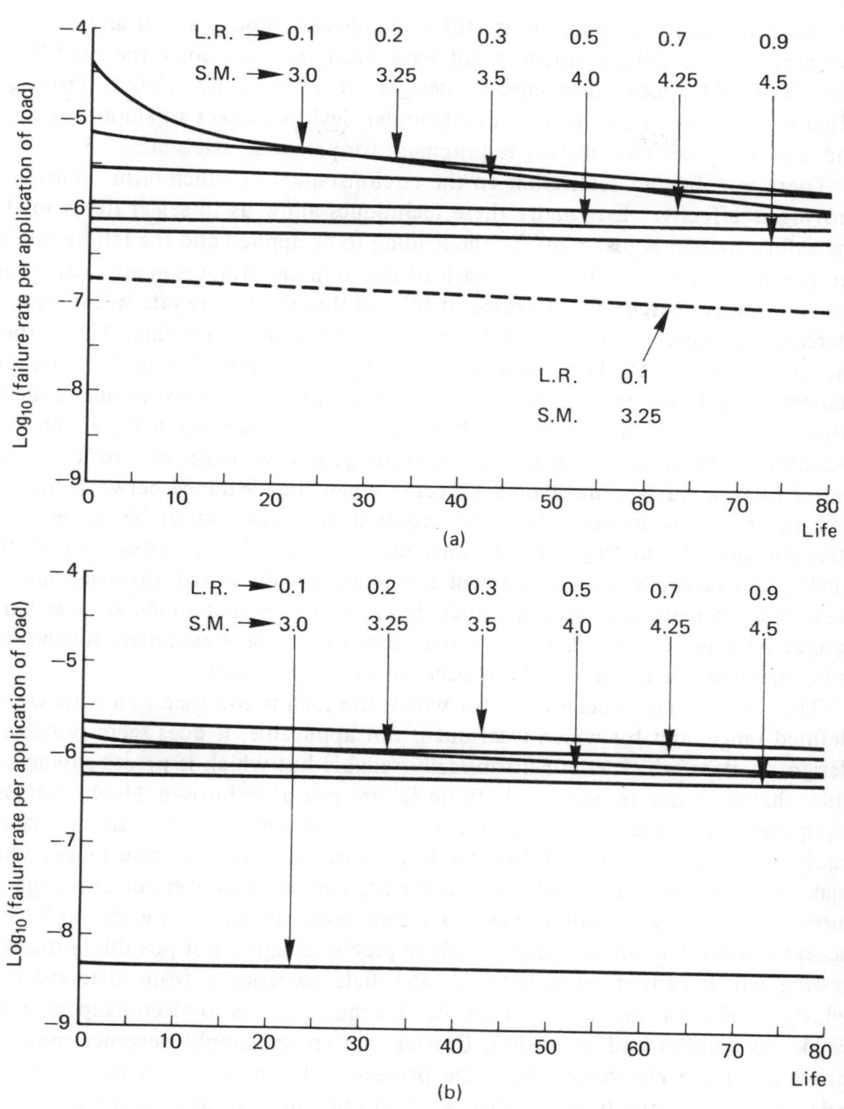


**Figure 4.19** Distribution of failures during life: (a) during normal operation for burn-in and (b) with one application of 20 per cent overload for proof test (L.R. = loading roughness, S.M. = safety margin)

adequate reliability of guns. It is still so employed, though so far as the author can ascertain no failures under proof have been recorded since the First World War. It would appear that modern designs are intrinsically reliable. Proving is often adopted for pressure vessels and similar devices subject to smooth loading, and is in many cases a statutory requirement for pressure vessels etc.

There is a further restriction to the circumstances in which either burn-in or proving is effective. Essentially these techniques allow us to select items so that the safety margin is just right for the loading to be applied and the failure rate we are prepared to accept. If the strength of the item deteriorates in any way during use, the safety margin will decrease. It follows that the failure rate will increase – increase markedly – above the value set during burn-in or proving, that is, above the acceptance level. The techniques are therefore ineffective in these circumstances. It is, however, important to note that an item can wear in one mode for which these techniques are invalid, but not wear in another mode for which these techniques are valid. For example, hitherto guns have worn by erosion of the barrel internal surface until failure occurs when the clearance between the shot and the barrel is so great that the required accuracy cannot be achieved. Its strength against bursting remains unchanged during that process, and so that quality can be established by a proof test as already described. However, modern wear control methods can sometimes delay failure in that mode so that barrel fatigue becomes a possibility before the clearances become excessive. A proof test is no guarantee that failure will not occur in the fatigue mode.

Thus in the very special cases for which the load is confined to a narrow well-defined range, and for which wear-out is not applicable, it does seem possible to design an item which is not intrinsically reliable but which is practically acceptable; that is, it has an acceptable finite failure rate after burn-in. Much electronic equipment falls into this category and it could also be true of some mechanical equipment subject to smooth loading. In the vast majority of circumstances which make up the general run of mechanical engineering, however, in exchange for intrinsic reliability we should have to accept severe penalties: we should have to accept substantial indeterminism in which precise design is not possible, prototype testing which cannot be quantified, and field experience from different users which could be at marked variance. All this must lead, as so often happens in this work, to conflict and confusion. Costing and spares supply becomes more the province of the clairvoyant than the professional expert. It can be a very high price to pay for the benefits, that is, small reductions in weight or gains in performances, resulting from a minor reduction in safety margin.

The reader should recall that the whole of the mathematical analysis and general argument of this chapter have been based on the assumption that the strength of an item under review does not deteriorate with time or repeated application of load. In spite of this restriction we have been able to make a number of significant deductions. However, in the last but one paragraph we were forced to acknowledge that strength may deteriorate with time – indeed, as we shall see, it virtually always does deteriorate with time. Thus we have really reached a point in our study of reliability when we can no longer postpone an examination of this phenomenon. Consequently the next chapter is devoted solely to that topic. However, since the effects of non-intrinsic reliability are so many, so varied and so great, and generally unquantifiable, we shall restrict our treatment to the intrinsically reliable product, which must be the target for all high-quality design.

# 5 Wear-out

## 5.1 General representation of the process

Some form of wear takes place with nearly all real mechanical items. Wear may physically be a phenomenon corresponding to the colloquial use of the term in which material is worn away so that clearances or stresses become too great for satisfactory use. It may additionally be due to physical or chemical deterioration, to ageing, or to corrosion, erosion and so forth. Alternatively, wear may be due to fatigue of the material leading to failure, or to creep leading to loss of clearance or possibly rupture. The term 'wear-out' thus embraces a wide variety of phenomena, the common feature being a reduction of any failure-controlling strength attribute of the component with time.

In this book we shall distinguish between the terms 'wear' and 'wear-out', though some writers seem to use them interchangeably. The word 'wear' is used to imply any deterioration in the strength, however assessed, of an item. Some wear will be acceptable and not result in any loss of quality. The term 'wear-out' will be used to describe the situation where wear has reached such proportions that failures occur, often at an ever-increasing rate. Some consideration must be given to discarding the item in these circumstances.

*Exercise 5.1*

Give examples of the following

(a) wear starting immediately a component is used
(b) wear starting sometime after a component is first used
(c) wear starting before a component is used.

Ideally wear may easily be introduced into the model we are using to study reliability. It is only necessary to displace the strength distribution to lower values as the age increases, accompanied probably by some change of form of the distribution as illustrated in figure 5.1. There is likewise no difficulty in introducing the process into the mathematical model, at least in mathematical terms. The

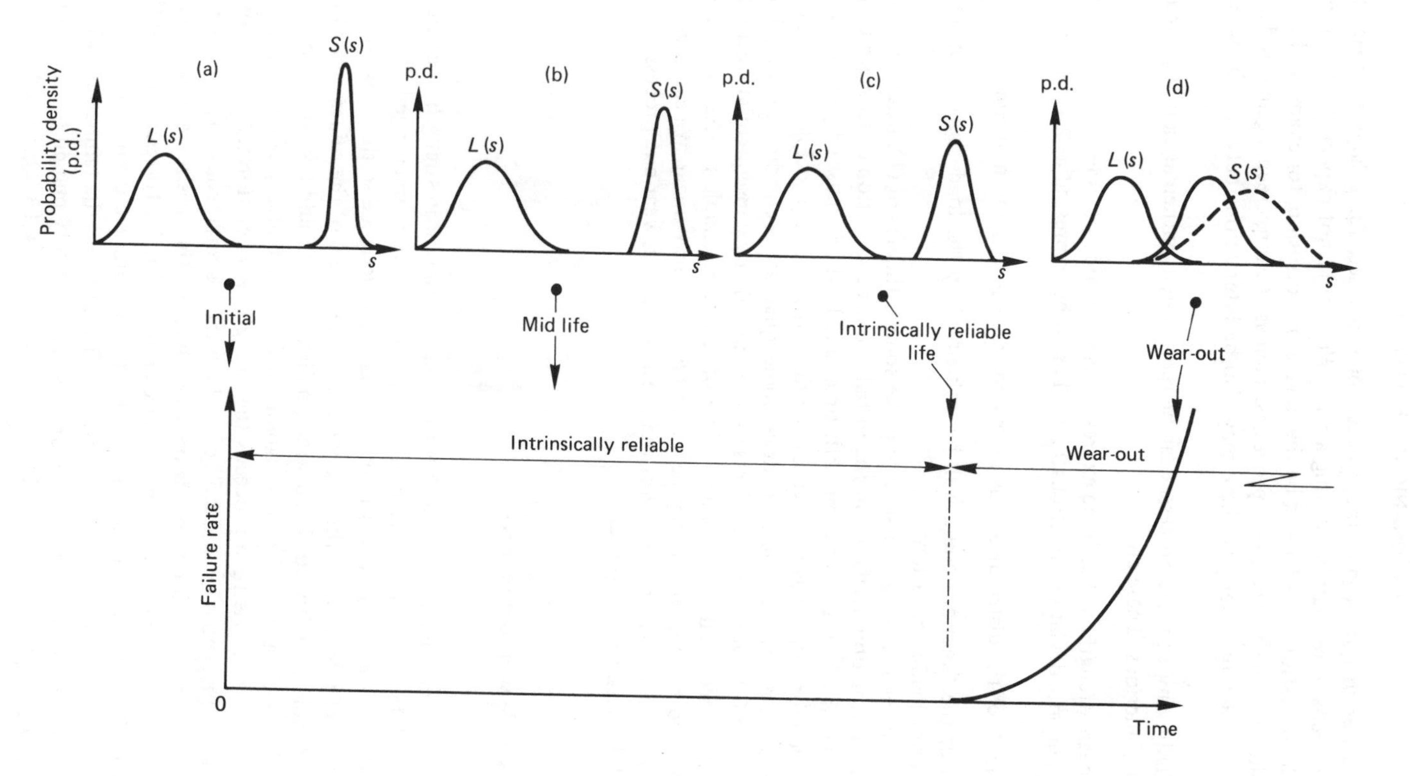


**Figure 5.1** Development of strength distribution with age and consequent effect on the failure rate

difficulty arises in quantifying the process, or processes, for a large number of unrelated mechanisms may be playing a part. We should not expect the form of strength deterioration in fatigue to be the same as in corrosion, for example. Likewise we should expect other wear processes, such as erosion, creep, ageing, and so on to be completely different. There may also be interactions between the wear processes.

Essentially, any of these processes can be represented mathematically, by three parametric functions. These are

(1) the mean strength, $\overline{S}$, has to be expressed as a function of time
(2) the standard deviation of strength, $\sigma_S$, has to be expressed as a function of time
(3) the shape of the distribution has to be expressed as a function of time.

The form of the functions could well be influenced by the load distribution and the number of load applications.

In a few cases it may be possible to make some estimates of (1), but if there is very little information available on the initial standard deviation of strength then there is none at all on its variation with time, and load. We must emphasise most strongly therefore that quantitative calculations from first principles are impossible. Nevertheless we are able to draw some qualitative conclusions and thus deduce some of the very general effects of wear and of any consequential wear-out on reliability. We shall first examine some of the more simple processes such as erosion and corrosion, which may be expected to be fairly linear. We can then go on to study fatigue which is perhaps the most important wear-out process of all, but which is markedly non-linear.

## 5.2 The simple wear-out process

This process is characterised by a linear decrease in the mean strength of an item with age, while the standard deviation of strength remains unchanged. We shall suppose that an item wearing in this manner has been designed so that it is initially intrinsically reliable; that is, the safety margin is in excess of the values quoted on figure 4.18. Such an initial load and strength distribution could take the form indicated on figure 5.1a. Initially the reliability will be substantially 100 per cent and the failure rate zero. As time proceeds the strength will deteriorate, until at some age it can be represented by figure 5.1c in which the instantaneous safety margin is just equal to the value plotted on figure 4.18. Between the ages represented by (a) and (c) on figure 5.1 the reliability will have continued to be 100 per cent and the failure rate zero. As the item becomes older the instantaneous safety margin will drop below the minimum value for intrinsic reliability and the failure rate will increase in response to the curve of failure rate against safety margin for the appropriate loading roughness, such as those shown on figure 4.13. One would expect the failure rate to increase very rapidly with time as the effective instantaneous safety margin continued to decrease. The process is illustrated diagrammatically on a skeleton failure rate–safety margin diagram in figure 5.2.

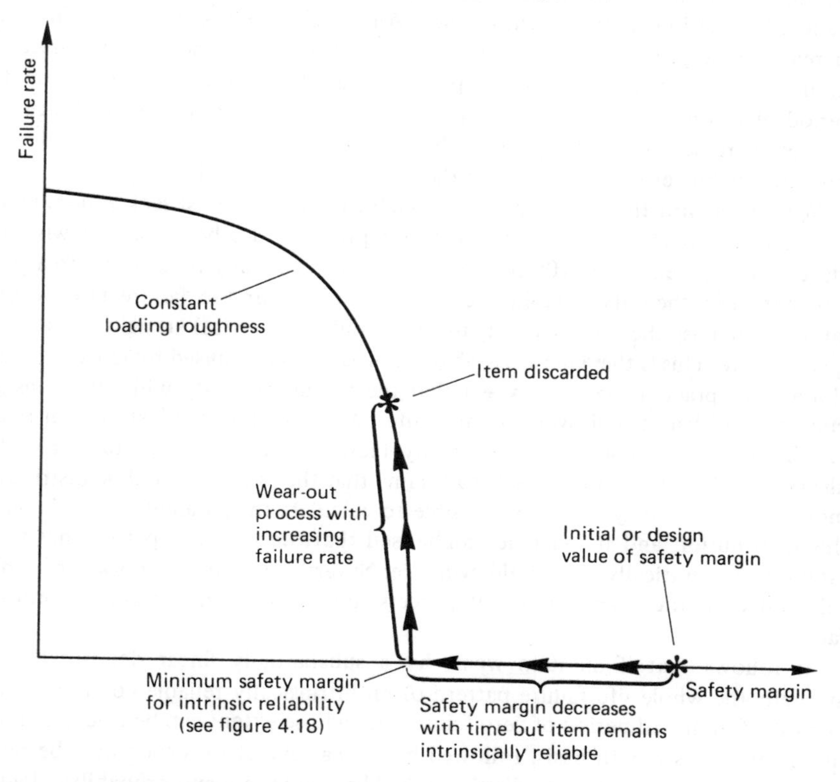


**Figure 5.2** Diagrammatic representation of whole life failure rate when mean strength deteriorates but its standard deviation remains unchanged

The failure pattern deduced qualitatively above is confirmed by the curves on figure 5.3. These have been calculated on the basis that the mean strength deteriorates linearly from an initial safety margin of 7 to a safety margin of 3 at full life. The standard deviation of strength was assumed to be independent of the strength and remained constant, apart from any distortion brought about by failures themselves. The calculated failure pattern confirms that deduced qualitatively. The onset of wear-out, that is, the life of intrinsic reliability, can be estimated on a quasi-constant strength basis if the law representing the deterioration of strength with time is known. In many practical instances this is all we wish to do, since the failure rate increases rapidly with age thereafter and the item would be discarded. Even so, it may be desirable on economic grounds to set the life somewhat higher than the age for intrinsic reliability. In those circumstances it is doubtful if any calculated values can be used to estimate the life, since, for example, it is not valid to assume the standard deviation of strength remains constant, to quote just one assumption. (One would expect the weaker items to wear faster so that the standard deviation increases with time.) All we can state is that the failure rate increases rapidly. Calculations have been carried out for a wide range of rates of strength deterioration. The same pattern was always revealed: following the period of intrinsic reliability the failure rate increases continuously with age, the rate of increase always being high, though naturally the more rapid the loss of strength the higher the rate at which the failure rate increased.

Without a firm theoretical base on which to build, we must seek a more global empirical approach to the simple wear-out process. Ideally we would wish any piece of equipment to be 100 per cent reliable (intrinsically reliable) up to a given moment, and then its reliability could fall to zero or the failure rate become infinite: that is, the curves on figure 5.3 would be vertical straight lines at the specified life. This is the 'One-Hoss Shay' approach immortalised by Oliver Wendell Holmes. In practice we may expect that the fall in reliability will not be instantaneous, but distributed over a more finite period of time as illustrated in figure 5.4. Taking into account the very many factors that cause departure from the ideal, it would not be unreasonable to assume that the failures would be distributed more or less normally. It is not possible to calculate the parameters which define this distribution and it must be emphasised that the wear-out pattern has to be established empirically from field evidence. Nevertheless this is a workable empirical model for the simple wear-out process, involving the minimum of empirical factors.

It follows that if an item wears by a substantially linear deterioration of strength the whole life failure pattern of an intrinsically reliable component will consist of an initial period of zero failure rate whose extent can be calculated on a quasi-constant strength basis, followed by a wear-out phase which may be represented by a normal or similar distribution. The failure rate and probability density function are shown in figure 5.5. It may be compared with the standard bath-tub curve which appears frequently in reliability literature. This is shown dotted on figure 5.5. It consists of a period – the early life – during which the failure rate drops with age, followed by an extended period of constant failure rate: the so-called useful life. This pattern is consistent with a safety margin below that for intrinsic reliability. The constant failure rate is followed by one which increases rapidly with age, designated as wear-out. This 'classic' curve is the one we should expect for an intrinsically unreliable component which wears linearly with age. The curve derived for the intrinsically reliable component with simple wear and the classic bath-tub curve are essentially the same; the former is a special case of the latter, arising from the choice of safety margin.

*Exercise 5.2*

It is reasonable to expect that the roof tiles used in good-class housing construction in temperate zone countries would have a negligible failure rate from natural

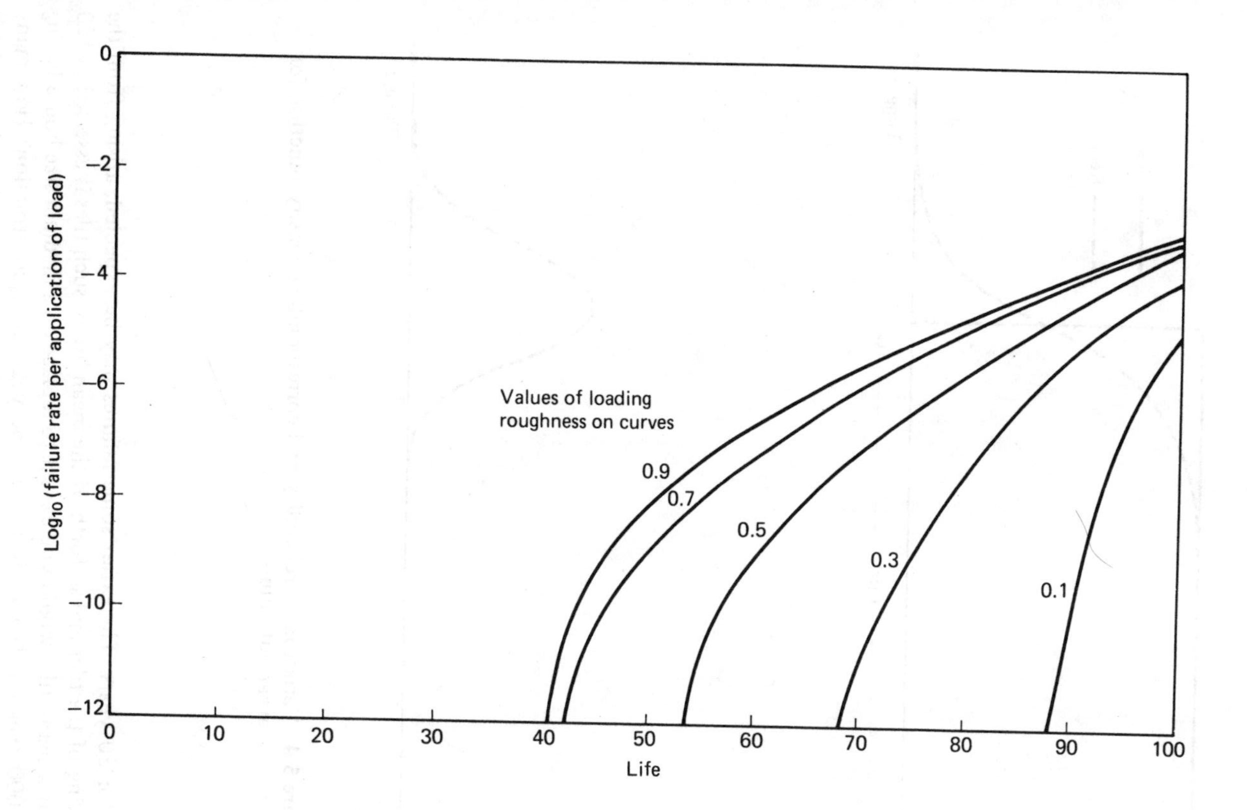


**Figure 5.3** Failure rate-life curves at various loading roughnesses for items whose strength deteriorates linearly with time. Initial safety margin = 7. Final effective safety margin = 3

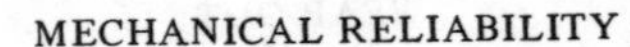

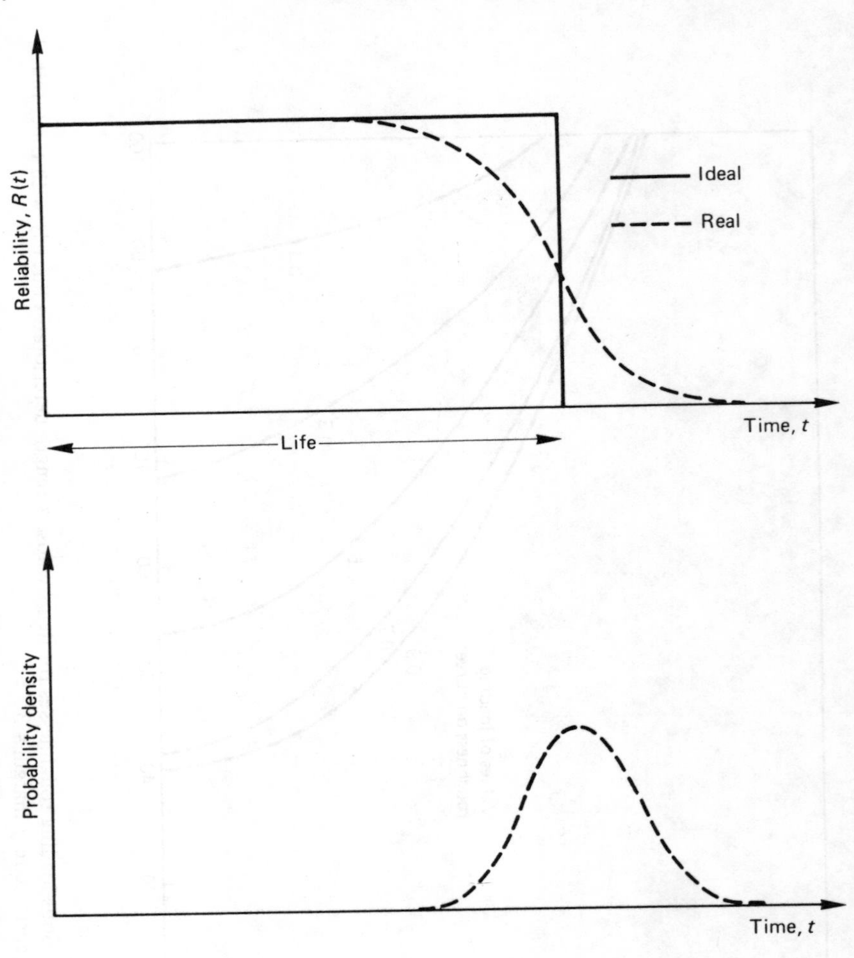


**Figure 5.4** Variation of reliability and corresponding density function for wear-out failures

causes for 30 years. Thereafter some failures may be expected, arising from the weathering of the tiles and so forth. If the mean life of such tiles is assessed as 150 years, how many tiles would you expect to replace on a medium size house having some 5000 tiles on its roof in the first 60 years after construction? How many would you expect to replace in 90 years? Explain what approximation you have had to make before calculating the second figure.

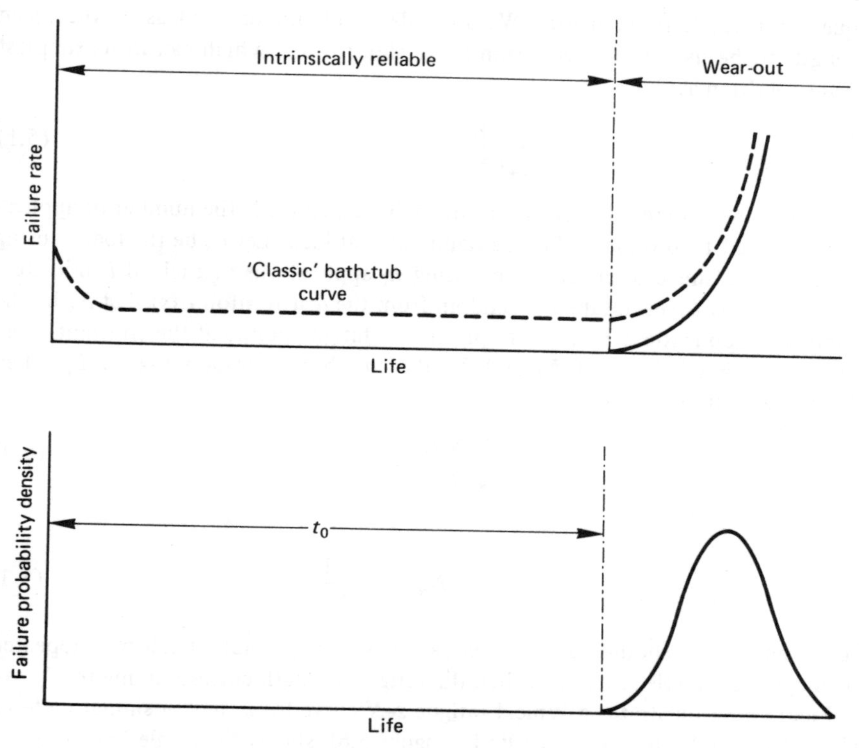


**Figure 5.5** Schematic bath-tub curve for an intrinsically reliable component compared with the 'classic' bath-tub curve of reliability literature

## 5.3 Wear-out by fatigue

It has been claimed that fatigue is responsible for at least 80 per cent of all mechanical failures, and since some claim that 80 per cent of all electrical failures are mechanical it may be responsible for most electrical failures as well. Without wishing to stand by such assertions, most people would nevertheless agree that fatigue is extremely important. Furthermore it is markedly non-linear, so the treatment just presented in section 5.2 is not applicable. A new approach is necessary.

The mean strength in fatigue of an item to withstand $N$ repetitions of a fixed load will be given by the well known $s$–$N$ curve, such as that illustrated in figure 5.6a. However, below the number of applications causing failure this does not give the strength used in the development of our theory: that strength is the ability to withstand a single further application of load, since this is the assumption upon which equation (4.7) has been derived, recognising that on account of the damage done by previous applications of load the strength may be less than its initial value.

Equation (4.7) is fundamental. We are able to obtain some ideas of the mean strength to be used in this equation from Miner's rule, which can in its simplest form be written as

$$\sum \frac{n}{N} = 1 \tag{5.1}$$

where $n$ is the number of applications of each load and $N$ is the number of applications required to produce a fatigue failure at that load. Let $L_s$ be the load causing a failure in a single application following $n_i$ applications of each of $i$ individual loads of values $L_i$, each taken at random from the distribution $L(s)$. Let $N_s$ be the number of applications required to promote a fatigue failure at the constant stress corresponding to the load $L_s$ and $N_i$ the number corresponding to $L_i$. Then Miner's rule can be written as

$$\frac{1}{N_s} + \sum \frac{n_i}{N_i} = 1 \tag{5.2}$$

or

$$N_s = \frac{1}{1-\Sigma\, n_i/N_i} \tag{5.3}$$

The strength that should be used in equation 4.7 for general reliability in repeated loading with variable strength is then the fatigue strength corresponding to $N_s$.

Figure 5.6a illustrates a typical fatigue $s$–$N$ curve for a steel or similar material which has a definite endurance limit. Figure 5.6b shows the single load strength plotted against the number of load applications. This has been derived from the $s$–$N$ curve using equation (5.3).

The characteristic shape of the single load strength curve at (b) follows directly from equation (5.3). Thus if the value of the cycle ratio summation, $\Sigma(n_i/N_i)$, in that equation is as high as 50 per cent, $N_s$ is only 2 and the strength at 50 per cent life is virtually equal to the original strength. Likewise it follows that the strength at 90 per cent of the life is that corresponding to the fatigue strength at 10 applications, and that at 99 per cent of the life is that corresponding to the fatigue life at 100 applications. Even at 99.9 per cent of the fatigue life the instantaneous strength is that corresponding to the fatigue strength at only 1000 applications. However, at 100 per cent of the fatigue life, that is, when the cycle ratio summation equals 1, the single load strength is the fatigue strength at an infinite number of applications, that is, the fatigue or endurance limit. Although using Miner to predict the single load strength is pushing the rule beyond its accepted limit, general practical engineering experience would tend to confirm the picture so deduced: nothing very much is thought to take place during the major portion of an item's life, during which the strength is thought to remain more or less constant. This is the crack nucleation or dormant period referred to in the solution to exercise 5.1A. Then cracks appear and the strength deteriorates rapidly; indeed

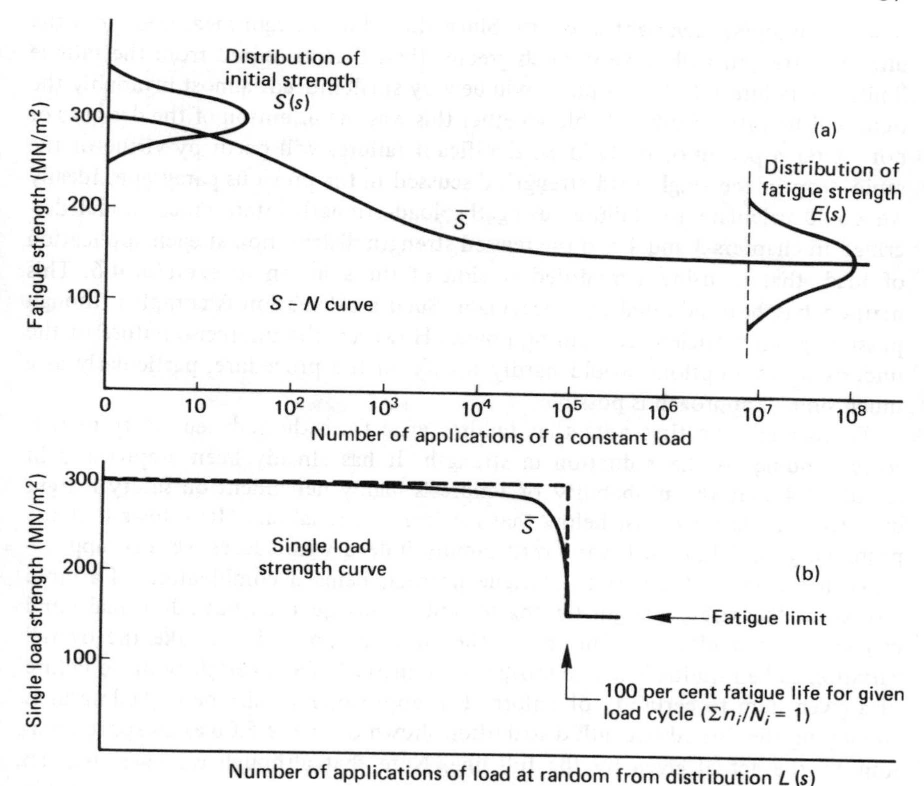


**Figure 5.6** Standard *s–N* curve for fatigue failure at (a) and the derived single load strength curve at (b)

until recently an item with cracks would have been immediately withdrawn from service. The predicted behaviour has been confirmed more scientifically in ref. 15, where it is shown that the distribution of static properties after fatigue damage to near failure is the same as that of undamaged test pieces. Thus it would seem that the above deductions are a reasonable basis on which to proceed.

It will be appreciated that only the mean values have been plotted in figure 5.6a and 5.6b and the strength of the total population will be distributed about these mean values. This implies that $N_i$ will be distributed and hence that $\Sigma(n_i/N_i)$ must also be distributed. It follows that at failure, when $\Sigma(n_i/N_i) = 1$, $n_i$ must be distributed in a corresponding manner, and that the sudden drop in the single load strength depicted in figure 5.6b will be distributed with respect to the number of load applications.

We can now envisage qualitatively the failure mechanism in fatigue. During the initial stages the strength distribution will remain unchanged and the reliability or failure rate can be calculated from the standard load–strength interference model employing invariant strength. Since the safety margin measured from the ultimate strength will be very much greater than that measured from the fatigue limit, the failure rate in this phase will be very small indeed: almost invariably the item will be intrinsically reliable whether this was the intention of the designer or not. After a period of no failures, significant failures will occur by virtue of the rapid drop in the single load strength discussed in the previous paragraph. Ideally we could calculate the failures using the load–strength interference model described in chapters 3 and 4 and the revised strength distribution at each application of load, that is, using a modified version of the solution to exercise 4.6. This method has been adopted by Kececioglu. Such a calculation is complex, though possible given sufficient computing power. However, the imprecise nature of the underlying assumptions would hardly justify such a procedure, particularly as a much simpler approach is possible.

To that end we first note that failures arise from the reduced safety margin corresponding to the reduction in strength. It has already been emphasised in section 4.4 that the probability of failure is highly dependent on safety margin over the critical range just below that for intrinsic reliability. It follows that the probability of failure will vary ***very*** rapidly indeed with successive load applications in the later stages of the fatigue process, being a combination of a rapid change in the safety margin (strength) and a consequential but additional rapid change in probability of failure. In the simpler approach we take the overall variation to be infinite. Thus we move instantaneously from certainty of no failure of a given item to certainty of failure. This approach may also be looked upon as employing the dotted strength distribution shown on figure 5.6b as an approximation to the actual given by the full line. Note that although we have used an extreme application of Miner to justify this approximation, we are not using that extreme application in the actual solution of the problem. With the above approximation the distribution of fatigue failures with time, or number of load applications, will correspond with the distribution of the instantaneous loss of strength.

For quantatitive evaluation we must estimate this distribution which stems from the distribution of the $s$–$N$ curve, or more precisely, the $s$–$N$ relationship. Now the distribution in the $s$–$N$ relationship arises from the fact that each item has its own $s$–$N$ curve. This curve is not usually referred to in the standard literature, but it clearly must exist. The curves making up the relationship are distributed in the same way that all other material properties are distributed, as shown on figure 5.7. If these curves were known the calculation would be relatively straightforward, but unfortunately individual $s$–$N$ curves cannot be determined in the absence of a non-destructive test for fatigue. We have therefore at this stage to make some assumptions. A plausible assumption is that the distribution of fatigue strength at any arbitrary number of load cycles is the same as the distribution of the fatigue or endurance limit stress, which is denoted by $E(s)$. This assumption does not necessarily imply that the distribution of fatigue strength $E(s)$ is the same as the distribution of static strength $S(s)$ – see figure 5.6a. It is well known that for materials in general, $E(s)$ does not correlate with $S(s)$, though whether

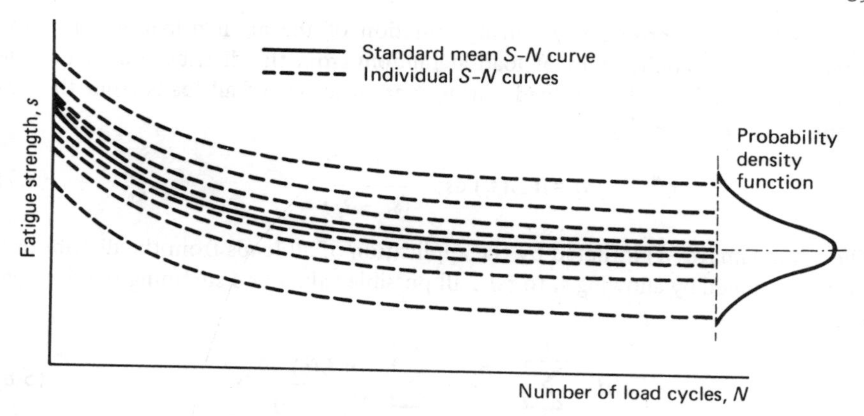


**Figure 5.7** Relationships between the $s$–$N$ curves for individual items and the standard mean $s$–$N$ curve

this would be true of variations in the same material is less certain. However, the correlation of $E(s)$ and $S(s)$ is only likely to be of significance in very low cycle fatigue, and would play no part in the general run of designs. It becomes part of our assumption, then, that weak items are as weak at all numbers of load applications. It follows that all the individual $s$–$N$ curves are parallel, and parallel to the mean curve as in figure 5.7. In this case the $s$–$N$ curves for all the individual items of a population reduce to a single function

$$N = N(s - s_f) \tag{5.4}$$

where $N$ = the number of cycles to failure at stress, $s$

$s$ = stress

$s_f$ = the fatigue or endurance limit of any particular item of the population. The set of $s_f$ values will, of course, be distributed so that $E(s_f) = E(s)$.

$N(s - s_f)$ = a function giving the cycles to failure at stress, $s$. Putting $s_f$ equal to its mean value gives the standard mean $s$–$N$ curve, so this function is known from standard data.

Before failure the application of a load does fatigue damage, which depends on the load and the number of cycles to failure at that load. Consider an individual item drawn at random from a large population. The damage done by the application of one load of magnitude $s_i$ is obtained from Miner's rule as

$$d_i = \frac{1}{N(s_i - s_f)} \tag{5.5}$$

where $d_i$ is the small damage inflicted. The expected number of times that a load lying between $s_i$ and $(s_i + \mathrm{d}s)$ is applied will be given by

$$n_i = n\,L(s_i)\,\mathrm{d}s \tag{5.6}$$

where $L(s)$ is the probability density function of the applied load and $n$ is the total number of applications of load at random from the distribution. Hence the damage inflicted by the $i$th load during $n$ applications of all loads from the distribution $L(s)$ is given by

$$d_i = n\, L(s_i)\, \mathrm{d}s \; \frac{1}{N(s_i - s_f)} \tag{5.7}$$

The total damage ($d$) inflicted by the application of all loads from the distribution $L(s)$ is obtained by allowing $s_i$ to take all possible values and summing the damage. Thus

$$d = \sum_{0}^{i} \frac{n_i}{N_i} = \sum_{0}^{\infty} \frac{n\, L(s)}{N(s - s_f)}\, \mathrm{d}s \tag{5.8}$$

Using Miner's rule in its simplest form, which ignores any effects arising from the sequence of loading, at failure

$$d = 1 \tag{5.9}$$

Hence at the failure of all items whose fatigue limit strength is $s_\mathrm{f}$

$$\int_0^\infty \frac{N_\mathrm{f}\, L(s)}{N(s - s_\mathrm{f})}\, \mathrm{d}s = 1 \tag{5.10}$$

where $N_\mathrm{f}$ is the number of cycles to failure of an item whose fatigue limit strength is $s_\mathrm{f}$, when subject to loads taken at random from the distribution $L(s)$. Rewriting

$$N_\mathrm{f} = \frac{1}{\displaystyle\int_0^\infty \frac{L(s)}{N(s - s_\mathrm{f})}\, \mathrm{d}s} \tag{5.11}$$

Equation (5.11) gives the life of every member of the population in terms of its own fatigue strength. The probability density function of population failure expressed in terms of the variate time (number of cycles) is obtained by transforming the probability density function $E(s_\mathrm{f})$ from the variate $s_\mathrm{f}$ to the variate $N_\mathrm{f}$. Thus the probability density (p.d.) at time $N_\mathrm{f}$ will be given by

$$\text{p.d.} = E(s_\mathrm{f})\, \frac{\mathrm{d}s_\mathrm{f}}{\mathrm{d}N_\mathrm{f}}$$

$$= E(s)\, \frac{\mathrm{d}s_\mathrm{f}}{\mathrm{d}N_\mathrm{f}} \tag{5.12}$$

Equations (5.11) and (5.12) then give the fatigue–age pattern of an item of distributed fatigue strength subjected to a distributed load, on the assumption that Miner's rule applies to each item of the population individually. Given the probability density function, the cumulative distribution and the failure rate are defined by equations (2.6) and (2.8).

If the material has a true fatigue limit then the integrand in equation (5.11) will be zero for all values of $s$ less than $s_f$, since $N(s - s_f)$ will be infinite. In that case equation (5.11) can be written as

$$N_f = \frac{1}{\displaystyle\int_{s_f}^{\infty} \frac{L(s)}{N(s - s_f)}\, ds} \tag{5.13}$$

Besides being easier to use for calculation when a true limit exists, it brings out the powerful influence of that limit.

It is implicit in equation (5.8) that all items see every load defined by the distribution $L(s)$. This is valid because the number of cycles to failure is very large, so every load will be encountered in the expected proportion. To that extent the solution may be considered deterministic, though semi-deterministic is a better description. Given the solution for a specified load distribution (load cycle) there is no difficulty in obtaining the solution for a distribution of distributions. Thus if the mean of the load cycle were normally distributed, the failure pattern for the various cycles comprising the family can be calculated and the probability density of the whole population will then be the weighted sum of the probablity densities of the individual cycles. This gives rise to a marked contrast between fatigue, indeed all wear phenomena, and stress rupture. For the latter the whole load spectra can be lumped together, for only those which cause failure are of any interest: those which do not cause failure are irrelevant. But with fatigue and other wear processes a load which does not cause failure may still do damage. We need to know the load spectrum of each operator in order to assess the damage done by that operator's load distribution. For an overall assessment encompassing all operators we need to know the distribution of distributions – far more information than was necessary to assess stress rupture. A serious problem is practically determining these distributions. Contemporary data on load cycles is very rudimentary indeed, that on variations of load cycles non-existent.

If the distribution of distributions is known, the probability density of failure may be deduced as follows

Probability of failure in time $dt$ at time $t$ from $j$th load distribution acting alone $= f_j(t)dt$ (5.14)

Hence probability of failure in time $dt$ at time $t$ from $j$th load as a constituent of distribution of distributions $= P(j).f_j(t)dt$ (5.15)

where $P(j)$ is the probability of obtaining the $j$th distribution from the whole family of distributions.

Whence probability of failure from all distributions at time $t$ in time $dt$ $= \Sigma P(j).f_j(t)dt$ (5.16)

$$\text{So that probability density of failure} = \frac{\text{d(probability)}}{\text{d}t} \tag{5.17}$$

$$= \Sigma P(j).f_j(t) \tag{5.18}$$

The effect of introducing distributed distributions would be to make the failure patterns we shall quote more widely distributed. In the subsequent treatment in this book we shall ignore this aspect and assume that all items see the same load distribution.

Only one further factor may need to be resolved before equations (5.11) and (5.12) can be solved. In many cases the load cycle comprises a mean stress as well as a fluctuating or alternating stress. In these circumstances the relevant *s–N* curve must be deduced from the standard rotating beam or push-pull *s–N* curve by one of the Goodman, Gerber or Soderberg criteria. Details are given in any good text-book on materials, and are not discussed further here as it is only incidental to our main studies.

Using the above theory the failure-age patterns in fatigue wear-out can readily be calculated. Since the number of load and strength distributions, their relation one to another, and the forms of *s–N* curves are very large indeed, only a few representative examples can be quoted here. In spite of this limitation it is believed that most of the salient features can be demonstrated. All the patterns presented were derived for near normally distributed loads and strengths. The *s–N* curve was expressed as a *s*-log *N* curve and linearised. Above $N = 10^6$ the value of $s$ was taken as a constant equal to the fatigue limit. The linear curve below that value was arranged to reach twice the fatigue strength at $10^3$ cycles. The resulting curve is shown on figure 5.8, where it is compared with a typical *s–N* curve. While considerable numerical deviations from this formalised representation can be expected in practice, the main features of real curves are present.

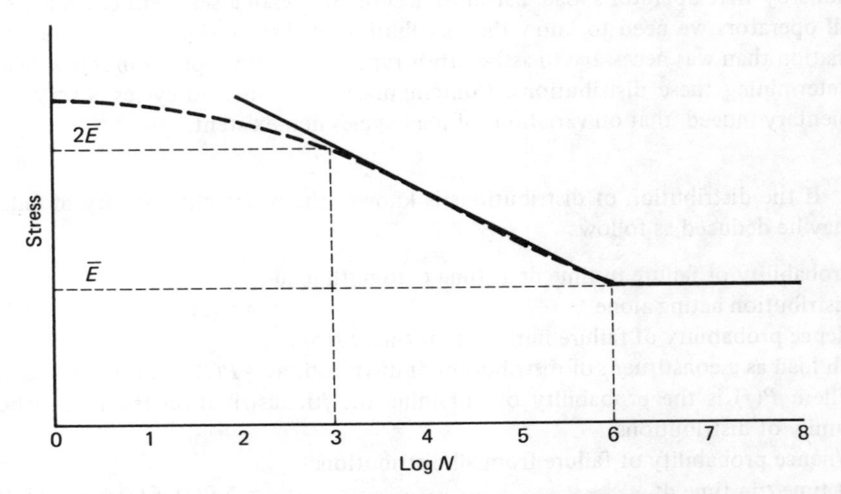


**Figure 5.8** *s–N* curve used for calculations plotted on log-linear axes. – – – – Typical *s–N* curve. ——— Linearised approximation

For stress rupture phenomena any marked interaction between the load and strength distributions is totally unacceptable because of the resulting high failure rate, but this is not necessarily true of fatigue, since a load in excess of the fatigue limit strength does not imply failure, only that fatigue damage is done. If only a short life is required, considerable interaction may be acceptable. The relative locations of the load and strength distributions for low cycle fatigue and high cycle fatigue are shown diagrammatically on figure 5.9. Medium cycle fatigue lies between these, and very high cycle fatigue outside the high cycle range indicated on the figure. Different load and strength distributions to those shown are possible in all cases, giving rise to different loading roughnesses.

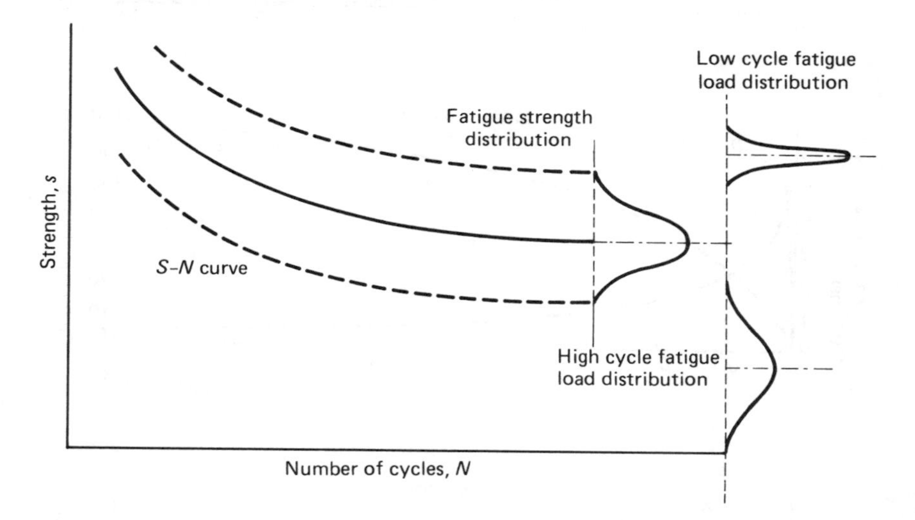


**Figure 5.9** Relative location of the fatigue strength distribution and the load distribution in low cycle and high cycle fatigue

The calculated distributions of failure rate and cumulative failures for rough, medium and smooth loading have been plotted against number of cycles for low cycle fatigue, high cycle fatigue and very high cycle fatigue in figures 5.10, 5.11 and 5.12. The last-mentioned characteristic has been defined by safety margins of –0.5, 3.0 and 4.5 respectively, referred to the fatigue limit. If, as in one or two cases, the fatigue safety margin is greater than that for intrinsic reliability, no significant fatigue damage is done and hence the fatigue life would be infinite. This is, of course, a common design target; but since if it is actually achieved fatigue plays no part in the failure process, it need not concern us here. What does concern us is the failure pattern when interaction is present.

The first thing to strike us is the very unusual pattern shown by all the curves; at least unusual in the sense that it is not quoted in standard reliability texts. In all cases there is the not unexpected initial failure-free period when fatigue damage is being done by the successive applications of load, but the damage is insufficient

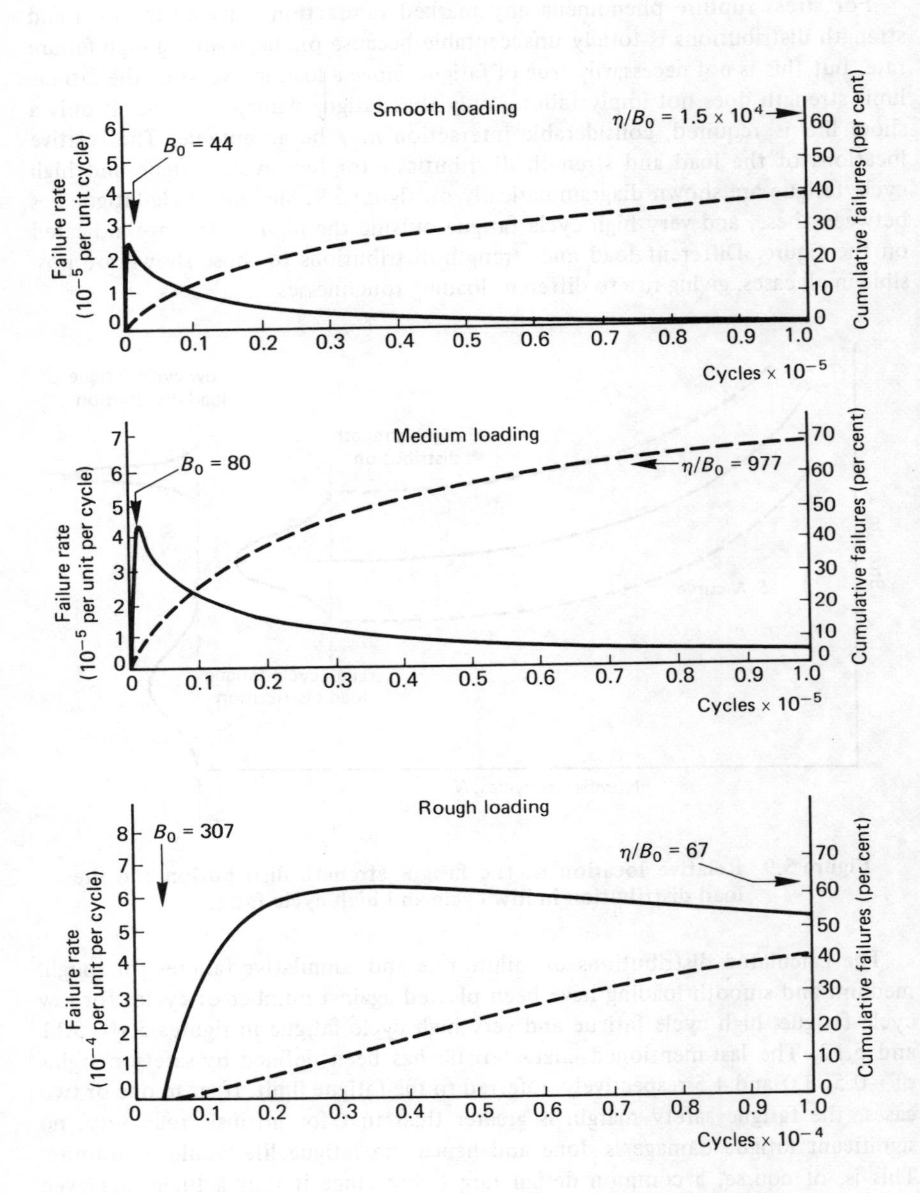


**Figure 5.10** Variation of failure rate and cumulative failures with age in low cycle fatigue at various loading roughnesses (——— failure rate, – – – – cumulative failures)

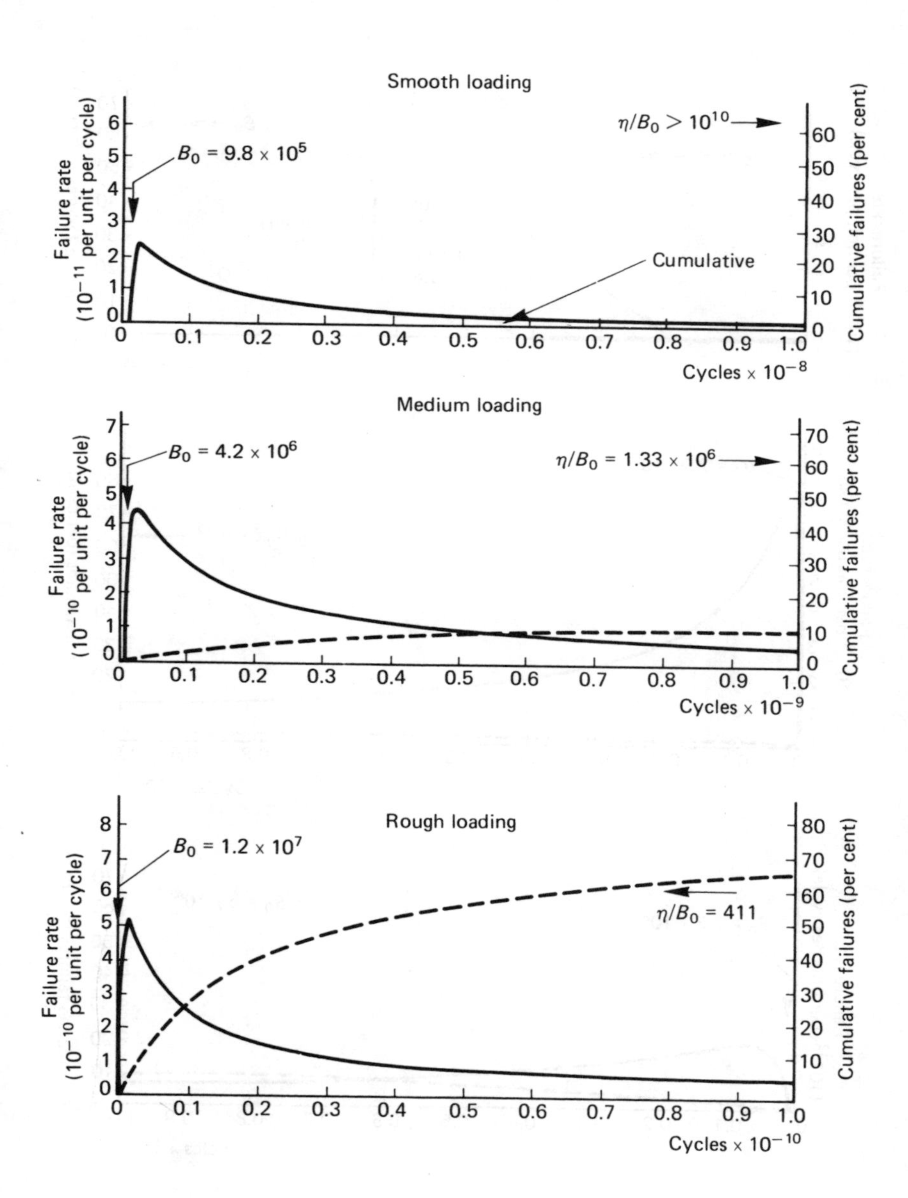


**Figure 5.11** Variation of failure rate and cumulative failures with age in high cycle fatigue at various loading roughnesses (——— failure rate, – – – – – cumulative failures)

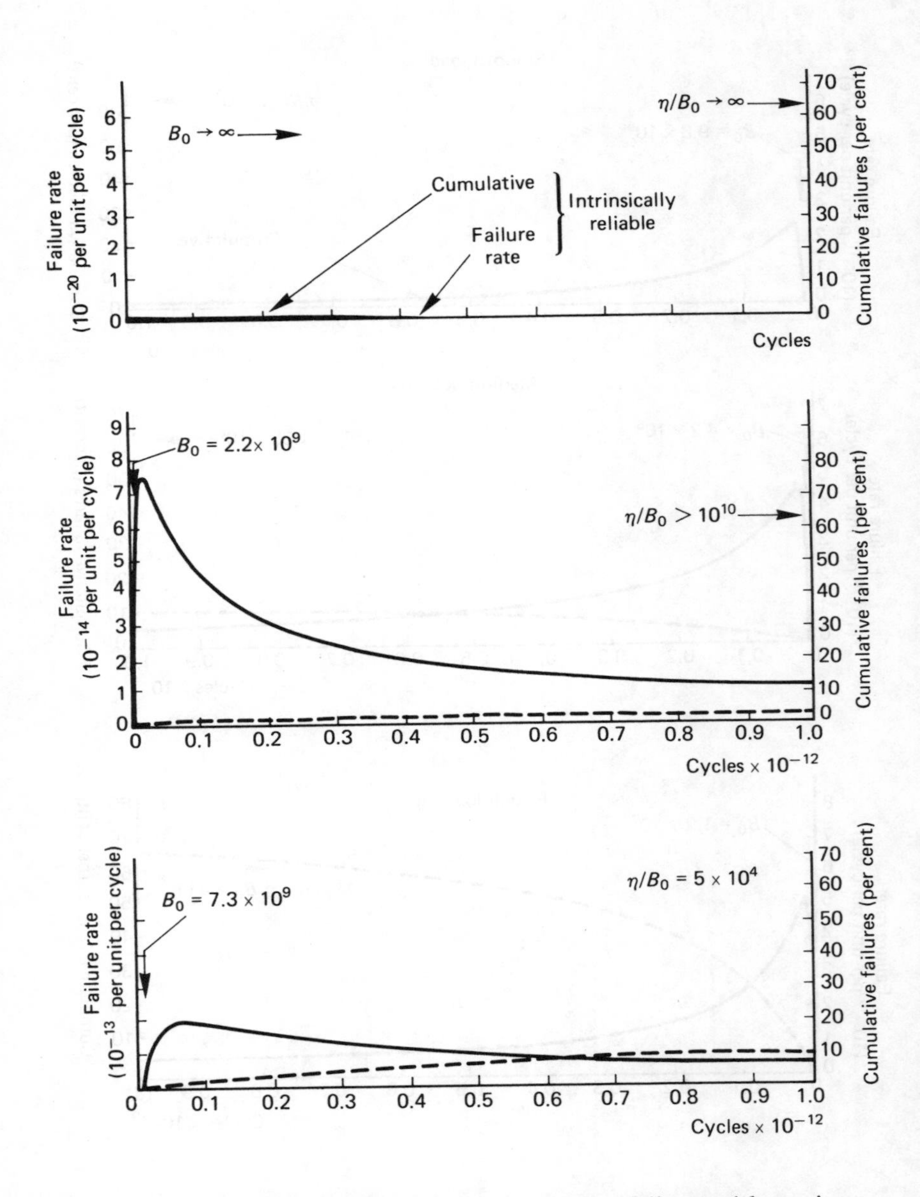


**Figure 5.12** Variation of failure rate and cumulative failures with age in very high cycle fatigue at various loading roughnesses (——— failure rate, – – – – – cumulative failures)

to cause failures. Once the damage is sufficient to cause failures the failure rate mounts very rapidly, almost instantaneously in some instances, and quickly reaches a maximum. Thereafter the failure rate ***drops*** slowly over a period which may be several hundred times the failure-free life for low cycle fatigue with smooth loading, to a value several orders of magnitude greater for high cycle fatigue with rough loading, before the characteristic life (that is, 63.2 per cent cumulative failures) is reached. Thus after the rapid rise of failure rate there is a very long period indeed of falling failure rate. Although it is most likely that this extended period will dominate the practical behaviour of the item, it should be noted that the 'sudden' rise in failure rate is only sudden by comparison with the very extended period, and plotted on a larger scale as in the bottom diagram of figure 5.10 it is seen that the rise is structured. However, this structured rise is confined to such a narrow range that it would only have meaning if the input data were accurately known. For practical purposes the pattern consists of a sudden rise followed by an extended period of falling failure rate. It would seem totally incorrect, therefore, always to associate wear-out with an increasing failure rate, as is generally done in contemporary literature and the classic bath-tub curve. It would certainly not apply to fatigue. The physical reason for the behaviour noted above is the extreme sensitivity to strength (safety margin) of the 'failure rate' (load exceeds strength), which becomes the damage rate in the initial stages of fatigue. Strong items suffer very little damage whereas weak ones suffer considerable damage. This sensitivity has already been noted for invariant strengths, and applies also to fatigue. The sensitivity is in fact exaggerated when dealing with fatigue because the fatigue limit strength presents a very strong discontinuity. It follows that an even only slightly stronger item can have a much longer life. This accounts for many of the very extended distributions. Indeed, it is possible for the weaker items of a population to have a very limited finite life while the stronger items have a sufficiently high fatigue limit strength to ensure an infinite life. Strength variations are therefore more significant for fatigue than for stress-rupture or erosive type phenomena. In fatigue an isolated excessive load will most likely only cause extra damage which may still be small in relation to the total damage, whereas with stress rupture an isolated excessive load must be expected to cause failure.

To obtain an overall view of the fatigue failure pattern we look first at the dependence of the failure-free life on safety margin and loading roughness. It is shown on figure 5.13. For fatigue the parameters have been evaluated with respect to the fatigue limit strength. It is possible to achieve an infinite fatigue life by choice of these parameters in the same way that it is possible to achieve zero stress rupture failures when those parameters are referred to a stress-rupture strength criterion. Figure 4.18 may be used for design purposes if infinite life is the objective: and of course it will be if the number of load cycles is very high, as may arise with high-speed machinery, or perhaps when the stress is a consequence of vibration, and so on. It would be essential if safety were involved. In this case rough loading demands a higher safety margin than smooth loading. However, a

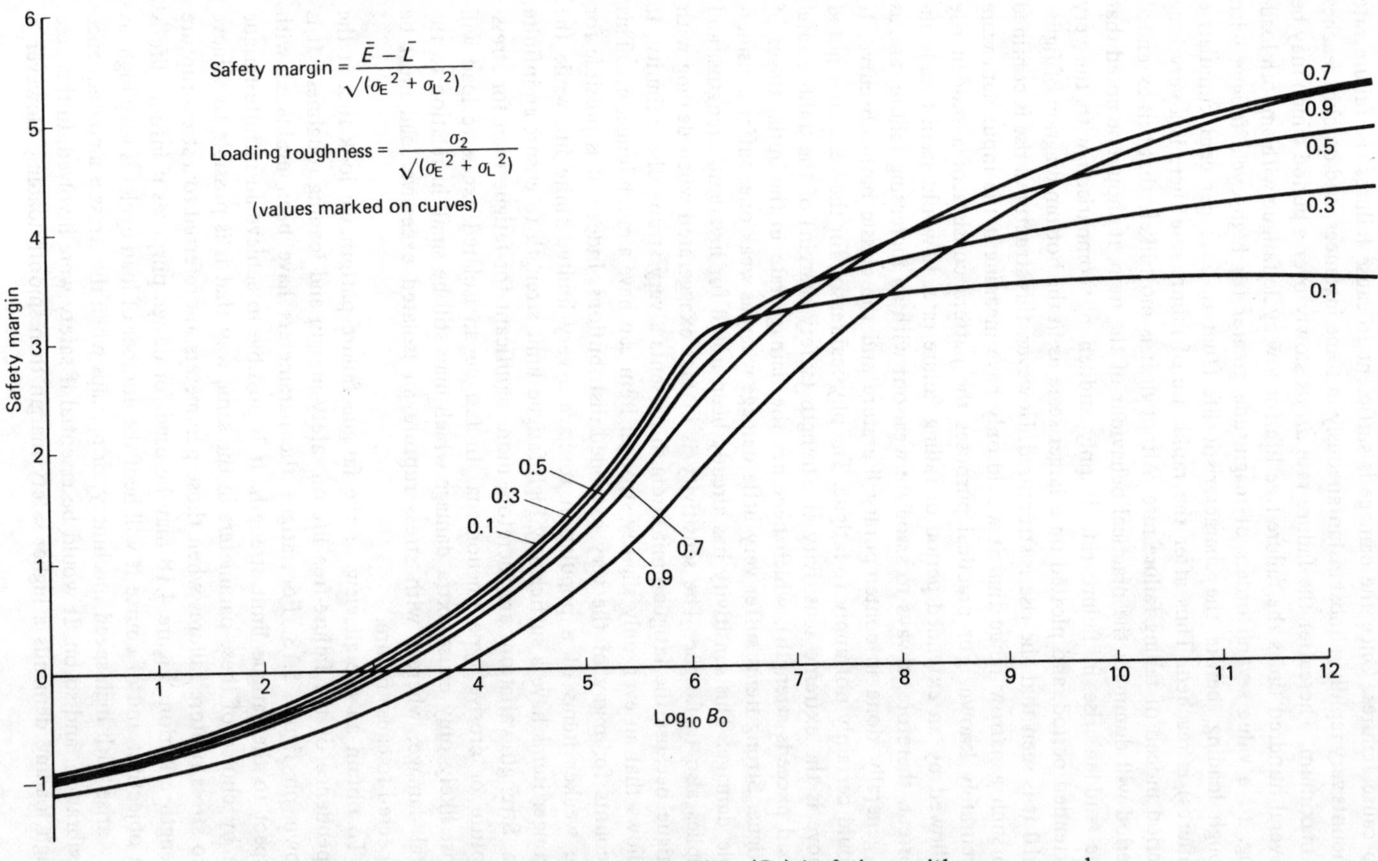


Figure 5.13 Life of intrinsic reliability ($B_0$) in fatigue with near normal load and strength distributions (Weibull $\beta = 3.44$)

very long, but not infinite, life may be obtained using somewhat smaller safety margins. This is attractive because the subsequent failure rate will be very low, as may be seen from figure 5.12. The difficulty in these circumstances is the sensitivity of both the failure-free life and the subsequent failure rate to safety margin and loading roughness. Sensitivity is always associated with the intersection of the tails of distributions. It is doubtful if full advantage can be taken of this design technique for the reasons that were advanced in connection with intrinsic reliability in section 4.4, though the penalties of missing the target are not nearly so severe, and 'risks' may be more sanguinely taken. By contrast with the case just discussed, however, it may well be possible to make use of smaller safety margins when only a moderate life is required. In these circumstances the main body of the populations will be interacting and hence the sensitivity to the controlling parameters will be much reduced. Figure 5.11 represents these circumstances. Even though the life expressed in terms of cycles is still reasonably high, the subsequent failure rate is low. This may not always be considered an advantage of course – design to a specified life with a minimum operating time after that life (that is, a high failure rate) being a more efficient use of the item's capabilities. Nevertheless, the lack of sensitivity problems makes this a much easier region in which to work. It appears that medium safety margins are practically feasible in fatigue even though such values would be unacceptable for invariant strength design. Absence of adequate data is the main problem here. As one moves to low cycle fatigue the behaviour becomes more predictable on account of the greater load–strength interaction, assuming that adequate data is available. More precise design is therefore possible. It is interesting to note from figure 5.13 that rough loading now requires a lower safety margin because many of the extremely low loads are doing no damage. Once the failure-free life has expired the failure rate is much higher, of course, but this is not seen as any drawback. In this type of application one is usually only interested in the failure-free period and the subsequent failure rate is of little interest. Indeed, a high subsequent failure rate implies an effective design since useful life is not being wasted. Before leaving this general description of fatigue failure patterns it must be reiterated that the possible combination of parameters and values is very high, and so no credence should be placed on the precise numerical values given on figures 5.10, 5.11 and 5.12 or 5.13 in relation to specific instances. It should also be remembered that we have only examined one shape for the *s–N* curve. Instead of the log-linear curve adopted here, it is often claimed that the *s–N* curve follows a cube law for bearings and a fifth power law for gears between about ten thousand and ten million cycles. Such modifications will substantially change numerical values. However, the salient feature common to most fatigue failure patterns is the extended life pattern after the failure-free period. Since this life cannot all be lightly thrown away, it presents some difficult design problems which will be taken up later.

Although attention has been concentrated on generally distributed loads and strengths as the practically important ones, it must be recorded that some strange failure patterns can arise in particular cases; for example, when the mean load is narrowly distributed and is slightly above the mean fatigue limit strength. Some typical curves are given on figure 5.14. The curves seem to have two parts. In all cases the first part of the curve is dominated by the failures of those items whose strength is less than the lowest load (this can be taken as the mean minus $3\sigma$ for a normal distribution). Every load therefore does damage. The second part of the curves is more influenced by the failure of those items whose fatigue strength is greater than the minimum load, and are therefore able to resist some of the loads without damage. The failure rate is thus much less, dramatically so in some instances – see figure 5.14b. Low slopes of the *s–N* curve give rise to a gradual transition from one regime to another: high slopes give rise to more rapid transition, even discontinuities. Another form of load distribution, which is an extension of that just discussed, consists of a small number of discrete loads, each of which

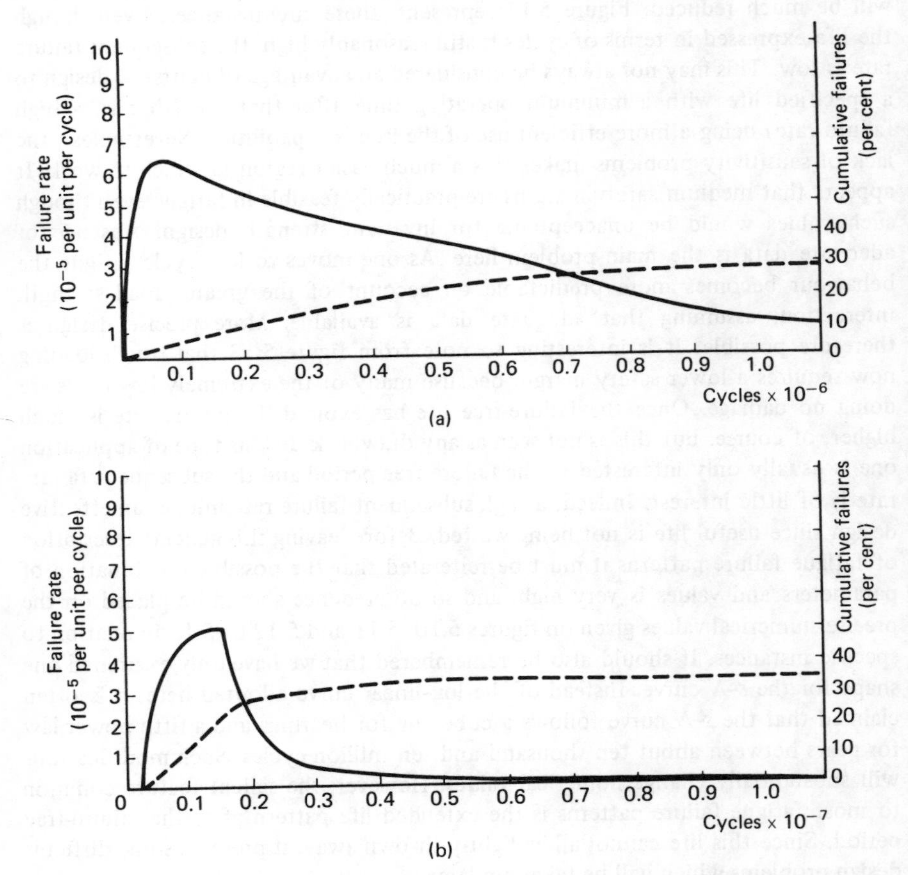


**Figure 5.14** 'Kinks' in fatigue failure patterns (——— failure rate, – – – – – cumulative failures): (a) mild kink, (b) pronounced kink

may, however, be distributed over a small range. This is a not uncommon load pattern though the mixture of levels, and consequently failure pattern, is varied. It should be remembered that unlike stress-rupture situations, fatigue can sustain considerable interaction between the load and strength distributions. Particularly if these distributions are allowed to take different shapes, and the *s–N* curve is also allowed to take different shapes, any number of fatigue failure patterns can be theoretically anticipated. Furthermore, we have only considered *s–N* curves for which a true fatigue limit exists. Many materials, which include the important light alloys, do not have a limit. If this is so, all loads do damage and hence infinite lives are not possible. The failure rate does not then fall as fast as when a true limit is present. Indeed, it is possible to create calculated curves having no falling failure rate, though the pattern is always far removed from the ever-increasing rate discussed in section 5.2. Whether all of these are practically significant is another matter, and each case has to be treated individually in its own particular context. It is thought to be not worth while pursuing the many theoretical possibilities at this stage, and further discussion of fatigue failure patterns is deferred until after field evidence has been examined, when it is taken up again under maintenance and design.

The reader may well ask why fatigue has been approached via standard *s–N* curves and Miner rather than through more modern fracture mechanics. There are a number of reasons for this which can be listed as follows

1. Fatigue calculations involving fracture mechanics invariably use Paris's law, which expresses the rate of crack growth as a single power term of the stress intensity range. It is recognised that this law is in error at low stresses: for example, it predicts that any stress, no matter how low, does fatigue damage. This cannot be correct, but because the predicted damage is very low at low stresses it is acceptable for deterministic work. It would, however, introduce unacceptable errors into the stochastic approach being followed here, since it is neglecting a significant source of the life distribution. Unfortunately there is no accepted modification to Paris's law and the author was not prepared to invent one because any derived phenomena could be attributed to the unsupported assumptions.
2. Even if 1 above is surmountable, the evidence the author has seen suggests that the constants in the Paris equation are statistically distributed. For deterministic work mean values may be acceptable, but again not for stochastic work. Some workers have correlated the constants with the yield strength, but it is not clear if this applies to individual members of a population and so further unsupported assumptions would have to be made.
3. It is usual in the fracture mechanics approach to fatigue to assume a maximum initial crack length which applies to the whole population. The actual values may be associated with a particular configuration and its method of manufacture. Subsequent life is based on this length in a deterministic calculation which gives the minimum life. However, the real initial crack formation consists of a

population whose lengths are obviously distributed, and once again an assumption would have to be made for a stochastic approach.

4. It seemed to the author that making assumptions for so many distributions, for which no evidence existed, would make any deductions extremely doubtful, if not completely valueless. On the basis of 'the devils you know. . .', it seemed much more satisfactory to stick with the *s–N* curve and Miner.
5. The approach in 3 can only apply to initially cracked structures, and although these form an important class there are many mechanical machine parts which are effectively initially crack free. Fracture mechanics does not encompass the crack nucleation period of such items, so the treatment adopted in this book would have to be followed in these instances anyway. Many items fall in this class.
6. There was a further not unimportant reason. The effects of many other wear-out processes, besides fatigue, are calculated on the basis of linear proportional damage (that is, Miner's assumption). For example, the creep process with varying load and temperature is estimated in this way. It does not seem unreasonable, then, to suppose that Miner's rule applies to many wear mechanisms in addition to fatigue. There are also other failure mechanisms which are very similar to fatigue but which do not involve cracking. As an example one can quote the loosening of a fastening under vibration. Most engineers would agree that there is a vibration level below which the life of a fastening is infinite, but at levels of vibration above some critical value the life is finite and the higher the level the shorter the life; that is, the vibration level–life curve is a typical *s–N* curve. A similar method of calculating the failure pattern is therefore probably valid.

For those interested in the application of fracture mechanics to stochastic systems refs 16 to 21 may prove a useful starting point, though fracture mechanics has progressed considerably since these papers were written. Unfortunately for one reason or another this work was not followed up, probably because the advances being made in fracture mechanics lead away from the thesis of these papers. Nevertheless, it should be recorded that Birnbaum and Saunders were the first to suggest the possibility of a falling failure rate in a wear process. It seems to have been completely disregarded at the time, doubtless because such a phenomenon would have been considered impossible in the electronic ambience which then dominated reliability studies.

In the United States the lead in the treatment of stochastic fatigue has come from Kececioglu (see, for example, refs 22 and 23). As stated, he follows the load/strength interference technique adopted for stress-rupture in this book. It is potentially the most accurate method, except that Kececioglu does not specify how the instantaneous, that is, single load, strength is to be evaluated. Some workers use the fatigue strength directly, but this is clearly in error above about 1000 cycles, with the error increasing significantly with cycles above that value. It is thus only valid for very low cycle fatigue when used in this way. An alternative method has been given by Gardner (ref. 24). His approach is based on the very plausible hypothesis that the numbers of load applications doing equal fatigue damage at different load levels are the same as those which result in equal cumulative fatigue failures at those levels. This hypothesis enables the damage at various load levels to be correlated, without using the controversial Miner's rule. The equivalent number of load cycles at any reference load level can then be estimated; from which the total cumulative failures and hence reliability can be calculated. The material input data is the distribution measured directly in any fatigue test. Gardner used a log-normal distribution, but Kinkead who has taken up the method in the U.K. (ref. 25) uses the normal distribution. However, neither of these distributions can be correct: there must be an initial failure-free period representing crack nucleation. If the cumulative failures are zero during this phase, Gardner's hypothesis leads to an indeterminate situation. Applied strictly, any number of load applications within the crack nucleation regime do zero damage – since damage is equated to cumulative failures – which is absurd. Like the Birmbaum approach, the Gardner method would appear to apply only to the crack propagating phase, which limits its application. It would, of course, be applicable to initially flawed structures. The less sophisticated method presented in this book was adopted on account of its basic simplicity and its straightforward interpretation of general fatigue failure distributions.

At this point it has to be emphasised that all existing approaches really only allow comparisons to be made. The calculation of absolute values is much more difficult, if not impossible, with existing knowledge. The failure patterns quoted in this book should only be treated as illustrative of general tendencies because of the many unknown factors involved. For fatigue these include: the doubts concerning Miner's law itself and the even greater doubts regarding its application to the *s–N* curves of individual items; the relationship between the *s–N* curves of individual items of a population when the influence of the manufacturing process, the component shape, surface finish etc. is taken into account; the known fact that the *s–N* curve can be a function of the loading sequence; the known influence of loads just below the fatigue limit; the known modifications to Miner when loads exceed the yield stress; and so on. Reference 26 shows how some of these factors may be introduced when calculating fatigue lives with variable loading, though it is confined to deterministic strengths only. Reducing the summation in Miner's law to less than unity is one method of approach. It may also be appropriate to recall here that if the intersecting tails of both the load and the strength distributions are at all extended, then the assumption of an instantaneous loss of strength may not be wholly valid. Hence the failure patterns quoted must be regarded as a qualitative indication of general behaviour rather than quantitative assertions. Nevertheless, we are able to draw a distinction between the various failure mechanisms, which can be most marked, and assess the consequences. In spite of the disclaimers it may still be possible to use the theory to make small modifications to field data to take into account changes in operation conditions or other factors, but this does require specialised experience in the particular area involved and is outside the scope of this book. For nearly all real estimates of fatigue failure patterns, extensive field data and experience is necessary.

*Exercise 5.3*

It is proposed to use a gas turbine designed and used successfully as a helicopter power unit as the power pack in a heavy earth-moving machine. Amongst other factors, some doubt exists concerning the reliability of the turbine disc for this purpose. You are asked to produce outline failure rate–age curves for the disc in both applications, and hence comment on its suitability in the earth-moving role. You may make use of the following simplifying assumptions and data in your assessment

1. The design and evaluation of the disc is to be based on the peak elastic stress (see ref. 27). The cycles to first crack for the disc are plotted against peak elastic stress on figure 5.15.
2. The only load you need consider is the centrifugal load.
3. The maximum peak elastic stress (rated stress) allowable in the disc design is set at 1500 $MN/m^2$.
4. In the helicopter role the mean stress due to the centrifugal load is 80 per cent of the maximum rated stress, and is normally distributed with a standard deviation equal to 6 per cent of the mean.
5. In the earth-moving role it is planned to run the turbine at a mean speed equal to 3/4 of that in the helicopter application. It is anticipated that the centrifugal stress will be normally distributed with a standard deviation of 15 per cent of the mean stress.
6. In the helicopter application there is effectively one cycle per mission, but in the earth-moving application there could be on average 125 cycles per mission.
7. The disc is made from high-class material under strict quality control, so you may assume that the strength is normally distributed with a standard deviation of the elastic endurance limit of 5 per cent of its mean value.

(*Note*: The calculations can be very tedious even with a hand calculator, and a microcomputer is really required to obtain numerical values readily. If such a machine is not available you can still make a qualitative assessment of the failure–age curves and hence make a qualitative comparison.)

**Exercise 5.4*

Using the assumptions for the *s–N* curve adopted in the preceding examples of this chapter, what would be the distribution of life in a standard fatigue test, that is, at constant load?

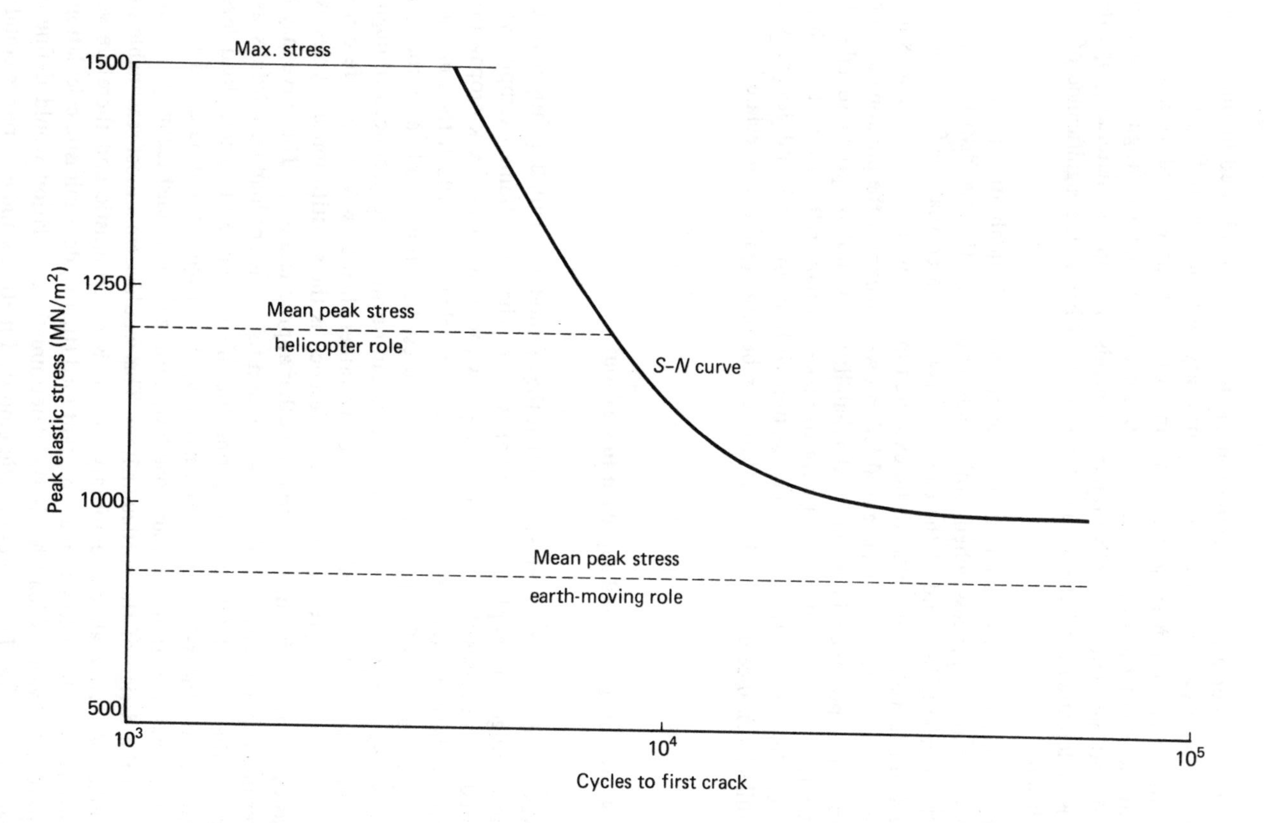


**Figure 5.15** The $s$–$N$ curve for a turbine disc based on peak elastic stress

**Exercise 5.5*

Using the mathematical representation of the *s–N* curve adopted in this chapter (that is, linear *s*–log *N* curve) and an arbitrary distribution of fatigue strength, show that if an item is subject to a number of concentrated loads (that is, deterministic loading) the ratio of life for 10 per cent cumulative fatigue failures to that for 50 per cent cumulative failures depends only on the material properties and is independent of the value of the loads. What is the significance of this conclusion?

It has been shown that it is possible to obtain a qualitative picture of the fatigue wear-out process when both loads and strengths are distributed. The results of calculations and of the exercises have indicated that the failure pattern differs substantially from that to be expected in pure wear. Since fatigue is held to be responsible for most mechanical failures we can expect the phenomena to be of special importance in mechanical reliability. It is not proposed to take the matter further here, but we shall have to make constant reference to the subject in later chapters, particularly in connection with maintenance and design, when the full significance of the failure patterns can be more readily appreciated.

## 5.4 A possible generalised approach to wear-out

It was stated in the previous section that the method of calculating fatigue failure patterns might be applicable to other wear mechanisms. Some examples were quoted of processes that resembled fatigue without involving crack propagation. However, here we want to look at any wear mechanism in which the same load does wear damage of some kind and also causes the ultimate failure. In these circumstances an equivalent of the *s–N* curve can obviously be produced, though it may have to be designated an *s*-life curve rather than an *s–N* curve. Life can be measured in any units. So the only difference is the substitution of cycles as a measure of life, as in the *s–N* curve, by other appropriate units. The curve may be determined practically by subjecting the item to a constant load until it fails, as is done in the case of fatigue. Measuring the life to failure at various load levels would define the *s*-life curve. Alternatively, the designer could be asked to estimate the life at various constant load levels, using the specialist techniques applicable to the design of the particular item. This would seem a most reasonable step to take. Virtually all contemporary design is deterministic and therefore very appropriate for life estimation at constant load (that is, deterministic calculations). Repeated for various load levels the deterministic calculation would define the mean *s*-life curve. Thus deterministic empirical methods and experience could be carried forward to a stochastic calculation of wear-life distributions. This is believed to be very important: reliability engineers cannot be experts in all branches of mechanical engineering, and it would be folly not to make use of all existing knowledge. Given such a curve a proportional damage law, for example, Miner's law, can be used to deduce the life, or what would be the distribution of life for any load distribution. Proportional damage laws are used so often in standard deterministic calculations that there is every justification for employing a similar assumption for this purpose. The equations of section 5.3 [except (5.13)] would then be equally valid, with the minor modification that the strength distribution would refer to the initial state rather than the fatigue limit. The significant difference between the two processes would be the absence of any limiting stress (the fatigue or endurance limit) so that an $s$-life curve may be expected to extend to zero stress.

It should be made clear that the author has no experience of applying such methods, other than in carrying out a number of theoretical calculations. It has not been found possible to make any direct checks, owing to the absence of practical data. Nevertheless, the method is such an obvious step to take that it is offered here for consideration by anyone who has to make an *ab initio* estimate of wear-out failure pattern in cases where no empirical techniques or field data are available.

*Exercise 5.6*

The wear load–life curve of a wood chisel was determined experimentally by cutting at constant load to failure for various load levels. It was found that the mean life increased with reduction of load over the working range so that the life was inversely proportional to the load. The cutting edge of the chisel had a mean strength of 25 N at 20 minutes. The standard deviation of strength at that point can be taken as 2 N. If the load applied during general workshop use has a mean value of 10 N and a standard deviation of 2 N, calculate the variation of failure rate and cumulative failures with age for this group of chisels. A failure in this context is taken to mean the chisel needs resharpening as a consequence of use; that is, all stress-rupture effects associated with mishandling, accidents etc. are neglected. How would the variation of the above quantities differ if the life was inversely proportional to the cube of the load? Compare the two curves and contrast both with the typical fatigue patterns presented earlier in this chapter and those you calculated in exercise 5.3.

The solution to exercise 5.6 has shown that in the absence of a fatigue limit the failure rate increases continuously during wear-out, as suggested in section 5.2 or by the classic bath-tub curve. If the life is inversely proportional to a higher power of the stress, however, the rate of increase is lower. One can imagine a whole series of wear $s$-life curves, from directly inverse through high inverse powers, to an infinite power representing the fatigue limit itself. Thus one can expect a range of wear-out curves characterised by different slopes of the failure rate-age curves according to the relevant wear mechanism. This would seem to exclude a simple empirical all-embracing representation.

The physical process modelled in this section differs from that described in section 5.2. The proportional damage law implies that the loss of strength is far from linear – see equation (5.3) and figure 5.6, and contrasts with the linear law assumed in section 5.2, and illustrated on figure 5.1. The proportional law obviously models some processes correctly. Thus any craftsman would confirm that the well-sharpened chisel of exercise 5.5, or any other tool for that matter, will keep its cutting edge and be just as effective for a long period of use, but once it starts to loose its edge it deteriorates very rapidly indeed. (In practice it will be re-sharpened on the first signs of deterioration, of course.) The sudden loss of strength is a phenomenon common to many wear processes (see section 9.4 for further evidence). It is represented in the model. This model is relevant to wear in which the same load does damage and also causes failure.

The linear damage model of section 5.2 is relevant to wear in which the load which eventually causes failure is not responsible for the damage. In these cases the deterioration of strength is gradual – if not truly linear as assumed in section 5.2, so that the likelihood of a failure increases continually as the strength is reduced. The load–strength interaction process applies to the whole life because there is no critical point at which the strength deteriorates suddenly and supersedes the interaction process. Linear and proportional damage are physically distinctly different phenomena, and should be treated as such. Philosophers may wish to consider the wider implications of linear damage leading to gradual decay but proportional damage to crisis. For example, the ill effects of rusting (and many other forms of corrosion) may fall into either category. Removal of the metal by rusting is a gradual process until the item fails because of inadequate strength, that is, the damage mechanism and the failure mechanism are separate. If we are concerned with aesthetic values, however, the same mechanism is responsible for damage and failure. Proportional damage would then seem more appropriate, and conforms with experience that suddenly the item 'doesn't look so good'. The division of wear mechanisms into these fundamentally different groups is generally ignored in reliability texts and literature, though it was recognised by Bomphas-Smith (see Reading list) as a significance factor in wear assessment.

## 5.5 The significance of wear-out

Virtually all real mechanical items wear in some form or other. This fact must be squarely faced if we are to approach mechanical reliability in a realistic manner. Although we can design items so that they are initially 100 per cent reliable to all intents and purposes, that is, intrinsically reliable, we must recognise that wear will take place so that eventually failures will occur. Intrinsic reliability does not mean unlimited failure-free operation. Ideally any component can be withdrawn from service and replaced, if necessary before wear has reached such a state that failures arise, that is, it is replaced before the wear-out phase. This council of perfection may be followed in some special instances, essentially so when safety is involved. Thus one would withdraw the helicopter disc referred to in exercise 5.3 well before the life of intrinsic reliability had expired. In such cases wear is of considerable importance since it determines the life, but the wear-out process and the associated failure–age characteristic is of no significance. However, in many circumstances it would be uneconomic to throw away the good useful life still available in many of the items during the wear-out period. This would be particularly true if the failure rate were low. In these circumstances a balanced decision when to withdraw or replace must be taken, and this defines the reliability ultimately achieved. Thus an assessment of mechanical reliability becomes an assessment of wear-out, and the more accurately this is defined the more likely will a valid conclusion be reached. But this is not the end of the story. Wear-out may be arrested, reversed or terminated by maintenance. So maintenance becomes an essential part of mechanical reliability. Clearly the optimisation of any maintenance plan requires a fairly specific understanding of the wear-out processes. These matters will be taken up in detail in chapter 9. Here we merely note the need for quantitative data on wear and wear-out. At the same time, it must be emphasised most strongly that this can only be obtained empirically at present. It is a great area of weakness in the general theory of reliability which has been neglected by most reliability workers, probably because they have concentrated on electronic reliability in which wear is of little significance. In the mechanical world, wear is all important.

In the absence of adequate fundamental knowledge it is inevitable, in some ways essential, that practitioners resort to empirical methods. There is consequently a number of empirical formulae which purport to estimate the reliability of mechanical items. For example, the reliability of bearings is often quoted as

$$R = \exp\{-(t/6.84\, B_{10})^{1.17}\} \tag{5.19}$$

where $R$ is the reliability and $t$ is the operating time in hours. Sometimes adjustment factors are used in conjunction with this equation. Thus one can write

$$Q = Q_L \times Q_S \times Q_T \tag{5.20}$$

where $Q_L$ = a load factor
$Q_S$ = a speed factor
and $Q_T$ = a tribological factor.

Then

$$R = \exp\{-(t/6.84\, Q\, B_{10})^{1.17}\} \tag{5.21}$$

There are in fact several versions of equation (5.19) or (5.21) in which the constants have slightly different values.

It is not our purpose to continue this theme here, although there are a number of such expressions for common mechanical components. Some of these are based on theoretically derived formulae with constants replaced by empirical values, while others are much more empirical in structure. In the hands of experienced engineers working in a known environment they appear to give results which satisfy those concerned. However the author must sound a cautionary note. It has been shown that the loading can exert a marked influence on the reliability behaviour when both stress-rupture and wear-out phenomena are involved. Furthermore, there are many loads which can affect the reliability, some of which may have been ignored in a particular formula. Such formulae are definitely for experts both in their knowledge of the derivation of the formula and its proposed field of application. Quite apart from these difficulties, it would be out of keeping with the object of this book to pursue the question further. The objective is to establish a basic understanding of the principles of mechanical reliability rather than to compile a handbook of empirical information.

The necessity to express wear-out phenomena in quantitative terms leads us to question and reconsider the definition of reliability in practical terms. Essentially we are returning to the interpretation of 'ability' in the British Standard definition – a problem first introduced in chapter 2, but then deferred for further consideration. If an item is not intrinsically reliable its reliability measured as a probability is practically indeterminate and need not concern us. If it is intrinsically reliable what we really need to know is how long, that is, to what age, it will remain intrinsically reliable and what is the failure-age relationship thereafter. As a first step we define 'ability' in the British Standard definition as the life of intrinsic reliability or zero failures. We shall denote this by the symbol $B_0$, in which the suffix zero denotes zero failures. In a similar vein we can define other 'abilities' as the age at which some arbitrary proportion of the initial population first fail. These ages, or lives, defined by percentage cumulative failures, will be denoted by the symbols $B_5$, $B_{10}$ etc., denoting 5 per cent, 10 per cent etc. cumulative failures respectively. Other proportions could, of course, be used. The symbol $B$ is a reflection of the fact that this nomenclature was first introduced by the bearing industry as a measure of the effectiveness of its products. In earlier practice, and often today, only one value is quoted, usually $B_5$ or $B_{10}$, but it is shown in the next chapter that if two or three values can be quoted (two plus mechanism or three without) it is possible to interpret the failure pattern by a statistical distribution and hence achieve generality.

It should be appreciated that there is no fundamental difference between ability expressed as a life or as a probability. It has already been stated that they are related through the failure-age distribution. The choice of the measure is one of convenience: that of expressing ability as a directly relevant quantity in a way which links directly with general engineering experience. Thus knowing the reliability at some arbitrary age is not likely to be very helpful. Because reliability can vary rapidly with age and in many different ways during the wear-out regime it is not possible to extrapolate (forwards or backwards) with any certainty. On the other hand, knowing the life for which the reliability never falls below some acceptable value is of direct importance: it expresses a limiting factor in the use of a component, and a definable target for design. It is therefore a superior practical measure, and the one we shall henceforth tend to adopt in this book. Unfortunately it cannot be used universally because users often prefer to demand a probability of success at a given time (or times) as a more direct measure of their requirement. An example would be the probability of completing a mission whose duration is only a small fraction of the equipment life, allowing maintenance and so on between missions. Furthermore, in a piece of complex equipment there will be a large number of components having different $B$ values and failure patterns. The relationship between these is best left to the designer. Most certainly so if the user has no technical knowledge. Thus it is not possible to formulate a universal definition other than in more general terms. This appears to be the course adopted by British Standards Institution, which the author would fully endorse. However, this is in apparent contradiction with practice in the United States where reliability has consistently been expressed as a probability in all published work. This may only be a reflection of the dominance of the electronic industry in that country. Thus it is worth noting that while NASA called for specified numerical reliabilities for the Apollo space programme, for the later and more sophisticated Space Shuttle programme no numerical reliability requirements or goals exist. This would imply some loss of faith in probabilistic reliabilities in a very advanced 'hi-tech' area when electronics is not the predominant technology.

The true nature of most mechanical unreliability becomes evident when reliability is expressed as a life. It may be assumed that intrinsic reliability is achieved, as it nearly always is in practice, but that it is lost as a result of wear after some use. For any particular wear-out mode the actual failure pattern will be substantially the same, but will set in at different ages for different designs as shown on figure 5.16, at (a) and (b) for example. Because of the general shape of the cumulative failure-age curve, the reliability at a subsequent time (at $t$ on figure 5.16) can be substantially different. The designer often has little control of the wear process – that is mainly dependent on the physical nature of the wear mechanism – but he does have control of the strength reserve, that is, the starting point of the curves (a), (b) or any other, on figure 5.16. Thus unreliability is more due to an incorrect choice of time at which to let wear-out begin, that is, choice of the appropriate $B$ life, than to an excessive value of the failure rate, whose variation with time is often outside the control of the designer. This explains why designers prefer to express reliability as a life which they can control, than as a probability or failure rate which they cannot control as such.

No matter how the wear-out process is measured, of one thing we can be certain – the strength or strength reserve of all mechanical components deteriorate with use (and sometimes even without use!). This has repercussions on our design philosophy. Even if the reader did not accept the arguments for intrinsic reliability advanced in section 4.6 and still hankers after the Will-o'-the-wisp advantages of the non-intrinsically reliable item, he must now face up to wear. Wear is an essential feature of virtually all mechanical items and some strength allowance

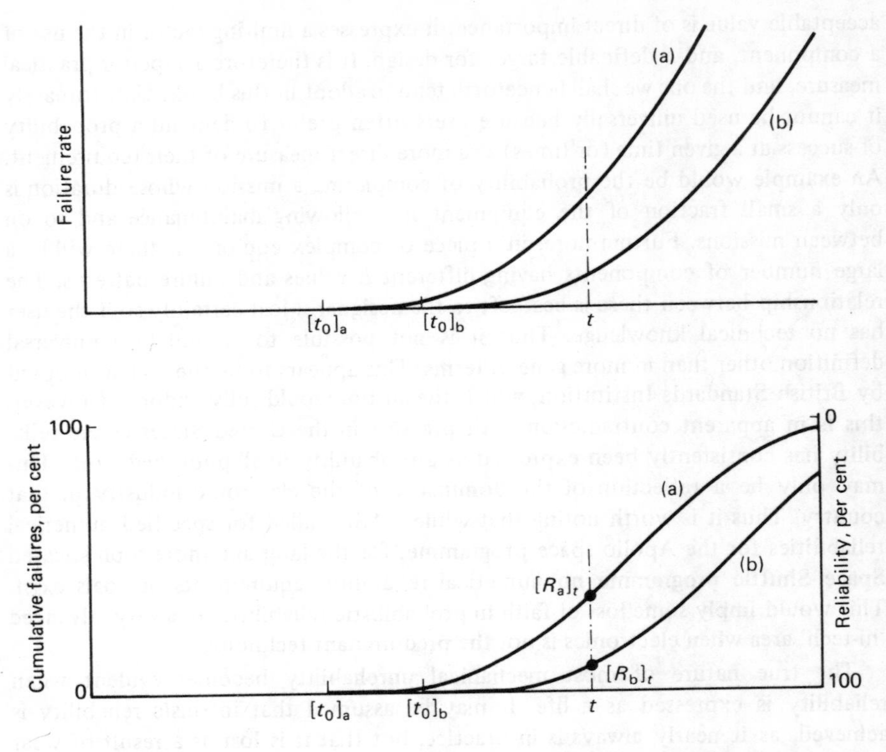


**Figure 5.16** Comparison of reliabilities of two components having identical failure patterns but different lives of intrinsic reliability

must be made to cater for it. Thus an item which could be designed for an invariant strength safety margin (that is, for a stress-rupture phenomenon) which is slightly less than that for intrinsic reliability, will almost certainly have to be designed for one that is greater when allowance for wear is made. As to all things connected with mechanical reliability, there will be exceptions: in this instance one could imagine that some static equipment is subject to negligible wear if proper protection against corrosion and so forth is taken. However, as a generality any consideration of wear only reinforces the conclusion that all items will be intrinsically reliable in the first instance.

A consequence of wear-out that may concern the mathematical modeller is that the failure-inducing load may no longer be the main load common to a series system of components. Indeed, no load causes failure – it is the cumulative effect of a sequence of loads which causes wear failure. The system is totally decoupled in the sense used in connection with stress-rupture phenomena. But the cumulative load cycle for each physical load will be common to all components, retaining dependence. Even so the cumulative effect of identical cumulative load cycles can differ for the different wear processes. So too can the susceptibility of different wear processes to different loads. Some degree of indpendence may thus be achieved, and must be correctly modelled. Unfortunately, the number of wear mechanisms is very large indeed. The list below shows some of the possibilities

Erosion (mechanical action)
- (i) abrasion
  - (a) sliding action
  - (b) rolling action
- (ii) rubbing
  - (a) sliding action
  - (b) rolling action

Corrosion (chemical or electro-chemical action)
- (i) solid/solid
- (ii) solid/liquid
- (iii) solid/gaseous
- (iv) mono-solid (owing to local material variation arising from impurities, stress, heat treatment, etc.)
- (v) stress corrosion
- (vi) surface protection breakdown by any wear process

Fatigue
- (i) low cycle fatigue
- (ii) high cycle fatigue
- (iii) thermal fatigue
- (iv) corrosion fatigue

Surface degradation
- (i) fretting
- (ii) pitting
- (iii) spalling
- (iv) cavitation action

Defastening
- (i) from vibration
- (ii) from repeated shock
- (iii) from thermal cycling

Creep
- (i) at normal temperature
- (ii) at high temperature

Ageing
- (i) thermal
- (ii) chemical
- (iii) structural
- (iv) environmental

Fouling
- (i) from dirt etc.
- (ii) by clogging
- (iii) from wear debris accumulation

Contamination
- (i) of liquids
- (ii) of gases

Leaking
- (i) past 'solid' joints
- (ii) through solids (permeability)

Thermal
- (i) overheating
- (ii) burning
- (iii) distortion

The above list is by no means comprehensive, for example, no radiation phenomena have been included. Furthermore, many of the processes have been described in name only and can vary substantially, quantitatively from one circumstance to another. Thus the local stress pattern will have a major control over sliding abrasion which can therefore have many different failure distributions. Even further, it will be modified by lubrication. The combinations of lubricants and wearing materials alone are infinite. Additionally, many common forms of wear failures are a combination of the basic mechanisms listed above; for example, binding and seizing usually include some erosive phenomena plus some overheating and distortion. Some of the processes may be statistically independent, some may be dependent, but it is of more importance that many may be physically interactive. In the latter case the reliability cannot be estimated by any statistical combination of the individual mechanisms – stress corrosion may be quoted as an example. Some strength degradation may be due to 'loads' which are irrelevant to the functional role and have no stress-rupture significance, such as incompatible materials in contact. The number of possible wear patterns is incalculable. Against all this, in many practical cases most of the mechanisms can be neglected because they are irrelevant or because the process is so slow that it has no effect on the practical life span.

The most significant feature is, however, the vast number of wear mechanisms that can possibly lead to mechanical failure. Combinations of the above basic mechanisms alone suggests well over a thousand different wear processes. There is no single treatment for them all. Conventional modelling of them all would be a formidable undertaking. On the other hand, those directly concerned with the various mechanisms usually know a great deal about them – they have to! Unfortunately most of it is qualitative in nature. The physics of the mechanism is often understood but the controlling variables are so numerous that quantification is almost impossible. As illustration, suppose we have two loaded steel panels spot-welded together, and protected by a given number of primer, undercoat and top coats of a given paint. The corrosive life distribution of the combination will depend on the effectiveness of the protective paint covering, including in particular its penetration of the joint, the effect of repeated loading in destroying the protective coating, the ability of the protective coating to resist the environment, local heating and consequent effects of the spot welding, residual stresses in the panels from forming during manufacture, lack of material homogeneity, and so on: to say nothing of the load cycle and environment. The standard explanation of corrosion contained in most textbooks is not going to help much in estimating the life distribution. What is going to help is the vast amount of practical experience and know-how which already exists in the hands of specialists. The way ahead is clear: this knowledge must be harnessed and expressed in a manner which will help reliability studies. There is a vast untapped source of specialist information which is neglected at present. The thousands of real failure patterns are unlikely to follow formularised statistical distributions, which cannot cater for the physical phenomena and on which so much effort is being dissipated. Reliability workers must forget the statistics for a while and get out into the field and find out what is going on. As an illustration, obtaining expert opinions of the lives of, say, 10 per cent, 50 per cent and 80 per cent of an original population of a given item under specified load and environmental conditions (that is, obtaining information in a way to which experts can respond) enables significant statistical deductions to be drawn regarding its reliability behaviour and required maintenance. For a long time such knowledge will be substantially subjective. Even so it will be authoritative, and thus can be made the basis for dependable quantitative progress.

In summary, then, we note that wear is all important for mechanical items. Its consequences are seven-fold

1. It reinforces the necessity to design for intrinsic reliability during the early and working life of an item.
2. It becomes the prime factor which controls the mechanical reliability (or unreliability) of all products.
3. It emphasises the role of maintenance in achieving any level of mechanical reliability.
4. It strongly suggests that reliability should be expressed in terms of life rather than as a probability, for practical engineering purposes.
5. It precludes the use of a constant failure rate in mechanical reliability work, since the failure rate is not constant in any of the wear-out mechanisms.
6. It introduces a vast variety of physical phenomena which can only be treated on an empirical basis by specialists in each wear mechanism.
7. The concept of a simple bath-tub curve has to be abandoned. Many distinctive curves are possible (see figure 5.17).

Clearly we have to take our study a long way beyond the stage we have reached so far, but the reader, like the author, may feel that we have had enough of theory and that it is highly desirable to back up the theory already developed with some experimental or field data. If the theory is corroborated we shall have a firm basis on which to build an understanding of mechanical reliability in its widest interpretation. If it is not corroborated, further studies along these lines are pointless! In order to examine field data we shall have to use statistical techniques. The next chapter is therefore a brief summary of some relevant statistical distributions,

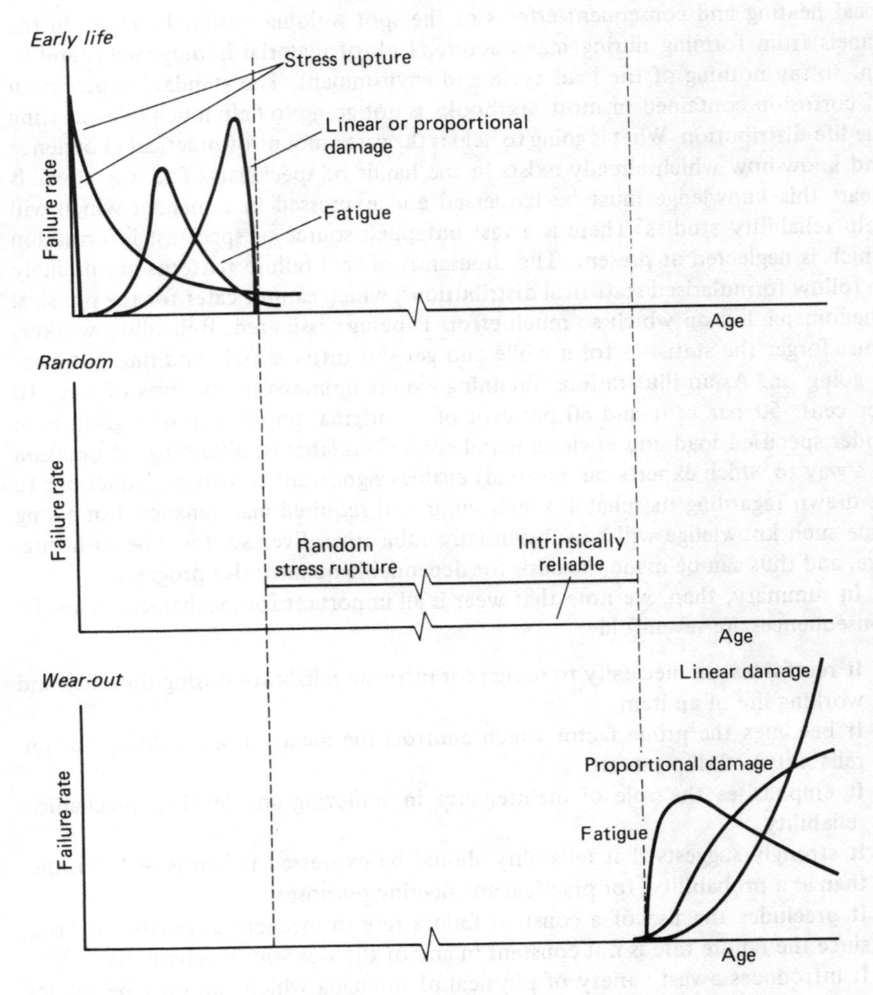


**Figure 5.17** Constituents of the bath-tub curve for mechanical components. Note that all combinations of early life, random and wear-out are possible

approached with application to reliability well in mind rather than as a treatise on statistics. It may be omitted by the reader who is familiar with statistics. With these weapons at hand we shall return to the main theme in chapter 7 with a discussion of field data, going on in chapter 9 to deal with repair and maintenance, which play such a vital role in controlling wear-out, and hence reliability.

# 6

# Some statistical distributions

## 6.1 Scope of study

A number of statistical distributions are used time and time again in reliability work, and need to be discussed particularly from that point of view before examining field evidence. Perhaps the one used most extensively is the negative exponential distribution. We have already seen in section 4.4 how it can represent the variation of reliability with time for infinitely rough loading. In addition any period of constant failure rate can be represented by this distribution. We shall therefore follow our brief study of the ubiquitous normal distribution with the negative exponential. However, it can be limited in application, particularly for mechanical products. Many real distributions do not quite correspond to the negative exponential, and significant errors could arise by forcing them into this mould. We therefore need something more flexible. The Weibull distribution fills this requirement rather well, better indeed than might be hoped for, because additionally it can be made to approximate very closely to the normal distribution. It is again very valuable for the same reason quoted in the case of the negative exponential distribution: many real distributions which bear superficial resemblance to the normal distribution are in fact sufficiently different as to exclude its use. Since it can 'stand in' for both the negative exponential and the normal distributions besides representing many other real ones, the Weibull distribution is very popular with reliability workers. It has a number of drawbacks as we shall see, but clearly merits close attention, and hence we shall follow our study of the negative exponential distribution by a study of the Weibull distribution.

Following our examination of the Weibull distribution we shall turn to the log-normal distribution. This has some importance in reliability work as it is claimed that it represents many repair processes. As we shall have to study the effect of maintenance on reliability, it is necessary that we be familiar with the properties of the distribution on this account. It is also quite likely that the log-normal distribution could be a much closer representation of other field data than, say, the Weibull distribution when more evidence has been accumulated.

Finally we shall take a look at the Pareto distribution. This is one of the lesser known distributions, and perhaps has little application in the general statistical field. However, the author believes it can be of great significance in the examination of field data. Since we shall be using it on several occasions in the subsequent chapters of this book, it is most appropriate that we take a look at its properties before examining field reliability data.

It must be emphasised that this chapter is not a study of statistics – good and adequate books already exist for that purpose (see the Reading list). The author would also record at this point that he is 'left cold' by debates surrounding the distribution which best fits data. First it must be realised that all distributions are empirical and there is no fundamental reason why one should be better than another. Secondly reliability data are often very scanty and the goodness of fit to a limited set is little guarantee of general application. Resort to the statistician's confidence limits is of little help, for if rigorously applied one would have precious little confidence in anything! What is important to the writer is that the distribution should adequately represent the data in engineering terms, and that any empirical constants should have some physical significance. In this way one can attach a conceptual confidence to the analysis, and apply it to other situations in which the engineering factors are readily understood, even if by so doing we are only achieving a first order approximation. At least the degree of approximation will be properly understood. We shall be looking at the distributions from that point of view and we shall assess their value for that purpose and that purpose alone.

The non-mathematical reader could take section 6.2 on the normal distribution in his stride. If he can manage section 6.3 on the negative exponential distribution he should do so, because it is used so widely in the literature: though not in this book. He must read section 6.4.1 on the general attributes of the Weibull distribution, section 6.4.3 which correlates Weibull distributions with failure patterns, and finally section 6.6 on the Pareto distribution. The remaining sections can be omitted, though this does mean some exercises cannot be tackled, but the general theme of this book can still be followed.

## 6.2 The normal distribution

The general form of the normal distribution, its properties and its application are fully covered in all standard statistical textbooks, and so require little attention here. Some standard data, which includes the tails, has been illustrated diagrammatically in figure 6.1, but full data can be obtained from standard statistical tables.

The probability density function of the normal distribution is given by

$$f(t) = \frac{1}{\sigma\sqrt{(2\pi)}} \exp\{-t^2/2\sigma\} \tag{6.1}$$

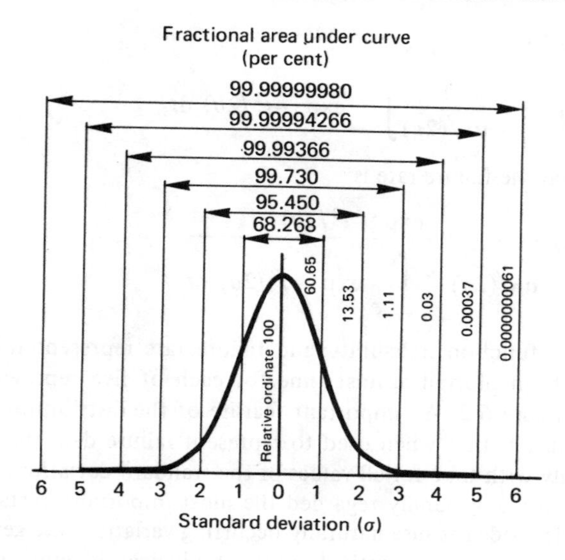


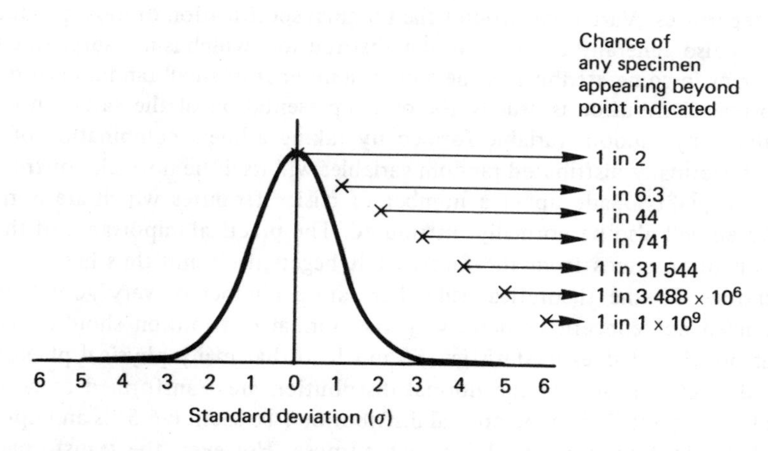


**Figure 6.1** Some properties of the normal distribution carried to high values of the standard deviation

where, following standard practice, $t$ is measured from the mean time. The corresponding cumulative failure distribution is then given by

$$F(t) = \frac{1}{\sigma\sqrt{(2\pi)}}\int_{-\infty}^{t} \exp\{-t^2/2\sigma\}\, dt \tag{6.2}$$

and hence the reliability by

$$R(t) = 1 - \frac{1}{\sigma\sqrt{(2\pi)}}\int_{-\infty}^{t} \exp\{-t^2/2\sigma\}\,dt \tag{6.3}$$

It follows from (6.1) that the failure rate is

$$\lambda(t) = \frac{\exp\{-t^2/2\sigma\}}{\sigma\sqrt{(2\pi)} - \int_{-\infty}^{t} \exp\{-t^2/2\sigma\}\,dt} \tag{6.4}$$

The probability density function, reliability and failure rate represented by the above equations have been plotted against time for each of five representative standard deviations in figure 6.2. An important feature of the distribution which is illustrated on the figure is that when used to represent failure data the failure rate increases continually with time for all values of the standard deviation.

The normal distribution is generally regarded the most important in statistics. This arises on the practical side because naturally occurring variations are generally normally distributed. This applies particularly to biological phenomena but equally to the properties of natural materials. The more natural they are the more likely this is to be true, but quality control of man-made materials can introduce some departures. Variations around the nominal specification of most production processes also generally follow a normal distribution, which is not surprising since if the error in some attribute is the sum of a number of small random errors then the normal distribution is usually the best representation of the variation in that attribute. Any random variable formed by taking a linear combination of independent, normally distributed random variables will itself be normally distributed; so that a quality made up of a number of basic attributes which are normally distributed will also be normally distributed. The practical importance of the distribution often stems from the fact that it begets itself and thus has very wide application. On the theoretical side there are a number of very good reasons, which need not concern us here, why the normal distribution should hold the central position it does in statistics. So much so that many physical phenomena which do not conform to the normal distribution are transformed so as to be normally distributed. The log-normal distribution (see section 6.5) is an important general distribution introduced for that purpose. However, the transformations involved in reliability representation are so complicated and imperfectly understood (see for example section 5.3) that this is not feasible when examining reliability field data. We have to resort to other distributions which are more directly applicable and which will be discussed in later sections. The normal distribution does not therefore play such a prominent part in representing mechanical reliability data as it does in general statistics. Nevertheless it can represent some failure patterns and it is of course the first choice in the representation of material properties for the reasons given above. However, one does need to adopt a sceptical view of its universality. Materials cannot have negative strengths as is predicted

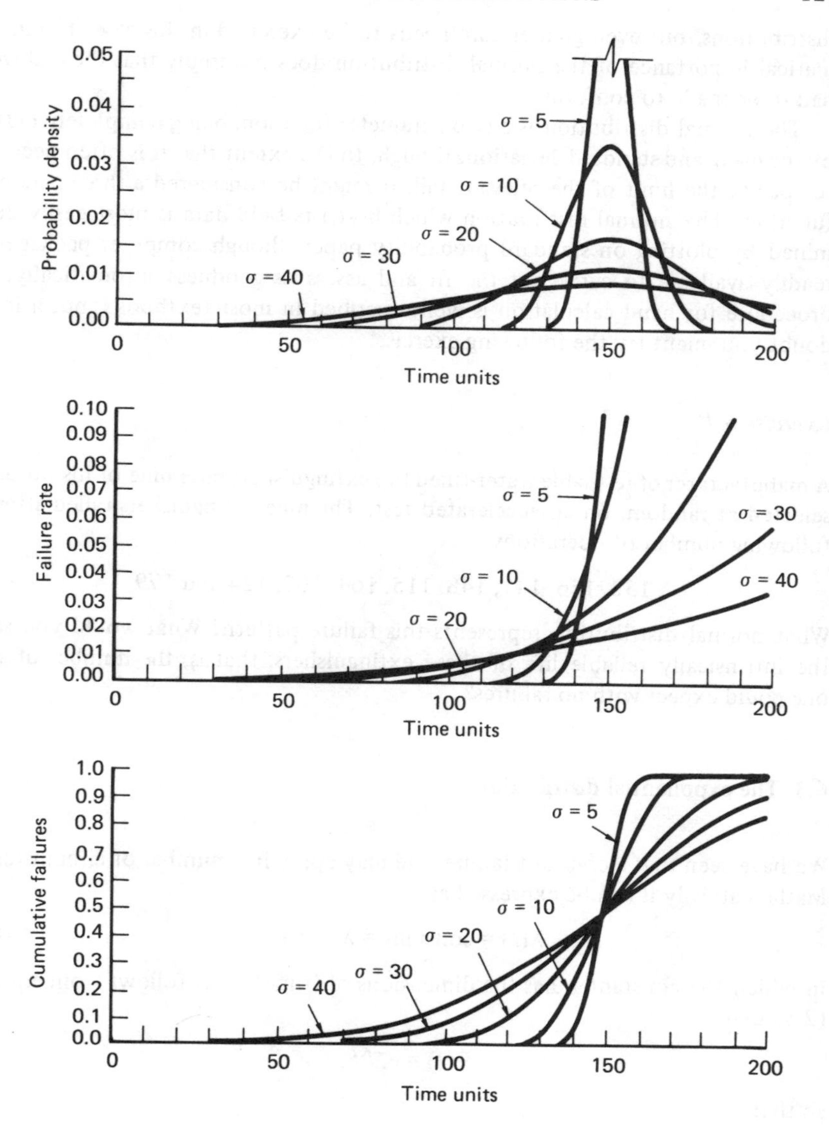


**Figure 6.2** Probability density, failure rate and cumulative failures (= 1 − reliability) for the normal distribution at the standard deviations ($\sigma$) marked on the curves

by the normal distribution, which is clearly invalid as it stands. This is particularly so as we are mostly interested in the tails. A simplistic belief in the normal distribution can lead to errors – see solution to exercise 3.2. It can be used for load distributions, but even greater care needs to be exercised in this case. The mathematical importance of the normal distribution does not imply that the real world had to be made to conform.

The normal distribution is a two parameter function, being completely defined by the mean and standard deviations though, to the extent that it is often necessary to specify the limit of the relevant tail, it might be considered a three parameter function. The normal distribution which best fits field data is most easily determined by plotting on standard probability paper, though computer packages are readily available to carry out the fit and assess its goodness automatically. The procedure for hand calculation is well described in most textbooks, but if in any doubt you might try the following exercise.

*Exercise 6.1*

A manufacturer of re-usable water-filled fire extinguishers puts nine of his products, selected at random, on an accelerated test. The nine extinguishers failed after the following number of operations

132, 156, 141, 148, 115, 164, 102, 124 and 179

What normal distribution represents this failure pattern? What would you say is the intrinsically reliable life of these extinguishers, that is, the number of shots one could expect with no failures?

## 6.3 The exponential distribution

We have seen that a constant failure rate may apply in a number of circumstances. Mathematically it can be expressed as

$$\lambda(t) = \text{constant} = \lambda \text{ (say)} \tag{6.5}$$

in which the constant $\lambda$ has the dimensions of $(\text{time})^{-1}$. It follows from equation (2.5) that

$$R(t) = e^{-\lambda t} \tag{6.6}$$

so that

$$F(t) = 1 - e^{-\lambda t} \tag{6.7}$$

and hence the probability density function of failure is given by

$$f(t) = \lambda e^{-\lambda t} \tag{6.8}$$

Equations (6.7) and (6.8) define the negative exponential distribution. Its probability density function, the cumulative failures and the failure rate are plotted against time in figure 6.3. The mean life, $\bar{t}$, is given by

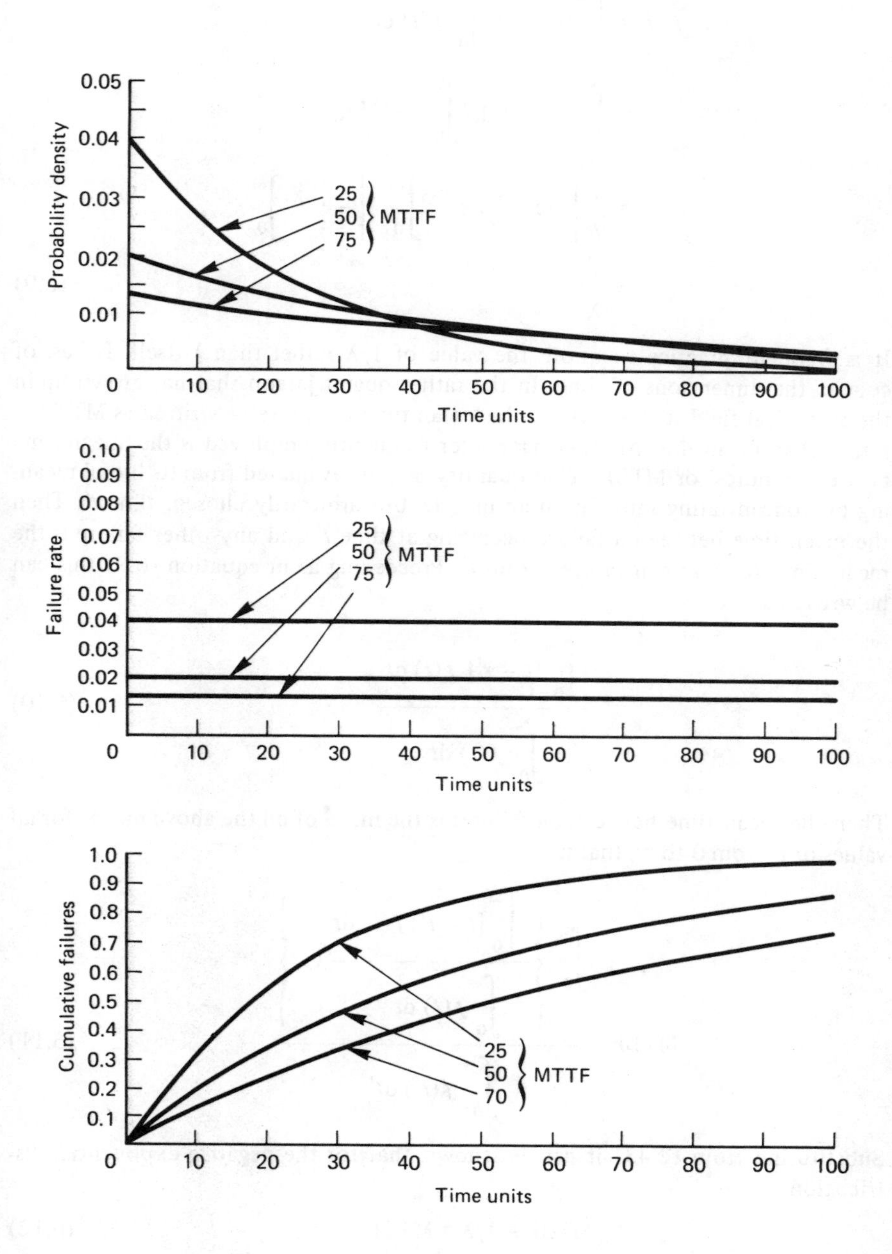


**Figure 6.3** Probability density, failure rate and cumulative failures (= 1 − reliability) for the negative exponential distribution at the mean times to failure (MTTF) marked on the curves

$$\bar{t} = \int_0^\infty t f(t)\,\mathrm{d}t \Big/ \int_0^\infty f(t)\,\mathrm{d}t$$

$$= \int_0^\infty \lambda t \mathrm{e}^{-\lambda t}\,\mathrm{d}t \Big/ \int_0^\infty \lambda \mathrm{e}^{-\lambda t}\,\mathrm{d}t$$

$$= \frac{1}{\lambda}\left[\lambda t \mathrm{e}^{-\lambda t} + \mathrm{e}^{-\lambda t}\right]_0^\infty \left[-\mathrm{e}^{-\lambda t}\right]_0^\infty$$

$$= \frac{1}{\lambda} \tag{6.9}$$

It is frequent practice to quote the value of $1/\lambda$ rather than $\lambda$ itself. It has, of course, the dimensions of time. In the rather quaint jargon that has grown up in the statistical field it is known as the 'mean time to failure' – written as MTTF – rather than mean life. Another parameter frequently employed is the 'mean time between failures' or MTBF. That quantity may be evaluated from its literal meaning by concentrating initially on an unique, but arbitrarily chosen, time $t'$. Then the mean time between a failure occurring at time $t'$ and any other failure is the mean time to failure measured from $t'$. Proceeding as in equation (6.9) this can be written as

$$\frac{\int_0^\infty [t - t']\, f(t)\,\mathrm{d}t}{\int_0^\infty f(t)\,\mathrm{d}t} \tag{6.10}$$

Then the mean time between all failures is the mean of all the above means for all values of $t'$ from 0 to $\infty$, that is

$$\mathrm{MTBF} = \frac{\int_0^\infty \left\{ \dfrac{\int_0^\infty [t - t'] f(t)\,\mathrm{d}t}{\int_0^\infty f(t)\,\mathrm{d}t}\,\mathrm{d}t' \right\}}{\int_0^\infty f(t')\,\mathrm{d}t'} \tag{6.11}$$

Substituting from (2.43) it can be shown that for the negative exponential distribution

$$\mathrm{MTBF} = 1/\lambda = \mathrm{MTTF} \tag{6.12}$$

It should be noted that it is only coincidental that for this distribution MTBF equals MTTF. It is not true for other distributions. Careless practice uses MTBF and MTTF interchangeably for the negative exponential distribution, and because that distribution is so universal the relationship is sometimes believed to be universal, causing confusion. As a physical interpretation of the significance of these terms in a case when they are certainly not equal we can again refer to Oliver Wendell Holmes famous 'One-Hoss Shay', for which

MTTF = 'one hundred years to a day'
MTBF = zero (all the components failed simultaneously).

This extreme example illustrates the role and importance of MTTF and MTBF. The former is more important as a measure of the reliability of components which are discarded from the population on failure. The latter is more significant for multi-component maintained equipment, when the mean life of the fully maintained equipment is infinite, but failures of maintainable components can be expected at the MTBF. However, both figures are necessary to fully represent any situation, particularly if MTTF and MTBF are different, as the above example so clearly brings out. The seal of confusion was affixed for the author when he heard MTBF referred to as the mean time before failure! A better nomenclature would resolve confusion, but established practice, however confusing, is difficult to avoid.

A more serious misinterpretation arises if the MTBF is identified with the mission time, $T_m$. Since

$$R = e^{-\lambda t} = e^{-(t/\text{MTBF})} \tag{6.13}$$

identifying $T_m$ with MTBF implies that the mission reliability is $e^{-1} = 36.8$ per cent. Most frequently this is *not* what is intended. From equation (6.6) it is seen that MTBF must have the values indicated in the table below to achieve the mission reliabilities quoted.

| *Mission reliability (per cent)* | *MTBF or MTTF* |
|---|---|
| 90 | 10 × mission time |
| 95 | 20 × mission time |
| 99 | 100 × mission time |
| 99.9 | 200 × mission time |

It is readily seen from this table that the MTBF is well in excess of the mission time. Furthermore, failure data often applies to the mission time only or to a relatively small number of missions, so that it would be quite unjustified to apply the relationship to the whole age up to MTTF. As a practical example, the death rate of young people is random at about 1.25 per thousand per year, giving a mean time to failure (MTTF) of 800 years. But that does not imply that anyone is going to live to that ripe old age! Both MTTF and MTBF are essentially statistical parameters, and need great care in physical interpretation.

Field data may readily be interpreted as the negative exponential distribution where it is applicable. If a piece of equipment is maintained so that failures are replaced to keep the number constant, its approximation to the constant failure rate law may be tested by plotting the number of failures per time interval. (An appropriate time interval can be estimated from Sturges' Rule which states that the number of intervals should be equal to $(1 + 3.3 \log_{10} F)$ where $F$ is the total number of failures.) The number of failures in each equal time interval should, of course, be roughly constant for the constant failure rate equation to apply. An estimate of the failure rate for the equipment will then be

$$\text{estimated } \lambda = \frac{F}{T} \tag{6.14}$$

where $F$ is the total number of failures in $T$ equipment hours of operation.

If the failures are not replaced, equation (6.7) may be used to check that the failure rate is constant. Let $N$ be the initial number of items. If the proportion of failures (expressed preferably as $F/(N + 1)$ rather than $F/N$) is plotted against time on log-linear graph paper, the result should give a reasonable fit to a straight line (from equation 6.7)). The value of $t$ corresponding to $F(t) = 1 - e^{-1} = 0.632$ will be the mean time to failure or the mean time between failures. As an alternative to the above method, and preferably if there is any doubt, the approximation to a constant failure rate may be tested by plotting on Weibull paper as described in the next section.

*Exercise 6.2*

The mean distance between job reports on an armoured fighting vehicle is 271 miles. The vehicles are known to be in the random failure state. What is the probability that a vehicle will complete a mission of 100 miles without failure.
(*Note*: This is often referred to as the 'mission reliability' for the distance quoted – in this case 100 miles.)

*Exercise 6.3*

A manufacturer of sapphire styluses for a popular brand of record player in the low price bracket claims that random failures of the stylus occur at the rate of 0.05 per thousand styluses per hour. If you had one of these record players and had already played 100 sides of a long-playing record with one stylus, when would you expect to have to change the stylus? You may assume that one side of a long-playing record plays for 30 minutes. State all assumptions you have to make in reaching your conclusion.

**Exercise 6.4*

Equation (6.9) has been derived from first principles. Some textbooks start from the condition that the mean of a distribution is equal to the first moment, and others from the relationship

$$\text{MTTF} = \int_0^\infty R(t)\,\mathrm{d}t \tag{6.15}$$

Establish this relationship and hence define the conditions in which it is valid.

*Exercise 6.5*

The manufacturer referred to in exercise 6.1 bought a further nine extinguishers produced by a rival and subjected them to the same test, recording failures after the following number of operations

48, 100, 161, 226, 309, 406, 538, 710
(the 9th extinguisher was not failed after 750 operations)

Show that these follow the negative exponential distribution. Can you reach any practical conclusions from the comparative tests?

## 6.4 The Weibull distribution

### *6.4.1 General characteristics of the Weibull distribution*

It has already been shown that failure rate-age characteristics may rise or fall in addition to remaining constant. The negative exponential distribution can only represent the latter case. It is highly desirable therefore to have at hand a distribution which can represent any form of failure rate-age curve. The Weibull distribution is one possibility.

In chapter 2 it was shown that

$$\lambda(t) = -\frac{\dfrac{\mathrm{d}}{\mathrm{d}t}[R(t)]}{R(t)} \tag{6.16}$$

and integrating

$$\int \lambda(t)\,\mathrm{d}t = -\log_e R(t) \tag{6.17}$$

or

$$R(t) = \exp\left[-\int \lambda(t)\,\mathrm{d}t\right] \tag{6.18}$$

In 1951 Weibull suggested that the simplest empirical expression representing a great variety of actual data could be obtained by writing

$$\int \lambda(t)\,\mathrm{d}t = \left(\frac{t - t_0}{\eta}\right)^{\beta} \tag{6.19}$$

so that

$$R(t) = \exp\left[-\left(\frac{t - t_0}{\eta}\right)^{\beta}\right] \tag{6.20}$$

It has since been found that very many actual reliability patterns can be represented by this equation which, as we shall demonstrate, is very easily applied in practice. The Weibull distribution is usually expressed as

cumulative distribution:

$$F(t) = 1 - \exp\left[-\left(\frac{t - t_0}{\eta}\right)^{\beta}\right] \tag{6.21}$$

having a probability density function:

$$f(t) = \frac{\beta}{\eta}\left(\frac{t - t_0}{\eta}\right)^{\beta - 1} \exp\left[-\left(\frac{t - t_0}{\eta}\right)^{\beta}\right] \tag{6.22}$$

The failure rate for this distribution is

$$\lambda(t) = \frac{\beta}{\eta}\left(\frac{t - t_0}{\eta}\right)^{\beta - 1} \tag{6.23}$$

Equations (6.20), (6.21) and (6.22) only apply to values of $(t - t_0)$ greater than, or equal to, zero. For $(t - t_0)$ less than zero, $f(t)$ and $\lambda(t)$ are zero. The constants appearing in the above expressions can each be given a physical interpretation

(1) $t_0$ is a locating constant defining the starting point or origin or the distribution.

(2) $\eta$ is a scaling constant, stretching the distribution along the time axis. When $(t - t_0)$ is equal to $\eta$ the reliability is given by

$$R(t) = \mathrm{e}^{-(1)^{\beta}} = \mathrm{e}^{-1} \tag{6.24}$$

$$= 0.368$$

The constant therefore also represents the time, measured from $t_0 = 0$, by which 63.2 per cent of the population can be expected to fail, whatever value is assigned to $\beta$. For this reason it is often referred to as the 'characteristic life'.

(3) $\beta$ is a shaping constant which primarily controls the shape of the curve. The failure density distribution, cumulative failures and the failure rate are shown plotted against time in figure 6.4, for various values of $\beta$. For $\beta < 1$ the distribution represents a decreasing failure rate with age. For $\beta = 1$ the Weibull distribution reduces exactly to the exponential distribution and can thus represent a constant failure rate. For $\beta > 1$ the curve represents increasing

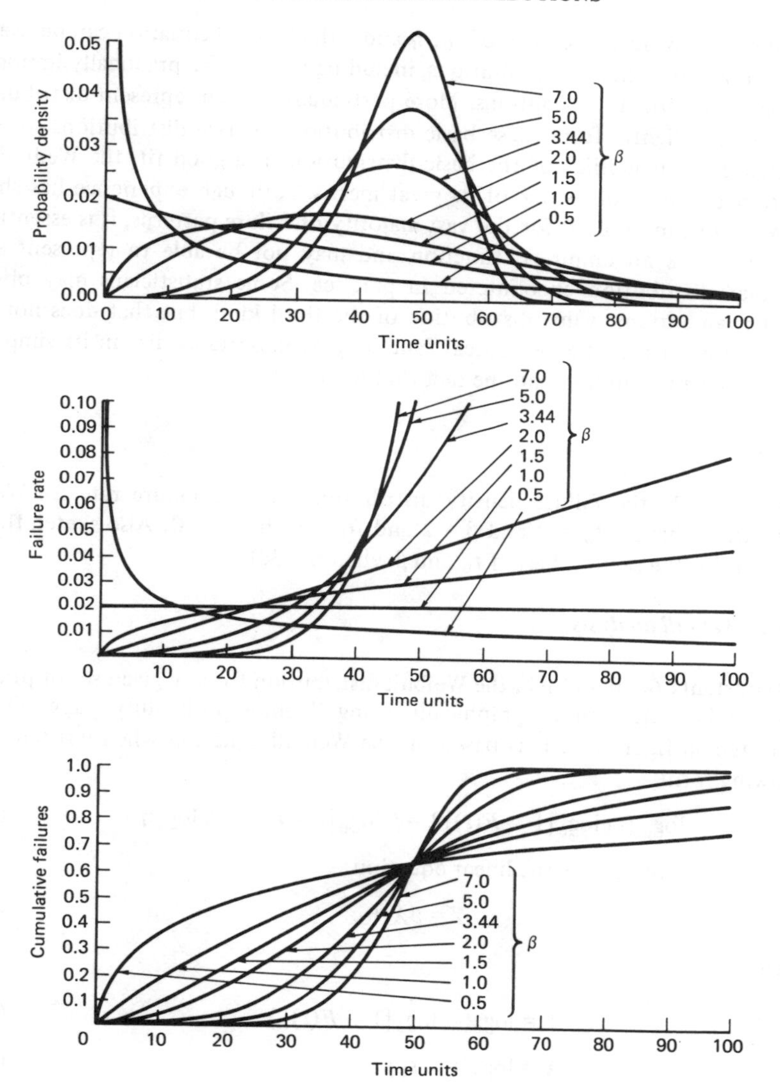


**Figure 6.4** Probability density, failure rate and cumulative failures (= 1 − reliability) for the Weibull distribution at the values of the shaping parameters ($\beta$) marked on the curves

failure rate with age. With $\beta = 3.44$ the Weibull distribution closely approximates to the normal distribution. The value of $\beta = 3.44$ is that which makes the median equal to the mean for the Weibull distribution as for the normal distribution.

Thus by appropriate choice of $t_0$, $\eta$ and $\beta$ the Weibull equation can be used to represent a wide range of distributions, including both of the practically important random and normal distributions. More particularly it can represent distributions which differ slightly from these basic distributions, as real distributions so often do. Hence when neither of the basic distributions is a good fit, the Weibull distribution can be. This is one of its great merits. Although experience has shown that Weibull can be used for the vast majority of failure patterns, it is essential to note that it is an empirical function and may not be able to represent some particular distribution encountered in practice. Some statisticians may observe that it is an extreme value distribution of the third kind, but that does not alter the fact that it is still an empirical function. Its importance lies in its simplicity and its wide adaptability, and the fact that it works!

**Exercise 6.6*

Sketch roughly the failure density distribution and the failure rate for Weibull distributions having $t_0 = 0$ and $\beta = 2$ and for which $\eta = 1.0$. Also sketch the distribution for the same values of $t_0$ and $\beta$ when $\eta = 3.0$.

### **6.4.2 Weibull analysis*

The constants $t_0$, $\eta$ and $\beta$ in the Weibull distribution for any given set of practical data are best determined graphically using Weibull probability paper. This is illustrated in figure 6.5. It is based on the Weibull equation when written in the following form

$$\log_e \{-\log_e [1 - F(t)]\} = \beta \log_e [t - t_0] - \beta \log_e \eta \tag{6.25}$$

This can be expressed as the linear equation

$$Y = \beta X + c \tag{6.26}$$

where

$$Y = \log_e\{-\log_e [1 - F(t)]\} \tag{6.27}$$

$$X = \log_e [t - t_0] \tag{6.28}$$

On Weibull paper the ordinate is a double natural log scale of $[1 - F(t)]^{-1}$ expressed in terms of $F(t)$, while the abscissa is a single natural log scale of the random variable $[t - t_0]$, as shown in figure 6.5 at (a) and (b). Any set of data which follows the Weibull distribution will plot as a linear curve on this paper. The slope of the line gives the constant $\beta$ and the intercept on the $Y$-axis $\beta \log \eta$. However, raw failure data are usually expressed in terms of the variable $t$, not $[t - t_0]$. Plotted in terms of $t$ the curve may be linear, in which case $t_0 = 0$ and the initial plot can be used for assessment purposes. However, the curve could take on one of the forms illustrated in figure 6.6. If the curve is non-linear it could

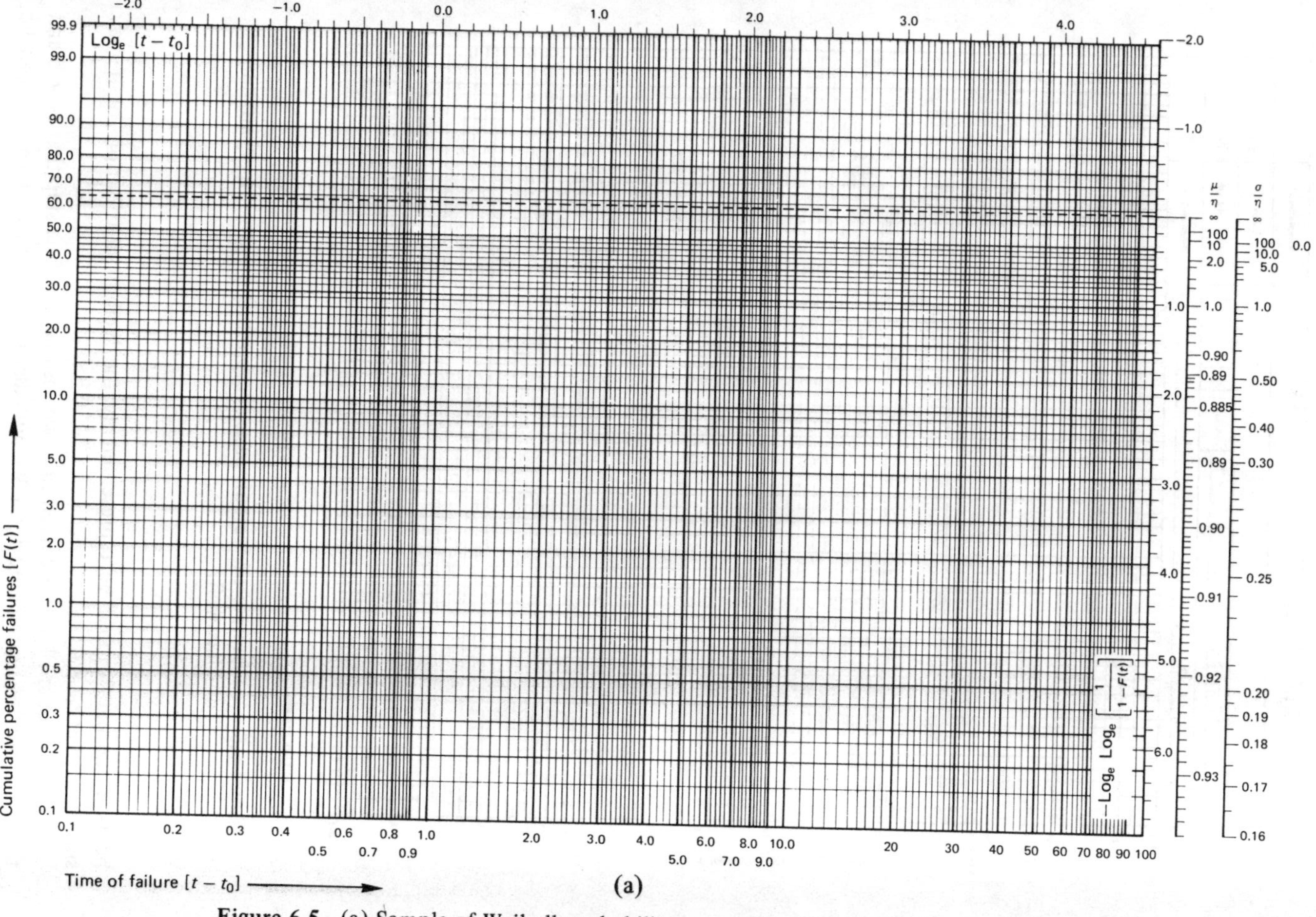


**Figure 6.5** (a) Sample of Weibull probability paper (from ref. 28) (*see overleaf for (b)*)

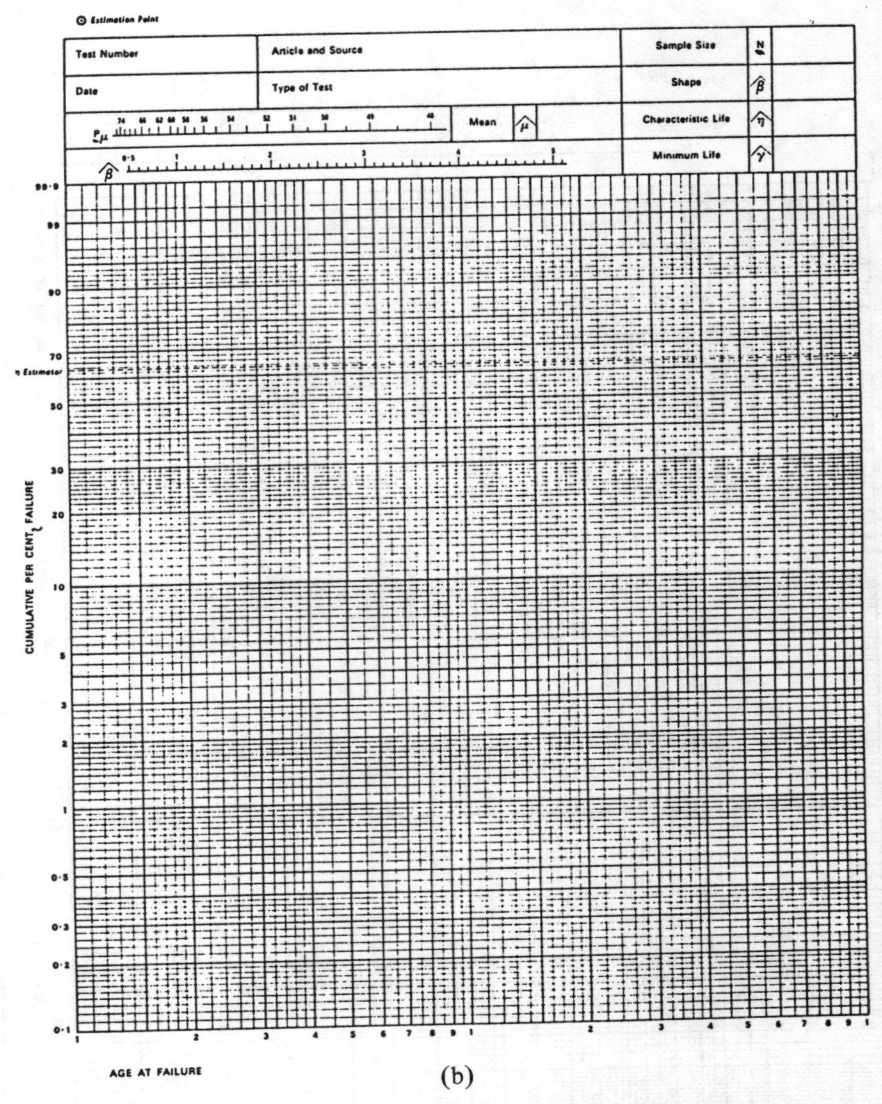


Figure 6.5 (b) Sample of Weibull paper (reproduced with the permission of Chartwell)

be that the data cannot be represented by the Weibull distribution, or it may simply be that $t_0 \neq 0$.

If the raw data plot as a curved line, we assume as a first step that $t_0 \neq 0$ and the data can be represented by a Weibull distribution. In that case $t_0$ can be

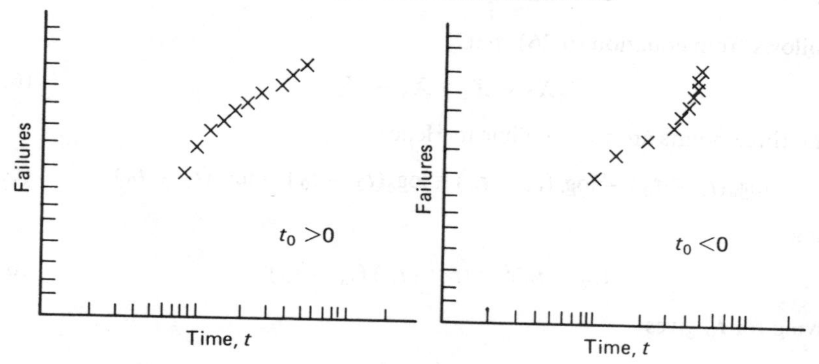


**Figure 6.6** Effect of $t_0 \neq 0$ on plot of raw data on Weibull paper

estimated (ref. 29) by selecting an arbitrary point roughly in the middle of the curve, and two other points both distant $d$ in the $Y$-direction, referenced subscript 1, 2 and 3 as illustrated in figure 6.7.

Since we have chosen the points 1, 2 and 3 so that

$$Y_2 - Y_1 = Y_3 - Y_2 \tag{6.29}$$

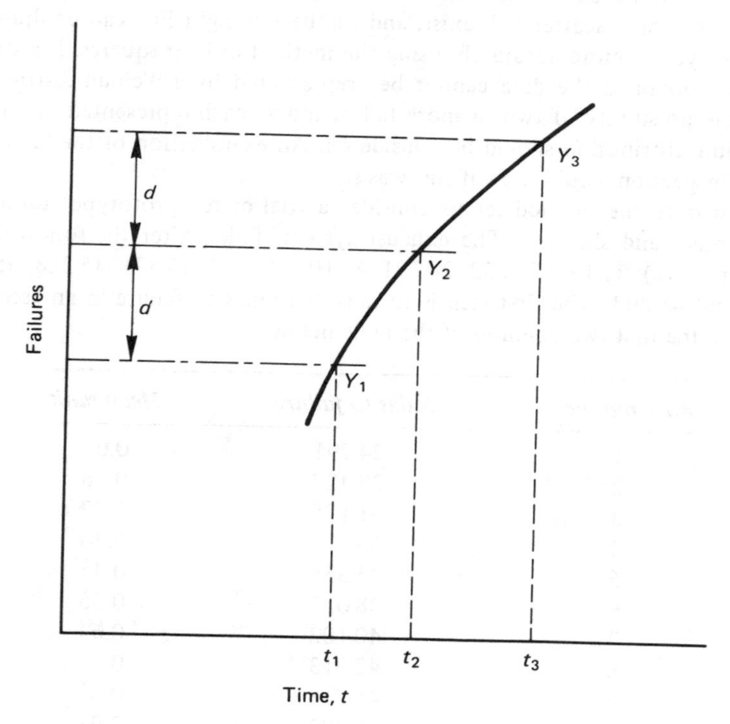


**Figure 6.7** Estimation of $t_0$

it follows from equation (6.26) that

$$X_2 - X_1 = X_3 - X_2 \tag{6.30}$$

if the three points are to be collinear. Hence

$$\log_e(t_2 - t_0) - \log_e(t_1 - t_0) = \log_e(t_3 - t_0) - \log_e(t_2 - t_0) \tag{6.31}$$

giving

$$(t_2 - t_0)^2 = (t_3 - t_0)(t_1 - t_0) \tag{6.32}$$

Solving for $t_0$ gives

$$t_0 = \frac{t_3 t_1 - t_2^2}{(t_3 - t_2) - (t_2 - t_1)} \tag{6.33}$$

which can also be written as

$$t_0 = t_2 - \frac{(t_3 - t_2)(t_2 - t_1)}{(t_3 - t_2) - (t_2 - t_1)} \tag{6.34}$$

The value of $t_0$ can now be calculated and the data replotted using $(t - t_0)$ as the variable. If the data follow the Weibull distribution the points will lie on a straight line. Some scatter will exist, and the best straight line can be drawn in, either 'by eye' or more accurately using the method of least squares. If a straight line is not obtained the data cannot be represented by a Weibull distribution, though the possibility of two or more failure modes each represented by a different Weibull distribution should be considered. An examination of the failures by physical inspection would show if this was so.

To illustrate the method let us consider a trial of ten prototype automotive exhaust pipes and silencers. The exhaust systems failed after the following distances (in miles): 31 175, 38 033, 24 791, 40 102, 42 913, 33 871, 45 218, 35 338, 28 427 and 48 203. The first step is to rank the miles to failure in an ascending order as in the first two columns of the table below.

| *Order number* | *Miles to failure* | *Mean rank* |
|---|---|---|
| 1 | 24 791 | 0.09 |
| 2 | 28 427 | 0.18 |
| 3 | 31 175 | 0.27 |
| 4 | 33 871 | 0.36 |
| 5 | 35 338 | 0.45 |
| 6 | 38 033 | 0.55 |
| 7 | 40 102 | 0.64 |
| 8 | 42 913 | 0.73 |
| 9 | 45 218 | 0.82 |
| 10 | 48 203 | 0.91 |

In the third column we enter the mean rank, calculated as in the solution to exercise 6.5. These data are then plotted on Weibull paper on figure 6.8 in which the cumulative failures equal the mean rank. It will be seen that the points do not plot on a straight line. Three points were arbitrarily selected as described to give values of $t_1$, $t_2$ and $t_3$ as

$$\begin{aligned} t_1 &= 2.55 \times 10^{-4} \text{ miles} \\ t_2 &= 3.50 \times 10^{-4} \text{ miles} \\ t_3 &= 5.02 \times 10^{-4} \text{ miles} \end{aligned} \tag{6.35}$$

Using equation (6.34) $t_0$ is found to be 9700 miles, and the data can be replotted to give the straight line shown on figure 6.8. The characteristic life $\eta$ is simply determined using the relationship established at (2) in section 6.4.1. It is the intercept of the straight line with the 63.2 per cent cumulative failure line, and in this case is equal to 30 000 miles. (Note the 63.2 per cent failure line is given by the horizontal dotted line on Weibull paper). The shaping constant $\beta$ is the slope of the straight line, and with this particular type of Weibull paper is easily determined by projecting a vertical line from the above intercept on to the upper [that is, $\log(t - t_0)$] scale as shown by the dotted line. Subtract one from this value and drop a perpendicular from this point on to the cumulative failure line, as shown again by a dotted line in figure 6.8. Draw a horizontal line from this point to the first right-hand vertical scale. The slope of the line is then the reading on the vertical scale divided by unity, that is, the intercept of the horizontal dotted line on the first right-hand scale gives the value of $\beta$ as a direct reading. In this example $\beta = 3.41$, representing a near normal wear-out failure pattern.

The mean value, $\mu$, and standard deviation, $\sigma$, may be found using the method of moments since $\mu$ is equal to the first moment about the origin and $\sigma^2$ is equal to the second moment about the origin less the first moment squared. The $r$th moment about the origin is given by

$$\int_0^\infty t^r f(t)\, \mathrm{d}t \tag{6.36}$$

Substituting the probability density function of the Weibull distribution it can be shown (see ref. 21) that

$$\frac{\mu}{\eta} = \Gamma\left(1 + \frac{1}{\beta}\right) \tag{6.37}$$

and

$$\left(\frac{\sigma}{\eta}\right)^2 = \Gamma\left(1 + \frac{2}{\beta}\right) - \left[\Gamma\left(1 + \frac{1}{\beta}\right)\right]^2 \tag{6.38}$$

where $\Gamma$ denotes the well-known gamma function. Thus $\mu/\eta$ and $\sigma/\eta$ are functions of $\beta$ only and can be plotted on Weibull paper from equations (6.37) and (6.38).

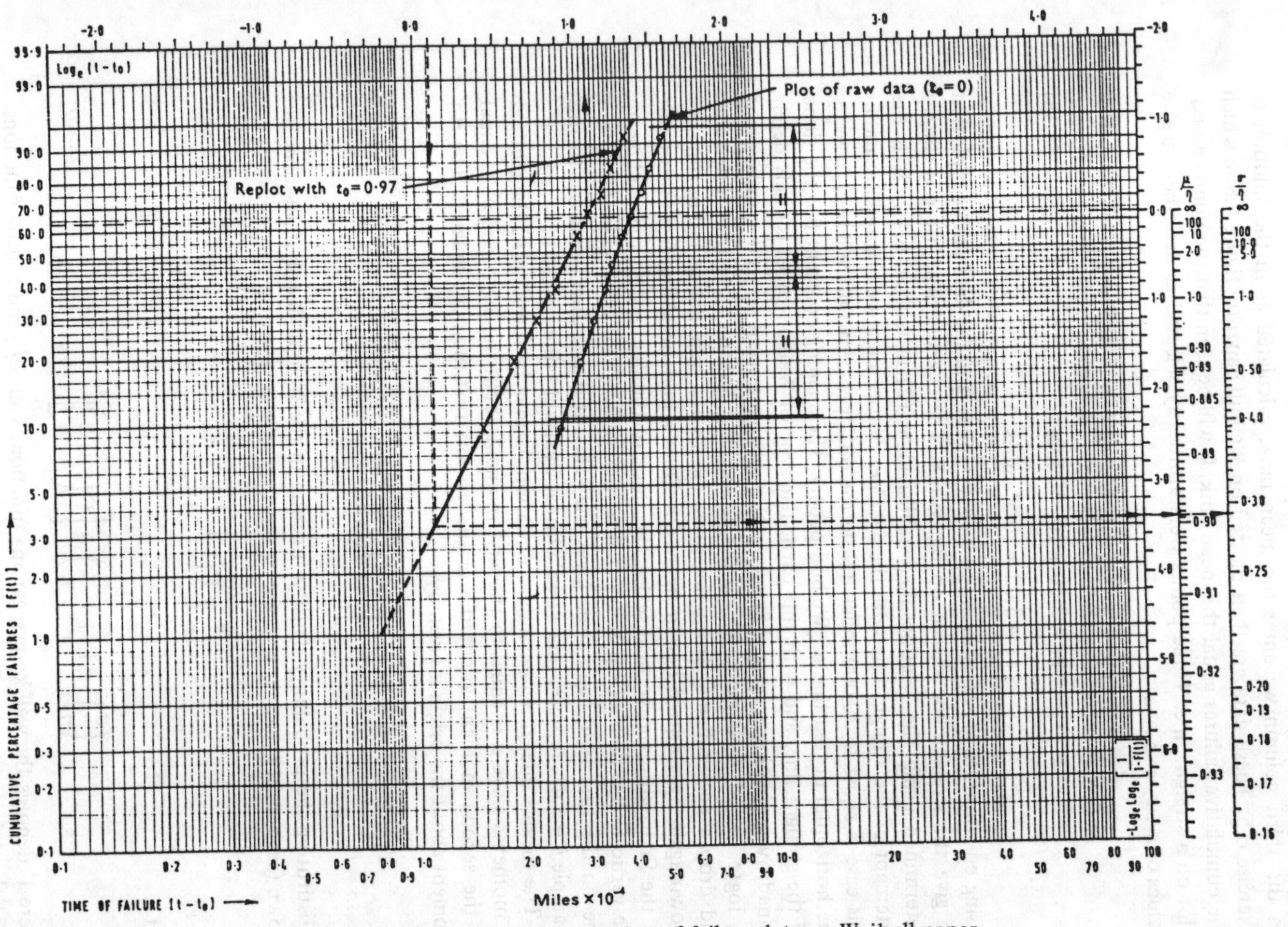


Figure 6.8 Example of presentation of failure data on Weibull paper

Nomograms are provided whereby an extension of the horizontal dotted line determining the shaping parameter enables $\mu/\eta$ and $\sigma/\eta$ to be read off directly. It should be noted that both $\eta$ and $\mu$ are measured from the origin $t = t_0$. As illustration, for this particular example $\mu/\eta = 0.899$, whence $\mu = 0.899 \times 30\,000 = 26\,970$ miles, and the mean value of $t$, that is, the mean life, is $26\,970 + 9\,700 = 36\,670$ miles.

With the type of paper shown on figure 6.5b, the nomogram scales are drawn at right angles to those on figure 6.5a. This makes the determination of the shaping parameter a little easier in that it is only necessary to drop a perpendicular from the 'estimation point', shown at the top left corner of the paper, on to the straightened Weibull line. Its intersection with the lower horizontal nomogram scale then enables the shaping parameter to be read off directly. At the same time the probability of the mean life can be read off the upper horizontal nomogram scale, and hence the mean of the distribution ($\mu$) determined by reference to the Weibull straight line plot. No scale is given for the standard deviation on this paper.

There is a third kind of Weibull paper in which the shaping parameter is obtained by drawing a line parallel to the Weibull straight line through an estimator point on separate enlarged axes. This enables the shaping parameter to be determined more accurately, given that the original data warrants it. This type of paper does not usually have scales for the mean and standard deviation. The axes on which the original data are plotted are exactly the same, of course, except perhaps for the number of log cycles on the time axis, on all forms of Weibull paper.

The example we have taken illustrates one big advantage of the Weibull distribution in analysing reliability data. Since the distribution does not exist for negative values of $(t - t_0)$, the failure rate is zero for all $t < t_0$. We can consequently identify $t_0$ with the life of intrinsic reliability ($B_0$). The absence of an infinite tail to the lower extremity of the Weibull distribution is an essential feature in this interpretation. It applies to some other distributions, of course. However, the reader must be reminded that there is no fundamental reason why the tail of the Weibull distribution, or any other, which fits the mass of the population should also fit the tail of that population. Sufficient data to justify that assumption must be obtained in the field. It may be of interest to record that the origin of the Weibull distribution when $\beta = 3.44$, that is, when it approximates to a normal distribution, is at 3.13 standard deviations from the mean. This should be compared with the tail of the normal distribution on figure 6.1. In the above example the exhaust systems were intrinsically reliable up to 9700 miles. Thereafter wear-out set in, and from the value of $\beta = 3.41$ we know that the wear-out was distributed in a near normal manner. This is readily seen from figure 6.4, from which it is also seen that the failure rate is continuously increasing. It follows from section 5.2 that the wear-out is taking place by a steady and continuous loss of strength: we may suspect corrosion, which is the obvious cause of failure in this item. Had the failure been in a strong fatigue mode (such as might arise from a resonant vibration) the value of $\beta$ would have been less than unity to give a failure rate which decreases with age – see figure 6.4 – in accordance with the description in section 5.3.

At this point the author must draw attention to some difficulties in Weibull analysis and its interpretation. It will be seen from figure 6.8 that the curvature of the raw data plot is very low. It would have been only too easy to have drawn a straight line through these points. In many instances the scatter is much higher than in the example quoted and this tends to obscure further any curvature. So that although the estimation of $t_0$, the age of intrinsic reliability, is ideally possible, it is often practically difficult. The extent of the problem may be seen from figures 6.9 and 6.10, which show plots of defined Weibull functions having different values of $t_0$, on Weibull paper. If it is remembered that we only see a small portion of these curves in any real analysis, the difficulty is appreciated. The reader may like to blank off with two sheets of paper the ranges represented by cumulative failure rates greater than 90 per cent and less than 10 per cent. This would be the only portion of the curves available with nine failures. These are often the numbers with which we are having to deal: after all is said and done, the whole object of the exercise is to reduce the number of failures! The difficulty in recognising curvature should now be apparent. It will also be seen from figures 6.9 and 6.10 that the difficulty of detecting curvature increases with higher values of $\beta$, and can be very difficult indeed for $\beta$ greater than about 3.

The reason for the difficulty is not far to seek. By replacing a curved line with a straight line on Weibull paper we are replacing the real distribution

$$R = \exp\left\{-\left(\frac{t - t_0}{\eta}\right)^{\beta}\right\} \tag{6.39}$$

by

$$R = \exp\left\{-\left(\frac{t}{\eta + t_0}\right)^{\beta}\right\} \tag{6.40}$$

Strictly, the $\eta$ values in equations (6.39) and (6.40) may differ, but the difference is very small. If $t_0$ is small in relation to $\eta$, as it usually is, we can write

$$\frac{t - t_0}{\eta} = \frac{t}{\eta} - \frac{t_0}{\eta} \tag{6.41}$$

$$\frac{t}{\eta + t_0} = \frac{t}{\eta}\left(1 + \frac{t_0}{\eta}\right)^{-1} \approx \frac{t}{\eta} - \left(\frac{t}{\eta} \times \frac{t_0}{\eta}\right) \tag{6.42}$$

Now if we have only a small number of data points they will be located near the 50 per cent cumulative failure point, that is, close to 63.2 per cent where $t/\eta = 1$. Hence the difference between (6.41) and (6.42) is small. Either distribution then adequately represents the limited data. The solution to this difficulty is clear. The analysis must concentrate on points for which $t/\eta$ is well removed from unity, so the data points will adequately define the curved portion of the Weibull line. This may be achieved either by increasing the number of data points so that for some of them $t/\eta$ is well removed from unity, or by paying more attention to points which are well removed from the characteristic life. The author tries to do this by analysing the points falling above and below the median separately, assuming the points in each group can be represented by a two parameter distribution. If the two straight lines do not agree *exactly*, curvature should be suspected. In these cases it is usually not too difficult to draw an adequate line through the points, though high-class draftsmanship is necessary. Particular attention must be paid to representing any low life points, for figures 6.9 and 6.10 show that these are the ones indicative of curvature. The author prefers a graphical analysis of this nature to a wholly numerical one, since it enables the analyst to see what is going on and to make appropriate judgements. Numerical methods are subject to the same errors in estimating the locating constant as the graphical ones (see, for example, ref. 30). Numerical methods therefore offer no advantage, tend to conceal difficulties, and are inflexible in use. As a substitute for graphical analysis it is, of course, very convenient to carry out the work on a computer with graphical display. It is then sometimes convenient to estimate the locating constant by successive approximations. The least squares straight line with the highest coefficient of correlation through the data points (adjusted for a range of assumed locating constants) is obviously the best fit and easily evaluated on a computer. It is still essential to maintain manual control of the process, however, since the method can give some very unpractical interpretations of some data. This arises from the very ill-conditioned nature of the maximum when scatter is present. In all cases it is also essential to inspect the raw data plot before proceeding with the analysis, to be sure that all the data points do belong to one distribution.

### *6.4.3 Representation of failure patterns by Weibull distributions*

Weibull distributions can be used to represent most of the failure patterns we have previously encountered in this book; the analysis is relatively simple, and when followed through, enables us to deduce a great deal about the failure mechanism. It cannot be over-emphasised, however, that the identification of the mechanism from the three Weibull parameters alone, without reference to the physical evidence, should be made with considerable caution. Nevertheless it has to be done on occasions, and from our study of the distribution we can identify the following combinations of Weibull parameters with particular failure mechanisms

(a) $t_0 = 0$. The item has no life of intrinsic reliability, and
  - (i) if $\beta < 1$ the failure rate falls with age without reaching zero, so we should suspect an early life characteristic arising from a low safety margin, giving rise to stress rupture failure
  - (ii) if $\beta = 1$ the failure rate is constant for all ages which represent a random or pseudo-random failure characteristic

or (iii) if $\beta > 1$ the failure rate increases with age at all ages which is indicative of wear-out commencing as soon as the item was brought into service.

(b) $t_0 > 0$. The item is intrinsically reliable from the time it was brought into service until $t = t_0 = B_0$, followed by

(i) if $\beta < 1$ a fatigue or similar type of wear-out in which the failure rate decreases with age after a sudden increase at $t_0$; low values of $\beta$ ($\sim 0.5$) may be associated with low cycle fatigue and higher values ($\sim 0.8$) with high cycle fatigue

or (ii) if $\beta > 1$ an erosion or similar wear-out in which the constant load life continuously decreases with increasing load; lower values of $\beta$ suggest a constant load life which is more dependent on load than would be the case at higher values.

(c) $t_0 < 0$. This could indicate the item was used or suffered failures before the data was collected, otherwise

(i) if $\beta < 1$ it could be an early life failure mode beginning before the item was brought into service, arising from a low safety margin

or (ii) if $\beta > 1$ wear-out by steady decrease of strength beginning before the item was brought into service, for example limited shelf life which had expired or was inadequate.

It will be seen from the above that Weibull analysis can give a good lead to the mechanism of the failure. The emphasis should, however, be on the word 'lead'. It can never be used as conclusive evidence. Great care should be taken to consider alternative interpretations that could be placed on the distribution, and indeed whether the data could not be represented by a different Weibull distribution if some of the data were re-interpreted. Corrupt data is a frequent source of difficulty, particularly when data is limited. In practice, this is often the real problem. Nevertheless Weibull analysis can often give a good insight into the mechanism involved, and when used with other information can provide a very powerful tool.

**Exercise 6.7*

A new type of protective coating for turbine blades was life-tested under normal operating conditions. Ten blades were randomly selected as a sample and life-tested. The hours to failure were recorded as follows: $3.2 \times 10^3$, $4.7 \times 10^3$, $1.8 \times 10^3$, $2.4 \times 10^3$, $3.9 \times 10^3$, $2.8 \times 10^3$, $4.4 \times 10^3$ and $3.6 \times 10^3$. At 5000 h the test was terminated, and two blades survived. Determine the type of failure and the characteristic life. What is the reliability for a life of 2000 h? Would you consider the coating satisfactory? (Pratt and Whitney data)

*Exercise 6.8*

In service failures of the cages of a bearing in a certain aero engine were recorded after the following engine running hours: 360, 525, 705, 925, 940, 1185, 1410,

1420, 1465, 1950, 2050, 2620 and 2730. At that time no failure in 51 of the bearings in other engines had been recorded. You can assume that all have completed at least as many running hours as the failed bearings. Determine the nature of the failure of these bearing cages. (Rolls Royce data)

*Exercise 6.9*

A record of the nine control valves in a process plant indicated failure after the following number of hours: 10 097, 7 622, 5 123, 6 000, 29 106, 18 997, 26 335, 14 780 and 10 211. Make whatever observations are possible from this data.

**Exercise 6.10*

A machine is known to fail by fracture of one of its brackets, which has a natural frequency of 1 hertz. Observation of the bracket shows that it vibrates only under certain conditions. These arise for 0.01 per cent of the total load cycle and for the purpose of this exercise you may assume that all the vibration takes place at constant amplitude, giving rise to a maximum stress of ± 200 MN/m$^2$. The bracket is made of mild steel having a nominal fatigue limit of ± 150 MN/m$^2$ and ultimate tensile strength of 300 MN/m$^2$. Sketch a typical *s–N* curve for the material and hence deduce the Weibull distribution which will represent the failures of this machine in the particular mode under discussion. State fully any assumptions you make.

**Exercise 6.11*

What other unbiassed estimate of the failure rank may be adopted in place of the mean rank quoted above? Indicate briefly how this may be estimated.

### **6.4.4 Weibull analysis with drop-outs (or censored data)*

In practice a further difficulty is often encountered in plotting data on Weibull paper, owing to suspended items or drop-outs, sometimes referred to as censored data. These may arise from failures in another mode to the one under investigation. Alternatively they may arise simply from the fact that many units will not have experienced a failure at the time the analysis is being carried out, although it will be known that they have survived a certain number of hours. This is a very common situation encountered in the analysis of actual service data. In this instance the amount of 'testing' is determined by service requirements and not by failure. Indeed, one hopes that failures would not be experienced on many items, and so any practically worthwhile analysis must be able to deal with data on 'non-failures'. (*Note*: The fact that an item has survived to a given age without failure is significant factual information in its own right which should be incorporated into the analysis.)

Another important application of censored data analysis arises when an item fails in more than one mode. Exercise 6.9 has shown that simple bulking of the data can be very misleading and is invalid. In this situation failure in one mode often causes removal of that item from the population and prevents the failure age in another mode being determined, that is, failure in the first mode censors the data in the second (or vice versa) though we do know that the item survived in the second mode up to the time of failure in the first (or vice versa). The correct method of analysis is then to identify the failure data which apply to each of the modes – by physical examination – and to treat all failures in modes other than the one under immediate analysis as drop-outs at the time of failure. This does, of course, often imply a considerable reduction in the number of analysable data points if many modes are present.

To illustrate the procedure in the circumstances envisaged above, let us return to the problem of exhaust pipes and silencers discussed in section 6.4.2 and suppose that the vehicle being used for what is in fact the seventh order number (giving an actual failure at 40 102 miles) were involved in a road accident and written off at 35 000 miles so that testing was discontinued before the exhaust pipe and silencer had failed. Its miles to failure will therefore be unknown. We could still produce the part of the table of failure orders up to the fourth failure, after which we must consider the drop-out, that is, the car involved in the road accident. It clearly would occupy one of the order numbers 5 to 10 inclusive, but we do not know which. Hence we assume it could occupy any of the six remaining positions. Since we cannot allocate it a firm position we have to adjust the order number of the remaining five known failures to cover uniformly the same range we would have obtained had we been able to include it. This is done by using a mean increment in order number given by the expression

$$\text{increment in order number} = \frac{(N_r + 1)\ \text{including drop-out}}{(N_r + 1)\ \text{excluding drop-out}} \tag{6.43}$$

where $N_r$ is the number of items remaining after the last certain failure. Equation (6.43) can also be written as in ref. 29

$$\text{increment in order number} = \frac{(N + 1) - (\text{previous order number})}{1 + (\text{number of items beyond present drop-out item})} \tag{6.44}$$

For the particular example we are considering, the increment in order number is (6 + 1)/(5 + 1) = 1.167. We can now complete the allocation of order numbers as in the table below. After the fourth order number, the next order number is given by 4 + 1.167 = 5.167 and the next by 5.167 + 1.167 = 6.334 and so on. It may be noted that we have introduced the concept of non-integral order numbers. It will also be noted that adding the increment to the last order number gives 11.002. This is equal to $(N + 1)$, allowing for rounding off, and serves as a useful check. The mean ranks can now be calculated as before to complete the table.

| *Order number* | *Miles to failure* | *Mean rank* |
|---|---|---|
| 1 | 24 791 | 0.09 |
| 2 | 28 427 | 0.18 |
| 3 | 31 174 | 0.27 |
| 4 | 33 871 | 0.36 |
| 5.167 | 35 338 | 0.47 |
| 6.334 | 38 033 | 0.57 |
| 7.501 | 42 913 | 0.68 |
| 8.668 | 45 218 | 0.79 |
| 9.835 | 48 203 | 0.89 |

The above values can be plotted on Weibull paper as previously described. As a matter of interest, the Weibull constants obtained are tabulated below and compared with those obtained from the original data.

| | *With drop-out at 35 000 miles* | *Original data* |
|---|---|---|
| $t_0$ | 12 000 | 9700 |
| $\eta$ | 28 000 | 30 000 |
| $\beta$ | 3.10 | 3.41 |
| $\bar{t}$ | 37 000 | 36 670 |

The values of $\beta$ obtained from the two data sets do differ, but the difference is negligible for all practical purposes. The major difference between the results from the two sets of data arises in the estimation of $t_0$. This does introduce a corresponding error into the estimation of the characteristic life, so that these errors compensate to give an estimate of the mean life, which is the same for all practical purposes from the two sets of data. If this is what is required, the method is reasonably accurate. However, we have identified $t_0$ with the life of intrinsic reliability, and this quantity is of considerable practical importance. We should therefore be very wary of estimates derived from a population containing drop-outs, particularly if the population is small. If at all possible, the drop-out should be returned to the population for further evaluation, but this is often not practical. Here we merely caution against another possible source of serious error. From figures 6.9 and 6.10 we can see that drop-outs occurring late in a data series have little influence on the analysis, because the curve is fairly linear at that point so that averaging of order numbers is justified, but drop-outs occurring early in a test sequence modify the curved portion of the Weibull plot where averaging of order numbers is not justified, thus introducing considerable doubt into any deductions, particularly of $t_0$. Errors are likely to be large if the total population is small, but to decrease with increasing population.

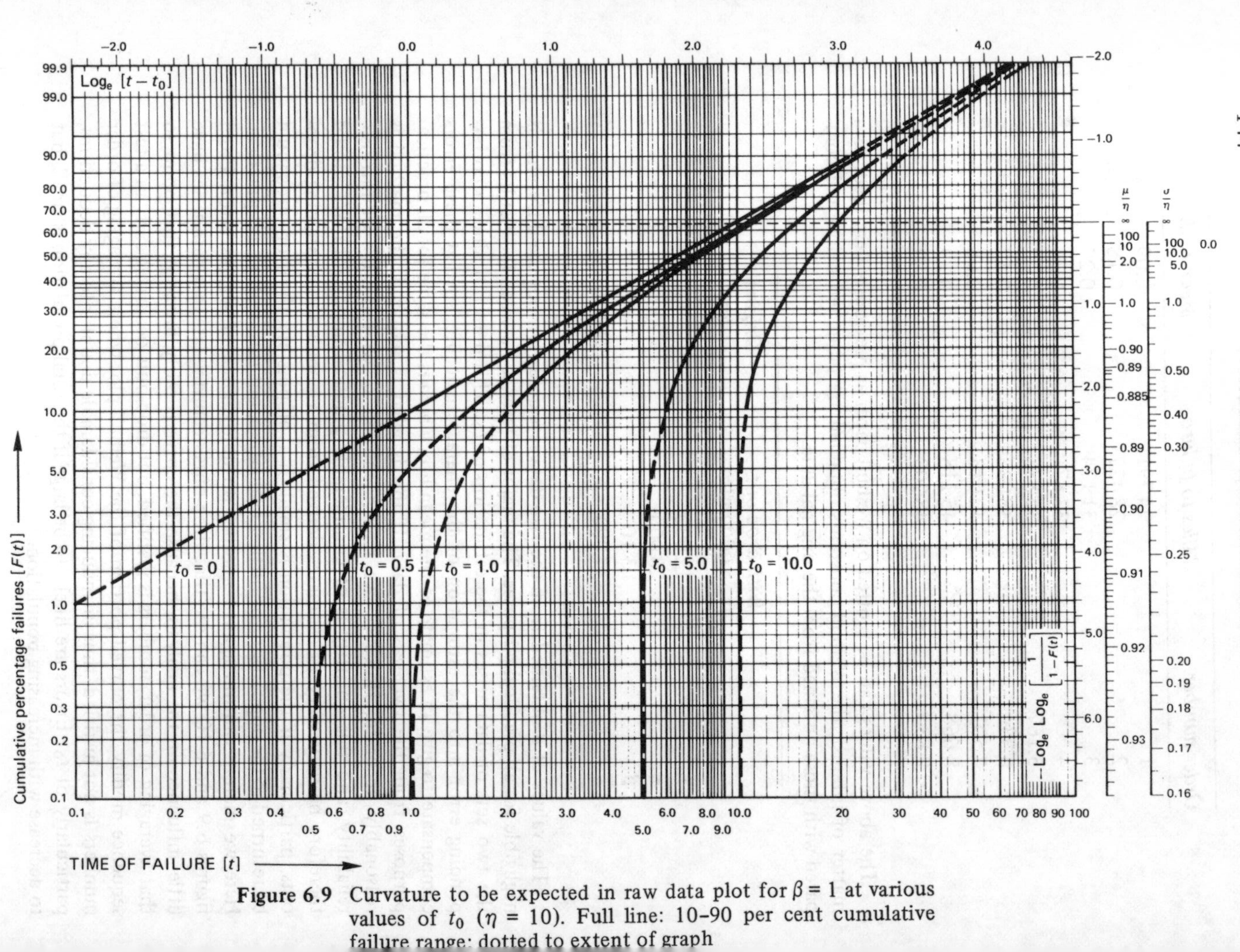


**Figure 6.9** Curvature to be expected in raw data plot for $\beta = 1$ at various values of $t_0$ ($\eta = 10$). Full line: 10–90 per cent cumulative failure range; dotted to extent of graph

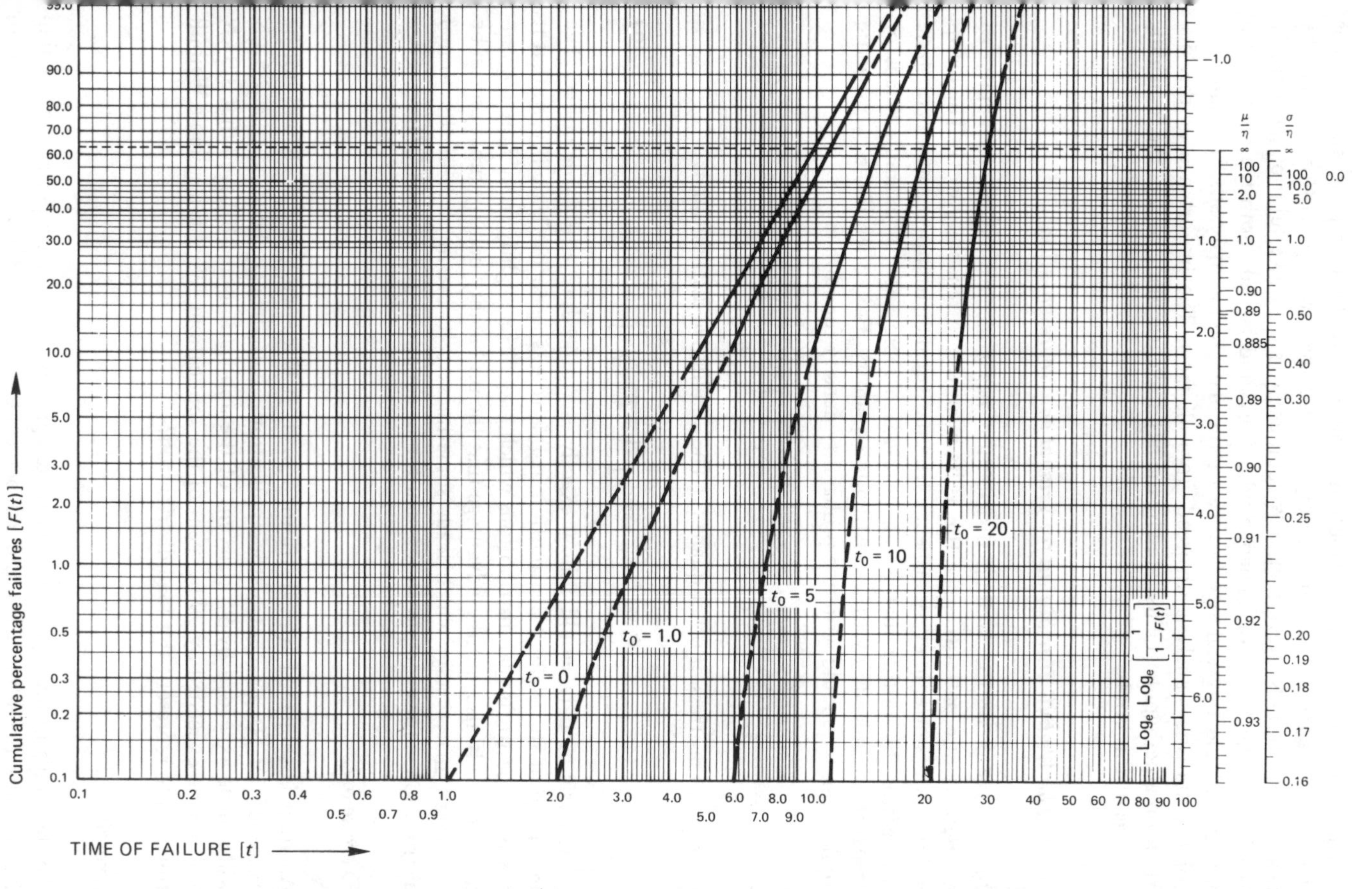


**Figure 6.10** Curvature to be expected in raw data plot for $\beta = 3.0$ at various values of $t_0$ ($\eta = 10$). Full line: 10–90 per cent cumulative failure range; dotted to extent of graph

The shortcomings of the Weibull distribution have been emphasised – perhaps over-emphasised – because it is firmly believed that every practitioner should be aware of the pitfalls. Nevertheless Weibull analysis is a powerful tool, particularly so when in the hands of someone who fully understands both the Weibull dis-bution and the physical significance of the data being analysed. The danger is isolated statistical analysis divorced from all facts. The importance of this distribution is perhaps best illustrated by the extensive use made of it in the remaining sections of this book.

### *6.4.5 *The Weibull distribution for general data representation*

Our discussion of the Weibull distribution has concentrated on its ability to represent failure patterns, since this is its main use in reliability work. The distribution is, of course, in every respect a true statistical distribution and we conclude this section by noting that it can be used to represent, for example, loads or strengths, or indeed any other distributed quantity. For either of the two specific applications we should write the probability density function as

$$f(x) = \frac{\eta}{\beta}\left(\frac{x - x_0}{\beta}\right)^{\beta-1} \exp\left\{-\left(\frac{x - x_0}{\eta}\right)^{\beta}\right\} \tag{6.45}$$

where $x$ is the stress or other quantity. The cumulative distribution is

$$F(x) = 1 - \exp\left\{-\left(\frac{x - x_0}{\eta}\right)^{\beta}\right\} \tag{6.46}$$

Analysis of data would follow exactly the procedure already described. The advantage of the Weibull distribution in this role is again its flexibility, so that it can be made to fit real distributions which may differ slightly from more conventional ones. It also has the advantage that the probability density function can be integrated analytically, which leads sometimes to its use in place of the normal distribution – for example, in the solution of equations (3.15) and (4.7).

*Exercise 6.12*

An item is subjected to a distributed load. It is known that 1.5 kN will not be exceeded by 50 per cent of the applications. It is strongly believed that 2.5 kN will not be exceeded by 95 per cent of the applications and believed that 3.0 kN will not be exceeded by 99 per cent of the applications. What is a suitable Weibull function to represent this load distribution?

*Exercise 6.13*

An item fails in low cycle fatigue. Its designer believes that 5 per cent will fail in 2000 hours operation ($B_5$) and 10 per cent will fail in 3500 hours ($B_{10}$). With what Weibull function would you represent this failure pattern? What proportion of the population can be expected to fail in 2500 hours?

### *6.4.6 *The modified Weibull distribution*

In spite of its great versatility, the basic Weibull distribution is unable to represent the patterns arising from the failures of a finite proportion of weak items such as may arise from improper assembly (such as the hammers quoted in chapter 4, section 4.5), or when the load distribution is of finite extent. The reasons the Weibull distribution is inadequate are three-fold.

1. It is most likely that failures from this cause would occur in the early life only and there would be no failures for a time in excess of a critical value, that is

$$f(t) = 0 \text{ for } t > t_c \tag{6.47}$$

   Although this implies a falling failure rate characteristic at some point, it does differ from early life failures which degenerate into a constant failure rate extending over the whole subsequent useful life.
2. It follows from 1 that the proportion of total population failures from this cause would be less than 100 per cent, no matter how long the items were used. Hence

$$\int_0^\infty f(t)\,\mathrm{d}t \neq 1 \tag{6.48}$$

3. Although it is an early life failure mode to the extent that such failures would occur in the early life, the failure rate could increase before decreasing to zero if the failure mechanism involved some forms of wear, that is, we could expect to find wear-out in the early life period.

The Weibull distribution is one of the most versatile, and is widely used in reliability work. There is every reason, therefore, why we should attempt to modify that distribution to meet the criteria set out above. As an obvious first step, we can identify the critical time ($t_c$) with the locating constant ($t_0$) of the Weibull distribution. However, the standard Weibull distribution only exists for $t$ in excess of $t_0$, whereas we require one to apply only for $t$ less than $t_0$. The mirror image of the Weibull distribution could meet this requirement. Furthermore, if we postulate that this distribution only exists for $t$ greater than zero, we satisfy the condition set out at 2. Optimistically, choice of the shaping parameter, $\beta$, will enable us to meet criterion 3. The distribution is illustrated diagrammatically on figure 6.11. It should be noted that that requirement could be met by applying the transformation to any of the standard distributions that do not extend to $-\infty$, for example, the log-normal distribution. However, since the Weibull distribution is one of the most versatile in its normal role, one would expect it to be equally versatile in its transformed role.

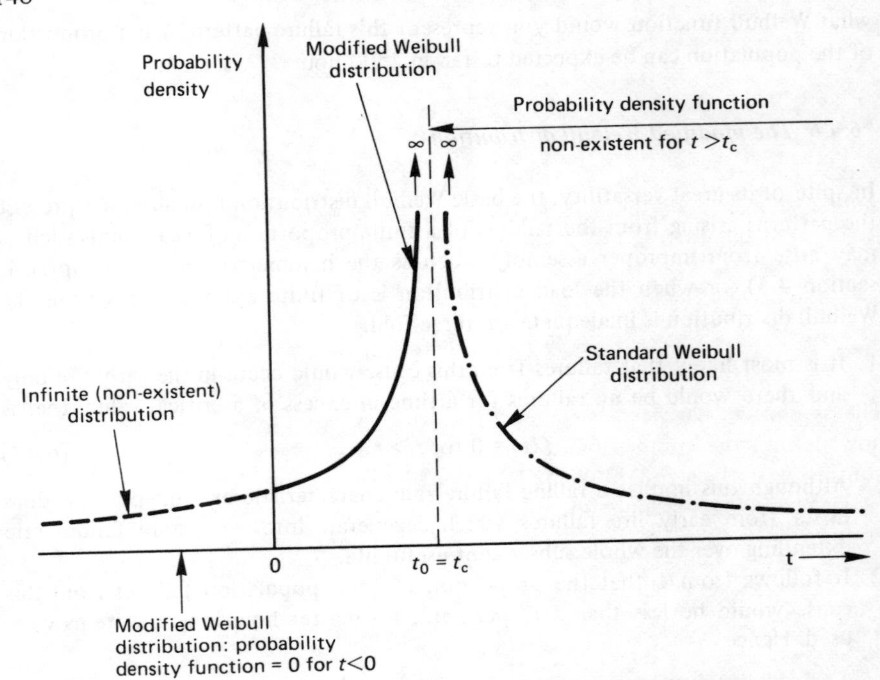


**Figure 6.11** Graphical representation of the modified Weibull distribution (full line) and its relationship to the standard Weibull distribution (chain dotted line)

If we write the probability density function of the general Weibull distribution as

$$f_W(\tau) = \frac{\beta}{\eta}\left(\frac{\tau - \tau_0}{\eta}\right)^{\beta-1} \exp\left\{-\left(\frac{\tau - \tau_0}{\eta}\right)^{\beta}\right\} \tag{6.49}$$

then its mirror image about $t = \tau_0$ is obtained by the transformation

$$t - \tau_0 = -(\tau - \tau_0) \tag{6.50}$$

or

$$\tau = 2\tau_0 - t \tag{6.51}$$

Substitution in (6.49) gives

$$f_W(2\tau_0 - \tau_0) = \frac{\beta}{\eta}\left(\frac{\tau_0 - t}{\eta}\right)^{\beta-1} \exp\left\{-\left(\frac{\tau_0 - t}{\eta}\right)^{\beta}\right\} \tag{6.52}$$

We therefore define a new probability density function, $f(t)$, to meet the criteria (1), (2) and (3) above, such that

$$f(t) = f_W(2\tau_0 - \tau) \text{ for } 0 \leqslant t \leqslant \tau_0 \tag{6.53}$$

and

$$f(t) = 0 \text{ for } t > \tau_0 \tag{6.54}$$

Identifying $\tau_0$ with $t_c$

$$\tau_0 = t_c \tag{6.55}$$

and hence from equations (6.49), (6.53), (6.54) and (6.55) the probability density function of the modified Weibull distribution will be given by

$$\begin{aligned} f(t) &= \frac{\beta}{\eta}\left(\frac{t_c - t}{\eta}\right)^{\beta-1} \exp\left\{-\left(\frac{t_c - t}{\eta}\right)^{\beta}\right\} && \text{for } 0 \leqslant t \leqslant t_c \\ f(t) &= 0 && \text{for } t > t_c \end{aligned} \tag{6.56}$$

The symbol $t_c$ has been used for the locating constant partly to denote that it now defines the completion of the distribution in time and partly to provide a contrast with the standard Weibull distribution.

The cumulative probability function for the modified Weibull distribution is given by

$$F(t) = \int_0^t f(t)\,\mathrm{d}t \tag{6.57}$$

that is

$$\begin{aligned} F(t) &= \left[\exp\left\{-\left(\frac{t_c - t}{\eta}\right)^{\beta}\right\}\right]_0^t = \exp\left\{-\left(\frac{t_c - t}{\eta}\right)^{\beta}\right\} - \exp\left\{-\left(\frac{t_c}{\eta}\right)^{\beta}\right\} && \text{for } 0 \leqslant t \leqslant t_c \\ F(t) &= 1 - \exp\left\{-\left(\frac{t_c}{\eta}\right)^{\beta}\right\} && \text{for } t > t_c \end{aligned} \tag{6.58}$$

It follows from equation (6.58) that the reliability will be given by

$$\begin{aligned} R(t) &= 1 + \exp\left\{-\left(\frac{t_c}{\eta}\right)^{\beta}\right\} - \exp\left\{-\left(\frac{t_c - t}{\eta}\right)^{\beta}\right\} && \text{for } 0 \leqslant t \leqslant t_c \\ R(t) &= \exp\left\{-\left(\frac{t_c}{\eta}\right)^{\beta}\right\} && \text{for } t > t_c \end{aligned} \tag{6.59}$$

Hence using equations (6.56) and (6.59) the failure rate, $\lambda(t)$, will be given by

$$\begin{aligned} \lambda(t) &= \frac{f(t)}{F(t)} = \frac{\dfrac{\beta}{\eta}\left(\dfrac{t_c - t}{\eta}\right)^{\beta-1} \exp\left\{-\left(\dfrac{t_c - t}{\eta}\right)^{\beta}\right\}}{1 + \exp\left\{-\left(\dfrac{t_c}{\eta}\right)^{\beta}\right\} - \exp\left\{-\left(\dfrac{t_c - t}{\eta}\right)^{\beta}\right\}} && \text{for } 0 \leqslant t \leqslant t_c \\ \lambda(t) &= 0 && \text{for } t > t_c \end{aligned} \tag{6.60}$$

The ultimate reliability of a system represented by equation (5.56) will be given by

$$R(\infty) = R(t_c) = 1 - F(t_c) = \exp\left\{-\left(\frac{t_c}{\eta}\right)^\beta\right\} \tag{6.61}$$

### *6.4.7 Evaluating the constants in the modified Weibull distribution*

The constants in the modified Weibull distribution may readily be determined in much the same manner as for the standard distribution, provided we can identify the data applicable to the failure mode. This is best done by physical examination. Let there be $n$ failures in this mode from a population of $N$ items. Then these failures can be interpreted as the last $n$ order numbers of the infinite cumulative distribution $F_\infty(t)$, given by

$$F_\infty(t) = \exp\left\{-\left(\frac{t_c - t}{\eta}\right)^\beta\right\} \text{ for all } t < t_c \tag{6.62}$$

that is, they will have the mean ranks

$$\frac{N-n+1}{N+1}, \frac{N-n+2}{N+1}, \frac{N-n+3}{N+1}, \ldots \frac{N}{N+1} \tag{6.63}$$

Since equation (6.62) can be written as

$$\log_e \log_e \frac{1}{F_\infty(t)} = \beta \log_e(t_c - t) - \beta \log_e \eta \tag{6.64}$$

standard Weibull paper can be used to represent the function provided the abscissa is relabelled $(t_c - t)$ and the ordinate $R_\infty(t)$.

It will be necessary to select, for the first approximation, a value of $t_c$ which is greater than zero – the value usually chosen as the first approximation for the locating constant in standard Weibull analysis. The resulting curve is then straightened by the usual techniques to give the true value of $t_c$. No difficulty should be experienced in deciding when to use the modified distribution. Standard analysis will result in an imaginary conventional Weibull distribution. This arises from the characteristic shape of these distributions when plotted on standard Weibull paper. Raw data which can be represented by the modified Weibull distribution plots in a manner which shows a concave upwards shape – suggestive of a negative $t_0$ – but in which the curvature increases with time. If $(t_2 - t_1) > (t_3 - t_2)$, then it follows from equation (6.34) that $t_0$ will be positive and greater than $t_2$. Since $t_2$ can be arbitrarily selected anywhere on the curve, it follows that $t_0$ will be greater than the $t$ value of all the data and hence a standard Weibull distribution is imaginary.

The method of calculation is best illustrated by an actual example. Figure 6.12a shows the failures recorded on a helicopter gear-box. The data has been

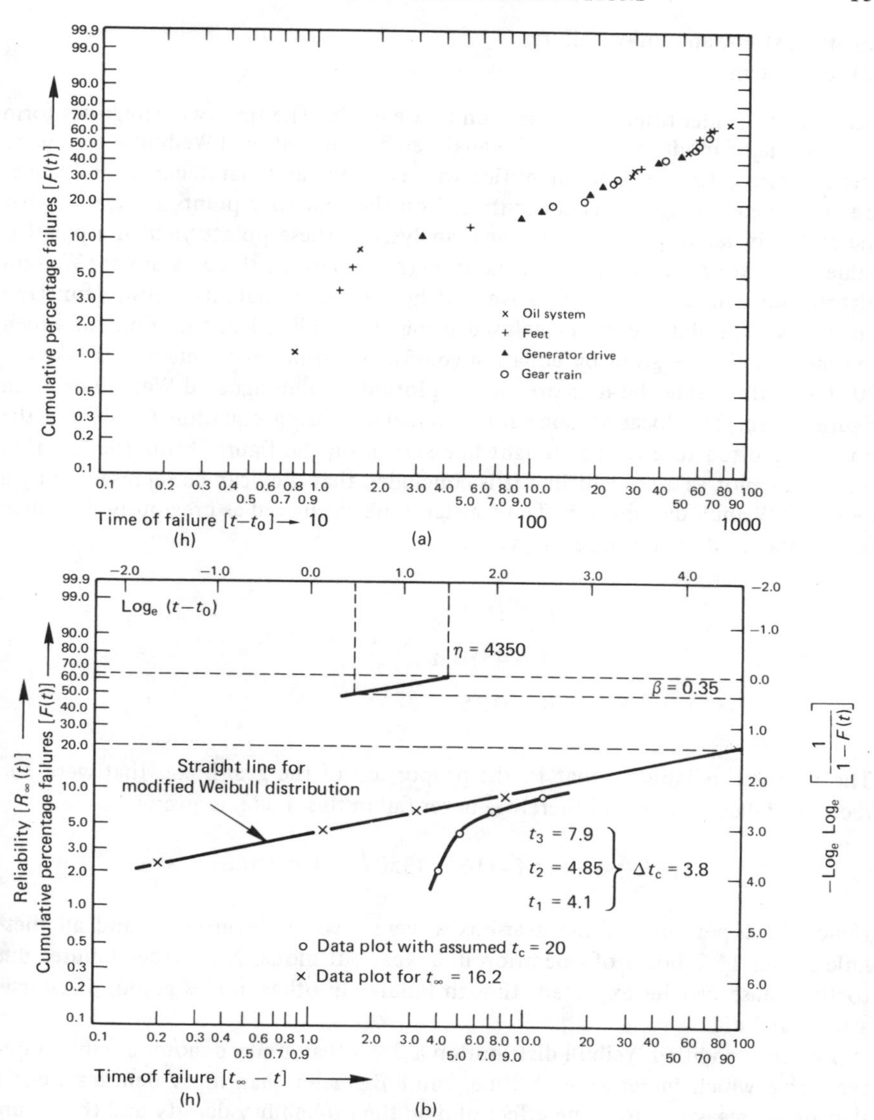


**Figure 6.12** Modified Weibull analysis of helicopter gear-box. (a) Raw data plot on Weibull axes. (b) Plot on modified Weibull axes.

taken from a comprehensive presentation in ref. 31. The failures were all categorised physically beforehand under four headings

(a) gear train failures
(b) generator drive failures

(c) feet, studs, and union failures
(d) oil system failures.

Each failure is identified graphically on figure 6.12a. The first two categories form the two major modes which can be analysed by conventional Weibull techniques. The last two categories were identified with early life and maintenance, and hence are combined. Analysis was concentrated on the first four points associated with the initial installation. A conventional analysis of these points yielded a positive value of 16 for $t_0$. It follows that the term $(t - t_0)$ in the three parameter Weibull distribution will always be negative and hence the probability density function imaginary. The data were re-analysed using the modified distribution for which the mean ranks are given by equation (6.63). As an initial estimate $t_c$ is taken as 20. Using this value the data are shown plotted on redesignated Weibull paper in figure 6.12b. The locating constant is calculated using equation (6.34) and the points replotted to give the straight line shown on the figure. From the fact that the points plot as a straight line it is concluded that they can be represented by a modified Weibull distribution. The constants are calculated as previously described for the standard distribution to give

$$t_c = 16.2 \text{ h}$$

$$\eta = 4350 \text{ h}$$

$$\beta = 0.35$$

The ultimate reliability, that is, the proportion of the gear-boxes that were correctly installed, and would therefore never fail in this mode, is given by

$$R(\infty) = \exp\{-(16.2/4350)^{0.35}\} = 0.868$$

Hence 13.2 per cent of the gear-boxes were incorrectly installed and all these failed after 16.2 hours of operation in a wear-out mode. No further failures due to this cause can be expected, though failures in other modes could, of course, arise – and did.

For the modified Weibull distribution a $\beta$ less than unity denotes a failure rate-age curve which increases with time, and a $\beta$ greater than unity denotes a curve that decreases with time. The effect of $\beta$ on the probability density and the failure rate is shown on figure 6.13. It is most likely that early life fatigue (such as may be triggered by imperfect finish on some item) could be represented by the normal Weibull distribution. The modified Weibull distribution would therefore mainly be used for the representation of early life erosive wear phenomena. In such situations the shaping parameter of the modified distribution would be less than unity. The quantity $\eta$ is solely a scaling parameter in the modified distribution, though as such it does control the ultimate reliability. It has no significance as a characteristic life.

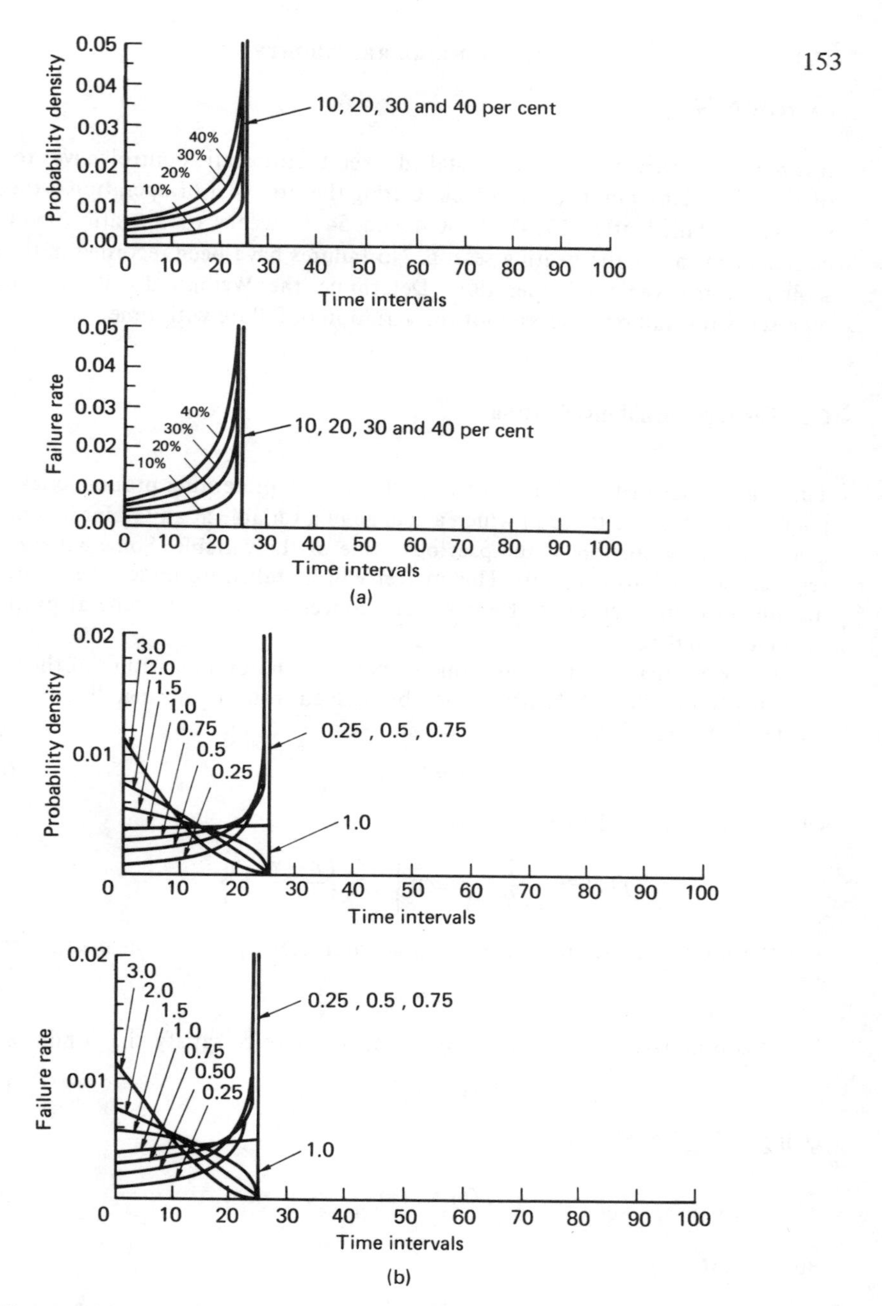


**Figure 6.13** Probability density functions and failure rates for modified Weibull distribution. (a) $\beta = 0.5$, various ultimate failures at time $t = 25$. (b) Different values of $\beta$ for modified Weibull distributions when 10 per cent of total population ultimately fails by time $t = 25$

*Exercise 6.14*

In order to conserve fuel a firm installed fixed thermostatic control valves to each of the 65 radiators in its office block. During the first year of operation 4 thermostat valves failed after 13, 42.2, 50.4 and 54.25 weeks of operation from the commencement of the heating season. No failures have been recorded in the subsequent three years of operation. Determine the Weibull distribution which represents this failure process. Plot the variation of failure with time.

## 6.5 The log-normal distribution

The log-normal distribution is of importance in representing proportional effect phenomena, that is, those in which a change in a variable at any point in a process is a random proportion of the previous value of the variable. Some believe maintenance falls in this category. This matter will be taken up under the heading of maintenance in chapter 9. Here we are concerned with the general properties of the distribution.

The log-normal distribution is one in which the natural logarithm of the variate is normally distributed. Hence it may be derived from the normal distribution by means of the transformation

$$\tau = \log_e t \tag{6.65}$$

where $\tau$ is normally distributed, that is

$$f(\tau) = \frac{1}{\sigma\sqrt{(2\pi)}} \exp\left\{-\frac{1}{2}\left(\frac{\tau - \tau_0}{\sigma}\right)^2\right\} \tag{6.66}$$

For the normal distribution the locating parameter $\tau_0$ is also the median, so that

$$\tau_0 = \tau_m \tag{6.67}$$

In transforming any statistical distribution we have to satisfy the condition that

$$f(t)\,dt = f(\tau)\,d\tau \tag{6.68}$$

giving

$$f(t) = f(\tau)\frac{d\tau}{dt} \tag{6.69}$$

But from (6.65)

$$\frac{d\tau}{dt} = \frac{1}{t} \tag{6.70}$$

Hence the log-normal distribution may be obtained by substituting (6.66) and (6.70) in (6.69) as

$$f(t) = \frac{1}{\sigma t \sqrt{(2\pi)}} \exp\left\{-\frac{1}{2}\left(\frac{\log t - \log tn}{\sigma}\right)^2\right\} \tag{6.71}$$

where $\sigma$ is the standard deviation of the normal distribution. To avoid any confusion we shall replace $\sigma$ by $\alpha$, and refer to it as the dispersion of the distribution. Then equation (6.71) can be rewritten as

$$f(t) = \frac{1}{\alpha t \sqrt{(2\pi)}} \exp\left\{-\frac{1}{2}\left(\frac{\log_e(t/tn)}{\alpha}\right)^2\right\} \tag{6.72}$$

It will be noted that the locating parameter of the original normal distribution has now become a scaling parameter in addition to being the median. The dispersion $\alpha$ is also a shaping parameter, as may be seen from figure 6.14 in which the probability density function for the log-normal distribution has been plotted for

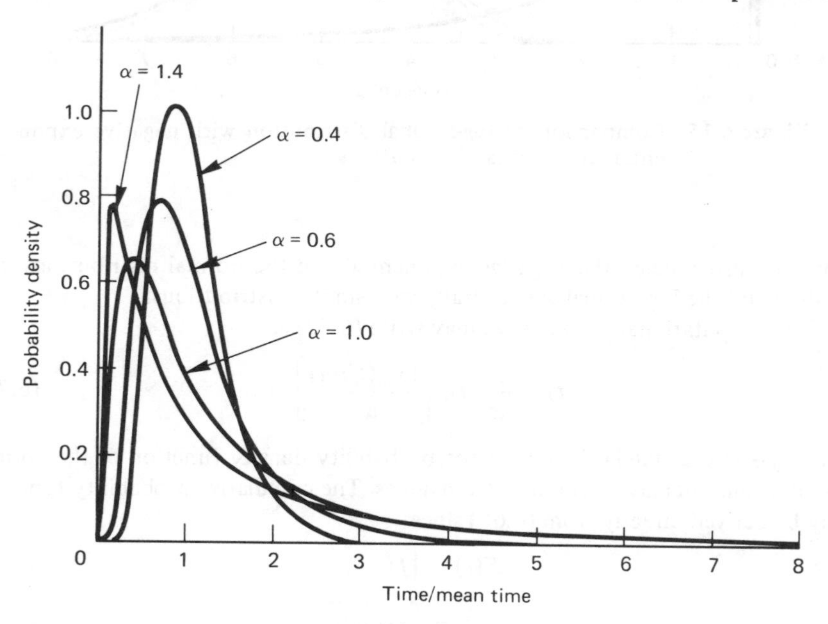


**Figure 6.14** Effect of the dispersion ($\alpha$) on the form of the log-normal distribution

various values of $\alpha$. For medium and large values of $\alpha$ the log-normal distribution is markedly skewed, but as $\alpha$ is decreased the distribution becomes more symmetrical. If $\alpha$ is close to unity the log-normal distribution is approximately equivalent to the negative exponential distribution, as shown by figure 6.15, though different origins must be chosen to obtain a reasonable fit. This figure also shows that for small values of $\alpha$, less than say about 0.2, the log-normal distribution approximates to the normal distribution. To the extent that the Weibull distribu-

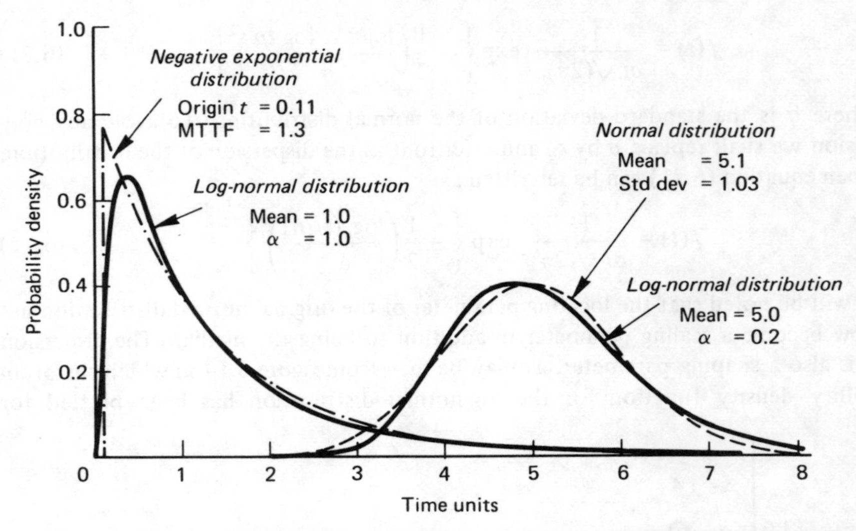


**Figure 6.15** Comparison of log-normal distribution with negative exponential and normal distributions

tion can also replace the negative exponential and the normal distributions, the Weibull and the log-normal are generally very similar distributions.

For computational purposes we may write (6.72) as

$$f(t) = \frac{1}{\alpha t} \ f_N \left[ \frac{\log_e(t/tm)}{\alpha} \right] \tag{6.73}$$

where $f_N$ is the standard form of the probability density function of the normal distribution, and may be obtained from tables. The cumulative probability function may be derived directly from (6.68) since

$$\begin{aligned} F(t) &= \int f(t) \, dt \\ &= \int f(\tau) \, d\tau \\ &= F_N \left[ \frac{\log_e(t/tm)}{\alpha} \right] \end{aligned} \tag{6.74}$$

where $F_N$ is the standardised cumulative function of the normal distribution, which is also available from statistical tables.

*Exercise 6.15*

Plot the cumulative distributions corresponding to figure 6.14, and hence deduce and plot the failure rate against time.

Unlike the normal distribution from which it is derived, the log-normal distribution does not extend indefinitely in the negative direction. The distribution begins at the origin, and like the Weibull distribution, does not exist for negative values of the variate. It is possible, of course, to move the distribution along the time axis, by introducing a locating constant into the distribution. If a locating constant does exist the raw data will plot as a curve on log-normal paper. Log-normal probability paper is very similar to standard probability paper except that the abscissa has a log scale instead of the linear scale of the standard paper. A curved line can be straightened in exactly the same way as described for the Weibull distribution. The parameters defining the distribution can then be determined.

Some workers always check whether the log-normal is a better fit to data than other distributions, such as the Weibull. It should be noted, however, that if one distribution is a better fit to a given set of data it does not follow that it is always the best fit, and each case must be examined separately.

## 6.6 The Pareto distribution

Pareto was a nineteenth century economist who first pointed out that in many real life problems a relatively small proportion of the independent variables nearly always accounts for the characteristic behaviour of the dependent variable. Nixon, Martin-Hurst and Shainin initially pointed out that this is invariably true of reliability: a majority of failures being caused by a minority of reasons. As an illustration of the Pareto distribution, figure 6.16 shows the faults that gave rise to failures in standby generators operated by the Department of the Environment (ref. 32). These have been plotted as a histogram in which the frequency of failure has been plotted against cause in decreasing order. The result is a typical Pareto distribution which is characterised by a very rapid drop in frequency over the first few items followed by a long tail of items recording a low frequency. A feature of many real curves is discontinuities in the distribution which enable items to be grouped more positively.

Referring to figure 6.16 as a typical example, we see that one item (instrumentation and control system) stands out like a sore thumb, accounting by itself for about 26 per cent of all failures. Then follow two items of equal importance (lubrication and cooling systems, and fuel injection equipment) each accounting for about 16 per cent of all failures. Thereafter follows a long list of causes of failure of roughly equal and low importance. The first three causes, accounting for 58 per cent of all failures in this case, make up Pareto's "vital few" which are responsible for the main behaviour of the system, while the remainder are Pareto's "trivial many", each of which plays only a small part in the overall behaviour. In assessing reliability, or any other subject which can be represented by a Pareto distribution, it is only necessary to concentrate on the 'vital few': the trivial many can be ignored as having little influence on the topic under review.

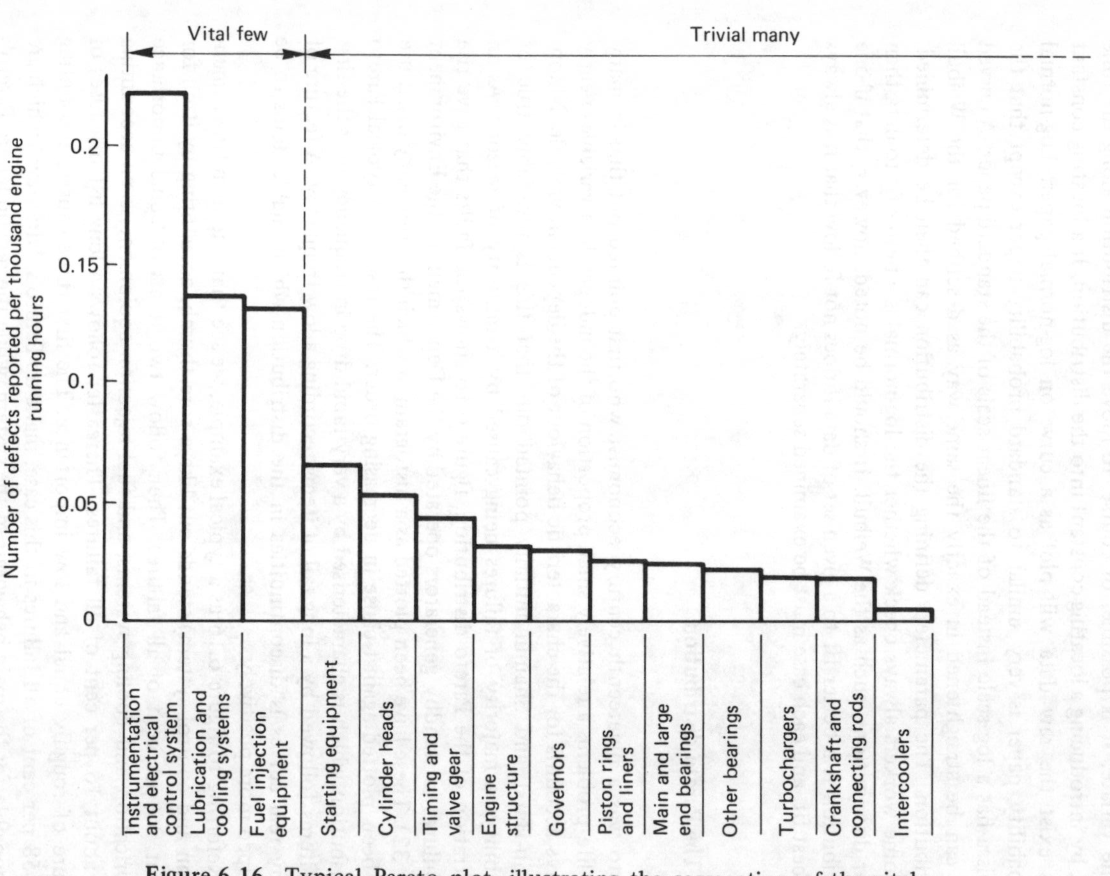


Figure 6.16 Typical Pareto plot, illustrating the segregation of the vital few from the trivial many

This is as far as we shall take the Pareto distribution in this book – indeed as far as it need be taken for most practical purposes. It will be observed that its application is very simple – the distribution is obtained from a straighforward numerical evaluation and its interpretation is mainly qualitative. In spite of this it is a very powerful tool in the assessment of field data and should not be pushed on one side in favour of more sophisticated statistical techniques on account of its simplicity.

By its very simplicity it is, however, open to pitfalls. Using the above example as illustration it will be noted that all instrumentation and electrical control system failures have been lumped together. It could well be that there is a single source of failure within the control system which is not revealed by this analysis. Of more importance the failures could be equally distributed over a large number of sources so that in fact they should all be treated as the 'trivial many'. There is some justification for the analysis in this instance in that the control gear is supplied as a complete sub-system. The supplier may therefore be regarded as the single source of failure, and the Pareto plot does pin-point action to be taken to reduce failures. The same is not true of the lubrication and cooling systems, though overheating could be due to either. If this were so, a revised title would be preferable: otherwise the writer strongly believes it would have been preferable to record failures due to lubrication and failures due to the cooling system separately. It is possible that one could become a member of the trivial many and attention could then be directed closer to the cause of the trouble. The same observation is not true of 'main and large end bearings'. Since the combined class is one of the trivial many, the constituents must also both be in the trivial many category. The overall conclusion remains unchanged. To sum up: the cause against which failures are debited must have a physical significance in the subsequent procedure, be the significance a technical one, an economic one or any other, if the Pareto distribution is to be a dependable guide. With this precaution we have a simple powerful tool at our disposal. We shall make use of it in the next chapter when we take up again the general theme of mechanical reliability by assessing field data.

# 7
# Failures in the field

## 7.1 Quality of performance

The first feature listed in chapter 2 in our definition of reliability is perhaps the one that presents the greatest difficulty. In deriving the Weibull distribution for a particular set of data, as in section 6.4.2, we listed failures but we did not state what constituted a failure. Disastrous failures, such as the breaking of an engine crankshaft or a turbine disc disintegration, are self-defining and uncontested; but incidents which are not disastrous and only impair the performance to a limited extent may be classed as a failure in one instance, and not in another. For example, the tune of a piano, which may be quite adequate for the majority of people in their own homes, could be considered totally inadequate for high-class public performance. The loss of performance (failure) due to loss of mechanical adjustment can be corrected by maintenance (retuning), but what constitutes a 'failure' is essentially a function of the audience expectations.

Differing requirements for the quality of performance for different operational roles can lead to great difficulty in evaluating reliability. The difficulty is not made any easier by the necessity to class any incident as a failure or success. This matter was first raised in section 3.1 and is a direct consequence of a statistical approach leading to equation (3.1). It may well be an accurate representation of many electronic items – they either work or they do not – but it is not always true of mechanical items, many of which, as we have already seen, gradually wear until at some point they are deemed to have failed. On what basis do you replace the windscreen wipers on your car? Sometimes they may fail disastrously, but more often they wipe the screen less and less effectively as time goes by. This cannot be represented in the classical statistical model. If we are to apply statistics we must first acknowledge the limitations imposed by the mathematical model and define in each and every case when a failure is to be reckoned as a failure.

A further complication, which this example illustrates, is that windscreen cleanliness is itself a feature which is measured subjectively by the driver. Quantitative manipulation of subjective data does not make it more objective. Additionally it must be recognised that although some failures can be defined objectively, in many cases even these are decided subjectively. To quote an example: a ball bearing may be allowed a permitted amount of 'play' beyond which it is deemed to have failed, but in practice it will often be replaced on the subjective assessment of a fitter without any reference to specified measurements. Statistical data on the replacement of these bearings could, in reality, tell us more about the behaviour of fitters than of bearings!

Many cases could be quoted in which the assessment of failure is completely objective, and many in which it is substantially subjective. Between these extremes lies a complete spectrum in which objective and subjective assessment are mingled in varying proportions. A full appraisal of the meaning of 'failure' is essential in every particular case. Many apparently conflicting data on reliability stem from this factor.

Additonally, before assessing 'failure' one must be certain of the maintenance given to the item. It is obvious that many failures can be averted by preventive maintenance or by unscheduled maintenance following inspection or operator observation – provided the operating programme allows it. Thus, in assessing failures of complex equipment, both the repairs made during preventive maintenance and the number of occasions when it received unscheduled maintenance must be taken into account, together with the quality of the maintenance provided. All this leads to the important conclusion that in assessing reliability data one must be extemely careful to allow for possible discrepancies in the interpretation of the term 'failure'.

Much of what is stated in this section may seem to be little more than a statement of the obvious, as indeed it is. However, it would be instructive if you tried the two amusing and quite short exercises given below before reading the suggested solutions.

*Exercise 7.1*

The washer of the cold-water storage tank of a normal domestic system allows water to leak past it. The water escapes down the overflow. Do you consider this is a failure of the system?

*Exercise 7.2*

Your wife/girl-friend wishes to borrow your car. She pulls out the choke and operates the starter. The engine fails to fire. She operates the starter again and again without success, the engine soon becoming over-rich. In a temper she calls you, who being a mechanical engineer, aspirate the engine and start it almost immediately. If you were making a study of the reliability of car-starting systems, would you class this as a failure?

Some authorities try to overcome the difficulty we have been discussing by use of two different words – 'failure' and 'defect'. A failure is defined as "the termination of the ability to perform a function" and a defect as "any non-conformance to specified requirements." In this nomenclature a defect is seen as a minor fault. In the author's opinion this only begs the question: when is a fault a minor fault? If a fault in a component does not impair performance, why is the component there? Other authorities retain the word failure, but prefix it with adjectives which define its relevance, for example: 'safety failures', 'major failures' and 'minor failures'. A more elaborate nomenclature would be

| | |
|---|---|
| over-critical failures | – involving safety |
| critical failures | – item unserviceable |
| major failures | – reduce usefulness |
| minor failures | – marginally reduce usefulness |
| incidental failures | – do not reduce usefulness |
| non-relevant failures | – do not matter. |

'Catastrophic' is sometimes used in place of critical. Care needs to be taken with such terminology since none has universal acceptance. The only agreed option appears to be that failures involving safety should be designated separately. This does seem most reasonable. The more elaborate of the terminologies are often replaced by a numerical scale ranging from 1 to 0 or 10 to 0 based on the consequences of the failure. The highest number usually implies safety is involved but the rest of the scale is based on arbitrary criteria – it may be cost, for example. We shall have occasion to use such a scale when looking at failure modes and effects analysis in section 11.3. Meanwhile we shall treat all failure data with scepticism, since for any numerical analysis we must still decide 'when a fault's a fault'.

## 7.2 Field evidence

As stated in the Introduction, it is not possible to conduct experiments in reliability, as in other branches of technology, to verify each theoretical step. We have therefore to rely on field data – which does have the advantage of being the real objective – but may include uncontrolled and unrelated effects avoidable in a laboratory test or worse still operating conditions of which we are not aware. The correlation cannot therefore be that which one would normally expect in scientific work, and must be thus assessed.

It will be recalled that our major postulate was that all mechanical items should be designed to be intrinsically reliable but that they would eventually fail, if not prevented by any maintenance action, in one of the wear-out modes discussed in chapter 5. This enabled us to devise a rational approach. Even so, it was recognised that some designs may not be intrinsically reliable, either because means were available for combating this state, burn-in and proof testing being the outstanding methods, or because the designer may wish to accept the disadvantages for other gains, or maybe because the design process had failed to achieve intrinsic reliability. We should first ask ourselves, then, how far do past designs conform to our postulations?

It seems that the best evidence could be obtained from the failure mechanisms themselves. If these were largely wear-out, then we may reasonably expect that the items were intrinsically reliable. If a significant number of failures were by mechanisms involving no wear process, often referred to as stress-rupture mechanisms, we could reasonably conclude that the items were not intrinsically reliable. Fortunately the author had available a good source of such evidence, for provision is made in the Army workshop reporting system FORWARD (Feedback Of Repair Workshop And Reliability Data) to record the nature of all failures. The records are by no means 100 per cent complete. This is not surprising when one considers the method of cellecting the data from job cards which may have to be completed in very adverse circumstances. However, it is far better that no record be made in cases of doubt than that subsequent analysis be based on guesses, and we shall have to use that which is available. On the other hand this defect is largely nullified by the vast amount of data available. To this end every failure experienced by the British Army over a period of two years was examined, for various classes of equipment, to establish the cause of failure. The British Army is one of the biggest users of a wide range of equipment and there is every reason to believe they are representative of most professional users, though perhaps not representative of untrained or casual users. If the authenticated causes can be treated as representative, and it is firmly believed they can, some interesting conclusions can be drawn.

We can look first at the frequency of occurrence of various failure modes in Army aircraft. These are thoroughly reported and provide the most comprehensive information available. The data has been plotted in the form of a Pareto distribution on figure 7.1. It conforms to the typical pattern. We may immediately identify Pareto's vital few with the first two modes, but to be more certain in this case we can extend it to include the first six modes, after which a discontinuity leads to the repetitive form of the trivial many. Of these six failure modes, every one has a wear-out characteristic. A number of the trivial many also have wear-out mechanisms, though we can note a small number such as 'broken' or 'bent' which are essentially of a stress-rupture nature. Hence we can deduce that all the significant modes of failure in aircraft items resulted from wear-out, though a few minor modes did exhibit stress rupture. This may be interpreted as strong indication that designs were initially intrinsically reliable, though there are a small number where intrinsic reliability was not achieved.

It may be argued that aircraft equipment is of high quality and not representative of the general run of mechanical equipment. The data for all commercial vehicles operated by the Army was therefore also examined. It is again presented as a Pareto plot in figure 7.2. In this case the first five items are identified as the vital few. However, a difficulty arises from the less precise reporting code used for the more run of the mill items. It cannot be established whether a 'broken', 'cracked', 'disintegrated' or 'sheared' part failed by stress rupture or fatigue. The

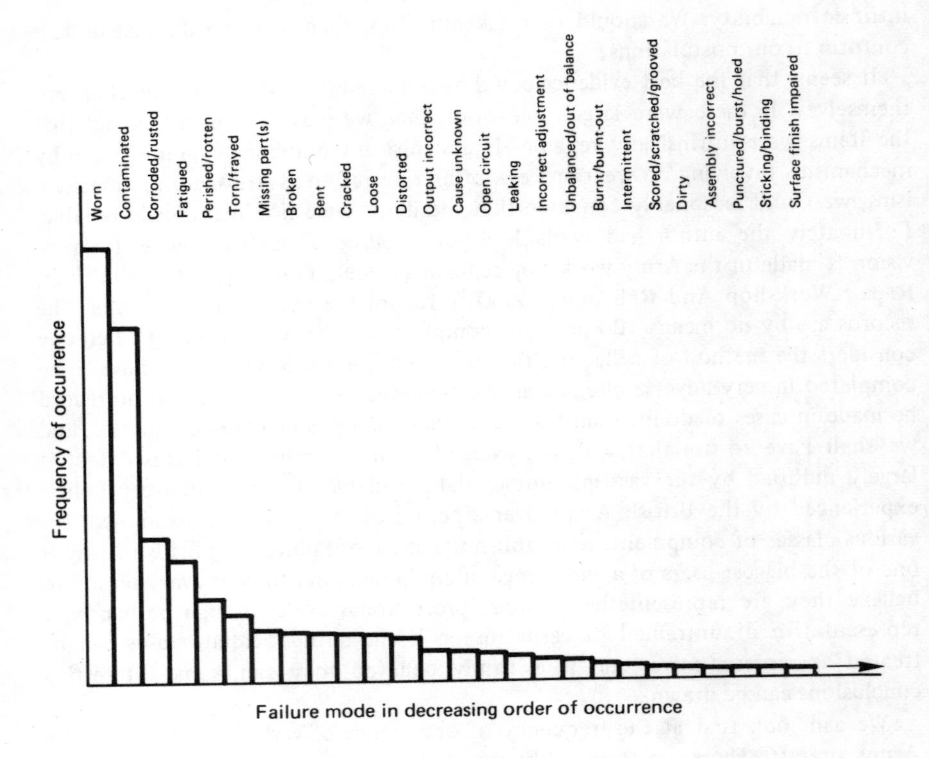


**Figure 7.1** Frequency of occurrence of various failure modes in Army aircraft

latter is not separately recorded. However, all the remaining vital few failure modes are of a wear-out nature. As for aircraft, a number of the trivial many are also wear-out. Thus we reach the same conclusion, though perhaps with a little less certainty, that the majority of failures are due to wear-out, but a few are due to stress rupture.

A number of special items of equipment were also examined for their failure modes: they all yielded virtually the same result. We therefore conclude that on the whole the design of mechanical items ensures that they are intrinsically reliable, even though this may not have been the specific intention of the designer. Good design ensures intrinsic reliability. We note, however, that stress rupture was responsible for a sufficient number of failures for us to conclude also that contemporary design does not always ensure intrinsic reliability.

It may be worth recording that fatigue was not the 'villain of the piece', as

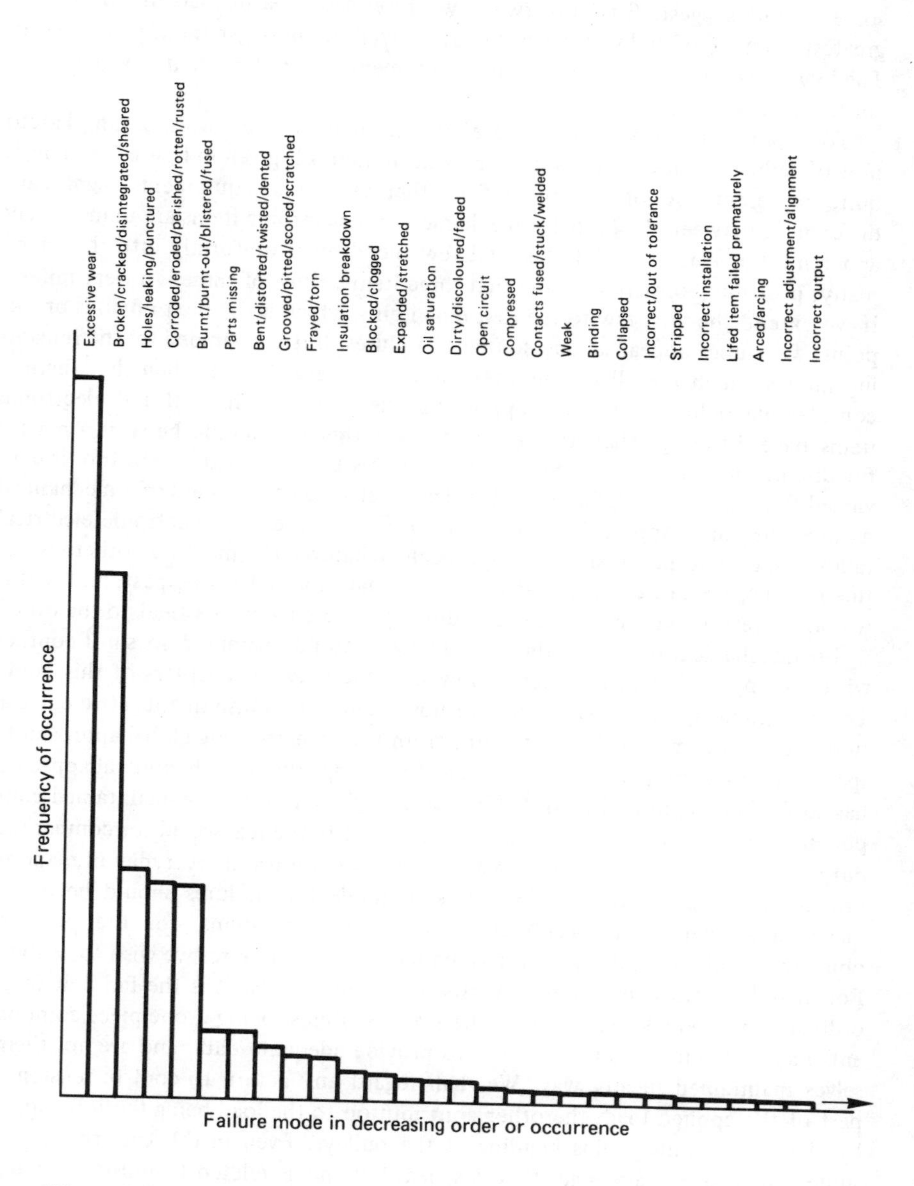


**Figure 7.2** Frequency of occurrence of various failure modes on 'B' vehicles operated by the Army

some would suggest. Straightforward wear (erosion) easily accounted for the greatest number of failures, with corrosion well up amongst the leaders. (*Note*: The low peace time utilisation of Army equipment may not warrant any generality for this conclusion.)

To compare the mechanical and electronic worlds, the corresponding Pareto plot of failure modes in Army communication radios is given in figure 7.3. This is quite representative of the plots for other electronic equipment. Significant differences between the plots for mechanical and electronic items are immediately apparent. For electronic kit the vital few blend more uniformly into the trivial many, probably consisting of the first three items. None of these is a wear failure. If we extend the vital few to the first nine items, which is the next obvious break-point, wear does appear as one source of failures, but is swamped by the remaining modes which are all of the stress-rupture variety. We see then that there is considerable field evidence to support the view that mechanical and electronic items have differing reliability behaviours. Whether this should be so is a matter for discussion which is outside the scope of this book, though the author is convinced that, in so far as failures of electrical or electronic items are of a mechanical nature, the same approach is relevant. It is only the true electronic/electrical failure that could lie outside that approach. Whatever the merits or otherwise of this may be, we can be sure that differences have existed in the past, and that it would be very unwise indeed to read directly across from one field to the other.

For mechanical reliability the field evidence so far examined does not conflict with the approach that has been followed in the previous chapters of this book. This is a good reassuring start, but we now need to establish quantitative correlation. Necessarily we shall have to turn from the all-embracing global approach to specific items. It may be worth recalling at this stage that our theoretical approach has so far been confined to the effect of a single load on a non-maintained component. We need therefore to seek its counterpart in the real world for comparison purposes. The item should be in reasonably common use in everyday life, and to provide adequate analysable data it is desirable that failures should be by no means unknown. This specification eliminates most items, for the practical objective of all reliability work is to eliminate the very failures we wish to analyse. Fortunately there is one item that fits our requirements. It is the fan belt of an ordinary road vehicle. It is one of the simplest items, being a one-piece component. Fan belts fail sufficiently often to provide adequate data, and are not themselves maintained in any way. We shall regard any action to control tension as part of the applied load, the other contribution to the load being tension applied by the driving pulley plus bending at the pulleys. Even in this case the item is subject to a secondary load, that is speed, but this is related to tension through the fan and pump characteristics. Although the relation between speed and tension may not be unique, we shall assume it to be so. (Both forward speed and generator load contribute to this departure.) How then does its failure characteristic compare with the concepts we have established so far?

Figure 7.4 shows a plot on Weibull paper of the fan belt failures plotted against

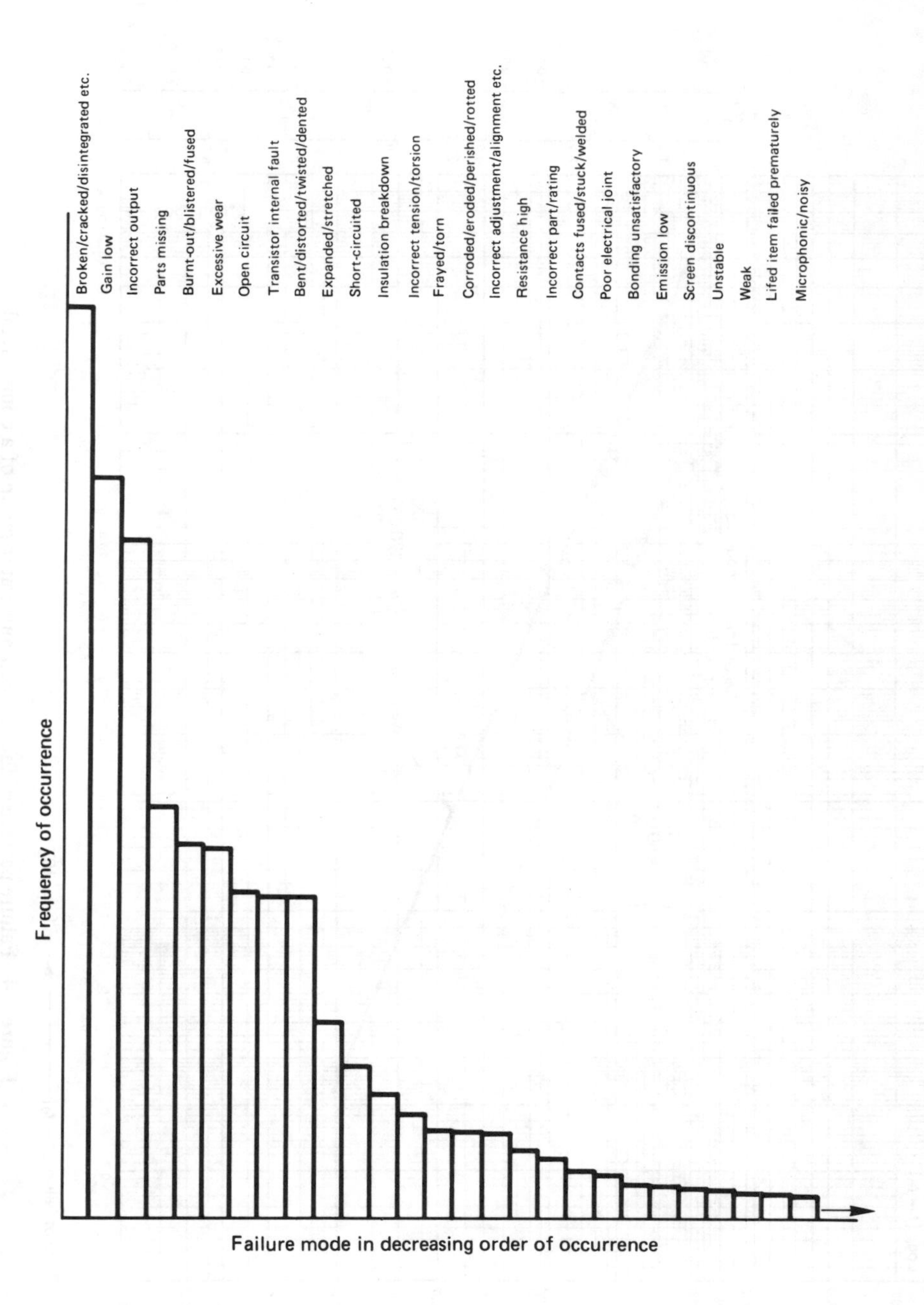


**Figure 7.3** Frequency of failure modes in Army communication radios

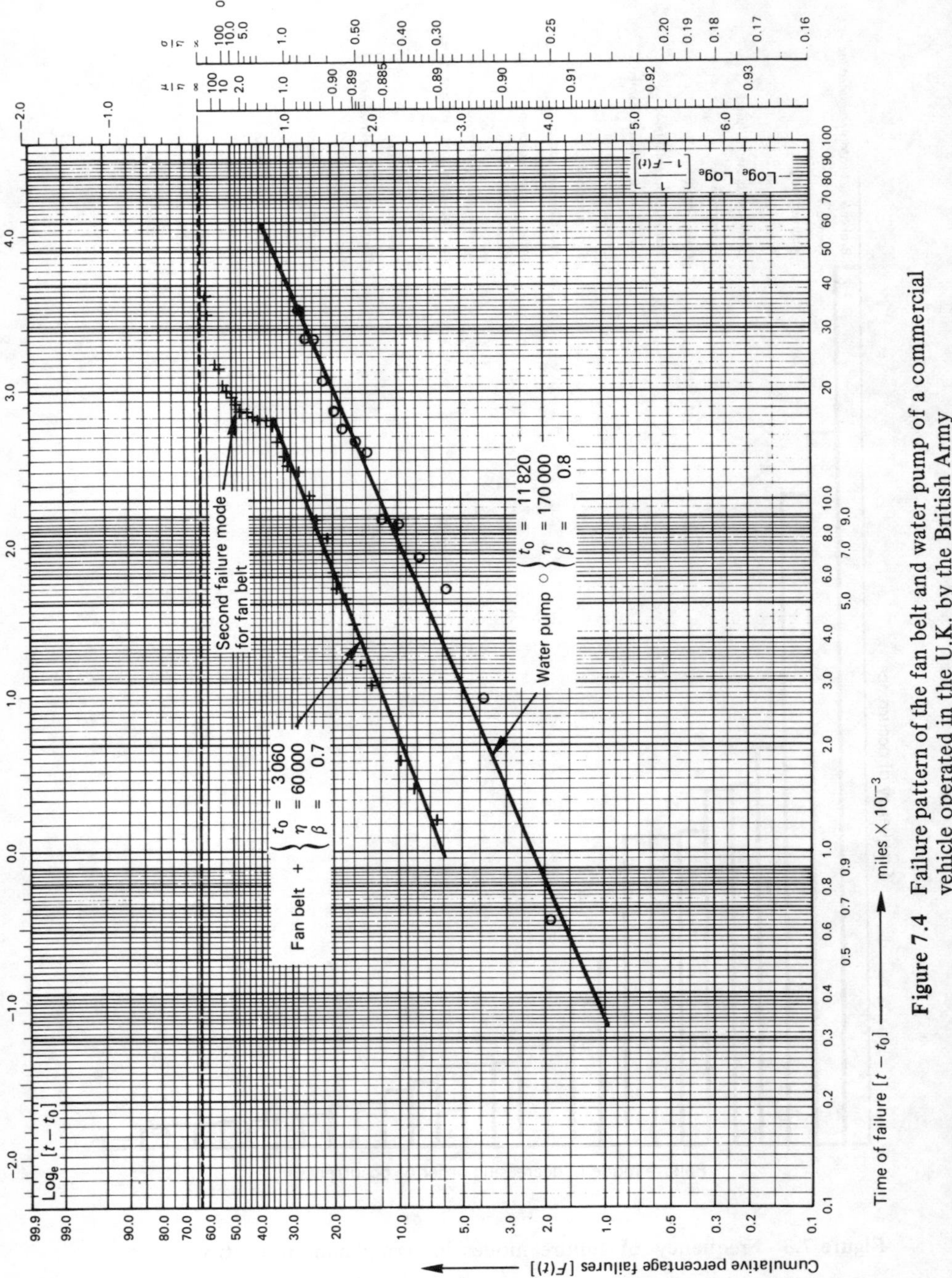


**Figure 7.4** Failure pattern of the fan belt and water pump of a commercial vehicle operated in the U.K. by the British Army

miles to failure as recorded on a commercial vehicle operated by the British Army in the U.K. Only the time up to first failure in any vehicle is plotted so that any significant maintenance is unlikely and plays no part. The results are interesting. It is seen that the failures can be represented by a Weibull distribution in which the locating constant ($t_0$) is 3060 miles. This conforms to our expectations, and indicates that the belts are intrinsically reliable for that distance, in that no failures can be expected under that distance. This evidence at least suggests that some items can be intrinsically reliable. The intrinsically reliable life ($B_0 = t_0$) is followed by the wear-out process represented by the Weibull distribution. This distribution has a shaping parameter of 0.7 and a characteristic life of 60 000 miles, giving a mean life of 87 000 miles. The value of the shaping parameter implies that there is a sudden jump in the failure rate at just over 3000 miles, followed by a long period of falling failure rate, strongly suggesting that failures arise from some fatigue process rather than by 'wear' in the sense of erosion. Any loss of belt material by scrubbing action will, of course, be taken up by retensioning during maintenance. As is well known by physical observation, most belt failures are preceded by cracking, that is, failure is by a fatigue process. It is interesting to note that even for such a 'simple' item as a fan belt a secondary mode can set in – for this particular case at about 20 000 miles. The failure patterns for belts in a number of different vehicles used in different parts of the world were substantially the same: a period of intrinsic reliability, some well in excess of the 3000 miles just quoted, followed by a fatigue type wear-out.

Two significant conclusions can be reached. First that a detailed analysis of a monolithic item confirms the conclusion that items can be intrinsically reliable before wear-out sets in. The wear-out is *not* preceded by random failures starting from first introduction into service. Secondly the hitherto unreported falling failure rate wear characteristic does exist. In view of its unusual nature this latter characteristic needs further confirmation by correlation with other known fatigue processes before it can be accepted.

Moving first to a more complex but still basic item, we can refer to the Rolls Royce failure data on bearings used for exercise 6.8. It was shown there that the bearings have a life of intrinsic reliability of 225 hours during which completely failure-free operation may be expected. This is followed by a wear-out mode represented by a Weibull distribution having a shaping parameter of 0.95. Although the shaping parameter is somewhat higher than generally encountered theoretically, it is not unduly so, and the field data indicates a jump in the failure rate at 225 hours followed by a long period of falling failure rate – typical of fatigue. It is the failure mechanism that engineering judgement would anticipate in this application. The data is positive additional evidence that the type of failure pattern predicted theoretically can, and does, occur in practice. It is apposite to record that similar evidence relating to bearings was produced as long ago as 1963 (ref. 33) but the analysis and interpretation was carried out in isolation. The full significance of the results was consequently not appreciated. Re-interpreted as a three parameter Weibull distribution, the results form very substantial backing for the theory as applied to this particular component, in which fatigue is known to be the failure mechanism.

As an example of a more complex component, the failures on the water pump of the vehicle used for the fan belt analysis were also examined. The results are shown on figure 7.4. Little need be said about them. They show that the pump is intrinsically reliable up to 12 000 miles ($B_0$), and that this is followed by a fatigue-type wear-out mode. The mean life is about 220 000 miles.

As a final example of failure in the fatigue mode of the failure data quoted by the British Standards Institution (ref. 34) for an electro-mechanical device will be examined. The results are plotted on figure 7.5a. These are very interesting data in that two failure modes can clearly be identified, and the appropriate curves have been drawn and labelled on figure 7.5a. One isolated point remains. Without a physical examination it is not clear whether this is one of the two identifiable failure modes or an entirely different mode. Assuming the latter to be the case, the first failure mode can be analysed (treating failures in the second mode as drop-outs) to give a Weibull locating constant ($t_0$) of 6.5 hours and a Weibull shaping parameter ($\beta$) of 0.45. The characteristic life is 90 hours. The low value of the shaping parameter combined with the finite value of the locating parameter ($t_0$) suggests that failure is arising from fatigue with a strong interaction between the load and the strength. Although the item is intrinsically reliable in this particular mode – no failures can arise under 6.5 hours – the presence of an unidentified mode could make it intrinsically unreliable.

This example is a particularly interesting one in that it was first used by the British Standards Institution to illustrate Weibull analysis. The curve they drew through the data plotted on Weibull paper is reproduced in figure 7.5b. It assumes that either only one failure mode is present or that all failure modes can be lumped. The possibility of several failure modes, which only set in after a period of operation, has been ignored. I would suggest that the curves of figure 7.5a are a far better representation of the field data. This is no criticism of the British Standards Institution. This actual example has been chosen from the literature because it fully represents established thinking of the time (1974) – indeed it is all too prevalent today – and because it is believed that the error is immediately apparent once the possible failure mechanisms are understood and electronic/electrical thinking is not allowed to exclude essentially mechanical phenomena. The prevalence of the negative exponential distribution in those fields does not mean it is omnipotent in mechanical applications. To assume so leads to confusion, since it excludes the possibility of intrinsic reliability and the existence of distinctive wear-out patterns. Much of the difficulty surrounding mechanical reliability arises from false tenets brought in from electrical, and particularly electronic, engineering. We have already seen that intrinsic reliability is rarely achieved in that application, so the random mode is considerably more prevalent. Whether this does or should apply to modern electronics is another matter. However, we must note that early reliability work was carried out in an atmosphere of all-pervading random failures, and hence it would be unwise to put much reliance on any data

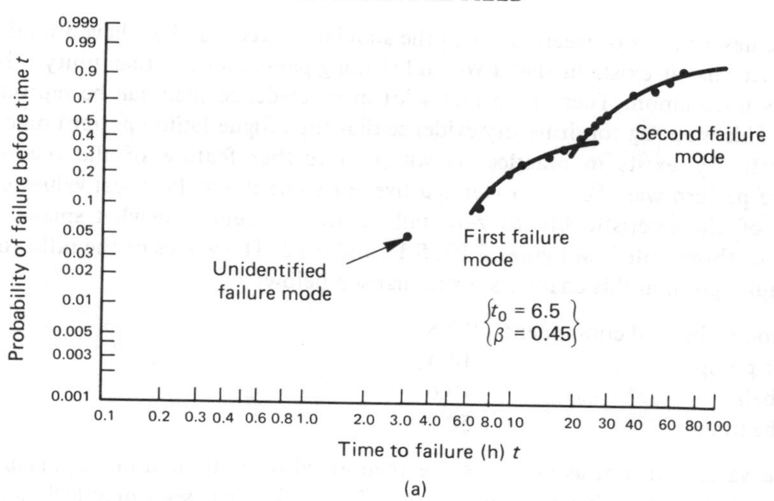


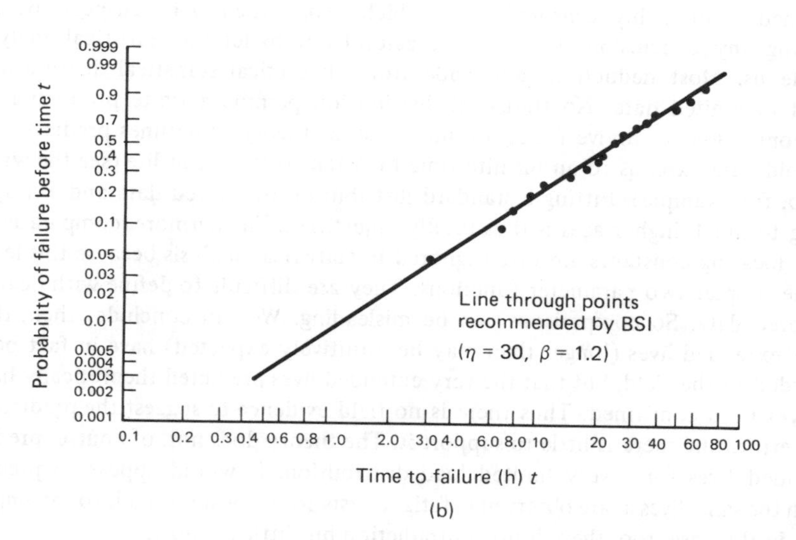


**Figure 7.5** (a) An example of multi-mode fatigue failure mechanisms. (b) Illustrating how an analysis of the two modes simultaneously can suppress essential features of the process

analysis of that time. The omission to establish by engineering examination that the data is self-consistent (that is, that it does not contain mixed failure modes) would appear to be a sin of many early workers in the field (see also exercise 6.9).

It seems we are on the right lines in our representation of fatigue: theory predicted a falling failure rate with age in this wear-out mode – which is contrary to the descriptions of wear-out in all the standard texts – and we have found clear evidence that it exists in that a Weibull shaping parameter less than unity is by no means uncommon. There is in fact a lot more evidence than can be reproduced here. This is strong confirmatory evidence that the fatigue failure pattern predicted theoretically exists in practice. However, a further feature of the theoretical failure pattern was the very extended lives of some items. Practical values of the ratio of characteristic life to zero failure life do seem somewhat smaller than some of those noted on figures 5.10, 5.11 and 5.12. The values of the ratio for the examples given in this chapter are summarised below:

| | |
|---|---|
| Hydro-mechanical component | 13.8 |
| Water pump | 14.4 |
| Fan belt | 19.6 |
| Ball bearing | 50.2 |

These values are not as high as some theoretical ones, though much higher than many practising engineers would quote. The author has seen practical values as high as 400 or 500, but this is still below the magnitude of the theoretical predictions with rough loading. Unfortunately, all the high practical values seem to be obtained from highly censored data, which makes them a bit suspect. Before drawing any conclusions we must be careful not to let the statistical analysis delude us. Most deductions are made from theoretical statistical distributions fitted to limited data. No statistical distribution permits a finite proportion of the population to survive for an infinite time, as theory sometimes predicts, and no field data extends to an infinite time or often to the mean life (see the water pump, for example). Fitting a standard distribution to limited data and extrapolating to much higher ages is thus totally unjustified. Furthermore, comparatively small locating constants are often ignored in statistical analysis because this leads to the simpler two parameter functions. They are difficult to define with heavily censored data. So comparisons can be misleading. We can conclude, then, that some extended lives (longer than may be intuitively expected) have in fact been recorded in the field, but that the very extended lives predicted theoretically have not yet been confirmed. Thus there is no field evidence to suggest the prediction is in error, but there is little to support it. The theory does not, of course, predict extended lives for a very limited load distribution. It would appear to predict much the same lives as are observed in fatigue tests for a constant load, for example. Thus in this case, too, there is no contradiction but little support.

Support for the theory is provided by the unexpected kinks noted on some theoretical curves. Figure 7.6 shows some practical kinked curves. Once again an unusual feature predicted theoretically is confirmed practically. Bompass-Smith (see Reading list), from whose work these results have been taken, quotes a number of similar results, and is aware that they are a feature of some fatigue data. He correctly attributes the phenomenon to 'weak' items. Strangely, he is the only worker in this field to draw attention to this aspect of the practical curves. It is a little surprising that it has not been observed more often, but it could well be that

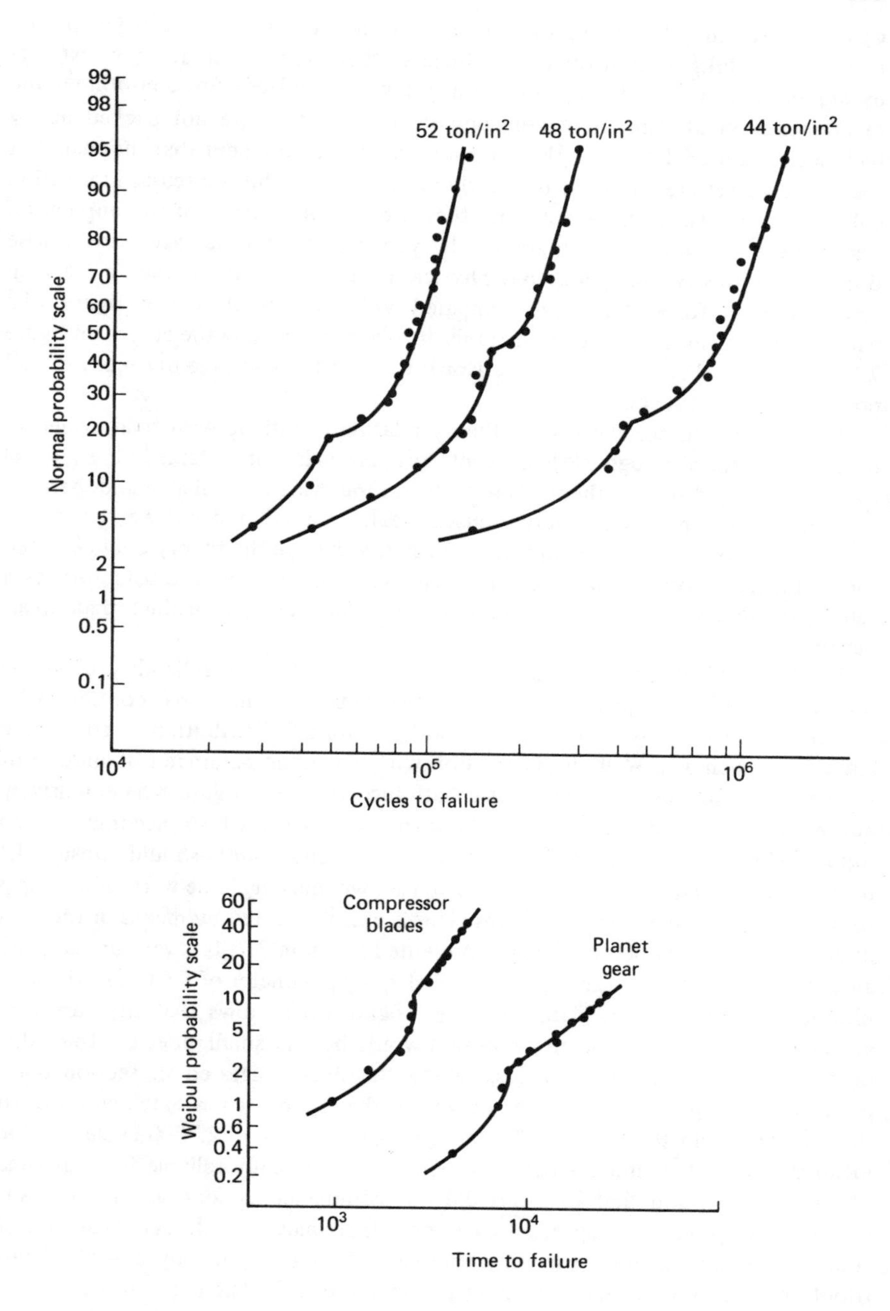


**Figure 7.6** 'Kinks' in fatigue cumulative curves (from Bompass-Smith – see Reading List)

with only a limited number of data points smooth curves have been plotted through the kinks. Indeed the author believes that the representation of test data by standard statistical distributions nearly always involves some smoothing, and recommends great care in making any deductions that are not backed up by technical evidence. For example, the reader may well comment that the shapes of the above curves are reminiscent of early life wear-out failures discussed in section 6.4.6. In some ways this is just what they are – a proportion of the population that have a low enough strength to always suffer fatigue damage – otherwise Bompass-Smith's 'weak' items. But physically we know that the failure mechanism is the same for all failures as compared with the data plotted on figure 6.12 where different modes are demonstrably involved. One thing the curves on figure 7.6 do show is that no standard statistical distribution can possibly represent all practical fatigue distributions.

To sum up the comparison of theory relating to fatigue with field evidence: it is noted that although 100 per cent correlation is not obtained, the general picture presented by the theory is reflected in the evidence available and, perhaps of more importance, some very unusual features predicted by the theory are found in practice. Much development is necessary before the theory could be used for quantitative predictions, but it would seem in every way satisfactory as a vehicle for understanding the general reliability characteristics of the fatigue wear-out process.

In chapter 5 it was envisaged that a period of intrinsic reliability could be followed by wear-out processes in which the failure rate increased continuously. The process was likely to be represented by a normal distribution or something close to it, such as a Weibull distribution with a shaping parameter of about 3 or so. This corresponds with the classic bath-tub curve (see figure 5.5) extensively quoted in all the literature. Thus abundant evidence has been accumulated to support this failure pattern. The reader who has any doubts should consult the many texts quoted in the Reading list. Even so, we must again be wary of attempts to force data into preconceived forms. This is best illustrated by means of the data quoted in ref. 35, which has been re-presented in figure 7.7. It is very understandable that a straight line (giving a Weibull shaping parameter of 5.6) was originally plotted through the raw data. However, figure 6.10 shows that the curvature arising from a non-zero locating constant would be very small indeed at this value of the shaping parameter. Analysis along the lines suggested in section 6.4.2 produces the superior straight line drawn on this figure. It is a much closer fit to the test data than the original. This analysis gives a value of 23 000 cycles for the locating constant. It implies that the item was intrinsically reliable for that time (23 000 cycles), and that wear-out did not commence as soon as the item was brought into service, as suggested by the original analysis – though wear did, of course. The subsequent wear-out pattern can be represented by a Weibull distribution, this time having a shaping parameter of 3.7. The characteristic life is 63 500 cycles. Strangely, this pattern would coincide with a picture drawn on classical statistical lines in which the item has a wear-out life which is normally

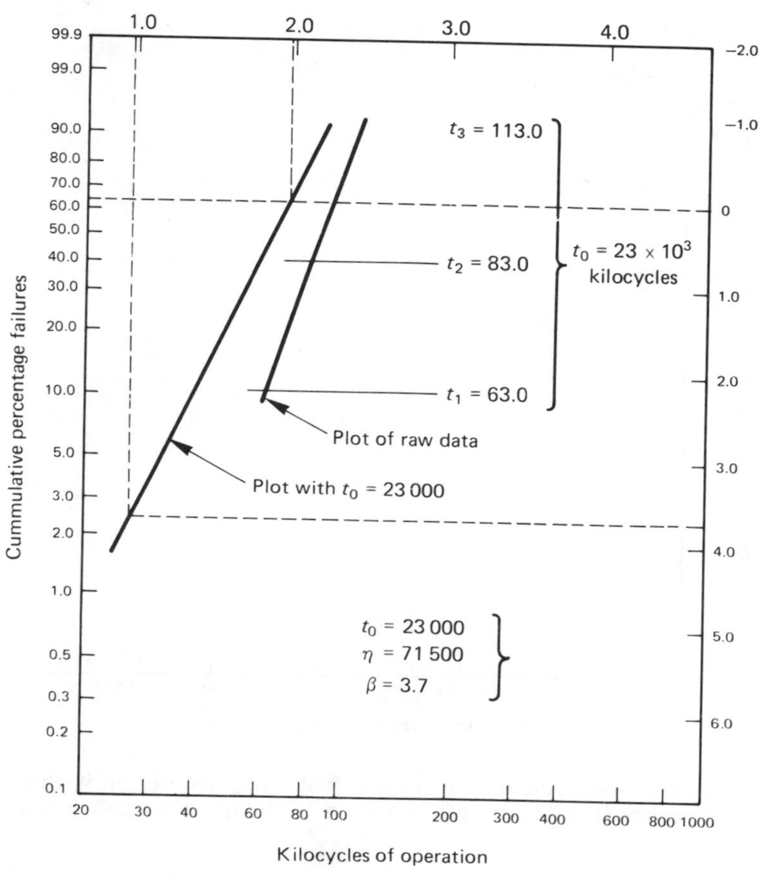


**Figure 7.7** An example of the erosive wear of some automotive switches

distributed about a mean value. Yet this failure pattern was not originally advanced, even though the lower $3\sigma$ point is markedly greater than zero, presumably because intrinsic reliability was discounted. The pattern fully conforms to the failure patterns we have envisaged. Thus not only is all the conventional bath-tub evidence available to confirm the practical existence of this type of wear, but supposedly less orthodox patterns can also fall into this category.

It is not possible to separate wear-out patterns in which the final failure-inducing load is responsible or not responsible for the prime wear process. The failure patterns are too similar and field evidence too inadequately reported. Very many of the field failure patterns have Weibull shaping parameters between 2 and 3, but whether this is of any significance or not is difficult to see without a specific investigation. As an example of a possible erosive wear in which the same load causes wear and failure, figure 7.8 shows the raw data plot of some aero engine fuel pump failures, taken from ref. 36. These pumps have a life of intrinsic

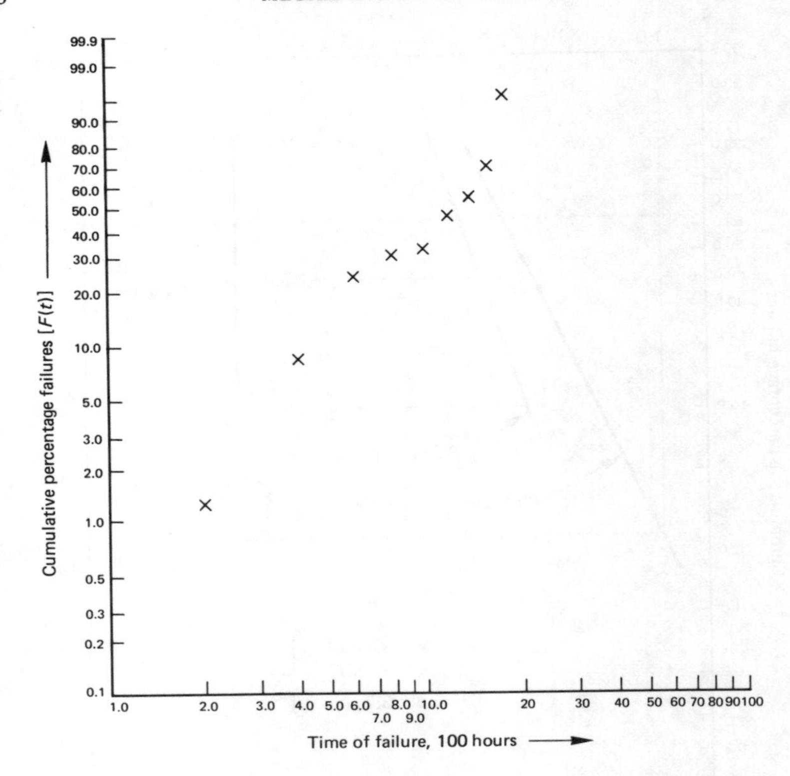


**Figure 7.8** Raw data plot of aero engine fuel pump failures on Weibull paper (data from ref. 36). $t_0$ = 82 hours, $\eta$ = 1283 hours, $\beta$ = 1.81

reliability of 83 hours, and the Weibull shaping parameter of the subsequent failure pattern is about 1.8. Erosive wear is a very likely failure mechanism in this case though ref. 36 gives no details. There are, of course, many drop-outs in the data. At least this data shows a wear-out pattern in which the failure rate is increasing, but less fast than for a normal distribution, with increasing time.

We conclude, therefore, that the general practical evidence for both the main wear mechanisms – fatigue and erosion – supports the ideas which have been advanced in the earlier chapters of this book, though it is not possible to account for detailed aspects of either process.

Turning to some less frequently occurring failure modes, the example used to illustrate the modified Weibull distribution is itself excellent evidence of the existence of early life failures of a limited proportion of the population, and also of the fact that mechanical early life failures can be in a wear-out mode with increasing failure rate, in addition to the well-established mechanism in which the failure rate decreases with age. Given only the raw failure data it would have been difficult to eliminate the possibility of fatigue failures applying to the whole population, the kink in the curve being not too untypical of some failures by that mechanism. However, the physical identification of the different mechanisms leaves no doubt, especially in this case when these particular failures are repeated at 300 and 600 hours, that is, at the maintenance interval. (There appears to be some difficulty in installing the gear-box, which, of course, only physical examination could reveal.) The other modes can be analysed by the techniques already discussed, when it can be established that both the gear train and the generator drive failed in fatigue modes, as may reasonably be expected. This example not only illustrates the failure mode under discussion, it also underlines once again the importance of separating failure mechanisms by physical examination before rushing into any statistical analysis.

There remains just one aspect of mechanical reliability behaviour, which is basic to the philosophy being developed in this book, for which there is on the one hand little evidence and on the other hand abundant evidence of a kind. It relates to the critical dependence of the failure rate on the environment, that is, the load distribution. It is extremely difficult to find published quantitative data on failure rates of the same component in different environments. One may note that on a sophisticated level such bodies as the Civil Aviation Authority employ only failure data that has been obtained 'in service', because that obtained from endurance testing can be in considerable error. Experience has shown that small environmental changes, such as temperature and vibration levels, can have a radical effect on component reliability. The author was told of some aircraft flaps which had been endurance tested for some 30 000 hours without failure; yet in service failures started to occur after only 15 000 hours. An environmental discrepancy in the test was later identified as the cause of the error. Again, after a study of pneumatic control systems in the nuclear power generation industry, which should be under close control and cover only a narrow range of loading roughness, T. R. Moss wrote "The widely distributed nature of mechanical equipment applications results in large variations in fault data for many normally identical items." From personal knowledge, the failure rate of commercial equipment used in a military environment often differs substantially from that experienced in the commercial world for which it was designed. Further justification may be obtained from the table below, taken from ref. 37. It lists the highest and lowest failure rates for some common engineering items, together with the ratio of highest to lowest rates. The data does not refer to identical items but does show that the range of failure rates can be very high indeed. The reader should remember that these are quoted as practical values that can be expected in normal service – they are not intended to represent good/bad design practice.

| *Item* | *Failure rate (failures per $10^6$ hours)* | | |
|---|---|---|---|
| | *Lowest* | *Highest* | *Ratio* |
| Mechanical components | $10^{-2}$ | $10^{1.4}$ | $10^{3.4}$ |
| Electro-mechanical components | $10^{-1}$ | $10^{1.6}$ | $10^{2.6}$ |
| Boilers and condensers | $10^{-0.8}$ | $10^{2.1}$ | $10^{2.9}$ |
| Turbines | $10^{-0.5}$ | $10^{2.6}$ | $10^{3.1}$ |
| Mechanical equipment | $10^{-0.5}$ | $10^{2.5}$ | $10^{3}$ |
| Pumps and circulators | $10^{-0.6}$ | $10^{3}$ | $10^{3.6}$ |
| Pneumatic equipment | $10^{-0.5}$ | $10^{3.7}$ | $10^{4.2}$ |

In addition to the fully reputable evidence just quoted, there is, of course, a lot of hearsay evidence which attributes widely differing failure rates to the same component. No one can have worked long in the mechanical reliability world without having run across it at some time. It may be true or it may not. However, it is not inconsistent with the theoretically predicted behaviour.

The field evidence presented has necessarily been restricted by space considerations, but it shows that a wide range of failure patterns does arise. Apart from Bompass-Smith (see Reading list and ref. 38), little attention appears to have been paid to the possibilities in the general run of literature, and virtually no attempt to correlate failure patterns with physical mechanisms.

*Exercise 7.3*

During the commercial operation of a fleet of 65 vehicles, failure of the propeller shaft universal joint was recorded at the following mileages: 2117, 5358, 7099 7750, 8265, 10 715, 11 172, 12 511, 13 876 and 20 584. No vehicle experienced more than one failure, and all vehicles have now completed more than 25 000 miles. You are asked to examine if this failure data agrees with the general concept of reliability being developed in this book.

*Exercise 7.4*

Figure 7.9 presents some failure data on ball bearings, taken from ref. 39, in which data for different levels of lubricant contamination are presented. It is stated in the summary to the reference that the evidence indicates that early failures are often initiated and caused by debris or an unsuitable surface topography. For comparison, figure 7.10 presents some different failure data on balls, taken from ref. 40. In that reference it is stated that all failures considered in the investigation appeared to be the result of classical sub-surface fatigue.

You are asked to examine all this evidence carefully, and to deduce how far it fits the theoretical picture being drawn in the previous chapters of this book. *Note*: Owing to the scale of the original drawings in the references quoted it is not possible to give numerical data and hence you will not be able to make accurate numerical deductions. However, you should be able to infer values to sufficient accuracy for the purpose of this exercise. All the information is plotted on Weibull scales.

*Exercise 7.5*

How far do you think it is possible to use equations (5.19) and (5.21) as a general representation of ball bearing reliability?

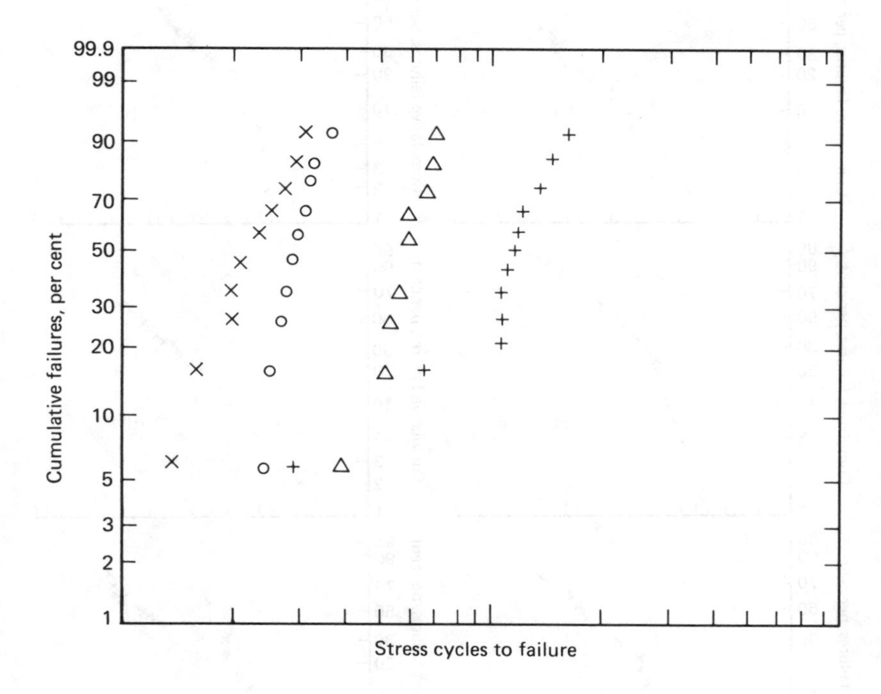


**Figure 7.9** Cumulative distributions of RHPN 1005 bearing failures for various levels of debris filtration plotted on Weibull scales (data from ref. 39). x 40 $\mu$m filtration, ⊙ 25 $\mu$m filtration, △ 8 $\mu$m filtration, + 3 $\mu$m filtration

*Exercise 7.6*

Figure 7.11 shows the failure distribution recorded on a type of Naval main propulsion Diesel engine. The data (from ref. 41) has been plotted on Weibull axes. In the author's experience such plots are not at all uncommon, though in many cases a straight line is drawn through the data and random failures are assumed. The authors of ref. 41 have, however, presented us with a definite curve which excludes random failures. In your opinion does the information conform or disagree with the thesis so far developed in this book?

How can field evidence be summarised? In the first place it must be recognised that there are no conclusive laboratory-type tests, and that true field evidence is

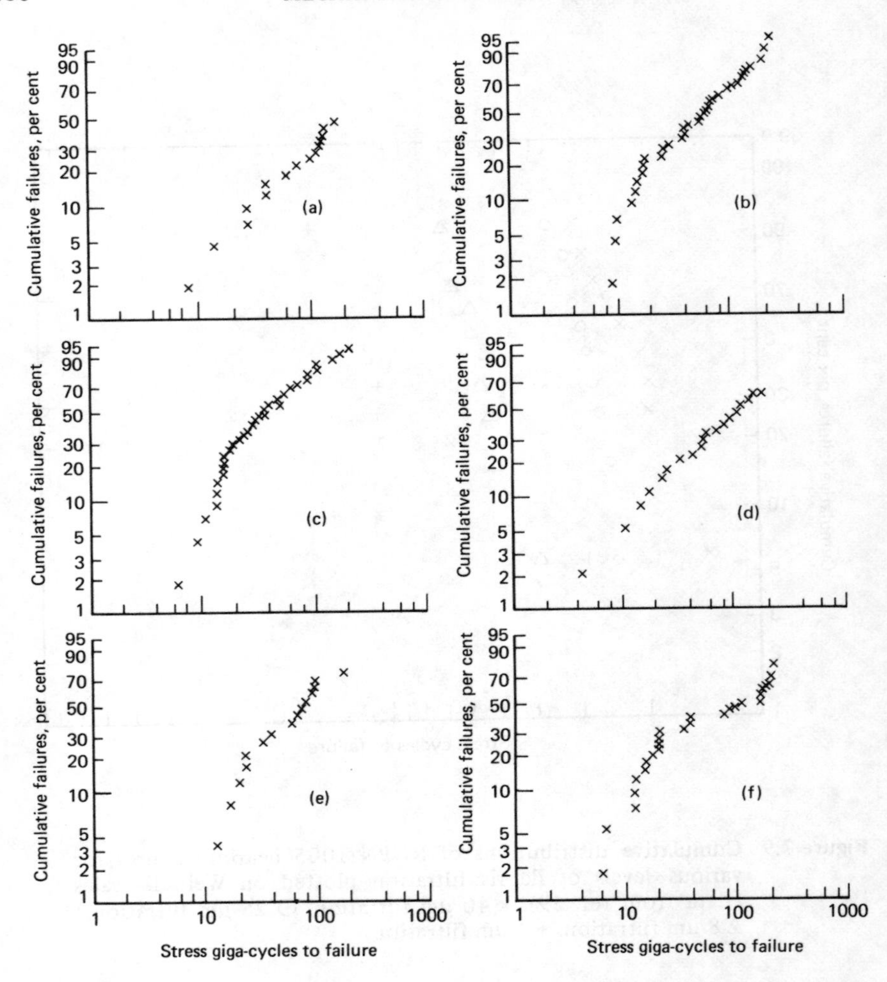


**Figure 7.10** Cumulative distribution of AISI 52100 ball failures with various lubricants, plotted on Weibull scales (data from ref. 40). (a) Paraffinic mineral oil. (b) Formulated tetraester oil. (c) Tetraester base. (d) Traction fluid 1. (e) Traction fluid 2. (f) Synthetic paraffinic oil

more circumstantial. Furthermore, much of the earlier data has to be treated as corrupt, because no attempt was made to separate failure modes, or alternatively establish that only one mode existed. Additionally, a great deal of early analysis unjustifiably forced the data into a constant failure rate regime. What remains strongly supports the philosophy developed in the preceding chapters of this book. In particular it has been shown that

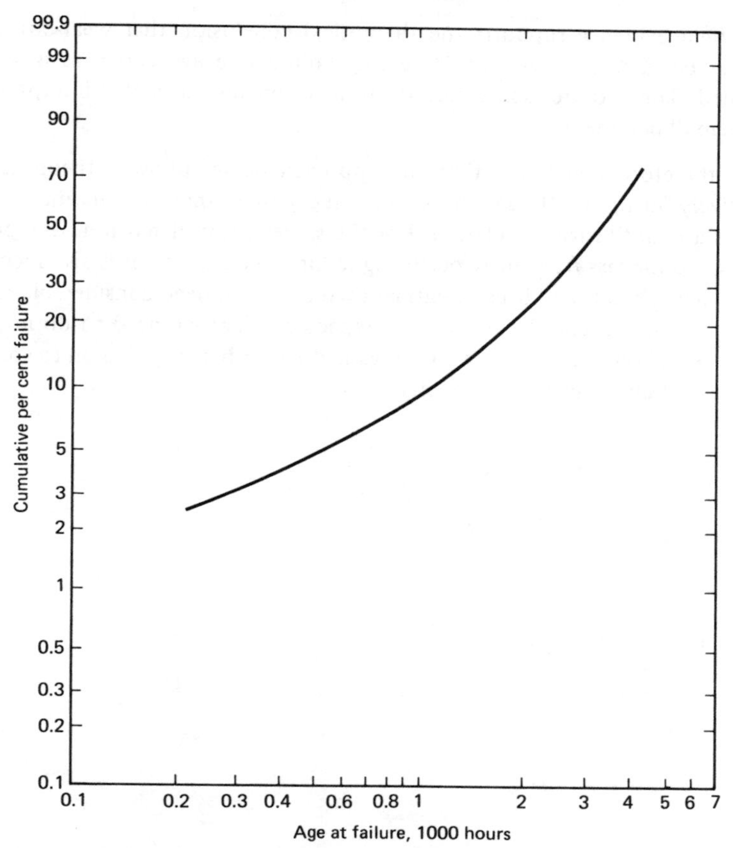


**Figure 7.11** Weibull plot of failures in Naval main propulsional diesel engines

1. Most components have a period of intrinsic reliability during which no failures arise. There are some exceptions to this general finding, showing that intrinsic reliability is not always achieved.
2. The subsequent failures in the limited cases examined occur by one or more of a number of well-established physical wear-out mechanisms.
3. The fatigue wear-out pattern experienced in the field confirms that expected on theoretical grounds, which differs from that usually postulated in the literature. Additionally, some peculiarities noted in the theoretical results have been confirmed by field evidence.
4. Erosive wear produces a different failure pattern whose existence has also been substantially confirmed by field evidence.
5. Early life failures which conform both to the type usually reported in the literature, and also to the case in which the early failures take place through a wear mechanism, have been isolated.

6. The field evidence supports the theoretical conclusion that wear-out cannot always be identified with an increasing failure rate-age curve, as is so often claimed. This requires some modification to generally accepted interpretations of Weibull parameters.

It is therefore concluded that the approach being followed theoretically is qualitatively justified, though it is sadly recognised that the absence of data prohibits a quantitative correlation. For the same reason it has not been possible to follow up the less frequently occurring failure mechanisms. It is also recognised that the theory is a broad brush treatment which would need considerable refining before quantitative correlation can be expected. What we need to do now is to 'take stock' of the knowledge we have gained so far before going on to study the influence of maintenance on reliability.

# 8

# An appraisal of component reliability

With the support of the field evidence quoted in the previous chapter we can summarise our whole findings on the reliability behaviour of non-maintained items or components as follows. Failure occurs when a load on any item exceeds its strength at the instant that load is applied. We have shown that this can arise even when there is no strength degradation with age, because of the distribution of individual loads and strengths about their means. Such failures are easily identified physically: they are of a stress-rupture nature in which no wear-dependent or time-dependent phenomenon is involved. Excluding, in some cases, a short early life regime in which the failure rate decreases with time, these failures are essentially characterised by a constant failure rate which applies over the major part of the life before being terminated by wear-out. They are commonly referred to as random failures. It has been shown theoretically, and substantiated to some extent by field evidence, that this failure rate is critically dependent on the loading. Failure rates may vary by several orders of magnitude in different loading environments. Because of this sensitivity, it is postulated that no rational approach is possible if the existence of random failures is allowed. Consequently it is a fundamental axiom of the philosophy we are following that all random failures must be eliminated by good design. The conditions to achieve this have been set out in section 4.5. Items which have zero random failures (pedantically and exactly – not 'zero', but so small that they can be regarded as an 'act of God' and are neglected) are said to be intrinsically reliable. There is considerable field evidence to show that a high proportion of existing mechanical components are intrinsically reliable. Whether this is achieved by accident or intent is difficult to say, but if achieved by intent it was most likely done so as a result of experience and empirical design. As a result we have found that there is a significant proportion of items which are not intrinsically reliable. So room for improvement still exists. It is recognised that for some loading conditions a non-intrinsically reliable design may be acceptable, though it may not be inappropriate to note at this point that if any allowance is made for wear, the design will most likely be initially intrinsically reliable, regardless of intent.

Since in the random failure phase, items can only be classified as either reliable or unreliable with any practical degree of confidence, it is preferable, and often of more meaning, to specify reliability as the life of intrinsic reliability ($B_0$) rather than as a probability of success. If it is contemplated that operation will be continued into the wear-out regime, that phase can be adequately and consistently defined by quoting the $B_5$ and/or $B_{10}$ ages, preferably both, or in some cases a higher $B$ life.

In this context, wear is defined as loss of a strength attribute with age. It is an inevitable feature of nearly all mechanical items. This loss of strength results in a loss of safety margin and eventually an increase in failure rate, even for an initially intrinsically reliable item. Thus the whole life failure pattern of an intrinsically reliable item will consist of a period of zero failures followed by a wear-out period for which the failure rate is finite. The various forms of wear-out have been fully discussed theoretically in chapter 5 and supporting field evidence given in chapter 7. What we should particularly note here is that some wear-out characteristics follow a substantially normal distribution which may be expected intuitively, while others follow a significantly different distribution which would not be expected intuitively, but which can be explained theoretically. However, a feature common to many distributions is that the characteristic life is almost always equal to, and very often considerably in excess of, the intrinsically reliable life. It would hardly appear economic, therefore, to regard the working life as that for intrinsic reliability. It is most likely that the extent to which we would wish to operate in the wear-out regime will differ substantially from one failure mechanism to another and from one application to another. Although we have been able to associate some classes of wear-out failure patterns with specific failure mechanisms, the correlation is to all intents and purposes a qualitative one. In no instance has it been found possible to evaluate ***a priori*** the parameters which define the pattern on a theoretical or even semi-empirical basis. The values can only be assessed on a wholly empirical basis ***a posteriori***, though experience of similar devices may enable some quantities to be intelligently anticipated (guessed!). This must be regarded as the most serious limitation in our development of the subject, and has significant practical consequences.

The treatment in the preceding chapters has assumed that a component is subject to one load and fails in one mode. This is hardly ever true, of course. Most components are subject to a variety of loads and mechanisms; all have to be taken into account and their effects aggregated. However, in most cases one load dominates the failure behaviour and only that load need be considered – we have referred to it as the failure-inducing load. If this is not so, the several contributing loads are most likely to give rise to different mechanisms. In these circumstances a physical assessment of each resulting mechanism and its statistical representation is called for, rather than any attempt to combine the effects statistically. Indeed this is an absolutely essential requirement. When loads combine to produce a failure mechanism, the combination is most often a physical one whose relation to the effects of the individual loads can be very complex, and statistical analysis should be confined to individual identifiable mechanisms. It has been shown over and over again that mixing failure modes can be totally misleading. This cannot be over-emphasised. It is to be hoped that statisticians and reporters of field data will keep this in mind in future. One important spin-off is that the statistical treatment we have adopted so far, which has been confined to single mechanisms, need not be extended. In our subsequent treatment we shall assume that a controlling failure/load mechanism has been identified.

Assuming intrinsic reliability can be achieved, and there appears to be no fundamental reason why it should not be, wear mechanisms control nearly all mechanical reliabilities. This is important, for its implications are far reaching. To assess mechanical reliability we must assess wear. To achieve mechanical reliability we must control wear. The control of wear is initiated at the design stage but is ultimately dependent on maintenance. Thus maintenance becomes a prime factor in controlling mechanical reliability, and hence it is to that subject that we now turn our attention. We do so, however, in the knowledge that we have built up a sound theoretical understanding – maybe not always quantifiable – of the reliability characteristics of non-maintained components, and this understanding is well supported by field evidence.

# 9 Maintenance

## 9.1 General introduction

Our study of reliability has so far completely ignored the all-important factor of maintenance. However, we have been able to build up a fairly detailed understanding of the behaviour of the non-maintained component and to substantiate this knowledge with field experience and evidence. Since the non-maintained item is the building brick of all more complex maintained equipment we are in a strong position to further our studies by introducing maintenance.

As a preliminary observation we may note that most maintenance involves the replacement of a component that has failed, or is about to fail, by a new one. Prior to such maintenance action we shall assume that the component is not modified in any way (other than by wear arising from normal usage), though it may have been inspected. We are therefore able to apply the theory already developed for a non-maintained component to any component before maintenance action, and again to the new component after it has been installed to replace the failed or worn-out one. We note that, of course, the age of the parent equipment and the age of the maintained (that is, replaced) component will no longer be the same. In addition there is a second form of maintenance in which only adjustments are made to the component – for example, to restore wear. The cleaning and adjustment of sparking plug points in the standard automotive engine is a good example. We shall initially assume that any such action restores the component to the 'as new' condition so that it may again be regarded, for theoretical purposes, as a replacement. The number of times such action may be taken will, of course, be limited. It should be noted that the above assumption is made purely for simplicity in assessing the broad effects of maintenance, and that there is no fundamental difficulty in using different reliability or failure-age patterns for components corresponding to each stage of adjustment. Assumptions must always, of course, represent actual circumstances. While this may sometimes present difficulties in particular cases, it need not impede our general investigation, for we can break down any equipment, however complex, into a number of non-maintained but replaceable/adjustable components which can be treated in the manner discussed in the earlier chapters of this book. We are then left solely with the problem of studying the consequences of replacement.

Although components may need replacement, the equipment never does: it goes on and on, even if in the course of time very little, if any, of the original may be left. We must all have heard of great grandad's axe which is still as good as new – it has just had three replacement heads and two replacement shafts! On the other hand the equipment may well become obsolete or uneconomic to run. Who would want to use great grandad's axe to cut down a tree when lightweight power chain saws are available? We cannot be concerned with such matters in this book. Our objective is to study the effect of maintenance on the reliability of the equipment, or alternatively to establish the level of maintenance required to achieve a given reliability; though we shall introduce some economic aspects in section 9.7. In some instances it is possible that the reliability level could be specified by the customer, in which case our objective is a clear one. Alternatively, reliability and maintainability may be regarded as ingredients of a more comprehensive quality sought by the operator. As discussed in chapter 2, this quality may be cost or it could be availability – possibly both. We shall be forced therefore to consider ultimately rather more than reliability. Maintenance and reliability often go hand-in-hand: they can be alternatives to the same objective.

Maintenance may be split into two general categories: scheduled maintenance (sometimes referred to as preventive maintenance) and unscheduled maintenance. Unscheduled maintenance may be further split into two sub-categories – repair maintenance and on condition maintenance – to give the three basic maintenance policies from which nearly all actual maintenance plans are constructed. The three policies may be defined as follows.

(1) Preventive or scheduled maintenance is carried out to keep an equipment in a satisfactory operational condition by providing systematic replacement of components before they are expected to fail – it may include some inspection activity.
(2) Repair maintenance is carried out on a non-scheduled basis to restore an item to a satisfactory condition by providing immediate correction of a failure after it has occurred.
(3) On condition maintenance is carried out before an item fails, but only when its condition, established by continuous monitoring, indicates that failure is imminent.

To a large extent the choice of policy, or mix of policies, is as we have indicated an economic one, but there are technical constraints within which the choice can be circumscribed. If an item is intrinsically reliable the choice is unrestricted: one can either wait until an item fails and then replace it, replace it on a scheduled basis which is set to retain intrinsic reliability, replace it on a scheduled basis which allows wear to take place to the extent that it is no longer intrinsically reliable, or alternatively replace it when a defined amount of wear has taken place. The choice is an economic one which we shall examine later. However, if the item is not intrinsically reliable so that random failures are present, then neither a scheduled nor an on condition maintenance is possible. It should be recognised that maintenance is basically a modification of the strength distribution. If the strength has deteriorated, as in figure 5.1b or 5.1c, then it can be restored to its original value, or something approaching that value, as in figure 5.1a: this is what maintenance is all about. But random failures occur with an invariant strength, so maintenance (that is, reversing a non-existent strength degradation) is physically not possible. Failures can only be rectified after they occur, that is, a repair maintenance policy is the only one possible, irrespective of the economics. The level of repair maintenance necessary to counter random failures is determined by, and via the reliability is very sensitive to, the loading environment once an item is in service. There is no way in which the reliability of a random failure process can be controlled by maintenance. The reliability, maintenance policy and level of maintenance is sealed once the design is sealed, and no further studies are necessary.

A cautionary note is necessary at this stage. The previous paragraph applies to the true random failure. Although such failures are associated with a constant failure rate, it would be a mistake, as a corollary, to identify a constant failure rate with random failures when dealing with complex maintained equipment. In a multi-component item the lives of intrinsic reliability of components will most likely be distributed: they will certainly be different. As failures occur the replacement procedures lead to an even wider distribution of component age, so that in the absence of any scheduled maintenance, failures of the equipment will arise randomly. Thus the occurrence of failures, which is often interpreted as the failure rate, will be constant. This could happen even if all the failures of the components themselves arise in a wear-out mode. This is illustrated on figure 9.1. It

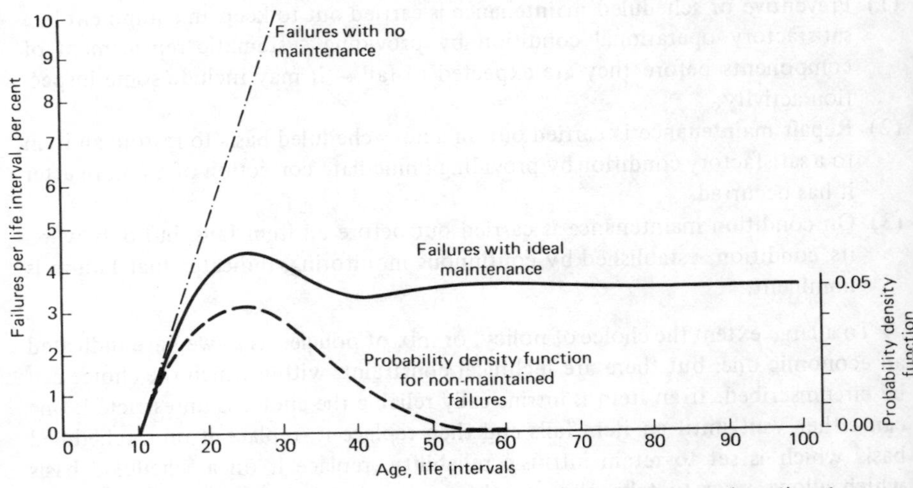


**Figure 9.1** Development of a constant failure incidence (rate!) when failures in a wear-out mode are repaired and the item returned to service

shows the probability density of the failure pattern and the ever-increasing failure rate of an item with no maintenance. The curve of failures per unit time interval with ideal repair maintenance is also shown. After the initial transient, the failure rate settles down to a constant value. Such a failure pattern is amenable to scheduled or on condition maintenance: the strength has deteriorated by wear-out and can be restored. To be absolutely clear we shall use the description 'pseudo-random' to refer to this type of failure regime, giving rise to a constant occurrence of failures (failure rate!) in wear-out modes. We shall restrict the description 'random' to the adventitious occurrence of failures arising from invariantly distributed loads and strengths.

The basic tenet of this book ensures all components are designed to be intrinsically reliable. Both scheduled and on condition maintenance are then possible. At one extreme all failures could be eliminated by scheduled replacement before the component's life of intrinsic reliability had expired. In some cases this may be done. No one, for example, would consider deferring the change of their automotive engine oil until it had become intrinsically unreliable. The cost of replacement oil and filter is so small compared with the potential damage arising from a lubrication failure that it is obviously economic to operate solely within the life of intrinsic reliability of the oil and filter. There is no need to do sums in this case: the economic optimum course of action is obvious. In other cases it is not at all obvious. In such circumstances we still demand intrinsic reliability of our components, so that we are not presented with an uncontrollable failure rate and repair maintenance bill, but we may continue operating these intrinsically reliable components beyond their life of intrinsic reliability; so some failures would occur. How many occur depends on the maintenance interval, that is, the replacement schedule. It follows that with intrinsically reliable components the reliability of the equipment depends very much on the maintenance. Maintenance is therefore a factor of prime importance, and the determination of the level of maintenance to optimise costs or achieve a specified level of equipment reliability or availability is a subject to which we must now pay considerable attention.

## *9.2 The mathematical model

The mathematical model we shall use to calculate failures with maintenance is essentially that proposed by Hasting (ref. 42). It will be assumed that any failure mode can be represented by a Weibull distribution. We then proceed by calculating the failures which arise in a small interval of time, $h$. In chapter 2 it was shown that the proportion of survivors at time $t$ from a large initial population that fail during the next interval, $dt$, is $\lambda(t)\,dt$. Replacing $dt$ by $h$ the numbers failing in the small interval of time, $h$, will be $\lambda(t)h$ where for a Weibull distribution $\lambda(t)$ is given by

$$\lambda(t) = \frac{\beta}{\eta}\left(\frac{t - t_0}{\eta}\right)^{\beta-1} \tag{9.1}$$

Now let the component age be given by $ih$ and the parent equipment age be given by $jh$, where $i$ and $j$ are integers, and additionally, in order that all times can be measured in terms of $h$, let

$$t_0 = ah \tag{9.2}$$

where $a$ is also an integer. Now if $p_i$ is the probability that a component will fail in the $i$th interval of component life, assuming it has survived to the start of the interval, then we can write

$$\begin{aligned} p_i &= \lambda(t) \times h \\ &= \frac{\beta h}{\eta}\left(\frac{t - t_0}{\eta}\right)^{\beta - 1} \\ &= \frac{\beta h}{\eta}\left(\frac{(i-a)\,h - \frac{1}{2}h}{\eta}\right)^{\beta-1} \quad \text{for } i > a \end{aligned} \tag{9.3}$$

where $(i - a)h - h/2$ is approximately equal to the mean value of $t - t_0$ in the $i$th interval of component life. If the component can fail in a number of completely independent modes for which the individual probabilities in modes I, II, III etc. are $p_i(\text{I})$, $p_i(\text{II})$, $p_i(\text{III})$ etc., then

$$p_i = p_i(\text{I}) + p_i(\text{II}) + \ldots \tag{9.4}$$

or

$$p_i = \frac{\beta_1 h}{\eta_1}\left\{\frac{(i - a_1 h) - h/2}{\eta_1}\right\}^{\beta_1 - 1} + \frac{\beta_2 h}{\eta_2}\left\{\frac{(i - a_2 h) - h/2}{\eta_2}\right\}^{\beta_2 - 1} + \ldots \tag{9.5}$$

where each term has the value zero if $i$ is not greater than the value of $a$ for the particular mode.

Given $p_i$ from equation (9.5) above, our objective is now to calculate $F_j$, the probability that component failure will occur in the $j$th interval of the equipment life. Hence let us consider the $j$th interval. In general a component may be in any interval of its own life, because failures and replacements may have occurred at any time before the $j$th interval. We must therefore consider all possible component ages, that is all values of $i$, from $i = 1$ up to some maximum permisible value which can be either the total life of the parent equipment, or the time at which a scheduled replacement is carried out, which would be less. Let the maximum number of component time intervals be $Y$, that is, $Y$ is the scheduled maintenance interval expressed in terms of $h$. Now to account for differences in component and parent equipment ages, let the probability that the component age is $i$ time intervals when the parent equipment age is $j$ time intervals be $q(i, j)$. Then the probability of a component in its $i$th interval failing when the parent equipment is in its $j$th interval is

$$q(i, j) \times p_i \tag{9.6}$$

Summing over the whole component life (from $i = 1$ to $i = Y$), the total probability of component failure in the $j$th interval of equipment life is given by

$$F_j = \sum_{i=1}^{i=Y} q(i, j) \times p_i \tag{9.7}$$

Now consider the $(j + 1)$th interval of parent equipment life. All the components which failed in the $j$th interval will be replaced by new ones, so that

$$q(1, j + 1) = F_j \tag{9.8}$$

All components which were formerly in the $i$th interval of their life and have survived proceed to the $(i + 1)$th interval. Now the probability that a component is in its $i$th interval and did not fail during the $j$th interval of parent equipment life is given from (9.6) as

$$q(i, j) \times (1 - p_i) \tag{9.9}$$

Hence

$$q((i + 1), (j + 1)) = q(i, j) \times (1 - p_i) \tag{9.10}$$

If one starts with all new components in a new equipment

$$q(1, 1) = 1 \tag{9.11}$$

otherwise one has to set up equation (9.11) appropriately. Equations (9.7), (9.8), (9.10) and (9.11) enable the values of $F_j$ to be calculated from $j = 1$ upwards using the value of $p_i$ from equation (9.5). The calculation proceeds up to some predetermined value of $j$ which represents the total life of the equipment in time intervals.

The above calculation enables us to evaluate the failures to be expected in any small time interval, $h$, when all components are replaced at a given age before failure (the scheduled replacement life) or are replaced on failure before that life has expired. The calculation is tedious and time-consuming by hand, but is readily carried out on a microcomputer. Hastings gives the listing of a program in ref. 42 which is readily adaptable to all micros. The only restriction in the published program is that the input failure patterns have to be expressed as Weibull distributions. The output gives the unscheduled failures in each time interval and the total scheduled replacements. As written, the program applies to individual replacement at the scheduled age, but can easily accommodate block replacement schemes.

In following Hastings' treatment we have adopted a numerical solution. His solution has in fact been based on standard theory (see ref. 42 for details and further references). However, virtually all standard published replacement theory deals with the asymptotic state, because it is impossible to obtain an analytical solution for the transient stage. This is considered to be a wholly unrealistic restriction for mechanical systems. Although equipment life can theoretically be infinite, it is shown in section 9.7 that there is an economic limit to the life span of most equipment, quite apart from technological obsolescence. The asymptotic state is rarely reached in practice, and solutions of this nature are of little help. Analytical solutions of the transient state would require numerical integration. The completely numerical approach developed by Hastings was preferred. It is simple and straightforward, and is indeed a step by step simulation of the actual process. The accuracy of the results are in every way commensurate with the accuracy of the input data. Of prime importance, it enables the transients in the early stages to be readily evaluated.

## 9.3 Reliability with scheduled and unscheduled maintenance

The non-mathematical reader who has skipped section 9.2 must accept that it presents a step by step calculation from which it is possible to calculate the expected failures in any time interval when any unexpectedly failed components are replaced by new components, and all components are replaced at any specified age if they have not already failed. To do this it is necessary to know the non-maintained failure pattern of the components and the scheduled time at which they will be replaced if not previously failed. Given this ability we are in a position to examine quantitatively the effect of maintenance on reliability. Our objective is to reveal underlying tendencies rather than to evaluate particular cases, and this is best done by examining uncomplicated representative examples. We shall therefore start by looking at a one component equipment and the simple ideal replacement of this otherwise non-maintained component on failure. Components are assumed to have the same failure–age characteristic, be they originals or replacements. Furthermore, for simplicity we shall assume that each component fails in one mode only, but we shall consider components which have different failure-age characteristics. Each will be represented by a Weibull distribution. In the first category of representative failure–age characteristics for the components we must cover all the failure mechanisms, which can be identified with the shaping parameter $\beta$ in the Weibull distribution. Thus fatigue and some vibrational effects may be represented by $\beta$ values of 0.5, 0.75 and 1.0, where the lowest value represents low cycle fatigue (LCF), and higher values cover the fatigue pattern more generally encountered. We must also represent the form of wear in which the strength decreases continually, such as corrosive/erosive wear, for which we can expect a much higher value of $\beta$. We shall use values of $\beta$ equal to 2 and 3.44 for this purpose. The latter provides normally distributed wear while the former represents a skewed wear distribution with a longer tail.

In the second category of representative characteristics we must cover the age spread over which we can expect failures, that is, the standard deviation of the failure pattern. In the case of the Weibull distribution this is measured by the characteristic life. It has to be related to the life of intrinsic reliability and to the planned life of the component. We first identify characteristic lives ($\eta$) which are roughly equal to the life of intrinsic reliability, and those which are considerably greater. We shall use values of $(\eta/t_0)$ equal to 1 and 10. In each case we shall take the planned or rated life of the equipment as datum. Then if $\eta/t_0$ equals unity a convenient value of $\eta$ and $t_0$ is 40 per cent of the planned life. It follows from the definition of characteristic life that 63.2 per cent of the components will have failed at 80 per cent of the rated life, whatever the failure mechanism, that is, whatever the value of $\beta$. By the time the rated life is reached, most of the first generation components may have been expected to fail. In order to represent the case where the life of intrinsic reliability is considerably less than the planned life of the equipment, so that all subsequent generation components can also be expected to fail, we shall investigate the case when $t_0$ is equal to 10 per cent of the rated life, the characteristic life having the same value of 40 per cent, so that $(\eta/t_0)$ equals 4. If the characteristic life is many times the life of intrinsic reliability – we have chosen 10 – it follows that the life of intrinsic reliability must be a small proportion of the planned life, otherwise a good deal of useful component life will be thrown away. We shall therefore restrict our investigation to the shorter of the two lives of intrinsic reliability studied in the earlier case, that is, $t_0$ = 10 per cent of the rated life.

The Weibull distributions representing the basic non-maintained failure patterns can therefore be summarised.

| *First category characteristics* | | | | *Second category characteristics* |
|---|---|---|---|---|
| Fatigue | $t_0 = 40$ | $\eta = 40$ | $\beta = 0.5$ | Good life of intrinsic reliability with Medium dispersion |
| | $t_0 = 40$ | $\eta = 40$ | $\beta = 0.75$ | |
| | $t_0 = 40$ | $\eta = 40$ | $\beta = 1.0$ | |
| Erosion, etc. | $t_0 = 40$ | $\eta = 40$ | $\beta = 2.0$ | |
| | $t_0 = 40$ | $\eta = 40$ | $\beta = 3.44$ | |
| Fatigue | $t_0 = 10$ | $\eta = 40$ | $\beta = 0.5$ | Short life of intrinsic reliability with Medium dispersion |
| | $t_0 = 10$ | $\eta = 40$ | $\beta = 0.75$ | |
| | $t_0 = 10$ | $\eta = 40$ | $\beta = 1.0$ | |
| Erosion, etc. | $t_0 = 10$ | $\eta = 40$ | $\beta = 2.0$ | |
| | $t_0 = 10$ | $\eta = 40$ | $\beta = 3.44$ | |
| Fatigue | $t_0 = 10$ | $\eta = 100$ | $\beta = 0.5$ | Short life of intrinsic reliability with Wide dispersion |
| | $t_0 = 10$ | $\eta = 100$ | $\beta = 0.75$ | |
| | $t_0 = 10$ | $\eta = 100$ | $\beta = 1.0$ | |
| Erosion, etc. | $t_0 = 10$ | $\eta = 100$ | $\beta = 2.0$ | |
| | $t_0 = 10$ | $\eta = 100$ | $\beta = 3.44$ | |

($t_0$ and $\eta$ are expressed as a percentage of the planned life)

The probability density, the failure rate and the cumulative failures for the first and last of these groups are given in figures 9.2 and 9.3 at (a), (b) and (c) respectively. The corresponding curves for the second group are, of course, identical with those of the first, translated along the time axis to commence at $t_0$ = 10.

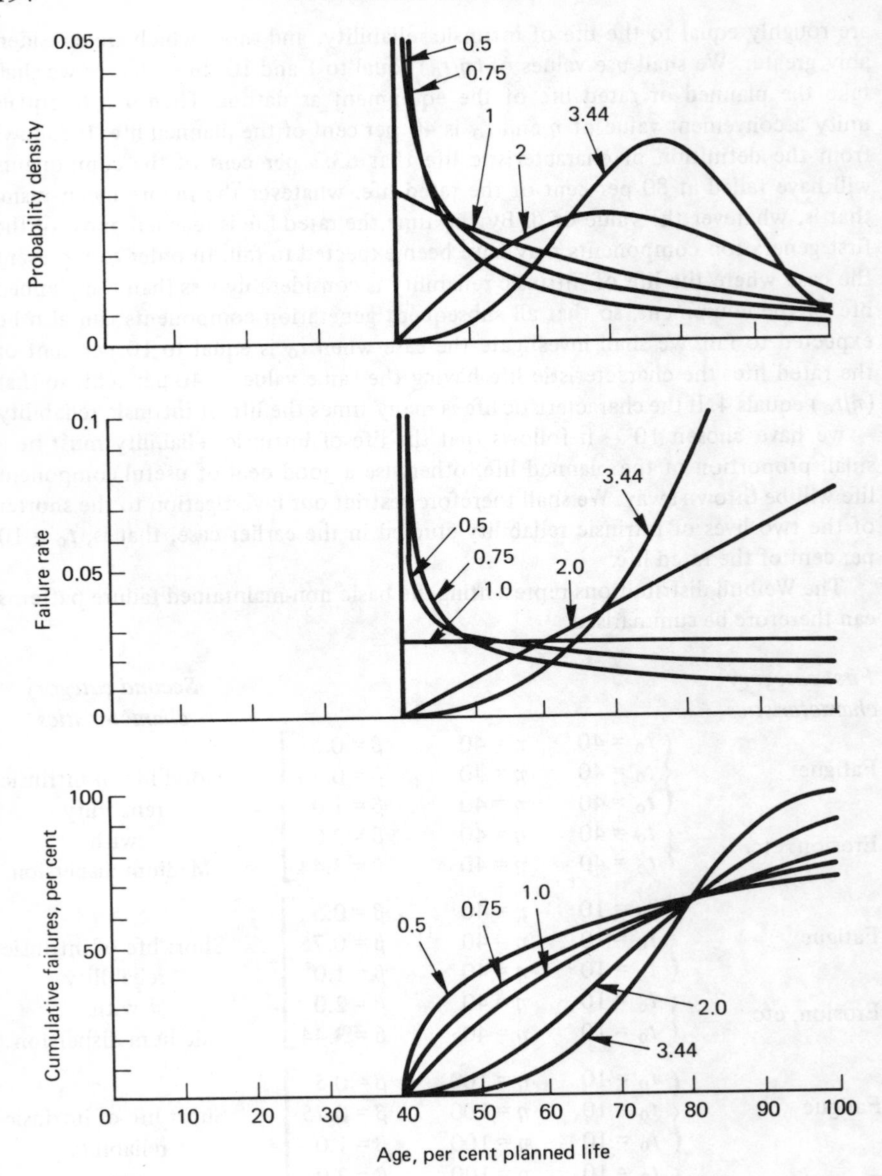


**Figure 9.2** Distribution of probability density, failure rate and cumulative failures of a non-maintained component having the following characteristics: (a) life of intrinsic reliability = $B_0$ = $t_0$ = 40 per cent planned life; (b) characteristic life = $\eta$ = 40 per cent planned life; (c) wear-out pattern defined by Weibull shaping parameter ($\beta$) marked on curves

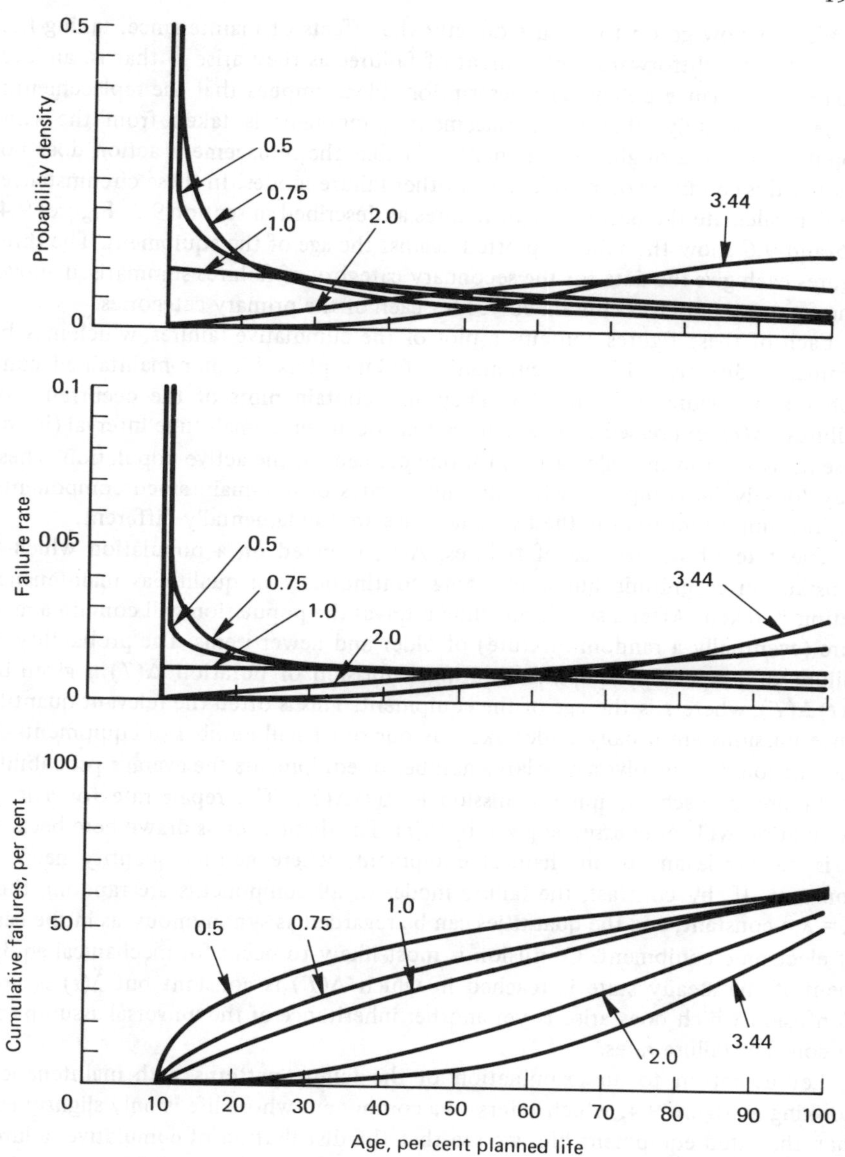


**Figure 9.3** Distribution of probability density, failure rate and cumulative failures of a non-maintained component having the following characteristics: (a) life of intrinsic reliability = $B_0 = t_0$ = 10 per cent planned life; (b) characteristic life = $\eta$ = 100 per cent planned life; (c) wear-out pattern defined by Weibull shaping parameter ($\beta$) marked on curves

We can now go on to discuss directly the effects of maintenance, taking first of all the straightforward replacement of failures as they arise – that is, an ideal repair maintenance policy. The description 'ideal' implies that the replacement is made immediately, that the replacement component is taken from the same population as the original component, and that the replacement action does not destroy this condition or introduce any other failure modes. In these circumstances we can calculate the occurrence of failures as described in section 9.2. Figures 9.4, 9.5 and 9.6 show the failures plotted against the age of the equipment. The three figures each give the data for the secondary category of failures summarised above. The figures themselves contain curves for each of the primary categories.

Each of these figures contains a plot of the cumulative failures, which may be compared directly with the cumulative failure plots for non-maintained components on figures 9.2 and 9.3. They also contain plots of the occurrence of failures, $\Lambda(t)$, expressed as the failures that occur in a small time interval (in this case $h$, as used in the calculation) for one per cent of the active population. These may loosely be compared with the failure rates of non-maintained components, but it is important to note the two quantities are fundamentally different.

The rate of occurrence of failures, $\Lambda(t)$, is based on a population which is constant in magnitude but is changing continuously in quality as maintenance action is taken. After a significant time interval the population will contain a mixture (eventually a random mixture) of older and newer items. The probability of failure of a specific equipment during a mission of duration $\Delta(T)$ is given by $\lambda(t)\Delta(T)$, where $t$ is the age of the equipment. This is often the relevant quantity since missions are usually undertaken by one or a small number of equipments. If the mission does involve a very large number of equipments the average probability of failure of each equipment mission is $\Lambda(t)\Delta(T)$. The repair rate for a large population will, of course, be given by $\Lambda(t)$. The distinction is drawn here because it is most relevant to mechanical equipment, where neither quantity need be constant. If, by contrast, the failure modes of all components are random, then $\Lambda = \lambda =$ constant, and the quantities can be regarded as synonymous, as in the case of electronic equipment. Confusion is most likely to occur for mechanical equipment if the steady state is reached in which $\Lambda(T)$ is constant but $\lambda(t)$ is not. Confusion which does arise is yet another inheritance of the universal assumption of constant failure rates.

Let us return to an examination of the failure patterns with maintenance. Looking at figure 9.4, which refers to a component whose life is only slightly less than the rated equipment life, we see that the distribution of cumulative failures bears superficial resemblance to the cumulative failure pattern of the non-maintained component shown on figure 9.2. However, if the occurrence of failures is compared with the failure rate for the non-maintained component, we can see some significant differences. Thus at higher values of $\beta$, representative of erosive and similar wear, the failure rate of the non-maintained component increases continuously while that of the maintained component increases at first, but then decreases as maintenance takes effect and the population includes an increasing

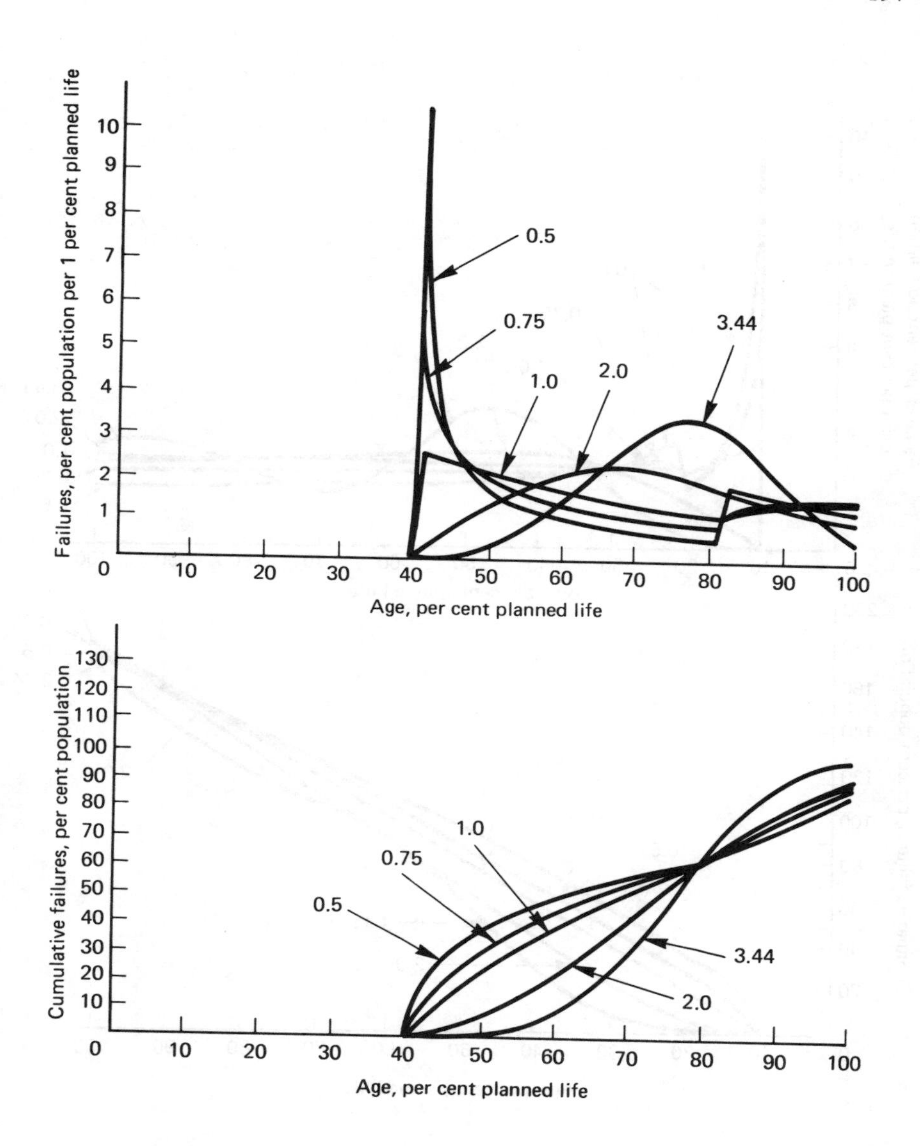


**Figure 9.4** Failures in each 1 per cent planned life interval and cumulative failures with ideal repair maintenance for a component having the following characteristics:

(a) life of intrinsic reliability = $B_0$ = $t_0$ = 40 per cent planned life;
(b) characteristic life = $\eta$ = 40 per cent planned life;
(c) non-maintained wear-out pattern defined by Weibull shaping parameter ($\beta$) marked on curves

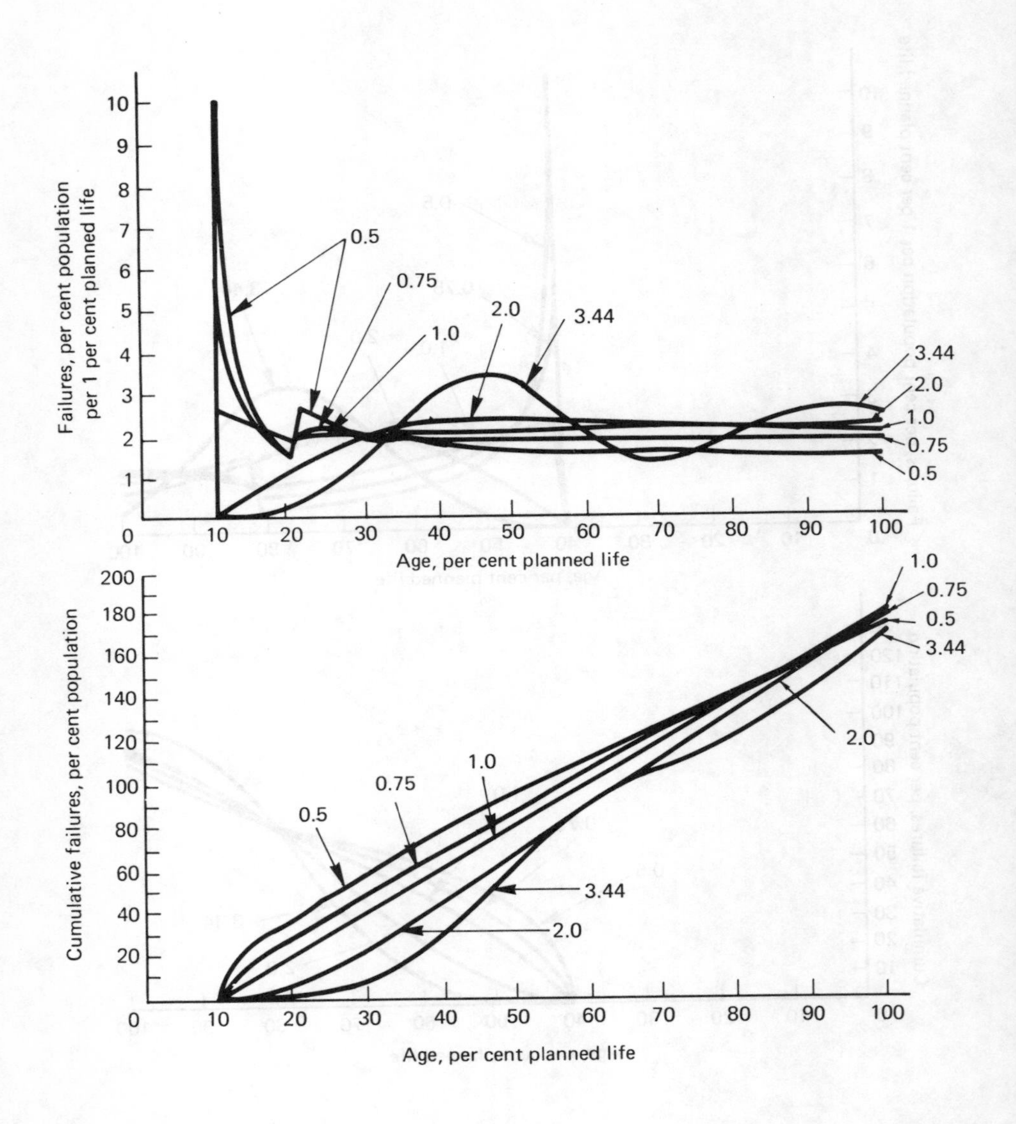


**Figure 9.5** Failures in each 1 per cent planned life interval and cumulative failures with ideal repair maintenance for a component having the following characteristics:
(a) life of intrinsic reliability = $B_0$ = $t_0$ = 10 per cent planned life;
(b) characteristic life = $\eta$ = 40 per cent planned life;
(c) non-maintained wear-out pattern defined by Weibull shaping parameter ($\beta$) marked on curves

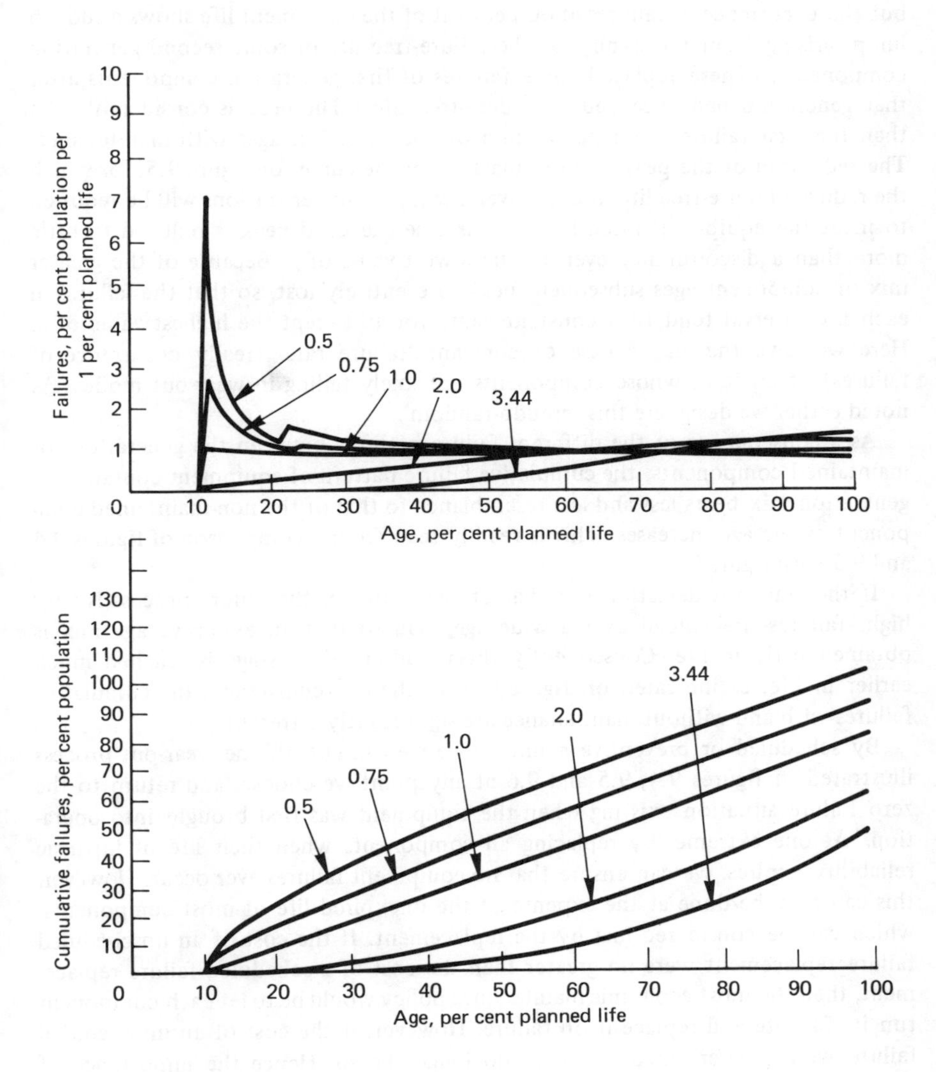


**Figure 9.6** Failures in each 1 per cent planned life interval and cumulative failures with ideal repair maintenance for a component having the following characteristics:
(a) life of intrinsic reliability = $B_0$ = $t_0$ = 10 per cent planned life;
(b) characteristic life = $\eta$ = 100 per cent planned life;
(c) non-maintained wear-out pattern defined by Weibull shaping parameter ($\beta$) marked on curves

proportion of second generation or newer components. At lower values of $\beta$, representative of fatigue or similar wear, the curves are initially of the same shape, but the occurrence of failures at 80 per cent of the equipment life shows a sudden jump, arising from the expiry of the failure-free life of some second generation components. (These replaced some failures of first generation components after that generation had exceeded its failure-free life.) The peak is considerably less than the first failure grouping because of the spread of ages with maintenance. The reduction of the peak is more marked for the curves on figure 9.5, for which the reduced failure-free life implies several component generations will be required to meet the equipment rated life. In this case the third peak is reduced to little more than a discontinuity even at the lowest value of $\beta$. Because of the greater mix of component ages subsequent peaks are entirely lost, so that the failures in each life interval tend to a constant value for all except the highest value of $\beta$. Here we have the classic case of constant 'failure rate' (really occurrence of failures) of an item whose components are solely failing in wear-out modes. As noted earlier we designate this 'pseudo-random'.

As a consequence of the different failure patterns between the generations of maintained components, the cumulative failure pattern of equipment containing a generation mix bears less and less resemblance to that of the non-maintained component as the age increases. This is readily seen from a comparison of figures 9.4 and 9.5 with figure 9.2.

If the standard deviation (or characteristic life) of the failure process is very high, failures are spread over a wide age span so that an extensive age mix is obtained early in life. Consequently the pseudo-random stage is reached much earlier in life, as indicated on figure 9.6. In these circumstances the cumulative failures with and without maintenance are significantly different.

By scheduled or preventive maintenance we can cut off the wear-out process illustrated in figures 9.4, 9.5 and 9.6 at any point we choose, and return to the zero failure situation existing when the equipment was first brought into operation. At one extreme, by replacing all components when their life of intrinsic reliability expires, we can ensure that no equipment failures ever occur. However, this can only be done at the expense of the unexpired life of most components, which can be considered lost by the replacement. If the cost of an unscheduled failure replacement were no greater than the cost of a scheduled failure replacement, then the most economic maintenance policy would be to let each component run its full life and replace it on failure. However, if the cost of an unscheduled failure were greater, then this may no longer be so. Hence the importance of scheduled maintenance; for in general we should expect the cost of an unscheduled replacement to be higher than that of a scheduled one. This arises from any secondary damage that may be done, from any loss of revenue that may be incurred while the unscheduled maintenance is being carried out, and from any additional expenditure involved in the failure. To assess the choice quantitatively

Let $N_s$ be the number of scheduled replacements in the life span or other accounting period

$N_u$ be the number of unscheduled replacements in the life span or other accounting period
$C_s$ the cost of a scheduled replacement
$C_u$ the cost of an unscheduled replacement.

$$\text{Then total cost of maintenance} = N_s C_s + N_u C_u \tag{9.12}$$

If $N_i$ be the number of replacements required to achieve continual intrinsic reliability, that is, to achieve zero failures

then we can express the cost as

$$\frac{\text{total cost of maintenance}}{\text{cost of retaining intrinsic reliability}} = \frac{N_s C_s + N_u C_u}{N_i C_s}$$

$$= \frac{N_s + K N_u}{N_i} \tag{9.13}$$

where

$$K = \frac{\text{cost of unscheduled replacement}}{\text{cost of scheduled replacement}} \tag{9.14}$$

The retention of intrinsic reliability throughout the whole planned life is taken as an intuitive council of perfection against which to compare other maintenance costs. To make this comparison the number of scheduled and unscheduled replacements which would be required for each of the failure distributions listed earlier can be calculated for any scheduled maintenance interval, and from that data the total cost of maintenance expressed as a proportion of the cost required to retain intrinsic reliability can be evaluated by means of equation (9.13). This cost ratio has been plotted against the scheduled maintenance interval for various values of $K$ on figures 9.7, 9.8 and 9.9. If the ratio (for any given $K$) is always greater than unity then a scheduled maintenance policy to retain intrinsic reliability, that is, to replace every component on the expiry of its life of intrinsic reliability, is the right one, since it is cheaper and has clear practical advantages. Values greater than unity are therefore irrelevant, and the curves have been shown dotted where this applies. Values of the ratio (for any given $K$) less than unity indicate that scheduled maintenance schemes allowing for some wear-out are more economic. The age at which the minimum occurs gives the optimum scheduled maintenance interval. If no minimum occurs when the ratio is less than unity then a repair maintenance policy (equivalent theoretically to scheduled maintenance with an interval equal to the planned life) is the cheapest.

It is evident on first glance at each of the above figures that substantial differences exist between the different failure mechanisms represented by $\beta$ values of 0.5, 1.0 and 3.44. The second feature to strike us is that the value of $K$ has a very strong influence in defining the optimum. In this respect the curves confirm the common-sense approach in that for all the cases examined a policy retaining intrinsic reliability is the right one if $K$ is high enough. As $K$ is reduced, other

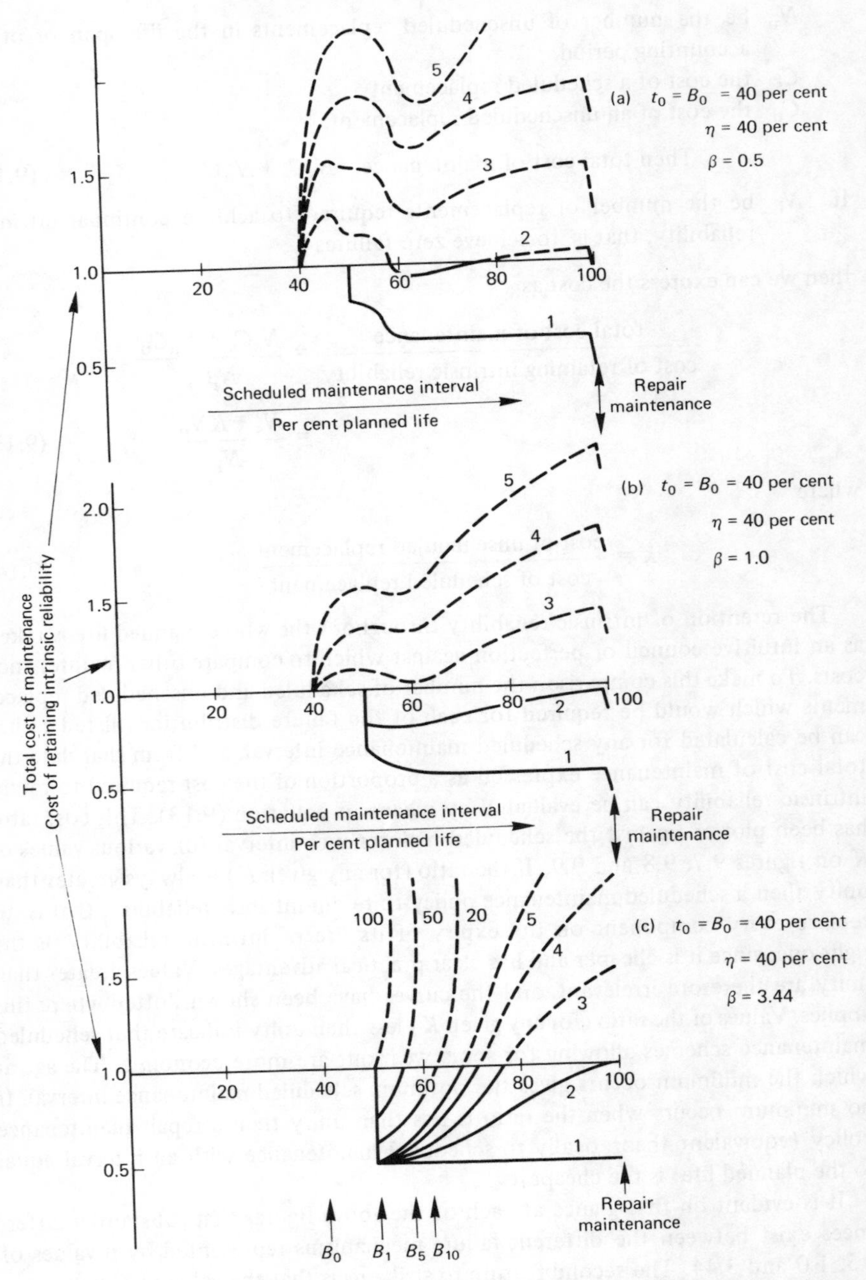


**Figure 9.7** Costs of scheduled and unscheduled maintenance plans for representative wear-out patterns. Values of cost ratio, $K$, marked on curves

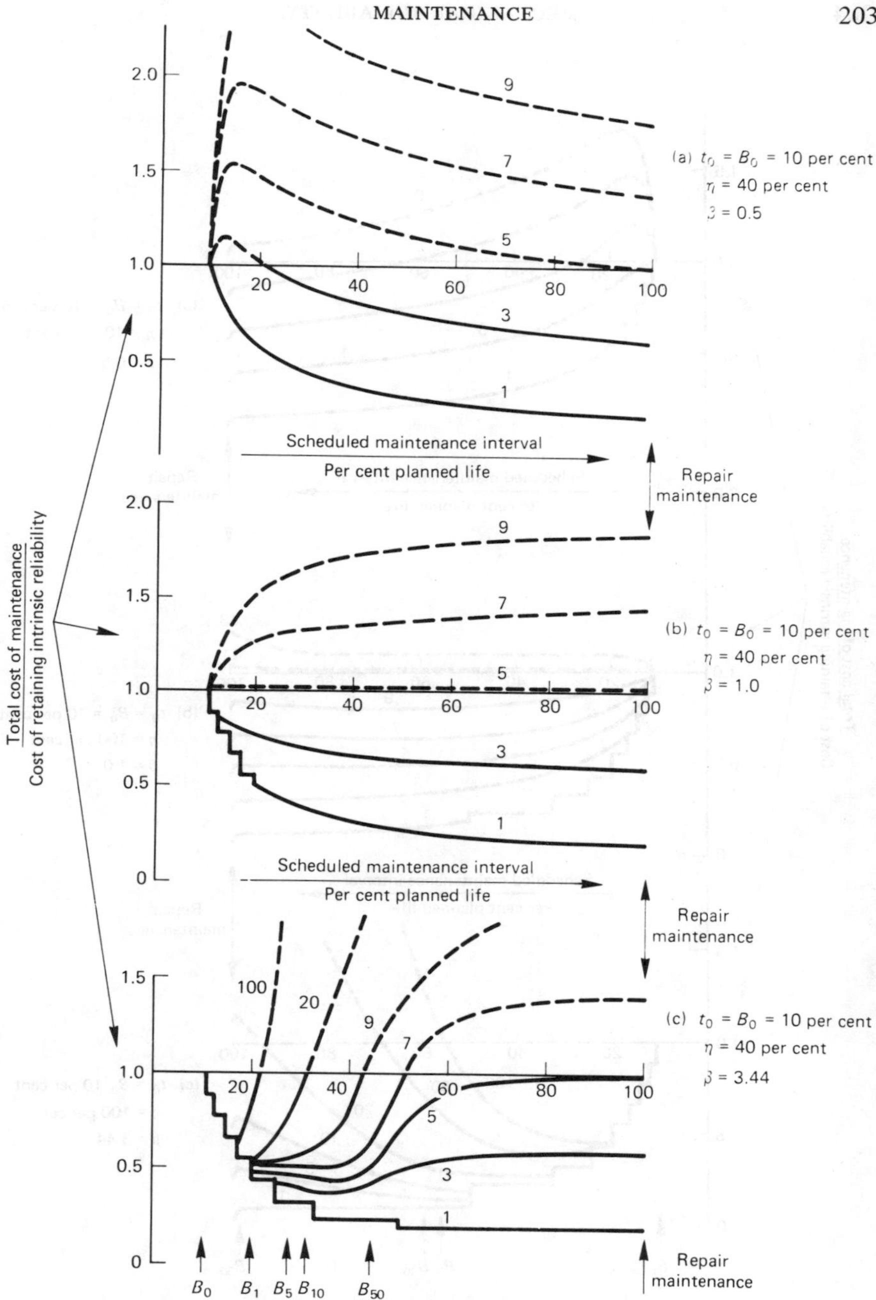


**Figure 9.8** Costs of scheduled and unscheduled maintenance plans for representative wear-out pattern. Values of cost ratio, $K$, marked on curves

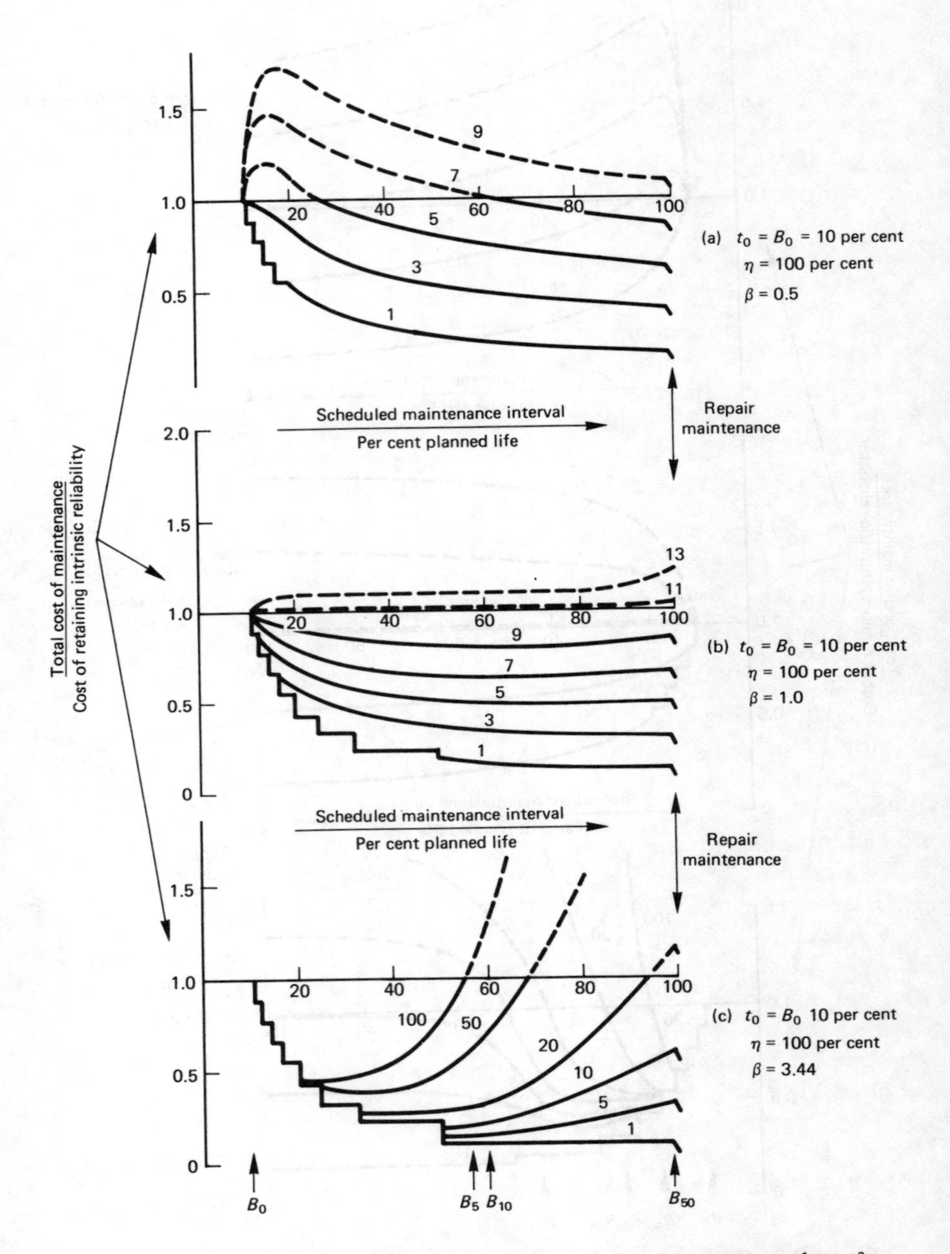


**Figure 9.9** Costs of scheduled and unscheduled maintenance plans for representative wear-out patterns. Values of cost ratio, $K$, marked on curves

policies become viable but the value at which this is possible depends on the failure mode. For the particular case in which $K = 1$, repair maintenance is always the cheapest, as already observed. Thus some generalisations are possible, and are maybe worth while exploring further.

So far as fatigue, or any fatigue-type failures, are concerned we are generally faced with a stark choice between a repair maintenance policy or a scheduled maintenance policy which retains intrinsic reliabilities. Scheduled maintenance between these intervals is usually more costly. There are some exceptions; see for example the curve on figure 9.7 for $\beta = 0.5$ and $K = 2$, giving an optimum scheduled maintenance at 65 per cent of the rated life, but in these instances the cost difference is marginal. The reason that one of the extreme policies is the optimum is not far to seek. In this failure pattern the bulk of the failures arise immediately after the failure-free period has ended and thereafter failures are distributed at a lower rate over a long period. This can be seen for non-maintained components in figures 9.2 and 9.3. Any scheduled maintenance carried out after the initial batch of failures discards some long life components, replacing them with a sample of the original population containing a proportion of short life components. The net effect is to replace some long life components by short life ones and hence increase the total number of failures. Any scheduled maintenance should therefore be carried out as early as possible to catch the bulk of the failures, or delayed as long as possible to gain from the long life element of the population. The choice between these two extreme policies is determined by the value of $K$: the greater the spread of the failure distribution the larger the value of $K$ necessary to justify scheduled maintenance. This is well illustrated in figures 9.7, 9.8 and 9.9, for which the $\eta/t_0$ values are 1, 4 and 10 repsectively and which give critical values of $K$ equal to 2, 5 and 9 when $\beta$ is equal to 0.5, and 2, 5 and 11 when $\beta$ is equal to 1.0. As a very rough and ready rule, one can plot these values as in figure 9.10.

The position on erosive wear, represented by a failure pattern having a Weibull shaping parameter ($\beta$) substantially greater than unity is quite different. Figures 9.4, 9.5 and 9.6 all show that initially only a few failures arise after intrinsic reliability has expired but that, as the equipment ages, failures arise with ever increasing severity. It is therefore possible to exceed slightly the intrinsically reliable life without too great a penalty, but if preventive maintenance is delayed too long the cost penalty becomes excessive as shown on figures 9.7, 9.8 and 9.9. Except for the very lowest values of $K$ a repair maintenance policy is consequently uneconomic. The extent to which the scheduled maintenance can be delayed beyond the intrinsically reliable life depends very much on the ratio of that life to the rated life. It will be noted from many of the graphs on figures 9.7, 9.8 and 9.9 that the cost arising solely from replacements at an early stage is a discontinuous curve. Thus for the failure pattern used in figure 9.7 it pays to increase the scheduled maintenance time beyond the intrinsically reliable life – with negligible increase in unscheduled failures – until a step reduction in scheduled replacements is achieved at 50 per cent rated life. Any further delay of scheduled maintenance only incurs increased unscheduled failures and cost. The optimum scheduled maintenance time is therefore 50 per cent of the rated life for this particular example. It should be noted, however, that even in this case the 'negligible' failures immediately after the intrinsically reliable life can become important if $K$ is high enough. This is illustrated by the curve for $K = 100$ on this graph. The same pattern is repeated for the other 'erosive' wear-out failure patterns on figures 9.8 and 9.9. ($\beta = 3.44$).

It can thus be concluded that for erosive-type wear-out a scheduled maintenance policy will be the most economic. The time at which scheduled maintenance should be carried out depends on the precise wear-out failure pattern, and of course, on the value of $K$. The oft quoted $B_5$ or $B_{10}$ life may be taken as a reasonable first approximation for average values of $K$ in the absence of firm data. They could, however, be in serious error if the failure pattern has a low standard deviation and/or a high life of intrinsic reliability (when these are expressed in terms of the rated life). For high values of $K$ an earlier replacement is always necessary to achieve minimum costs. Strictly, the optimum maintenance interval has to be assessed for each particular case.

In very general terms we can sum up our findings on scheduled maintenance as follows.

(a) If $K = 1$, no advantage is to be gained from scheduled maintenance, and hence a repair maintenance policy is the best whatever the failure mechanism.
(b) If failure is by an erosive wear-out process or any process represented by a Weibull distribution having a shaping parameter significantly greater than unity, and $K$ is greater than unity, then a repair maintenance policy will be more expensive than a scheduled maintenance policy. The optimum time of scheduled maintenance depends on the precise failure pattern. This optimum policy is very unlikely to retain intrinsic reliability unless $K$ is very high indeed.
(c) If failure is by a fatigue wear or similar process represented by a Weibull distribution with a shaping parameter less than unity, and $K$ is only slightly greater than unity, then a repair maintenance policy will give the lower cost. If $K$ is well in excess of unity, then a scheduled maintenance policy set to retain intrinsic reliability will give the lowest cost. The critical values of $K$ separating these policies are indicated roughly on figure 9.10.

The reader may have noted that the very high values of $\eta/t_0$ associated with some fatigue processes have not been used in this appraisal of maintenance policies. There were two reasons for this. Firstly, the possibility that different policies should be associated with different wear mechanisms required that the same values of $\eta/t_0$ should be used for comparison: the very high values are only associated with fatigue. Secondly, the very high values for fatigue only represent an exaggeration of the results presented here, leading to a quicker establishment of the pseudo-random state, and reinforcing the case for a repair maintenance policy.

It must be recognised that only a rough guide of overall trends can be obtained from a broad brush treatment. Nevertheless, an appreciation of the factors involved and their influence on the final outcome is invaluable at the feasibility and design

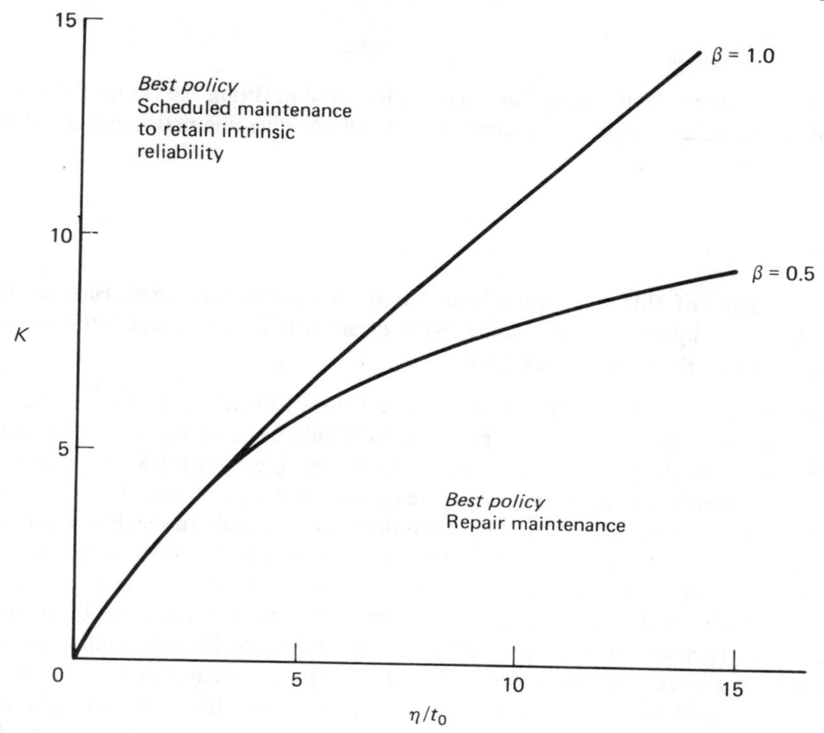


**Figure 9.10** Optimum maintenance policy determined approximately by values of $K$ and $\eta/t_0$ for fatigue wear out

stages, when a tentative optimum has often to be adopted to allow the design to proceed.

*Exercise 9.1*

Justify the current policy of replacing the water pump whose failure pattern is described in chapter 7 only when a failure has occurred (that is, a repair maintenance policy). In what circumstances would a scheduled maintenance policy be justified? If one is justified, what would you recommend for the scheduled maintenance interval?

*Exercise 9.2*

The treatment in this section has been based on the achievement of minimum total cost. How would the treatment be modified if the criterion determining the maintenance policy were to be based on maximum availability? (Availability is defined in chapter 2.)

*Exercise 9.3*

Examine in more detail the factors that are involved in fixing the ratio of the cost of an unscheduled repair to the cost of an identical schedule maintenance action ($K$).

*Exercise 9.4*

The treatment of this section has been based on ideal maintenance. How would you allow for imperfect maintenance? What effect would you expect it to have on the conclusions reached in this section?

The treatment of maintenance as so far developed in this section is, of course, grossly over-simplified. In the first place we have only dealt with a one component equipment, though this is not believed to be of any great significance. Clearly it would be a simple matter to extend the treatment to an equipment of any number of components, although in that case it would not be possible to reach any general conclusions regarding the maintenance policy, for the optimum would depend on the mix of components. There is, of course, no reason whatsoever why the maintenance of different components of the same equipment should not be treated differently. This is often done in practice: it is most common to find only a limited number in a given equipment subject to a scheduled maintenance plan, the remainder being treated under a repair maintenance procedure. This is entirely right and proper if it minimises cost. All the conclusions reached above would apply to the components taken individually, and assessments can be made on that basis.

To define the appropriate maintenance policy it is first necessary to determine $K$. We have already noted that $K$ has a major influence on the overall cost and hence must be known accurately if costs are to be evaluated. Unfortunately it takes into account a great many factors which have been brought out in the solution to exercise 9.3. Most of the features discussed there can be modelled mathematically – relatively easily in most cases – but the variations in procedure are so great that no generalisation is possible. A study of these aspects would take us well away from mechanical reliability and into the fields of management and operation research. They are outside the scope of this book. Carrying out such an evaluation is itself costly and could only be justified, if at all, by a mass user. For the general run of equipment, an intelligent estimate of $K$ based on past experience would seem every bit as adequate – the more so since a second difficulty in defining the failure–age pattern has to be dealt with. It would be idle to pretend that this will be known at the design stage. The best one can hope is that, on the basis of past experience, a designer can give a reasonable estimate of the values of $B_5$ and $B_{10}$, and perhaps $B_{50}$. $B$ value estimates are not likely to be very accurate at the design stage, so an accurate estimate of $K$ by itself is of little use. All estimates at the design stage must be treated as very, very tentative. It is only when more precise knowledge of the failure pattern becomes available during the development process (see section 10.4) that detail maintenance procedures can really be formulated. Very often they have to be regarded as tentative even at that stage and can only be 'firmed up' as more information becomes available during the useful life of the equipment. The more sophisticated techniques then come into their own.

All is not gloom, however. Fatigue and fatigue-like processes account for a large number of wear failures, and we have seen that for this mechanism there are effectively only two choices: scheduled maintenance to maintain intrinsic reliability or a repair maintenance policy. In most cases the decision is clear cut, requiring only rough values rather than detailed knowledge of $K$ or the precise failure–age pattern. Even with erosive and similar wear the optimum scheduled replacement interval is not likely to exceed $B_5$ or $B_{10}$, which are fairly good first order approximations for the scheduled replacement interval, with medium and low $K$ values respectively. So the type of policy can be resolved at the design stage with reasonable hopes of success. Furthermore, achieving compatible replacement intervals in the case of multi-component equipment is seen as the first target. Compatibility is often more important than any absolute life target, since it reduces overall maintenance costs by sharing maintenance activity amongst the several components requiring attention at the same time. If that can be achieved, it is not difficult to formulate a maintenance policy that is not unacceptably far from the optimum so far as individual components are concerned. The true optimum can then be evaluated in the light of operational experience. It may be worth adding that although it may not be too difficult to determine the appropriate type of policy because precise values of the input information are not required, the absolute cost of that policy does depend on accurate values of $K$ and the failure pattern. Thus, although policies can be reasonably confidently decided to allow design to proceed, a great deal of uncertainty surrounds whole life cost estimates until field data becomes available.

## 9.4 On condition maintenance

The cheapest possible maintenance activity is repair maintenance when the cost of an unscheduled replacement is no greater than the cost of a scheduled one ($K = 1$). The reason for this is that we are then making full use of the actual life of all components, and are not paying any penalty for the unplanned interruption of the equipment's output. We may thus regard this as the ideal (that is, the lowest possible) cost. The influence of the cost ratio, $K$, on the actual cost is considerable: even the most economic scheduled maintenance policy can be 2 to 5 times the ideal cost for very moderate values of $K$. At higher values of $K$ the figure is much greater. It could be very attractive indeed therefore if we could recoup some of the difference. This could be achieved if each component were operated beyond the statistical optimum group replacement life but not to the time of failure. If actual failure can be averted the maintenance bill approximates to the ideal maintenance cost, the approximation becoming closer the more accurately we are able to predict failure, and delay the maintenance action. Extended component life is 'profit'.

The importance of this approach is accentuated by the observation that identical products may be operated in significantly different ways. The statistical approach to maintenance optimisation must ignore these engineeringly significant differences, and bulk the small discrete operational groups into a single larger population having a greater standard deviation in its characteristics. Hence either a lower reliability may be achieved or a higher maintenance level may be demanded than if only essential maintenance were provided.

Although a great deal has been done to understand the mechanisms of wear and fatigue, it is not at present possible to assess accurately the effects of different usage on the failure pattern. The only realistic assessment is obtained from an examination of each and every component. As a practical illustration we may take an automotive tyre. Tyre wear varies considerably with speed, braking and acceleration, cornering, road surface and so on. If tyres were replaced on a scheduled basis, that is, after a given distance, the scheduled distance between replacements determined from the total population would be so low that many serviceable tyres would be discarded, and the cost of motoring increased. If the distance between scheduled replacements were increased to reduce the monetary cost to the individual, a higher total cost to the community (and some individuals) arising out of accidents would have to be accepted. In such circumstances the obvious step is to reject any scheduled maintenance plan, and replace the tyre when inspection shows it to be necessary. This is the basis of on condition maintenance. The same considerations apply to the majority of mechanical components, and it is only the difficulty of assessing the 'condition' that has held up this technique. The development of miniature sensors and reliable associated electronic accessories to measure and assess the wear process has given great impetus to the technique, and in the author's opinion it represents a most powerful tool to minimise costs. It is necessary, of course, that the cost of the monitoring equipment plus any additional monitoring costs involved should be (substantially) less than the difference between the cost of the optimum scheduled maintenance policy for the appropriate value of $K$ and the ideal maintenance cost.

Although on condition maintenance is closely akin to repair maintenance – the intention is that replacement should take place just before rather than just after the failure – a simple version of this policy may be assimilated into a scheduled or preventive maintenance plan. In this case the monitoring is carried out at significantly discrete intervals, at the time of some other scheduled maintenance activity. There are numerous examples: the inspection of brake pads or shoes during the routine servicing of a car, so that they are only replaced when their condition demands it, is a good example. This practice has been adopted for a long time and is a very valid procedure which has the advantage of avoiding withdrawal of equipment from service on an unscheduled basis. It implies that each component can operate successfully until the next inspection and that as a consequence some life is lost if these inspections are widely spaced. Modern on condition maintenance is an extension of this practice and has become associated with a much more frequent assessment than that which could be linked with standard scheduled maintenance activities. While some on condition monitoring may be continuous in the true mathematical sense, it need not be carried out with that rigour. The frequency has to be carefully planned in the knowledge of the elapse time between the first signal of impending failure and the failure itself.

Condition monitoring is based on the knowledge that wear-out is responsible for the vast majority of mechanical failures. It follows that mechanical failure does not often occur suddenly and is usually preceded by some change in the sensible behaviour of the machine. Thus the continuous (in the colloquial rather than the mathematical sense) recording of appropriate factors can give a clue to an impending failure.

The simplest form of monitoring is, of course, visual. Visual inspection has always played a large part in preventive maintenance schedules, and is indeed one of the most powerful techniques. Modern design is providing more access covers, hatches and so on through which the internal state of a machine may be visually examined, using intrascopes, miniature TV, and similar modern instrumentation. Powerful and useful though this technique may be, it is generally subjective, and relies on the experience of the inspector. Some of the subjective quality may be removed by demanding quantitative measurements of the observed item. For example, crack lengths may be measured, and the old established visual inspection of the condition of automotive tyres has given way to objective and defined measurements of tread depth by statutory decree.

Alternatively, or perhaps additionally, some readily accessible and easily measured feature of the machine's operation can be recorded. The simplest example is the standard oil pressure and cooling water temperature gauges fitted to most automobiles. Any departure from the average value is indicative of some malfunction. (Are gauges or warning lights to be preferred?) As an extension to this, the temperatures and pressures at many critical points in a machine can be measured easily by modern techniques and continuously recorded automatically if so desired. The recording of bearing temperatures is a long established practice in some high-duty machines, for example, or where high reliability is required as in Naval and power generation practice. Alternatively, and maybe additionally, the performance of the machine itself may be logged as a pointer. Practices along the lines sketched above have been used for many years in well-managed organisations, although it is only in recent years that modern instrumentation and automation has allowed extensive use.

The success achieved by the simpler techniques has raised the question whether more direct methods would not give better results. The majority of wear-out processes fall into two main classes. These are

(1) true wear, usually caused by the relative movement of adjacent surfaces, in which small quantities of material are worn away, and

(2) fatigue in which there is no removal of material, or indeed any outward sign at all during the early stages, but which causes cracking of the material before final failure.

The techniques for dealing with each of these classes is taken separately below.

Since the type of wear specified in (1) above is associated with relative adjacent motion, it follows that the surfaces will usually be lubricated, and that the material worn away must be contained within the lubricant. The quantity and size of the material particles contained in the lubricant may therefore be used to estimate the rate of wear, and their composition locate to some extent the wearing surfaces. The detail mechanics of the wear process are extremely complicated, and refs 43 and 44 may be used as a source of further information on this subject. However, looking at the grosser aspects, two types of particles are usually found in used oil. The first is small (0.025 mm) regular pieces resulting from normal wear. Their size and rate of production remains roughly constant throughout the life of the machine. The second type is produced by surface failure. They differ substantially in size (say 0.25 to 1.5 mm) and shape from normal wear particles. The shapes vary with the type of failure and the shape of the original part. Figure 9.11 shows some typical particles resulting from normal operation compared with those from the break-up of components just before failure. A typical curve of size and quantity of these particles is given in figure 9.12.

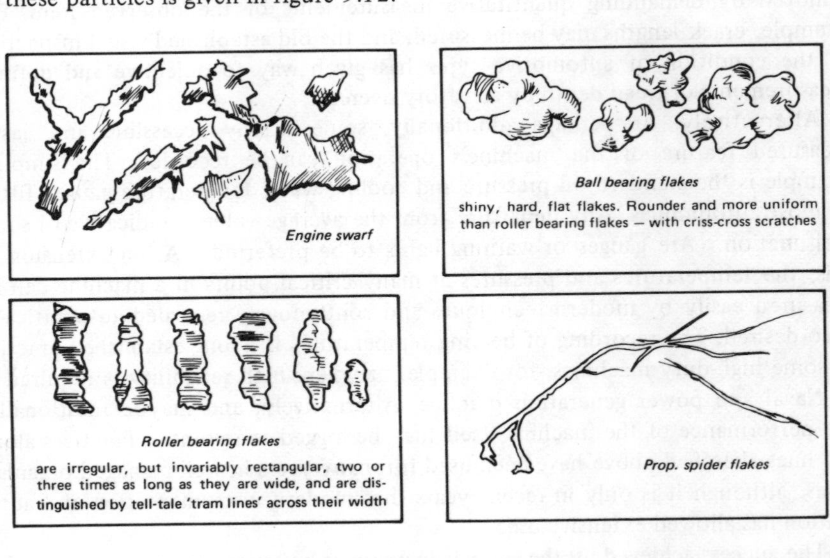


**Figure 9.11** Particles exposed on turbomag plugs (from ref. 45)

The objective of any 'health' monitoring' or 'early failure detection' kit is to detect the small change which precedes the actual failure. Several techniques are available. All the particles contained in the oil may be separated by a normal filter and subsequently examined, but this obviously requires a filter change. A simpler

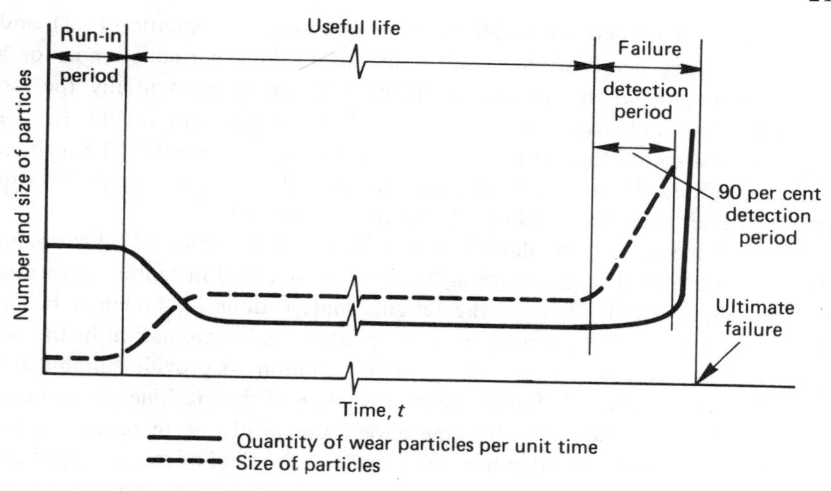


**Figure 9.12** Characteristic of wear particles with time of use

method where ferrous material is involved makes use of a magnetic probe to collect the debris. Self-closing magnetic probes enable frequent and regular inspection to be made. The particles collected by probe can then be examined visually, using a microscope or other suitable device. As an alternative, small samples of oil can be withdrawn from the system and examined in detail. This technique is simple and all the constituents of the oil can be examined using laboratory techniques. For example, spectrographic analysis (using either X-ray or arc emission) can provide information on the elements present in the oil and the quantity, to a high degree of accuracy. Samples taken every 10 to 50 h running can provide a complete record. British Rail, for example, claim a high success rate with this technique. Other users are less convinced. The difficulty is one associated with every sampling technique. The normal wear rate tells little of incipient failure, and the sample could well miss the small number of larger particles (see figure 9.12) indicative of failure. If the time between initial breakdown and complete failure is small – an average may be taken at 100 h – the failure could occur before significant warning is received. In spite of limitations it would appear that high success rates can be achieved provided the technique is matched to the machine in question, and its method of operation. Generally speaking, oil analysis is most effective when the oil remains clean, that is, when it is not contaminated by sources which give rise to spurious alarms, and more particularly when the operation of the equipment is on a regular basis and sufficiently extensive to warrant the cost of the operation. These conditions are fully met by commercial air lines and air forces, who use the techniques extensively.

A greater time between the warning and the failure would greatly extend the effectiveness and hence the use of these techniques. However, the short duration is a fundamental feature of the process when the same load causes both wear and failure. The instantaneous strength is then represented by equation (5.3), and it has already been demonstrated that it predicts a long period of more or less constant strength, when the wear particles will remain substantially the same, followed by a rapid loss of strength in the last few per cent of life. It is only during this last period that different wear particles can be expected. The technique is often confined, therefore, to the closely controlled operation, where regular samples can be taken, followed by rapid action if required.

For the same reason, the detection of fatigue failures under (2) above is more difficult in the initial stages: once again there is no loss of instantaneous strength or other physical indication of the fatigue damage done – though it becomes obvious, by virtue of the crack, on appropriate visual examination in the latter stages. For this reason it is becoming more common to provide suitable access covers, hatches, etc. through which the internal state of the machine can be visually examined. This technique has become a very powerful one in recent times on account of the big advances that have been made in the field of fracture mechanics. If the length of the longest crack is known, by condition monitoring, it is now possible to estimate fairly accurately the load which could cause brittle fracture in a given component, or the number of cycles of a given load magnitude that will cause a fatigue failure. It is not possible, or even desirable, to discuss the relevant fracture mechanics in this book, because adequate texts already exist. (The reader unfamiliar with fracture mechanics is referred to the book by Parker in the Reading list, which is a good starting point and contains many additional references.) What we note here is that such techniques do exist, but if maximum use is to be made of these new facilities it is necessary to make provision at the design stage for easy access to the critical crack locations. A comprehensive assessment of the failure modes is still therefore required to determine these locations. This form of conditioning monitoring should never be regarded as a means of trapping design errors. The assessment of the remaining life by this technique is subject to all the unknowns that bedevil conventional life assessment. Thus the material properties upon which the assessment is based will be distributed in the same way in which all other material properties are distributed, and must be taken into account. So too must the expected load cycle subsequent to the time of inspection – that prior to inspection is already evaluated in the measured crack length. Nevertheless, trustworthy assessments of remaining life are now possible using the considerable advances that have been made in fundamental understanding of crack propagation during the last decade, while on the ground the techniques of crack detection and appraisal are not too difficult or too expensive to implement (at least at the design stage). Crack monitoring therefore warrants consideration in many circumstances when the life is fatigue controlled.

Fracture mechanics is, of course, concerned essentially with crack propgation, and can tell us nothing about the crack nucleation process. Unfortunately this occupies most of the practical life of many items. During nucleation there is little that can be achieved by monitoring techniques, though some operators do keep a record of the actual load cycle from which to make theoretical life estimates: but this is not condition monitoring. A feature which may provide a useful monitoring signal is that many fatigue failures arise from extraneous forces which have not been properly – often cannot be properly – assessed at the design stage. Such forces may arise from progressive wear, for example. In this case it can be assumed that in addition to causing fatigue damage – or indeed any other damage – these forces will also cause vibration. Vibration and alternating stress can then be related as follows.

In any simple vibrating system the maximum kinetic energy is reached at the instant of maximum velocity, and is given by

$$\text{kinetic energy} = K_1 \frac{\rho V}{2} (2\pi f x) \tag{9.15}$$

where

$K_1$ = a constant depending on the shape of the deflection curve and the position of any point under consideration
$\rho$ = material density
$V$ = volume
$f$ = frequency of vibration
$x$ = amplitude of vibration at the point under consideration.

The maximum strain energy is reached at the instant of zero velocity and is given by

$$\text{strain energy} = K_2 \frac{VS^2}{E} \tag{9.16}$$

where

$K_2$ = a constant similar to $K_1$
$S$ = maximum stress at the point under consideration
$E$ = Young's modulus.

Equating the strain and kinetic energies

$$K_1 \frac{\rho V}{2} (2\pi f x)^2 = K_2 \frac{VS^2}{E} \tag{9.17}$$

so that

$$S = \frac{1}{\sqrt{2}} (2\pi f x) \sqrt{\left(\frac{K_1}{K_2}\right)} \sqrt{(E\rho)} \tag{9.18}$$

It is thus seen that the maximum vibrational stress, $S$, is proportional to the vibration velocity, $2\pi fx$. It follows that the vibrational stresses in two geometrically scaled systems made in the same material can be compared by measuring the vibration velocity at corresponding points.

The great attraction of this proposition is that the ratio $K_1/K_2$ and the product $E\rho$ do not vary greatly for a very wide range of engineering structures made in normal materials (see refs 46 and 47). On the basis of this proposition Yates (ref. 46) and Rathbone (ref. 47) have both devised scales in which the level of vibration determines the acceptability of the machine. These scales differ somewhat and are compared in figure 9.13 together with a scale proposed by VDI (ref. 48). However, we may disregard such universal scales yet still be able to specify a level of vibration, corresponding to an acceptable, even if unknown stress level, for any particular machine. The level of vibration has, in fact, long been accepted as a criterion of mechanical integrity. It is only recently, however, that instrumentation has made its routine quantitative measurement a practical proposition. Continuous monitoring of the vibration would show any deterioration during operation which is both

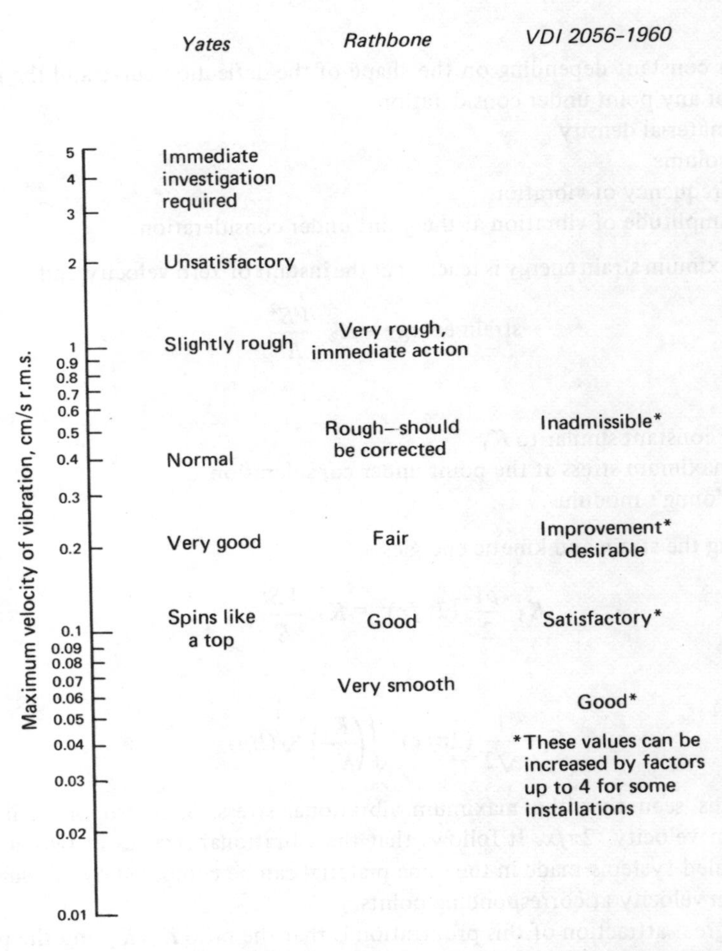


**Figure 9.13** Roughness–vibration level scale for mechanical equipment

directly indicative of wear and would warrant attention on that score alone, and indirectly indicative of additional cyclic stresses and hence possible fatigue damage.

Vibration is always manifest in noise. Although the very near noise level can be calculated from the vibration level in some very elementary instances (see refs 46 and 47) the general relationship between vibration and noise levels is too complex and involves too many unknown factors for practical use. In spite of this it should be appreciated that in a given situation there is a relationship between the vibration and noise levels and hence between the noise level and stress, as every practical engineer has known and used qualitatively for the last century or so (hence the term 'old banger'!). Modern equipment allows the noise to be measured and assessed quantitatively for both frequency and intensity. One may suppose, therefore, that monitoring the noise level from any machine is a potential method of detecting failures or incipient failures. Some of these techniques are described in some detail in ref. 41. Later papers in this reference contain accounts of the application of acoustics as a diagnostic tool in several specific fields. It is clear from the papers and discussion (ref. 49) that a great deal of progress has been made in the last few years. However, it is equally clear that the technique has met with mixed success. For example, Bertodo states, *à propos* of large diesel engines: "a relatively coarse amplitude-frequency spectrum can be used to good effect to monitor the behaviour and isolate changes", while Pride and Grover state that "the diagnosis of engine operational faults by simple frequency analysis of emitted noise is an extremely difficult task." Summing up the general experience, it would appear that acoustics can be a very valuable diagnostic tool in certain straightforward circumstances – for example, gear monitoring. For more complex equipment in complex surroundings, the interpretation of the data present difficulties. In spite of this, modern electronic equipment can be used to process noise data in the same way that it can process vibration data, so that various 'signatures' may possibly be interpreted in terms of machine performance. One can expect this technique to develop as computer noise recognition becomes more widely developed.

It is impossible to cover every possible monitoring technique in this book, and it will be apparent to the reader that many 'signals' can be used for monitoring. Besides the direct engineering signals which are indicative of any malfunction, it should be recognised that many indirect and more readily accessible signals may be employed. This could make monitoring a much more attractive proposition and is potentially a powerful technique if the signal pattern from several monitors is correlated in terms of malfunction or impending failure.

A detailed appraisal of the various current monitoring techniques appropriate to aero engine practice is given by Nutter and Kirby (ref. 50) who list the 'top ten' in order of preference as follows

(1) visual chip detectors (including oil filter)
(2) borescope (that is, internal visual inspection)
(3) vibration indication

(4) automatic flight data recording
(5) radiography
(6) cockpit instrument monitoring
(7) scavenge oil temperatures
(8) hot section life recorder
(9) radiation pyrometry
(10) trend analysis of cockpit instrumentation.

It is relevant to ask, however, in each case whether the introduction of an excessive amount of monitoring does not reduce the reliability as a result of 'false alarms'. Although each case must be assessed on its own merits, there is little doubt that monitoring is proving economic in an increasing number of applications. As a single example, the monitoring proposed for the Rolls Royce BS 360 is given in figure 9.14. This illustrates the extent to which condition monitoring has already been accepted in an application where high reliability is essential. Reference 51 is a good starting point for anyone wishing to study the technique in its modern powerful context.

One of the serious drawbacks of condition monitoring is the inopportune occurrence of signals predicting impending failure. If the equipment is in the pseudo-random phase the signals will of course arise in a random fashion. Withdrawal of the equipment for servicing then often entails many of the drawbacks of repair maintenance. It is a necessary concomitant of on condition maintenance, therefore, that special provision be made to rectify impending failures with minimum, if not zero, interference to the operating schedule. The greater the warning period offered by the on condition monitor the easier this becomes, of course. In addition to the normal techniques to facilitate maintenance, special construction, such as modular construction, allows complete sub-assemblies to be readily removed when failure is imminent and immediately replaced by standby serviceable modular units. Figure 9.15 shows how this is achieved in the Rolls Royce engine quoted previously in connection with monitoring techniques (figure 9.14). This not only involves special dismantling and assembly techniques, but demands complete interchangeability. Such construction can be expensive and the cost must be offset against some of the savings achieved by this maintenance policy.

It is obvious that the circumstances most favourable for on condition maintenance have high $K$ values, in which the large difference between the costs of scheduled and ideal maintenance can in part be used to pay for the condition monitoring equipment and special construction. High $K$ values are associated with much modern equipment, which either because it is 'universal' or because of ever increasing complexity to achieve a high level of performance, often coupled with more complicated automatic control, locks up a great deal of capital. The lost revenue and high interest on capital in such circumstances, alone accounts for a high $K$ value. When this is coupled with the increasing availability of reasonably priced and reliable transducers to measure condition, and microprocessors to assess condition, the trend to this form of maintenance is inevitable.

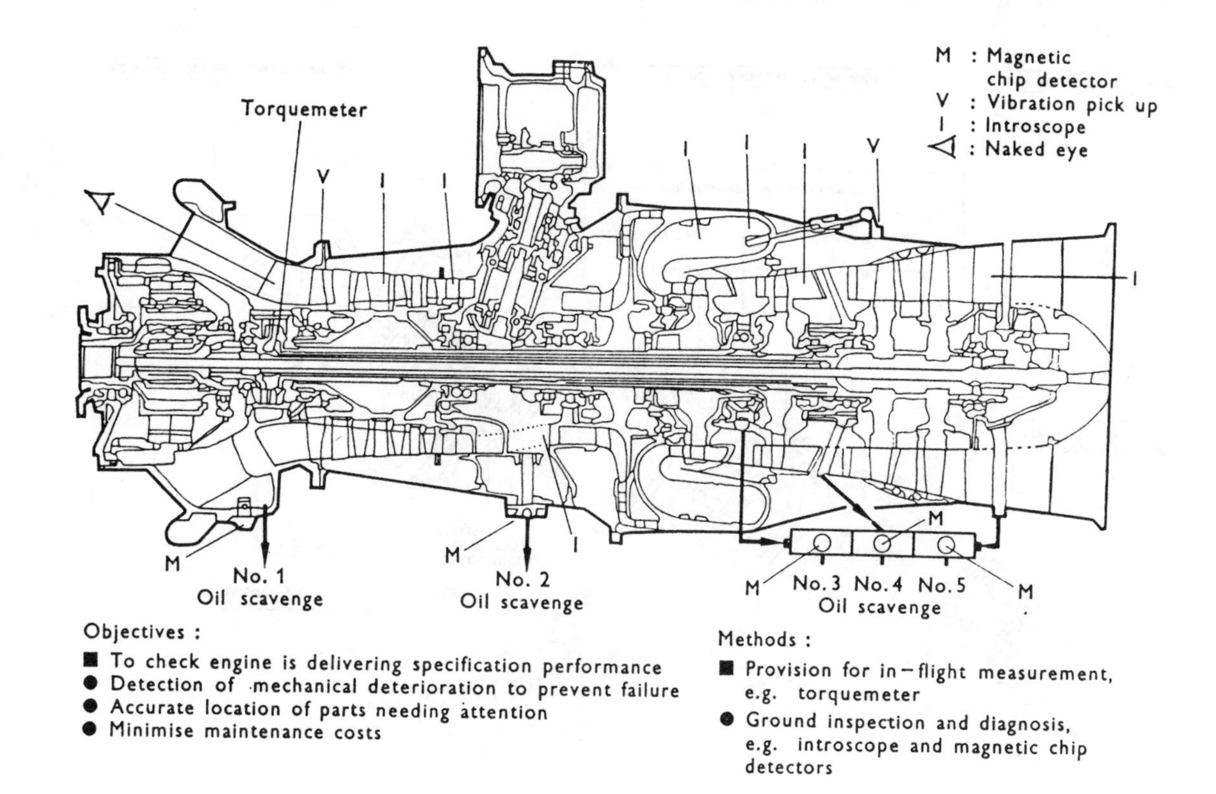


Figure 9.14 Condition monitoring in Rolls Royce GEM helicopter engine

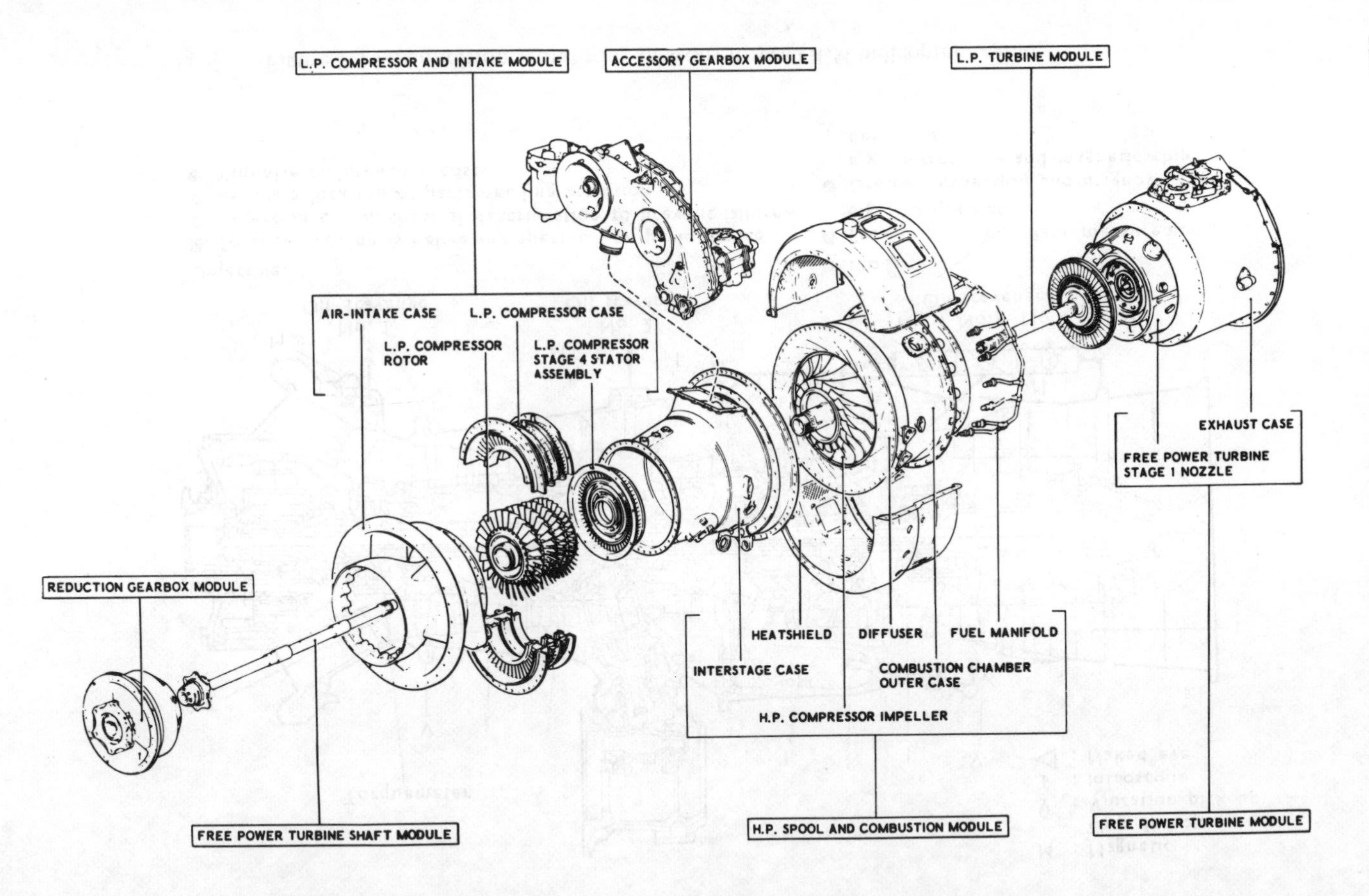


Figure 9.15 Modular construction of Rolls Royce GEM helicopter engine

*Exercise 9.5*

A typical modern small-bore domestic central heating system depends on a circulating pump rather than thermosyphon effects to maintain circulation. However, if such a pump fails it acts as an effective throttle so that the whole system fails. The reliability of domestic circulating pumps is not as high as it might be. What monitoring system would you suggest to give early warning of pump failure?

*Exercise 9.6*

It is essential that prior warning of failure should be received from the fan of an air conditioning unit which operates in a chemically polluted environment. What do you suggest be adopted to monitor its behaviour?

## 9.5 Maintainability

One of the major factors controlling the cost ratio, $K$, is the ease with which the maintenance can be carried out. This is measured by the 'maintainability' of the item, in which that quality may be associated with the complete maintenance action or with the active maintenance process. Qualitatively, maintainability has been defined as "The characteristic of design and installation which makes it possible to meet operational objectives with a minimum expenditure of maintenance effort (manpower, personnel skill, test equipment, technical data, and maintenance support facilities) under operational environmental conditions in which scheduled and unscheduled maintenance will be performed" (ref. 52). Maintainability has been defined quantitatively as "The probability that an item can be retained in, or restored to, specification within a stated period under the stated conditions of maintenance." It is usually denoted by the symbol, $M_\tau$, where $\tau$ is the stated period, sometimes called the time constraint. It is, of course, an unique probability at time $\tau$, which may be regarded as taken from a general cumulative distribution, $M_t$, giving the probability of achieving the aim within any arbitrary time, $t$. The time constraint, $\tau$, can be different for different applications of the same item, so $M_\tau$ has no unique value special to the item and should not be taken as a measure of its maintainability. Maintainability is measured by the distribution $M_t$. By analogy with reliability, the author would prefer to measure maintainability as the time in which a specified proportion of the items can be retained in or restored to specification under the stated conditions of maintenance. It may be denoted by the symbol $Z_x$, where $x$ is the specified proportion. The value of $x$ would normally be in the range 90 to 100 per cent, by contrast with the lower values quote for reliability. The median (50 per cent) value could, of course, be quoted instead. Both $M_\tau$ and $Z_x$ are obtained from the same distribution $M_t$ referred to above, and are alternative measures of the same concept –

maintainability – in that same way that $R$ and $B$ are alternative measures of reliability. The form of that distribution is open to some conjecture. Based very largely on experience and evidence from the electronic world, it is often stated to be log normal (see, for example, ref. 53), but justification by field evidence from the mechanical world is very sparse. A theoretical justification may be advanced as follows. If we take the very simplest of maintenance actions, such as undoing one of a number of nuts and bolts to release a component, then we may surmise that there would be a minimum time under which that job could not be done. In real life we could still expect that the job could be completed in the minimum time on a number of occasions, but on a substantial number of occasions it would take longer on account of adventitious difficulties. These could range from minor stubbornness to advanced corrosion, for example. It is reasonable to assume that the difficulties which resulted in the maintenance action exceeding the ideal minimum would be random. In that case we would expect the distribution of time to complete a small maintenance operation to take the negative exponential form with the minimum time as origin. Now a full maintenance job will be built up of a significant number of small activities of the type just described. Howard (see ref. 54) has shown that the distribution of the sum of a limited number of such jobs each conforming to the negative exponential distribution would be log normal. If the number were very large one would expect it to be normal. There are very good reasons for believing, therefore, that the active maintenance time for many mechanical maintenance activities should be represented by the log-normal distribution. Turning to field data, it does seem reasonable to conclude that where evidence does exist, it confirms, if somewhat pedantically, that the data is most closely represented by the log-normal distribution (see, for example, ref. 55). However, the evidence is extremely limited and is open to a variety of interpretations, so it would be unwise to draw any firm conclusions at this stage. What does appear to be certain is that wide discrepancies are usually experienced in the active rectification man hours for even the same reported defect on the same equipment, leading to a highly skewed distribution rather than to a normal distribution about an expected mean. It thus supports theoretical reasoning. This should not surprise practical maintainers – a relatively minor difficulty can often take a long time to overcome.

The log-normal distribution is not easily handled. Many theoretical statisticians prefer to replace it with the Gamma distribution for analytical purposes. In view of the uncertainties involved, the author prefers to replace it with the Weibull distribution – a comparison of the distributions has been made in section 6.5 – which would seem to represent field data as adequately, and allows a similar treatment of maintainability and reliability data. It is then found that man hours to repair can be represented by Weibull distributions having a shaping parameter ($\beta$) which can range between 0.8 and 1.5 with a mean value only slightly in excess of unity for first line repair, but which can range between 1.6 and 2.5 with a mean value of about 2.2 for second line repair. These values probably represent the simpler nature of first line repair and hence the closer approximation to the negative exponential distribution discussed above. Second line repair is more extensive and tends towards a log-normal distribution. It is generally believed that the time for major overhauls, which involve a very large number of jobs, is distributed normally. The reason for $\beta$ values less than unity for some first line jobs is not clear: it could have some fundamental significance or could be just a quirk of inadequate data. As may be expected, the values of the locating constant and the characteristic life vary enormously and it is impossible to give any guide on statistical grounds. It would seem far safer to estimate the parameters on an engineering basis. The parameter $t_0$ is the minimum time in which the repair can be expected to be carried out, so it should be possible to estimate it. The value of $\eta$ is the difference between this time and the time by which at least 60 per cent (63.2 per cent) of the repairs could be expected to be completed. This may be estimated, though much less readily than the minimum time. Rough and ready estimates of repair time distributions may thus be made.

It should be made clear that we are referring to a specific repair job, such as replacing a defective exhaust pipe and silencer in a given vehicle. Different distributions would apply to different maintenance activities – for example, replacing a clutch would have a different distribution. Electronic engineers tend to bulk all maintenance actions on an equipment into a single distribution: thus there would be a single distribution for all maintenance activities on a TV set. This may be justified because all repair actions are physically much the same, and more particularly are random. For mechanical equipment the repair activities are not random, but follow the particular wear pattern of the components making up the equipment, and the repair actions on different components can be of a totally different nature. If one does want an all-embracing maintenance time distribution for mechanical equipment, it must be constructed from the individual distributions, taking into account their expected frequency. This may have advantages in some cases: for example, first line maintenance which consists of a large number of different minor jobs often arising in a random manner may be treated in this way, but mixing major failure modes can be misleading.

*Exercise 9.7*

A vehicle is in the pseudo-random failure state so far as minor failures are concerned, having a constant failure rate, $\lambda$. It is deemed to have failed in its mission if its time of arrival is $\tau$ behind schedule. If the driver is able to rectify minor faults in a mean time of $\phi$ show that the mission reliability is

$$\exp\{-\tau x \exp(-\tau/\phi)\}$$

*Exercise 9.8*

The concept of maintainability, $M_T$, based on a time constraint has been largely developed by the electronics industry in the U.S.A. and subsequently taken over by some mechanical engineers. Discuss the validity of this measure of maintainability in mechanical engineering.

Following on exercise 9.8 and the preceding remarks, it cannot be overemphasised that maintainability is essentially an engineering matter, and furthermore a design matter; for it is at that stage the maintainability is achieved – or otherwise. It will be readily appreciated that the high standard deviation of maintainability often encountered in the field has far more influence on the cost of maintenance than the failure to achieve the statistical optimum policy. In accordance with Pareto's principle, that is where the effort should be made. The first step in achieving good maintainability is to list all the expected failures and their frequency of occurrence in the planned life of the equipment. (Note that this should be done for other reasons as part of the design process and is discussed fully in section 10.3.) Following from that, the maintenance procedure – the physical action – to rectify each failure should be laid down and assessed. Poor maintainability is usually a consequence of not taking this step. The key to good maintainability is good accessibility. The kind of atrocious design which leads to poor accessibility and hence low maintainability is illustrated in figure 9.16. It has been adapted from ref. 56. That reference quotes other examples. The situation described will be only too familiar to anyone who has carried out maintenance work. Reference 56 also notes that the average time taken to replace the defective part in the cases examined was only 1/16 of the time of the total operation. This highlights the scope that exists for improved design for accessibility. The difficulty in achieving access is believed to account for much of the variability in the time

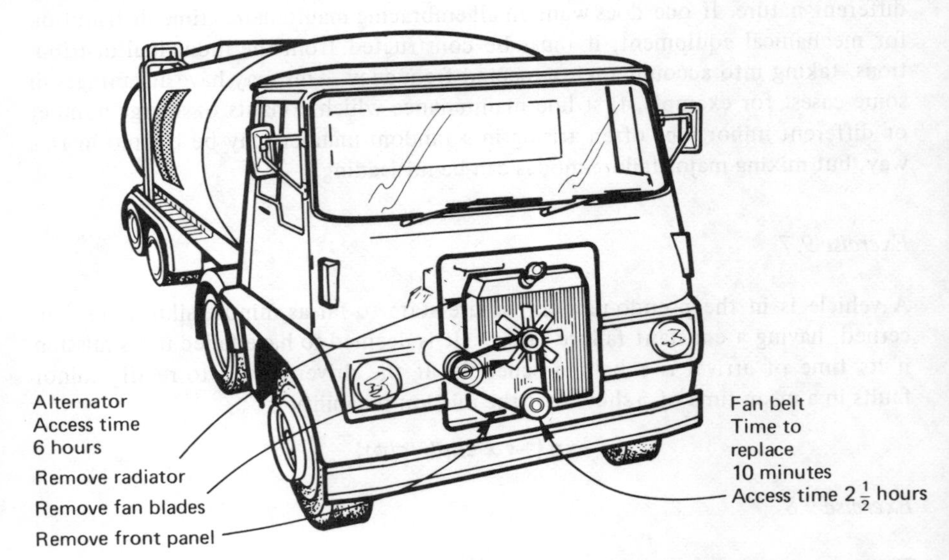


**Figure 9.16** An example of a comparison of the contribution of access time and replacement time to the total active repair time

taken to complete most mechanical repairs. Evidence available to the author suggests that the time to complete 62.6 per cent of the repairs on an item (the characteristic time) ranges from about 5 to 150 times the minimum time of repair, for first line maintenance. The average value is about 32. Such figures really are – or should be – totally unacceptable. Far too often design is optimised for production, not for maintenance, even though production is a one-off job (on each item!) while maintenance may have to be carried out repeatedly. It follows from this that, if good accessibility cannot be achieved, then it should not often be required – that is, the reliability, interpeted as the life of intrinsic reliability, $B_0$, should be commensurate with the life of the equipment.

At the other end of the scale, exceptional positive steps can be taken to facilitate maintenance. This has already been touched on in connection with on condition maintenance. An essential feature is very easy access to a component or sub-assembly together with simple removal so that any defective ones can easily be replaced. This obviously involves extra expense, which, one hopes, can be set against increased life when on condition monitoring is employed, or otherwise against the increased revenue from higher availability. To achieve this, the maintenance policy and its supporting facilities must be decided at the design stage. The level of replaceable items must be decided, that is, sub-assembly or sub-sub-assembly. In turn this requires decisions on repair procedure: for example, is the replaceable item to be repaired at second line (central depot) or base workshop (factory)? or if the replaceable item is at a low enough level is it to be repaired at all? Such factors may fall outside the accepted field of reliability, but they do influence the value of $K$ (the cost ratio introduced in section 9.3) and hence the optimum reliability. Furthermore, if exceptional methods are introduced to increase maintainability, it may be possible to offset the cost against savings accruing from a design concept of lower inherent reliability. The trade-off between maintainability and reliability is a possibility taken up in the next section.

## 9.6 Trade-off: reliability/maintainability

We have just observed that it is possible to compensate for low reliability by high maintainability, and of course the converse would be equally valid. This statement is often made in a loose qualitative way and must be true in a general colloquial sense, but it would be advisable to examine it a little more rigorously to define the circumstances in which the substitution is economically feasible. We shall base our analysis on the assumption that a user would wish to minimise total costs for a given level of availability. This may not always be the relevant criterion, but it is the one that would meet the majority of circumstances. Availability is defined in chapter 2 and the factors which control it are illustrated diagrammatically on figure 9.17.

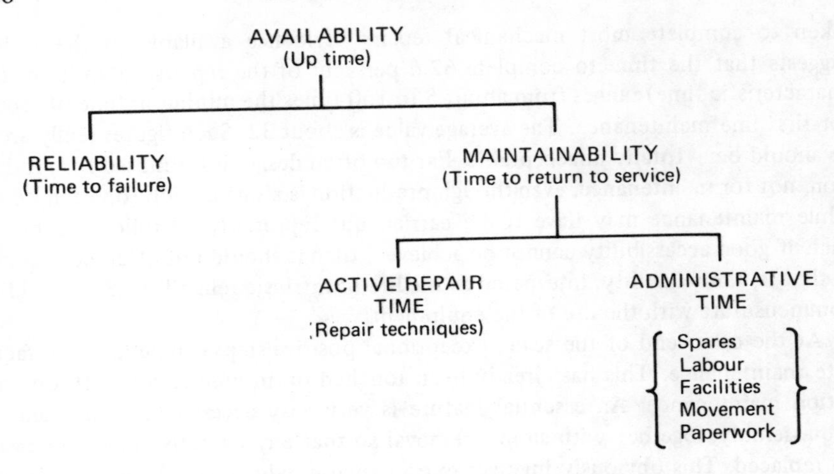


**Figure 9.17** Factors contributing to availability

At a given availability the basic running costs, which would be made up of fuel and/or material costs, operator costs, lost revenue and so on, will be constant. If that is so we can write

$$C = C_R + C_M + C_O \tag{9.19}$$

where

$C$ = total cost
$C_R$ = cost of equipment to achieve reliability $R$
$C_M$ = cost of maintenance over the planned life
$C_O$ = basic running costs over the planned life.

The minimum total cost will arise when $dC/dR = 0$, that is, when

$$\frac{dC_R}{dR} + \frac{dC_M}{dR} = 0 \tag{9.20}$$

Now we can write

$$dC_R = \Delta R \times \Delta C_R \tag{9.21}$$

where

$\Delta R$ = a small change in the reliability which is to be traded for maintainability
$\Delta C_R$ = cost of an unit improvement in reliability.

Likewise we can write

$$dC_M = -\Delta M \times \Delta C_M \tag{9.22}$$

where

$-\Delta M$ = a small change in maintainability which is to be sacrificed for improved reliability

$\Delta C_{\mathrm{M}}$ = cost of an unit improvement in maintainability.

Substituting in equation (9.20) and re-arranging

$$\Delta M \times \Delta C_{\mathrm{M}} = \Delta R \times \Delta C_{\mathrm{R}} \tag{9.23}$$

or

$$\frac{\Delta R}{\Delta M} = \frac{\Delta C_{\mathrm{M}}}{\Delta C_{\mathrm{R}}} \tag{9.24}$$

It should be noted that we should derive the same equations (9.23) and (9.24) if we sacrificed reliability for maintainability. The quantities $\Delta R$ and $\Delta M$ are trade-offs, or the marginal substitution rates of reliability and maintainability. It follows from equation (9.24) that the marginal substitution ratio ($\Delta R/\Delta M$) of maintainability for reliability must equal the inverse cost ratio in order to minimise costs.

According to equation (9.23) the cost of any added reliability must just equal the cost of maintainability that is replaced, if the total cost is to be a minimum, and vice versa. Now if $\Delta R \times \Delta C_{\mathrm{R}} > \Delta M \times \Delta C_{\mathrm{M}}$, the cost of the maintainability being replaced is less than the cost of the added reliability (that is, the substitution ratio is greater than the inverse cost ratio), and we may then reduce the total cost by planning for or developing more maintainability and less reliability for the same desired level of availability. If $\Delta R \times \Delta C_{\mathrm{R}} < \Delta M \times \Delta C_{\mathrm{M}}$ (that is, the substitution ratio is less than the inverse cost ratio), reliability can be added at a lower cost than the maintainability replaced, and we should proceed in that direction. (Both $\Delta C_{\mathrm{R}}$ and $\Delta C_{\mathrm{M}}$ will be functions of the reliability and maintainability respectively, and it is only justified to use equation (9.23) for small changes.)

A closer appreciation of the factors involved may be obtained by an examination of a simple illustrative example. We shall take the case of an equipment that has reached the pseudo-random failure stage, so that the mean time between failures can be taken as constant. In chapter 2 we defined availability as

$$\text{availability} = \frac{\text{up time}}{\text{up time} + \text{down time}} \tag{9.25}$$

which in this particular case can be written as

$$\text{availability} = \frac{\text{MTBF}}{\text{MTBF} + \text{MTTR}} \tag{9.26}$$

in which any standby or idle time has been ignored. Using this relationship it is possible to plot lines of constant availability on a reliability–maintainability map as in figure 9.18. On this figure, reliability has been measured as MTBF and maintainability has been measured as the repair rate per unit time, the units of time

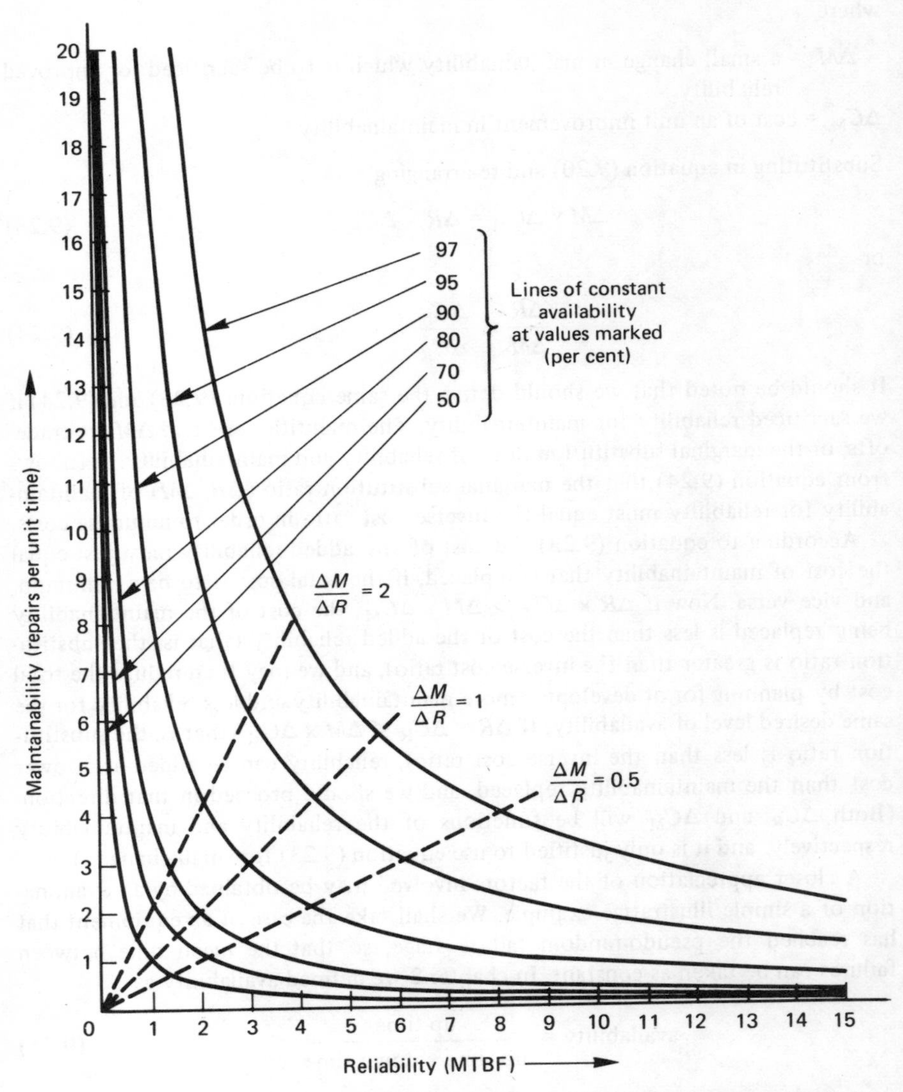


**Figure 9.18** Roles of reliability and maintainability in achieving availability

for reliability and maintainability being the same. Referring to figure 9.18, it will be observed that parts of the curves are substantially vertical or horizontal, particularly at low availabilities. In these areas the availability is insensitive to maintainability and reliability respectively, so that in qualitative terms it would pay to move the operating conditions towards the negatively sloped part of the curve, in as much as this would involve fewer resources. In terms of the mathematical analysis presented earlier, the slope of the constant availability curves is a measure of the marginal substitution ratio ($\Delta M/\Delta R$) which has values approaching infinity and zero in the regions concerned. This leads to the inequality discussed and the actions proposed.

Lines of constant marginal substitution ratio have been shown by the dotted curves on figure 9.18. These are curves connecting points of equal slope or inclination of the constant availability curves (for this reason they are often referred to as isoclines). The particular isocline corresponding to the inverse cost ratio (equation 9.24) will give the optimum combination of reliability and maintainability at each level of availability. It will be observed that the optimum combination differs for different availabilities. This assumes, of course, that the cost ratio remains constant over the range being considered – a fair assumption over small ranges. It follows from this that if we wish to improve the availability of a piece of equipment it is necessary to follow an isocline to maintain optimum conditions, that is, both reliability and maintainability must be increased together, in the proportion indicated by the relevant isocline.

It would be as well to note from figure 9.18 that at low levels of availability the isoclines are bunched closely together so that the optimum combination is insensitive to the cost ratio, and in these circumstances it may be possible to use rules of thumb and experience in deciding combinations of reliability and maintainability. As the availability increases the isoclines diverge and the constant availability lines are less curved: it follows that the optimum combination of reliability and maintainability is more sensitive to the cost ratio and hence a more exact knowledge of costs is necessary.

Figure 9.18 has been derived for a simple situation. Curves for more real situations will differ quantitatively, but will be generally very similar and the same reasoning applies. The difficulty here, as in so much mechanical reliability work, lies in attributing figures to the quantities involved in any particular set of circumstances.

In the above approach, which is in the form usually presented in the literature, it should be noted that $C_M$ is the total cost of maintenance over the planned life, as indicated under equation (9.19). It is therefore a function of the reliability (since higher reliability means a lower total maintenance cost, irrespective of the cost of each maintenance action), and the variables have not been separated as suggested by the form of equations (9.23) and (9.24). It might be better to define $C_M$ in terms of the cost of a maintenance activity rather than the total cost. Let $C_M'$ be the cost of a maintenance activity. Since only failed items need maintenance with a repair maintenance policy, equation (9.19) becomes

$$C = C_R + (1 - R)C_M' + C_O \tag{9.19a}$$

A number of versions of equation (9.19) can be set up to represent various maintenance policies. However, carrying out the analysis using the simple version at (9.19a) leads to the result

$$\Delta R\,(\Delta C_R - C_M') = \Delta M\,\Delta C_M \tag{9.23a}$$

for minimum total cost. Although less exact than equation (9.23), it does show that the left-hand side is likely to be much smaller than is immediately apparent from equation (9.23). Unless steps have already been taken to optimise, $\Delta R_R(\Delta C_R - C_M')$ will probably be less than $\Delta M\, \Delta C_M$.

Thus following the same reasoning as before it would seem more economic, in all probability, to build in additional reliability rather than maintainability, at least to a point when any further increase of reliability is markedly more expensive. It is a gross over-simplification to postulate that low reliability can always be compensated for by easy maintenance. It may be possible when the maintenance is itself very easy and cheap, but there is always a combination of reliability and maintainability which gives a minimum total cost, and generally this optimum involves high reliability. The claim that reliability and maintainability are interchangeable is, consequently, very much a conditional one.

## 9.7 Durability

It has been stated that maintained equipments, unlike components, never wear out. However, it is possible for such equipment to become uneconomic to maintain beyond a certain age. Suppose we have an item of equipment made up of intrinsically reliable components. It is unreasonable to expect that the lives of intrinsic reliability of all components will be the same, so that as the equipment ages an increasing number of components will need replacing. Instead, therefore, of the cumulative maintenance cost being a linear function of age, it will increase at an ever higher rate as shown on figure 9.19. The curvature will be further

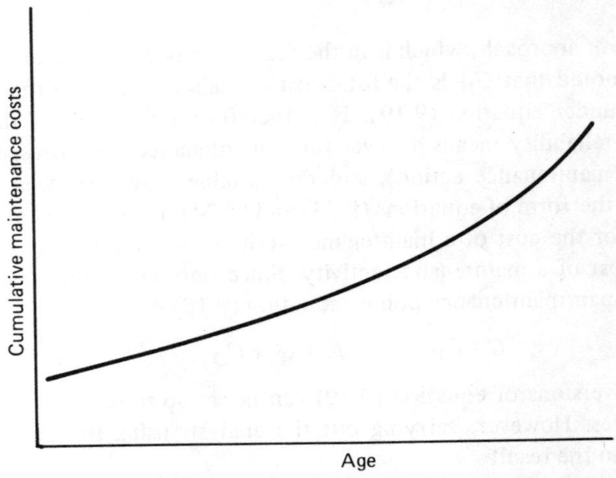


**Figure 9.19** Typical variation of the cumulative maintenance costs with age for reasonably complex equipment

accentuated by the natural endeavours of designers to achieve a longer life for the more expensive components. Clapham (ref. 57) has suggested that the logarithm of the mean cumulative maintenance cost will be a linear function of the logarithm of the age. A typical curve is illustrated in figure 9.20. In that case the mean cumulative maintenance cost will be given by

$$C_m = Bt^k \tag{9.27}$$

where $B$ and $k$ are constants. The value of $k$ generally increases with the complexity of the equipment (number of components). It would, of course, have a value of unity for a single component equipment in the pseudo-random phase. As a guide we could expect values in the range 1.3 to 1.6 or so for automobiles and up to about 2.2 for armoured fighting vehicles and locomotives. The total cumulative cost will be given by

$$C_c = A + Bt^k \tag{9.28}$$

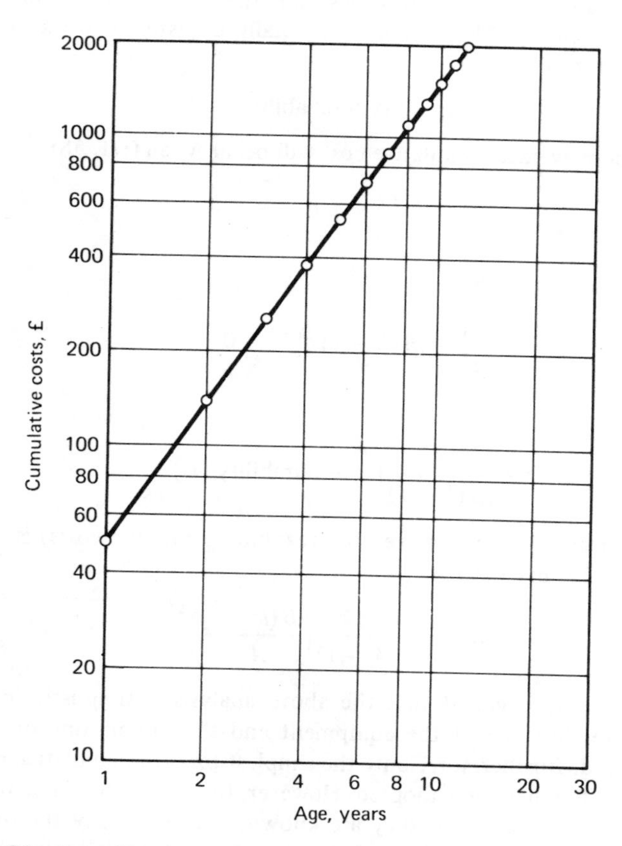


**Figure 9.20** Actual cumulative maintenance costs for Land Rovers operated by U.K. Army (1970 values). Data from ref. 58

where $A$ is the acquisition cost of the equipment. It follows that the average total cost per unit of operating time is

$$\overline{C}_c = \frac{A}{t} + Bt^{k-1} \tag{9.29}$$

The cost represented by (9.29) decreases with age until a minimum is reached, thereafter increasing continuously with age. One would expect the wise operator to continue using the equipment until this minimum is reached and then replace it. By so doing, his average cumulative cost will be reduced to a minimum. Extending use beyond the age for minimum average cost must incur a higher cost. The age at which minimum average cumulative cost occurs may be regarded as the durability of the equipment. We therefore define durability as "The ability to resist the adverse effects of environment, use, and repair with the progress of time."

A word of caution. Some authorities omit the words "and repair" from their definition of durability. This would seem totally unjustified to the writer, for in the absence of repair

$$\text{durability} = \text{reliability} \tag{9.30}$$

The minimum average cumulative cost will occur when (ref. 58)

$$\frac{\mathrm{d}\overline{C}_c}{\mathrm{d}t} = 0 \tag{9.31}$$

Hence from equation (9.29)

$$\frac{A}{t^2} - B(k-1)\,t^{k-2} = 0 \tag{9.32}$$

or

$$t = \left(\frac{A}{B(k-1)}\right)^{1/k} = \text{durability } (D) \tag{9.33}$$

The minimum average cumulative cost (excluding running costs) is obtained by substituting (9.33) in (9.29) as

$$[\overline{C}_c]_{\min} = \frac{Ak}{(k-1)}\left(\frac{B(k-1)}{A}\right)^{1/k} \tag{9.34}$$

It should be appreciated that the above analysis is simplistic in that it has ignored any resale value of the equipment and the discounting of cost has not been considered. Furthermore, only the simplest form of cumulative maintenance cost representation has been adopted. However, there is no fundamental difficulty in introducing these factors if they are known. That is usually the rub. They are generally not known in advance, and since the durability is very dependent on the numerical values its estimation becomes elusive. Nevertheless we should note that there is an economic limit to the time for which a piece of equipment may be operated and repaired. This is its durability.

*Exercise 9.9*

What is the durability of the roof constructed of tiles having the wear-out pattern described in exercise 5.2? For the purpose of this exercise you may assume that the owner replaces all the tiles that have failed during the previous ten years on a ten year cycle. The cost of replacing a tile is, on average, twice the initial cost and the cost of removing all the tiles from the roof is one-fifth of the cost of the original tiling. You should base the durability of the roof on that of the tiling alone and you may neglect the effects of inflation or the mortgage interest cost in your assessment.

*Exercise 9.10*

If the actual cumulative cost followed equation (9.27) they should yield a straight line when plotted against age on log–log paper. However, real data sometimes plots as in figure 9.21 (from ref. 59) in which two distinct patterns with different values of $k$ (in this case 1.65 followed by 1.15) can be identified. Explain why this should occur, and indicate any practical significance the phenomenon may have.

Durability has been defined in terms of the population average, but in practice the situation often arises in which a very expensive repair estimate to an individual item suggests it would not be worth while even during its durable life. This arises because individual repair costs occur at discrete intervals in discrete amounts, as illustrated diagrammatically on figure 9.22 as contrasted with the theoretical continuous curve of figure 9.19. It is suggested in ref. 60, therefore, that decisions should be made for individual items rather than for the population as a whole so as to avoid doing costly repairs to some items, and to take advantage of the fact that others run for long periods with comparatively little repair requirements.

When an equipment requires repair we are faced with a choice between two alternatives

1. Repair it
2. Scrap it and substitute a new one.

Now at any time, $t$, the notional value of any individual item will be the total cumulative cost (given by (9.28) above) less the optimum economic operating cost (given by (9.34) above multiplied by the time of use, $t$). If the repair cost is greater than the notional value, there is no way in which the cost of operating that item to the group life can be made equal to or less than the group average. In that case we should scrap it and substitute a new one. If the repair cost is less, then it is possible that its operating cost could be less than or equal to the group average. In that case we repair it, and continue to consider repair of each failure until such

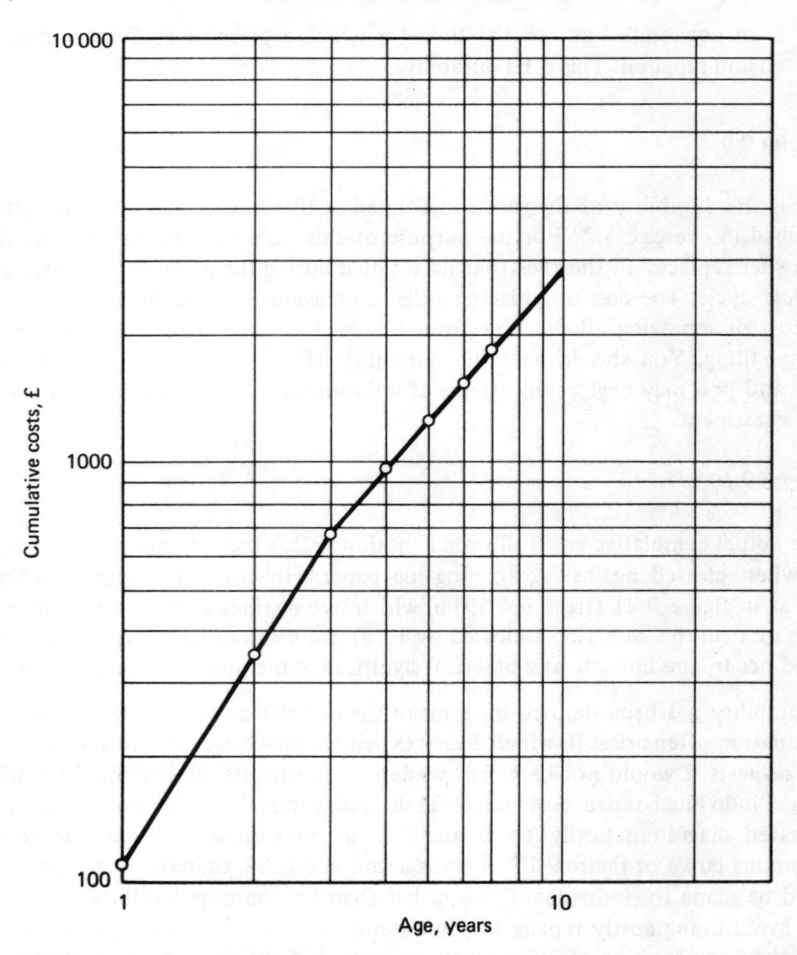


**Figure 9.21** Cumulative maintenance costs for a saloon car (1970 values). Data from ref. 59

time as the repair cost is greater than the notional value. The repair limit cost is therefore equal to the notional cost, or expressed analytically

$$r(t) = A + Bt^k - \frac{Akt}{(k-1)}\left(\frac{B(k-1)}{A}\right)^{1/k} \tag{9.35}$$

If at any time, $t$, the estimated cost of repair is greater than $r(t)$ then the item is scrapped. If at the time, $t$, the estimated cost of repair is less than $r(t)$, then the item is repaired and restored to service. In this way those items whose actual cumulative cost would be greater than the simple group average cumulative costs are retired early but the services of those items whose actual cumulative cost would be less than the simple group average are retained. This 'weighted' group

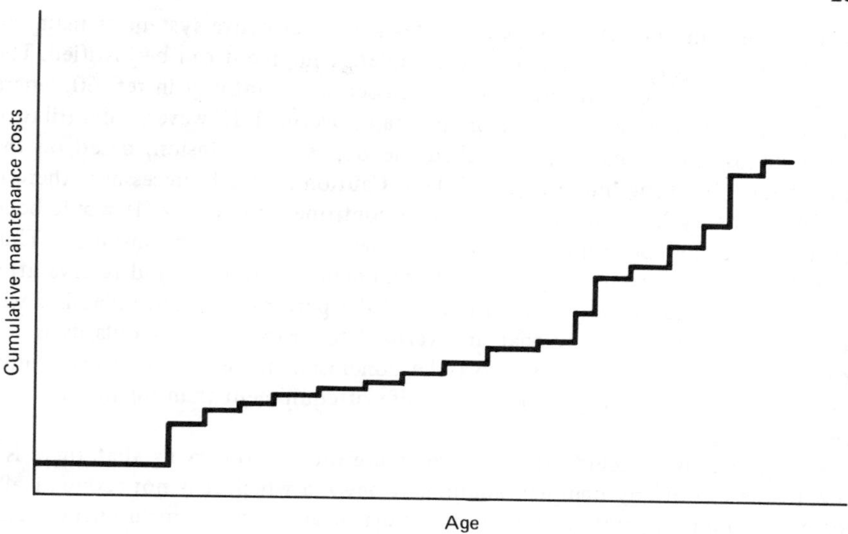


**Figure 9.22** A more realistic presentation of figure 9.19 showing how maintenance costs actually involve expenditure of discrete amounts at discrete time intervals

average must be equal to or less than the simple group average cumulative cost. A repair limit cost criterion therefore offers the possibility of maintenance costs which are less than the simple group average given by equation (9.34). It should be noted that operating a repair limit policy will of itself modify the cumulative cost curve. A second estimate can then be made on the basis of the revised data (see ref. 60) and the process repeated again, if desired. However, all the caveats relating to group replacement apply to repair limit cost replacement, and the higher accuracy giving lower cost may be even more illusory.

The concept of durability has been developed around the equipment as a whole. It may, however, have been equally validly developed around a major sub-assembly. Thus durability may apply to either sub-assemblies or to whole equipments, though it is essential that the sub-assembly should be reasonably complex, otherwise $k \to 1$ and the durability approaches an infinite life. Nevertheless, the concept is usually applicable to major sub-assemblies. If a number of these reach their durable life at roughly the same equipment age, it is possible to overhaul the items concerned to restore the equipment to the 'as new' condition. However, three factors make overhaul disproportionately more expensive than component replacement. First, there will be the large number of components with expected lives greater than the time between overhauls, but less than twice as great: all these must be replaced. Second, the process is usually a long one and its frequency and duration are major factors in determining the backing of complete equipments needed. Third, it is likely to involve the movement of equipments over large distances to appropriate workshops. It seems open to question then, whether, given a thorough knowledge of failure rates and an effective system of maintenance, the need for the overhaul of the complete equipment can be justified. The economics of the procedure have been discussed quantitatively in ref. 60, where some doubt is thrown on the value of any major overhaul. However, a contributor to the discussion of that paper reached the opposite conclusion, based on the same theory, by using 'more realistic' data. Caution is clearly necessary: there is no lack of theory but a marked absence of confirmed input data. It would be as well to take some cognisance of practical experience in such circumstances. Thus the U.S. Army relegates many kinds of equipments to training and reserve after one overhaul, suggesting that after overhaul the performance and reliability will be reduced, or alternatively that an overhaul to restore these standards is too expensive. It would seem to support the conclusion in ref. 60 that overhaul is less effective at extending the economic life of equipment than might have first been supposed.

We may safely conclude this section, therefore, by observing that there is a limit to the age of any complex equipment beyond which it is not economically worth continuing operation with the support of any form of maintenance. This defines the durability of the equipment. We note, however, that it may be very difficult to assess quantitatively because of a lack of technical data and of some economic factors.

*Exercise 9.11*

The central heating boiler in your home has been installed for eight years, and has developed a serious leak. Inspection has shown that several internal leaks have been present for some time and hence all gaskets and the faulty backplate will have to be replaced. Complete dismantling and re-assembly of the boiler will be necessary at an estimated cost of £150. The boiler originally cost £550 (adjusted for today's prices) and the amounts listed below (also at today's value) have been spent on annual maintenance. By comparison with the costs of colleagues who have the same type of boiler, you have good reason to believe your maintenance costs to date have been about average. Would you go ahead with the repair or scrap the boiler and install a new one?

| | | |
|---|---|---|
| End of 1st year maintenance cost = | £42.5 | £42.5 |
| End of 2nd year maintenance cost= | £49.5 | £92.0 |
| End of 3rd year maintenance cost = | £63.5 | £175.5 |
| End of 4th year maintenance cost = | £44.5 | £220 |
| End of 5th year maintenance cost = | £115 | £335 |
| End of 6th year maintenance cost = | £55 | £390 |
| End of 7th year maintenance cost = | £105.5 | £495.5 |

## 9.8 Review of maintenance aspects

It has been shown in the previous sections that maintenance has a major role to play in controlling mechanical reliability. Maintenance and reliability are inseparable, but have individual roles. The essential characteristic of maintained equipment is that in the event of a component failure the equipment is not physically thrown away, and we have to consider what happens thereafter. Yet this is only a part truth: if we are only concerned with mission reliability the item is effectively discarded from the active (though not total) population on first failure – unless it can be repaired and returned in time to complete the mission successfully, when the event does not constitute a failure (see exercise 9.7). In calculating mission reliability even the most complex equipment is modelled in the same way as a component having a large number of potential failure modes. Thus in rough loading the overall reliability is the lowest of the dependent mode reliabilities, whereas in smooth loading it is the product, and so on. So far as a mission is concerned, the first failure of a component masks any other potential failures of components, which cannot then arise on that mission. However, even if the equipment is discarded from the mission, it is not discarded from the population as a whole but will be returned to base or whatever for repair. It does not then become a new equipment because the non-failed components will be retained and hence have a failure rate appropriate to their age. Thus the state after repair is often called 'bad as old' – in contrast with 'good as new', although some improvement in capability must have been introduced by the new component – assuming the repair was carried out efficiently. The reliability for the next mission has to be calculated with the new age mix of components, and so on. If the state of the equipment is being represented by a global failure distribution, this is achieved approximately by continuing to use the same failure distribution – the 'bad as old' criterion.

The description 'bad as old' appears to have been first coined by H. Ascher, but has been used widely since without acknowledgement, and contains important concepts that warrant further examination. Suppose one has an item in which every component has only random failure modes. Then if the failure rate of the individual components is represented by $\lambda_i$, the initial failure rate will be given by $\Sigma\lambda_i$, assuming smooth loading so that the product rule is applicable. If component $j$ fails and is repaired, the failure rate after repair is still given by $\Sigma\lambda_i$, that is, the failure rate is equal to its initial value. The item after repair is therefore 'as good as new'. This would obviously apply to electronic equipment. If, however, the failure rates are not constant but are represented by $\lambda_i(t)$, then the initial failure rate is $\Sigma\lambda_i(0)$, and the failure rate after repair to component $j$ at time $T$ is given by

$$\Sigma\lambda_i(T) - \lambda_j(T) + \lambda_j(0)$$

This is clearly not the same as its initial value. If the difference between $\lambda_j(T)$ and $\lambda_j(0)$ is ignored, the failure rate is still not equal to its initial value but to the value just before repair – the 'bad as old' state. If the loading roughness is high, the above summations have to be replaced by the minimum values when a series system is being considered, or the maximum value for a parallel system, but the same conclusion is reached. It is concluded therefore that if a system consists of components having random failure modes (most electronic equipment) the state after repair is 'good as new', but if any components have a wear characteristic (most mechanical equipment), then the state after repair is 'bad as old'. Any mathematical modelling of systems which are maintained or subject to repair in any way must conform to this conclusion or else serious error can obviously be incurred. Note this implies that if only random failures are involved, repair action can effectively be ignored in calculating overall reliability but this is not possible with wear-out.

The conclusion in the previous paragraph tallies with experience. If I buy a TV set and the tube fails after say 2 years, the set is as good as new after replacement – perhaps a bit better because any early life failures of other components will have been eliminated. If on the other hand I buy a car and the clutch fails after 2 years, the car will certainly not be as good as new after replacement – the tyres will have some 20 000 miles of wear, to take just one item as example. Although in many cases some improvement will be achieved by the repair of worn components, this need not *always* be so. The improved reliability which is achieved by replacing one old component (that is, the failed one) by a new one may be cancelled or even negated by other consequences of the maintenance. For example, some replacements involve a modification of the tolerance stack-up and hence a disturbance of surfaces which have run in. This can accelerate subsequent wear. The failure pattern of disturbed but not replaced components can then show a higher failure rate than before maintenance. Note, this has nothing to do with the efficiency of the maintenance. It arises from the different 'fit' of worn parts when mixed with new. The actual failure pattern of components subject to wear after maintenance can be markedly different from unmaintained ones. 'Bad as old' is just one possibility.

On the other hand we may not be concerned with reliability, but with the maintenance process itself. In that case, every expected component wear-out failure will occur at some time and constitute an incident requiring attention. The total failures are then the sum of the component failures, whatever the loading roughness.

It has been shown that even for a single component, the occurrence of failures will tend to a constant value at high ages because of the random age mix which develops with repair over extended time. This is brought out on figure 9.1. It is even more likely to happen to a sub-assembly with a mix of components of different (though comparable) intrinsically reliable lives. It is then very attractive theoretically to replace the actual failure pattern of components, or sub-assemblies, by a constant 'failure rate', equal to the ultimate steady occurrence of failures, which can be considered to apply to all ages. It is a 'safe' approximation. The attraction is, of course, that the whole of constant failure rate theory which has been developed for electronic equipment can now be applied to mechanical items. Within limits this approach is valid, but it would be completely wrong to use it as a generality. As illustration, the time at which preventive maintenance is carried out to arrest some wear mechanism will control the ultimate steady occurrence of failure, and if the method is to be adopted the 'failure rate' appropriate to the maintenance policy must be used. There is no unique value which can be attributed to the component or sub-assembly itself. It would also be as well to recognise that the pseudo-random phase is only established over several generations with a repair maintenance policy. With scheduled maintenance the occurrence of failures can be far from constant, as illustrated by the full line in figure 9.23, which shows the life time occurrence of failures for an item failing by a wear mechanism when it is replaced on a scheduled basis at the 10 per cent cumulative failure age. Very many generations would be required to establish pseudo-randomness in this case, whereas the dotted line shows it is quickly achieved with repair maintenance. Additionally, the constant failure rate approach is only valid when applied to similar failure modes, and can be extremely misleading if applied to sub-assemblies, with multiple varied failure modes. As an equipment ages an increasing number of components pass out of their intrinsically reliable life and start to exhibit wear-out failures, so that increasing failures arise because more and more components

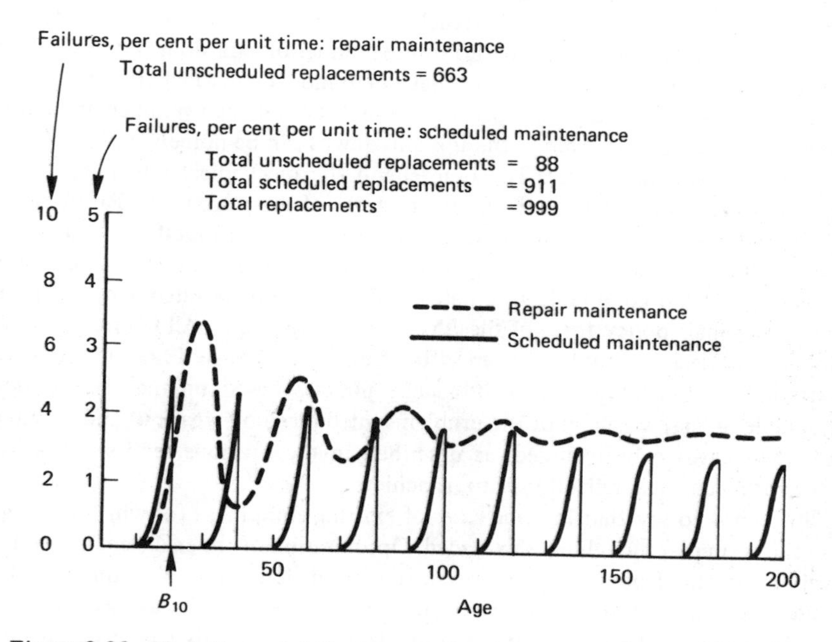


**Figure 9.23** Variation of failure rate with age for repair and scheduled maintenance policies. Non-maintained life pattern: Weibull ($\beta = 3.44$, $t_0 = 10$, $\eta = 20$). Scheduled maintenance interval: 20. Total life: 200.

are involved, even though each ultimately may have a constant pseudo-random failure rate of itself. Thus ever more maintenance is required with age. It gives rise to durability, as discussed in section 9.7.

For most purposes it is necessary to use the true failure rate (or occurrence of failures) appropriate to the age of the equipment and the maintenance policy being applied. This is unlikely to be constant for any other than the simplest of items subject to repair maintenance.

The author also has a conceptual difficulty, which appears to be shared by most mechanical engineers but not by most electronic/electrical engineers or most statisticians, in representing a multi-component sub-assembly by a single entity to which can be assigned a comprehensive failure rate, be it constant or not. To the author, the failure rate of such a sub-assembly ('component') must be made up of the constituent failure patterns (which can vary widely) and the maintenance policy applied to each. Take as an example the engine of a motor car: there can be no representative failure rate for such a sub-assembly, even if we confine ourselves to a given size bracket used in a given type of car. Design differences are so great that no universal figure can be quoted, other than on the basis that the ups and downs between components cancel out to give a workable unique value. This seems to imply that whatever designers do it makes no difference to the failures (that is, the reliability) – a palpably absurd conclusion. The concept of a sub-assembly as an entity can only be reached by ignoring the subjective nature of mechanical engineering design and the infinite number of variations in the solutions to any problem. Furthermore it does matter (to most people) whether the failure mode is a fan belt breakage or a piston burn-out, even if the overall engine failure rate is the same, that is, even if the ups and downs can be numerically set against each other model for model. To comprehend the mechanical reliability of any item it is necessary to establish the failure patterns of the vital (in the Pareto sense) mechanisms applicable to that item and the way they are modified by maintenance. As has been stated frequently before, each failure mode must be treated separately: it is no good collecting data on the rotting of wooden window sills to establish a repair policy for roof tiles! Yet this is often done. All failure modes on complicated pieces of equipment are often bulked (as I have done for a house to illustrate the absurdity), and maintenance policies based on that distribution. Intractable as may be many of the problems introduced by wear-out and maintenance, they have to be included, as described in the previous sections, if a true picture of mechanical reliability is to be achieved.

This is not to say that in some cases of existing equipment for which adequate data exists simpler modelling is excluded. Optimisation of the maintenance is then largely a matter of operational research, since that determines $K$ – and as such is outside the scope of this book. Even so, significant economic savings can sometimes be made on the basis of quite crude data. This is usually true if no previous optimisation has been undertaken. The calculation follows the methods outlined in section 9.3. Of course, it is often all one can do with existing equipment. It must be emphasised, however, that such modelling optimises an existing system, and is only valid when data applicable to that system already exists – reading across from one mechanical system to another is unjustified. It is an axiom of statistics, often ignored in comtemporary studies, that extrapolation is not valid. Such optimisation does nothing to the system itself, and if the system is second rate the optimum will be second rate. A first-class system can only be established by good initial design.

We have devoted a lot of attention to maintenance in this book on mechanical reliability, though we have concentrated on those aspects which affect reliability and have made no attempt to cover the subject as a whole (that would require another book). The inclusion of such material is considered fully justified. Any properly designed mechanical equipment should only fail in a wear-out mode and since all wear can be controlled by maintenance, all mechanical reliability is consequentially controlled by, or can be controlled by, and is hence dependent on, maintenance. This fact and its importance is often not recognised. Thus, for example, much of the poor reputation for the reliability of British cars is probably not so much due to poor manufacture (there are exceptions), which evidence suggests is little, if at all, worse than cars of foreign manufacture, as to the very poor servicing and maintenance by inadequately run and staffed garages. This contrasts with the usually keen and enthusiastic garages initially holding overseas franchises; but it is noticeable how, even with these garages, customer satisfaction is decreasing as they grow and adopt modern management. Design, manufacture and maintenance must go hand in hand if customer satisfaction is to be achieved. Much of good maintenance is good workshop practice. We have not discussed that aspect at all in this book, since it is assumed to be the background of anyone reading this book who is going to operate in the field of mechanical reliability. If it is not so, the first step becomes very clear!

What we have attempted to do is show that the maintenance policy can be optimised for the failure pattern and the costs attributable to various maintenance schemes. This involves maintainability, but also brings in all the administrative and managerial aspects of maintenance. They are most often not amenable to straightforward analysis, but can completely dominate the maintenance activity, and in turn control the reliability that is achieved and experienced. This is what makes mechanical reliability so intractable. If cost is the criterion, and it is difficult to see how that can be contested, we must accept the levels of reliability and maintenance which minimise that factor. This can lead to anomalies: for example, more effectively administered maintenance will lead to a lower value of $K$, and hence possibly to a switch from scheduled maintenance which retains intrinsic reliability to a repair maintenance policy for an item failing by a fatigue mechanism, or to scheduled maintenance at a later $B$ life for other mechanisms. This implies that unanticipated failures (that is an apparent lower reliability) will be introduced. There should be nothing surprising in this to a professional engineer: the optimum system does not necessarily comprise the best of all components. However, we should not lose sight of the fact that this economic optimum may be less acceptable. Unexpected failures can be deplored even when cost effective, and a psychological cost may be incurred. All such factors have been swept up in a global quantity, $K$, the cost ratio introduced in section 9.3 and defined by equation (9.14). The importance we have attached to $K$ is the importance we must also attach to these activities. Its determination can present some novel problems. Even so, it would be folly to pursue the matter to a depth that would enable $K$ to be estimated with great accuracy. It would profit us little because we cannot formulate the wear-out process with commensurate accuracy, certainly not at the design stage. Very often it is only possible when the equipment is obsolescent. Intelligent anticipation is the best we can hope for at the design stage. Let it also be recognised that numerical or analytical optimisation of maintenance such as presented in most of this chapter is often impossible in real life. We have only been able, by way of rigorous calculations of defined, simplified and representative cases, to illustrate that an optimum exists – and to isolate the controlling factors, together with their influence on reliability. By this means, though, informed decisions based on an understanding of the fundamental factors can be made in real situations. Our approach is an educational one. Although there may be some cases in which continuous functions can be used to represent the real thing, actual practice has often to deal with step functions. Thus if a certain bought-in component has a life of intrinsic reliability that does not quite match that of the other components of the equipment, it could well be impossible to extend it. The next best may be too expensive. More often than not, both reliability and maintainability come commercially in quantum steps. Hence we have often to compare the total cost – but it should be the total or whole life cost as far as it can be estimated – of a number of discrete possibilities. Quite crude calculations (of the slide rule variety) backed up by sound experience can be far more valuable than the best computer solutions based on unrepresentative modelling. If the first assessment is 'in the right street', there is every chance that a close approximation to the optimum can be achieved by fine tuning when development and early field data become available. A great deal depends on the designer's skill to achieve a close approximation to the optimum in the first place. More than a great deal – it is crucial to get this right, because once an item is in service it is often very expensive indeed to introduce modifications.

The intractability of reliability and maintainability may be discouraging, but if properly recognised it avoids unjustified decisions. It also lends support to the growing practice of adopting on condition maintenance. The unknowns are swept up in the monitoring process. Further growth of this technique is governed by the availability and cost of the monitoring equipment – and *its* reliability. Given satisfactory monitors, so many difficulties are overcome that it must be the way ahead. Even so, it is not the complete panacea, because its effectiveness and the provision of spares still have to be evaluated beforehand. It is also important to recognise that nearly all monitoring systems have to be introduced at the design stage. As in so much concerned with mechanical reliability, design is the key activity.

Perhaps the most important conclusion to reach from our study is that although items can be designed so that failures are negligible, or may be treated as 'an act of God', it is very often uneconomic to keep them in that state for the whole of their lives. Consequently, failures will arise because of wear, and may only be limited, rather than eliminated, by maintenance. It is the resulting 'failure rate' which is the source of 'unreliability' of mechanical equipment when the highest-class design and development are employed. It is the relevant quantity for all assessments and the one which is so difficult to evaluate because of the many disparate factors involved. Because maintenance plays such a big part in controlling the reliability, the author believes it must be included in the definition of reliability. Hence reliability is defined as:

> The ability of an item to perform a required function under stated conditions of use and maintenance for a stated period of time.

This chapter on maintenance is concluded cn a provocative note, stemming from one of the consequences of maintenance optimisation. It is very generally believed that fatigue is responsible for about 80 per cent of unscheduled failures: some authorities would put it higher, at 90 per cent (see ref. 61). This is usually quoted as an indictment of mechanical engineers and designers, or as a dire warning to designers to be ultra cautious when dealing with fatigue. Now at a conservative estimate we can assume that fatigue is the failure-controlling mechanism in 50 per cent of the wear processes. From our studies in section 9.3 it can be asserted that for at least 50 per cent of these the optimum policy would be repair maintenance. It follows that from this cause alone (fatigue), 25 per cent of the total unscheduled failures should be treated as planned failures. For the remaining 50 per cent of the population subject to non-fatigue failure-controlling mechanisms, section 9.3 suggests that replacement before not more than 10 per cent of that population had failed may be taken as a good generalisation – indicating a planned unscheduled failure rate of about 5 per cent of the total population from that cause. Hence about 30 per cent (25 per cent + 5 per cent) of ***all*** potential failures are planned failures, and the remaining 70 per cent should be prevented by scheduled maintenance. Of the total planned unscheduled failures we should expect 83 per cent (25 per cent/30 per cent) to arise from fatigue! The nature of the above approximations are such that it could be a little higher. This theoretical estimation of global optimum unscheduled fatigue failures tallies so well with general experience quoted above (80 per cent to 90 per cent) that one wonders what all the fuss is about. We should expect the current level of unscheduled fatigue failures. Indeed, according to this assessment we should be concerned if it were *not* as high as this. Apart from some well-known 'clangers', is then fatigue the bogey so often pictured? Can it be that mechanical engineers have got it right after all?

# 10 Design for reliability

## 10.1 The design process

Design is a vast subject with never-ending ramifications; a subject which often means very different things to different people. Yet is is probably the keystone to reliability. Presenting the 1974 James Clayton lecture to the Institution of Mechanical Engineers, which was intended to summarise eight conferences or symposia held by that body from the mid-sixties to the mid-seventies, the author concluded "the significant theme which has recurred over and over again at conferences . . . . . . is the matching of the design to the requirements of the user in its intended environment, and the necessity of good communications to achieve it. Common sense tells us this is so, and mathematical theory supports it by revealing the critical role played by the safety margin and by the loading roughness. . . . . . All my theoretical and practical studies of the subject, which fully support each other, have led in one direction only – the importance of design and development in achieving quality. This stands out above all else." Clearly, then, there are good and valid reasons for taking this subject seriously.

The milestones which mark out the life of a piece of equipment from the time it is first conceived until it is finally discarded will differ from organisation to organisation. However, a certain pattern does appear and may be represented by the following simplified ideal version of the steps from concept to casting.

(1) A general target is formulated.
(2) A feasibility study is carried out.
(3) A specific requirement is prepared.
(4) A project study is carried out.
(5) The project is accepted and approval for development is given.
(6) Detail design proceeds.
(7) Prototypes are manufactured.
(8) Trials are run and evaluated.
(9) The equipment is accepted by the originator of the project (user, producer, etc.).
(10) The equipment is manufactured.

(11) The equipment is effectively used.
(12) Equipment is retired or cast.

It will be appreciated that this is not a rigid process and that new ideas may well be generated at, say, stage (8) and fed back into the system. Furthermore, the results of pure, applied and directed research, and maybe service operational experience of similar equipment, will be accumulating throughout the whole process and will be fed in at all stages. However, under the heading of design we must include all the steps (1) to (9) inclusive, regardless of their precise relationship in particular cases.

It will be convenient to group these steps, or their equivalent, under three main headings. These are

(A) formulation of the requirement
(B) the conventional design process
(C) development.

Each of these is taken up separately in the next three sections. It is necessary to point out that we shall be concerned only with those aspects of these activities which relate directly to reliability. We shall in no way be considering functional design or aesthetic design. For example, if we were dealing with a machine for shelling peas we would assume that the machine is fundamentally capable of doing that job to the customers' satisfaction (and comes in the colour they like!). Our concern is to ensure that it will continue to shell peas at the same level of customer satisfaction whenever the crop is ripe for the desired number of seasons. This division of the design process cannot be absolute, but is generally a convenient approach. If the reader wishes to pursue functional design there are good textbooks available (see Reading list), dealing with that subject in its own right. The Engineering Design Guides published on behalf of the Design Council, the British Standards Institution and the Council of Engineering Institutions are also an excellent source of general design information which by virtue of ensuring good design generally set us on the right road to acceptable reliability.

## 10.2 The requirement

The phase of operations which defines the requirement is covered by steps (1) to (3) of the programme set out at the beginning of this chapter. This can, or should be, a relatively controlled process when the user, design authority and sometimes the manufacturer are all part of the same organisation. This can happen with military equipment and sometimes in the civilian world; for example, British Rail operate in this way for much of their specialised equipment. It is an ideal worth aiming for. More often than not, however, these operations are split between different concerns and even in the all-embracing organisation many components or even sub-assemblies have to be brought in, each with their own specification.

In the general commercial world, steps (1) to (3) would be taken by the user of either large and expensive equipment or of large numbers of equipment. The successive steps would then be implemented by contractors and sub-contractors. In other cases the requirement may be formulated, through steps (1) to (3), by the manufacturer; sometimes in close consultation with the user, on other occasions or circumstances through market research involving maybe little contact with the individual user. As an extreme procedure, the producer formulates his own requirement and creates the demand by advertising.

Whatever the basis controlling the procedure, it would generally be assumed that responsibility for steps (1) to (3) rests with 'higher management', since they largely control policy. At the same time it must be recognised that formulating the requirement essentially includes some specification of the environment or load; that is, $L(s)$ is largely defined. We have already seen that this quantity plays a dominant role in determining the reliability. So if any degree of reliability is to be finally achieved, the first actions towards its achievement must be taken in steps (1) to (3) of the above programme.

It will at once be recognised that it is one thing to set up a mathematical model in which some functions $L_i(s)$ represent each and every load, but it is quite another to interpret these physically. Clearly $L_i(s)$ will not be known in exact terms; but this should not prevent a representation being devised, based on intelligent anticipation (or even inspired guesswork!) – for leaving the load definitions to pure chance is even more reprehensible. We can usually allocate a value to the nominal mean load, and then make additions for the various adventitious effects which may be anticipated from a thorough appreciation of the requirement or the circumstances in which it is anticipated the equipment will be used. The allowances may well be crude and arbitrary, but at least they are better than no allowance at all. Perhaps because higher management scorns the scientific approach, there are often virtually no systematic data available on which to base these estimates. This is one of those areas where much progress could be made, for it may not be without significance that in the two fields where reliability is paramount – aircraft and nuclear engineering – many data do exist. It is recognised that it is difficult, and sometimes expensive, to assemble all the relevant information, but it is more expensive not to do so. It has been shown in chapter 4 that the form of the load distribution is far, far more potent than the form of the strength distribution. This must be regarded as one of the most important conclusions of theory, and can only be ignored at peril. Yet how often are such imperatives ignored! For example, environmental conditions are a source of continual difficulty – temperature and humidity together with physical and chemical pollution are all factors obviously to be assessed. So too must the vibrational environment due, say, to transport and other causes (including accidental dropping!). These side effects may well be as important as any load in conventional operation, and can only be formulated with full knowledge of the intended method of use. One of the major obstacles to better design in this respect is the failure to set up good working communications between the user and the specification writer and subsequently the designer and the manufacturer. More than good documentation and formal meetings is necessary. Those concerned must absorb the total atmosphere surrounding the use of prospective equipment. It cannot be over-emphasised that in whatever manner the specification is drawn up it must convey full details of the loading to be expected. Without such information, high reliability – or even any specified level of reliability – can never be achieved.

In addition to specifying the conditions of use, the reliability to be achieved by the equipment will be evaluated in steps (1) to (3). Now we have already seen that the optimum reliability (that is, that for minimum cost) is controlled by the maintenance, and hence any limitations that have to be unavoidably applied to the maintenance procedure must be set out at this stage. It would be wise to note that a particular operator's requirements will depend on circumstances. Thus mission endurance is the time (expressed in any units) for which the item is required to operate substantially continuously and independently of any logistic back-up. It represents a period when the value of $K$ is particularly high so that high reliability is required. To this end a mission reliability can be related to a mission endurance for specification purposes. Once the equipment has completed the mission, scheduled servicing and minor first-line maintenance may be acceptable, but major unscheduled first-line maintenance or any form of second-line maintenance could be unacceptable. The duration of this phase of operations is difficult to define. In a military context it is usually taken as the period for which men of a unit can operate efficiently without a complete stand down, which experience suggests should be about five times the mission endurance for land-based equipment. The periods of a day and a week are general civilian counterparts leading to the same factor of five. However, the duration, intensity and pattern of operation vary so greatly that no generalisation is possible and maintenance constraints arising from operational factors must be included in any specification. Beyond the medium term, a more flexible approach is possible in which the operational pattern may well play a secondary role to the technical requirements, recognising of course that the ultimate object is to achieve the optimum reliability ensuring minimum whole life costs.

It is also necessary to recognise during the specification stage that performance and reliability cannot be entirely divorced. It is self-evident that long life and high reliability are associated with conservative performance levels and operating conditions. On the other hand one assumes that the object of a new design is higher performance. The consequences of higher performance in terms of reliability must always be asssessed. While failing to specify a reliability in these circumstances may be one sin, over-specifying reliability requirements can be an even greater sin. Very high reliability has to be paid for, both in design and production. Furthermore, excessive reliabilities necessitate extensive development trials and testing before they are achieved. In spite of this many managers drawing up specifications think it 'good form' to ask for something 'a little bit better than we had last time' and, without due consideration, place on it values which have little hope of being achieved. Manufacturers do so in the apparent belief that glowing specifications will attract customers, and users on the simplistic basis that the more reliable the better (and it is too difficult to calculate the optimum anyway). The actual result can be either that the specification is treated seriously and the product priced out of the market, or the specification is ignored and the final reliability becomes a matter of chance. We repeat: the aim of every specification should be to achieve that reliability which gives the lowest whole life cost.

The same general principles apply to the smallest and largest items, though the specification for the reliability of a complex item will necessarily be more elaborate than that for a simple item. As illustration, the specification for a simple item of everyday use – a ball point pen – will be drawn up, but before doing this it is suggested the reader draws up his own specification. The solution to exercise 10.1 used for illustration is given in chapter 15. Then try exercise 10.2.

*Exercise 10.1*

Draw up a specification for the reliability of a ball point pen to be used by students of engineering at college.

*Exercise 10.2*

Draw up a specification for the reliability of demolition charges as used in building demolition etc.

The purchase of new equipment with which to carry out some function already performed by an existing item of equipment presents an additional problem. The reliability of new equipment is fairly certain to be initially lower than the existing equipment. This arises from manufacturing difficulties which must arise with new processes, untried items which may not live up to expectations, and a maintenance policy that is unlikely to have been optimised. All such factors can be rectified by time and experience, but the result is a short-term loss of reliability. It is for consideration whether the superior performance of the new equipment results in a reduction of costs which outweigh the increase of costs associated with the higher failure rate. In many cases 'stretching' or 'developing' the existing item may provide a more cost-effective solution, at least in the short term. The long-term demands are often more difficult to assess, but it should be recorded that the 'nettle' of new equipment has to be grasped periodically, otherwise stagnation must ensue. This is perhaps the one instance where the small consumer of mass-produced equipment is at an advantage over the large-scale user who has to place specific orders. Prestige considerations apart, the small-scale user can delay purchase until the product has proved itself (who would buy a new model motor car when it is first put on the market, for example?) whereas the large-scale user, particularly of special equipment, cannot avoid this short-term loss of reliability at some time or other. Planning should ensure that this occurs when it can most conveniently be accommodated – during an expected period of low demand, for example, when surplus equipment would be available to restore the reliability by parallel redundancy, or other suitable occasions.

Before leaving the subject of specification we should introduce the necessity for a design review. Design reviews are more closely associated with conventional design and are taken up in section 10.3.5. However, a number of organisations do like to carry out a preliminary review when the specification is taking shape. The object of a review at this stage is to provide a formal occasion when those who will subsequently be responsible for the project can voice any misgivings and highlight possible problem areas. The areas covered will be similar to those covered at the main reviews, though much of the treatment must be more superficial in the absence of specific design studies. Nevertheless a critical review of the feasibility study is essential in some form or other, and the conventional design review is a convenient vehicle, particularly in large organisations, or in circumstances in which the feasibility study group, the main design team and the manufacturing organisation are separated, or even different sub-contractors. It is essential if the design is at all innovative.

To sum up, formulating the requirement is an essential part of design, and it is at this stage that the foundations of high reliability are laid. In drawing up the specification both operational requirements and total or whole life costs should be taken into account, care being taken to ensure that all the actual operating conditions are specified as fully as is possible. Lines of communication outside the written specification should be set up between all those concerned with the design, testing and manufacture of the proposed item.

## 10.3 The conventional design process

Conventional design may be thought of as the processes covered by steps (4), (5) and (6) of the typical procedure outlined in section 10.1.

It has already been stated that this is not a design textbook, but it is possible to pick out certain aspects of design which have a particular impact on reliability. These can be itemised as follows

(1) First we must assess if the design is one for which failure would involve safety. If so, redundancy or fail-safe or other special techniques may be called into play.
(2) Consideration should be given to secondary functions that the item may have to perform.
(3) It is clear that a well-tried off-the-shelf item should be used whenever possible in preference to new design. Failing that, a modified off-the-shelf item may do the job. In either case it is essential to ascertain the prior failure history of the item and to establish whether the part is completely interchangeable with that of another manufacturer (which may involve proof testing). If an off-the-shelf item is selected it must be established that

its previously experienced environments and normal operating levels are not being exceeded – if indeed these limits are known.

(4) If a completely new design is necessary a search must be made for failures in models or similar research equipment. In any case the expected modes of failure should be listed, and the safety margin checked (see failure modes and effects analysis – section 10.3.5). An effort must be made to avoid any of the obvious defects notoriously responsible for unreliability – such as stress raisers; unsuitable or badly situated pipes, hoses or wiring; unsuitable or badly situated connectors; inadequate locking arrangements; and so on. It must be checked that the design is suitably stressed for environmental effects of vibration, temperature, humidity, dust, etc., and that all the components are physically and functionally compatible with each other. Finally, if selective assembly is not adopted it must be fully established that high and low tolerance components will properly mate. Checklists to ensure all these aspects are properly considered are discussed in section 10.3.4.

(5) In all cases is the design as simple as possible? What isn't there can't go wrong!

(6) Maintenance has a direct bearing on reliability. Full consideration must be given to the techniques that the operator should adopt – that is, repair, scheduled or on condition – and the physical methods by which a failure is to be detected and repaired. This can have a major impact on the design.

(7) Cost and ease of production must be treated in their own right, but it is essential that no 'value engineering' improvements be introduced at the expense of desired reliability. A real danger exists in the case of subsequent modifications to a proven design.

(8) The method by which an item is specified should receive adequate attention, especially where parts are made by sub-contracting organisations. It should be appreciated that only limited information can be conveyed by means of drawings and written words. Many items contain, even in this scientific decade, a great deal of 'know-how' in their manufacture which cannot be written into any specification. Proof tests and an 'approved' manufacturer may have to be written into a design.

(9) Arising from this it follows that facilitating inspection and laying down adequate inspection and test procedures are an integral part of design.

(10) Although operating conditions may be met, other adventitious loads due to handling, transportation and packing do sometimes exceed these. Storage time and conditions are additional factors which can induce wear-out failures even before the operating life proper has begun.

(11) Finally, one should try to give a numerical value to the reliability (realistically measured by a $B$ value) of each item and decide if it is adequate. It is worth considering how greater reliability could be achieved and what would be the consequences. The writer does not attach undue importance to numerical estimates themselves but they do stimulate critical thought and contingency design plans.

It will be noted that there is very little which is new in any of the above suggestions. Most of the points raised would be covered without any specific or conscious attention in all major designs. It is significant that a close examination of the causes of failure in many different types of equipment reveals, more often than not, that it is the straightforward error in the relatively cheap and often insignificant component that accounts for the majority of failures. The achievement of high reliability at this stage is mainly the application of engineering common sense coupled with a meticulous attention to details. While not a subject superficially of great academic attraction, it should be recognised that a systematic approach and a sound appreciation of the fundamentals of engineering are vital to success. It is therefore essentially an intellectual activity.

The nature of the design activity needs to be fully understood if we are not to stray far from our objective. Overall design is a creative activity, comparable with writing a novel or play or composing a symphony. It is an exercise in synthesis, and contrasts with science, which is concerned with analysis of the existing natural world. These activities involve diametrically opposed throught processes. The ability to create is largely inherent in a person's make-up and cannot be taught, though it can be encouraged and developed. This is best done by example and precept, in the same way that an appreciation of playwriting is best achieved in the theatre. A study of good (and bad!) designs 'in the flesh' is essential. Books are of limited value; real ironmongery is the essence of good training. Given the right human material, much can be achieved by example, practice and experience. Given the wrong material, there is virtually nothing that can be done. The first step in achieving a good design is therefore that of obtaining a good designer, and the second step is one of providing a sympathetic atmosphere in which he/she or they can flourish. We shall not pursue this matter further here. What we shall do is note that as the composition of a novel or a symphony must obey the basic rules of syntax and harmony, so too an engineering design must conform to certain technical rules. We shall be very much concerned with these, for in the same way that a great artistic creation can be marred by poor execution, a great engineering concept can be ruined by poor detail design. Very often, as we have already observed, this is the core of quality and reliability.

Outstanding among the technical demands is the necessity to provide adequate strength, using the word in its widest connotation, as in section 3.1. Conventionally, it is assessed in an idealistic calculation, and an arbitrary factor – the factor of safety – applied to bring the calculated values into line with experience. This is not good enough today. It is the height of absurdity to employ highly qualified staff equipped with modern expensive aids to estimate stresses to an accuracy of a fraction of a per cent, for example, only to multiply the result by an empirical factor which may not be known to an accuracy of better than 20 or 30 per cent. Some factors have more rationale than others, but the increased accuracy available from modern techniques can only be properly utilised by eliminating the empirical and unmeasurable, so far as we can possibly do so.

### 10.3.1 Design for stress-rupture phenomena

It has already been stated that the only route to high quality is intrinsic reliability, and figure 4.18 gives the minimum safety margins necessary to achieve this. Exercise 4.7 demonstrated how this concept may be applied practically. This would remove much of the empiricism, or at least limit it to the derivation of the load and strength distributions which can be quantitatively assessed. However, the calculation is a bit involved, and bears little relation to conventional calculations. We seek, therefore, in this sub-section a rational method which can readily be applied in the conventional design process and which can also absorb the wealth of existing empirical knowledge.

The strength of an item will be primarily related to the strength of the material from which it is made by

$$\frac{\sigma_S}{\bar{S}} = \frac{\sigma_P}{\bar{P}} = \gamma, \text{ say} \tag{10.1}$$

where $\sigma_P$ and $\bar{P}$ are the standard deviation and mean of the relevant material property in the as finished condition. Thus, in application to a simple tensile member, for example

$$\bar{S} = A\bar{P} \tag{10.2}$$

$$\sigma_S = A\sigma_P \tag{10.3}$$

where $A$ is the cross-sectional area of the member and $P$ is the proof stress or yield stress as the case may be. (It could possibly be the ultimate tensile stress or fatigue limit, depending on the nature of the design.) Similar considerations apply to other forms of loading. Some caution is necessary in interpreting $\gamma$ in physical terms, since it must include rather more than the basic material properties. It must be based on the material properties of the finished product, not a test piece, and must include any modification to the basic properties introduced by the manufacturing process, by the form of the item including tolerances, by the finishing process, and so on. However, no difficulty is seen in assigning a value to $\gamma$.

The condition for intrinsic reliability is then best handled by linearising the curve in figure 4.18. We can write each piecewise linearisation as

$$SM = m(LR) + c \tag{10.4}$$

where $m$ and $c$ are constants. Equation (10.4) can then be written as

$$\frac{\bar{S} - \bar{L}}{\sqrt{(\sigma_S{}^2 + \sigma_L{}^2)}} = m \times \frac{\sigma_L}{\sqrt{(\sigma_S{}^2 + \sigma_L{}^2)}} + c \tag{10.5}$$

Substituting $\sigma_S = \gamma\bar{S}$ from equation (10.1) into equation (10.5) and re-arranging gives the quadratic equation

$$(1 - c^2\gamma^2)\bar{S}^2 - 2(\bar{L} + m\sigma_L)\bar{S} + \{(\bar{L} + m\sigma_L)^2 - c^2\sigma_L{}^2\} = 0 \tag{10.6}$$

This may be solved for $\bar{S}$ to give

$$\bar{S} = \frac{(\bar{L} + m\sigma_L) + \sqrt{\{c^2 \sigma_L{}^2 (1 - c^2\gamma^2) + c^2\gamma^2 (\bar{L} + m\sigma_L)^2\}}}{(1 - c^2\gamma^2)} \tag{10.7}$$

where the positive sign has been taken for the square root term as being the one relevant to this problem. Equation (10.7) expresses the mean working strength, $\bar{S}$, in terms of known quantities, and solves our problem, but rather fascinatingly it can be rewritten as

$$\frac{\bar{S}}{\bar{L}} = \frac{\{1 + m(\sigma_L/\bar{L})\} + c\sqrt{[(\sigma_L/\bar{L})^2 (1 - c^2\gamma^2) + \gamma^2\{1 + m(\sigma_L/\bar{L})\}^2]}}{(1 - c^2\gamma^2)} \tag{10.8}$$

where $\bar{S}/\bar{L}$ is the conventional factor of safety (or simply replated to it).

Estimation of the working strength may be made directly from equation (10.7), but more attractively for most practising mechanical engineers, the old method of design using factors of safety may be retained by using equation (10.8). This analysis has shown that the factor of safety is just as valid a criterion as any other so long as it is recognised that the factor of safety is functionally related to the load and strength coefficients of variation (standard deviation/mean). The quantity, $\gamma$, $(= \sigma_P/\bar{P})$ is a fundamental property of the real material in its finished form and has physical significance and reality. The ratio $(\sigma_L/\bar{L})$ which appears in equation (10.8) is only a functional grouping and has no physical significance, since $\sigma_L$ and $\bar{L}$ are not related and may vary independently in an arbitrary manner. It is importand to note, however, that these two parameters, $\gamma$ and $\sigma_L/\bar{L}$, completely define the factor of safety, and no other factor can logically then be applied. Thus the so-called 'factor of ignorance' has no place in this approach. If there is any ignorance it applies to the contributing quantities themselves and any empirical correction should be applied at source, so that it is properly weighted.

The linearised curve used in subsequent calculations has been compared with the original safety margin – loading roughness curve on figure 10.1. It is readily seen that any error involved is well within practical limits. Using this linearised curve the minimum factor of safety for intrinsic reliability was calculated from equation (10.8). It is plotted against $(\sigma_L/\bar{L})$ for various values of $\gamma$ in figure 10.2. We should recall that the values apply to loads and strengths which have a Weibull distribution with $\beta = 3.44$, that is, near normal distributions. There is no difficulty in calculating these curves for any other known distributions. Even so, difficulty may be experienced in representing some tails. The most likely departure could be in the strength distribution, where quality control could well limit the tail. As an extreme case, therefore, the curve of minimum safety margin for intrinsic reliability against loading roughness was recalculated assuming an idealised quality control cut-off at $1.75\sigma$ from the mean. The method of calculation is given in the solution to exercise 4.4. The extent of this quality control is illustrated on figure 10.3, and represents the rejection of the weakest 5 per cent of the manufactured items. It is believed that a higher rejection rate would be generally unacceptable.

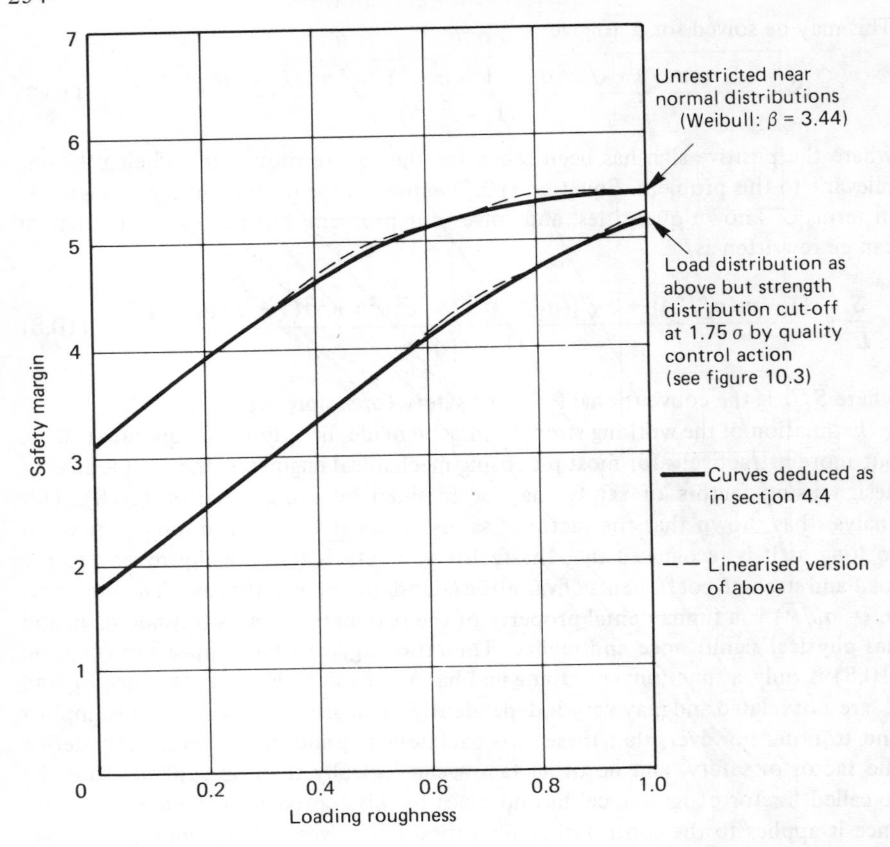


**Figure 10.1** The minimum safety margin required to achieve intrinsic reliability with unrestricted near normal distributions (Weibull: $\beta = 3.44$) and with restricted distribution to represent quality control action. Note that the values of the safety margins and loading roughnesses quoted apply to the distributions before quality control is exercised

The minimum safety margin–loading roughness curve and its linearised version is shown on figure 10.1, where it can be compared with that for the uncontrolled strength distribution. The corresponding factors of safety have been plotted on figure 10.4.

A comparison of figure 10.2 and 10.4 shows, as would be expected, that a curtailment of the strength distribution by quality control enables a lower factor of safety to be used while still ensuring intrinsic reliability. However, a comparison of the numerical values shows that the reduction is small if the strength distribution is inherently narrow and that quality control really pays off when the strength is more widely distributed compared with the load. The big savings are

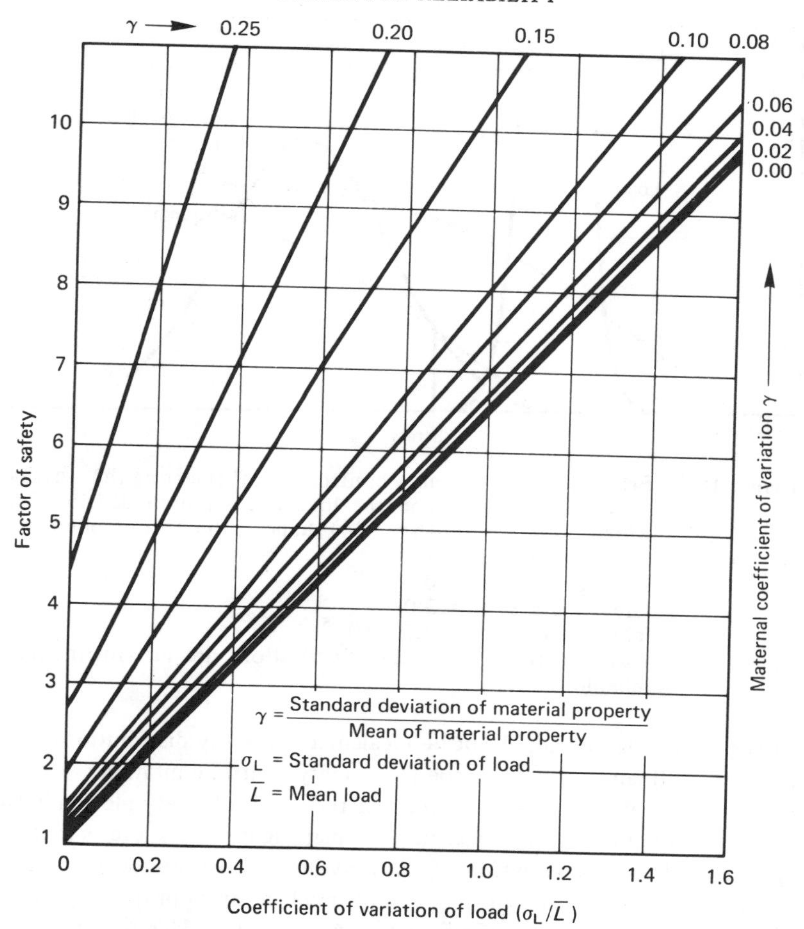


**Figure 10.2** Minimum factors of safety to give intrinsic reliability with near normal load and strength distributions (Weibull: $\beta = 3.44$)

obtained in smooth loading conditions. In those cases the gains can be significant, though it would be wise to recognise that they have been plotted for a reasonably heavy rejection of weak components by quality control. Additionally, an ideal cut-off has been assumed. We should note that quality control applied early in the manufacturing process is more likely to reduce the standard deviation of strength than produce the type of cut-off used in these calculations.

It would, of course, be unwise to use factors of safety applicable to a curtailed strength distribution unless it were known to be definitely applicable. We shall therefore refer henceforth to the curves on figure 10.2 for unrestricted near normal distributions for illustrative purposes. In real life there is no reason why

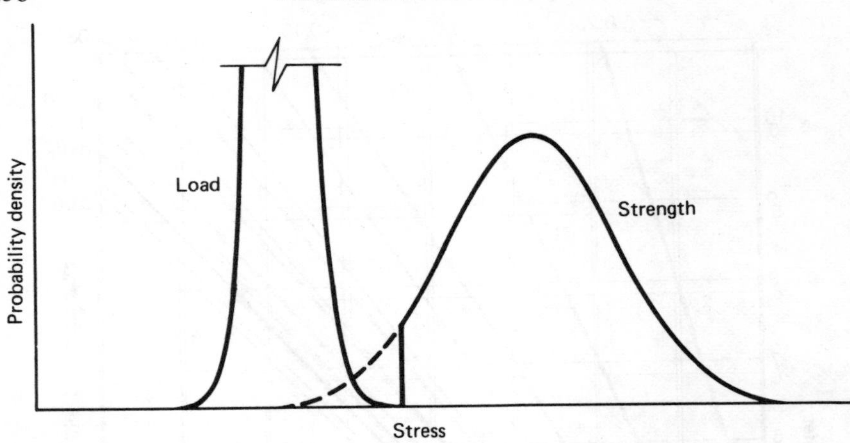


**Figure 10.3** Effect of a $1.75\sigma$ quality control cut-off rejecting the weakest 5 per cent of items on the load and strength distributions which have Weibull ($\beta = 3.44$) shape in the uncontrolled condition, for which
loading roughness = 0.3
safety margin = 3.0
failure rate = $2.5 \times 10^{-5}$
With quality control the distributions are just intrinsically reliable

the curves of this figure should not be recalculated for any distributions of loads and strengths with any perfect or imperfect quality control limitation.

One immediately notes that the values of the factor of safety plotted in figure 10.2 span the usual empirical range. In particular, figure 10.2 suggests that for an average mechanical item for which $(\sigma_L/\bar{L})$ may be equal to about 0.15, if it is not too much abused, and made in mild steel for which an appropriate value of $\gamma$ would be 5 per cent, a factor of safety of 2 would apply. This is just about the value one would adopt empirically. Incidentally, the loading roughness of this item is 0.83, so it is being treated reasonably. A more interesting example arises from the author's early experience as the designer of aero engine axial compressors. Here it was considered that the maximum load on a blade due to aerodynamic buffeting, vibration, etc. was about equal to the steady gas bending stress. Ideally, this could be covered by a factor of safety of 2 but a careful appraisal of successful compressors showed that a factor safety of 3 was necessary and was the empirical value used in design. Now let us adopt a more theoretical approach. Most of the variation in a quantity can be contained within ± 3 standard deviations, so in this case adventitious loading equal to the steady (mean) load would give a $\sigma_L/\bar{L}$ of 0.33. Reference to figure 10.2 shows that the value of the factor of safety should be exactly 3 for an average value of $\gamma$ of 5 per cent. Closer agreement could not be expected. The loading roughness in this case would be 0.91. It is roughish loading.

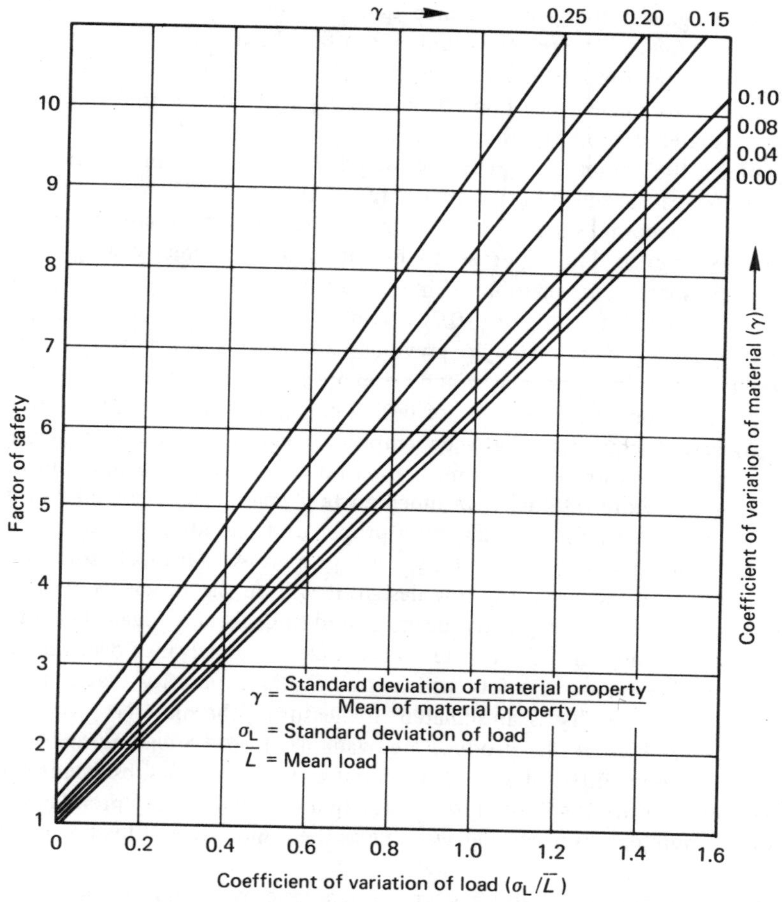


**Figure 10.4** Minimum factors of safety to give intrinsic reliability with near normal load and strength distributions (Weibull: $\beta = 3.44$) when a $1.75\sigma$ control cut-off is applied to the strength distribution

Another interesting example at the other end of the loading roughness spectrum concerns the gun. Here the Ordnance Board requires the charge to provide a gas pressure within 1 ton/in$^2$ of the design pressure in the gun tube. Taking an average value of the design pressure as 25 ton/in$^2$, and again assuming all practical variations are contained within three standard deviations, we may assume $\sigma_L/\bar{L}$ to be 0.013. Figure 10.2 then indicates that the value of the factor of safety should be 1.25 for $\gamma = 0.05$ compared with 1.2 adopted in practice. Because barrels are made from high-quality steel and because the manufacturing process itself (autofrettage) often provides a measure of quality control, the value of $\gamma$ could well be somewhat lower than assumed. Agreement is very good.

It is recognised that the evidence quoted above to justify the theoretical calculations *vis-à-vis* experience is limited (to say the least); but the necessary information for a comprehensive comparison is just not available. The examples quoted refer to single piece components of high quality so that estimates of $\gamma$ based on material property should be valid, or at least not so far in error as to compromise the comparison. It is considered that such examples as these justify the use of intrinsic reliability as a basis for calculating design working strengths. This can be accomplished through conventional factors of safety, allowing all existing experience and expertise to be retained, and directly weighed against theory: an aspect believed to be of immense value and importance. It is not, for instance, imagined that figure 10.2 would be used directly for design, except perhaps in a new field where no previous experience was available. Rather is it envisaged that past experience would be expressed in the form of figure 10.2 so that extrapolation and assessment of design changes could be made more rationally. Marked discrepancies between empirical values and those on figure 10.2 would suggest unknown factors must (one hopes profitably) be examined. Such an approach is more rational than much current practice by virtue of its theoretical basis, and the emphasis it lays on a proper understanding of the contribution made by loading conditions and material properties. Although some intuitive anticipation is still necessary during design, it will be the anticipation of physical quantities that can subsequently be measured and checked against assumption, thus ever improving future estimates. This is clearly an advance over the intuitive assessment of a factor of safety which is an abstract quantity that can never be measured. No difficulty is anticipated in measuring the material property, $\gamma$, at least when samples of the product are available. It is the loading factor $(\sigma_L/\bar{L})$ which will present difficulties both in initial estimation and subsequent measurement. This quantity is still so elusive that experience will be at a premium, though more and more data is becoming available as the importance of loading conditions is widely appreciated.

The blending of well-established empirical practice with a more rational approach is considered to be particularly important at this stage of development of design techniques. It must be appreciated that the old style factors of safety took in much more than the distribution of loads and strengths. This was sometimes recognised by multiplying the factor of safety by a 'factor of ignorance'. One major area of ignorance was stress analysis. Because of lack of computing facilities, relatively crude methods were used to derive stress values and this was allowed for in assessing the factor of safety. Stress concentration is a well-known example. However, modern computers enable exact values to be obtained almost as readily as the older and cruder methods. The allowance for this area of ignorance is consequently no longer required and the old established empirical values of many factors of safety are no longer appropriate. A more rational approach is necessary – indeed, is essential if design is to keep pace with modern techniques. Either the newer methods of stress analysis, fracture mechanics etc. must be ignored and design continue to be based on the consistent old style technique, or new factors of safety of commensurate accuracy must be found.

*Exercise 10.3*

What factor of safety would you recommend for the design of the crankshaft of the engine whose duty cycle is represented by figures 3.5 and 3.6?

*Exercise 10.4*

Recommend the factor of safety which should be used for the rungs of the wooden ladder which formed the subject of exercises 3.1c and 4.7. Hence deduce the rung diameter and compare your value with current practice.

*Exercise 10.5*

In the electronic industry it is customary to base design on the '$3\sigma$' rule. This assumes that all loads and strengths are contained within $3k$ standard deviations of the mean, where $k$ is some empirical constant equal to or slightly greater than unity. Design is then based on the old maxim that the minimum strength is greater than the maximum load, or mathematically

$$\bar{S} - 3k\sigma_{\mathrm{S}} \geqslant \bar{L} + 3k\sigma_{\mathrm{L}}$$

How does this compare with the design rule of figure 10.1? Account for any differences.

It must be noted that in this section we have really been concerned with the provision of an adequate strength to sustain a given load. The identification of the loads has not been covered in any way. While the main load may be dictated by the required duty the item has to perform and is thus outside the control of the designer, this is not true of secondary or internal loads, which are often very much under his control. In many cases these secondary loads can be the failure-inducing loads, and it has been well established in the earlier chapters of this book that the load has far more impact on the reliability than does the strength. This is particularly so for the stress-rupture phenomena under discussion in this section, but applies equally to many wear failure modes to be examined later. Design for reliability does therefore imply that full attention be paid to the source of internally generated loads, and particularly to those factors which lead to significant variations from the mean. The author has frequently stated that quality control of the environment is far more important than quality control of the manufacturing processes in achieving high reliability. The designer has thus a major duty to ensure that all internal loads stemming from his design are properly identified, evaluated and minimised. The most obvious example of an internal load which is under the designer's control is vibration. A full appraisal of all sources of excitation is a ***sine qua non*** of all design. Another example is stresses arising from thermal expansion and contraction: these can be minimised by good kinematic design. Further examples are inertia loads, loads due to possible misalignment, residual stresses, hydrostatic and hydrodynamic loads and so on. Achieving intrinsic reliability for a given load is but one step in achieving high reliability. Before that can be implemented, ***all*** the failure-inducing loads have to be identified. Furthermore, adequate strength to sustain those loads can be achieved in a number of different ways. The relative strength of overhead and buried cables has already been mentioned. As further illustration the author recalls the case of a hand-operated selector which was held in its various operative positions by a spring-loaded ball engaging in indents – a very common arrangement. However, in this particular case, the ball very occasionally jumped out of the indent, with what could be serious consequences. This was due to heavy shock loads which were an essential feature of the machine's duty. A stiffer spring would increase the safety margin and hence the reliability, though perhaps making operation more difficult. A better design would be a selector handle incorporating a squeezegrip lever which could operate a bolt engaging positively in latches – another common arrangement. This very simple example has been quoted to illustrate that just providing adequate strength is not necessarily the best way to achieve mechanical reliability. The best means of performing the required function has to be identified from a large number of alternatives, a process involving qualitative and subjective judgement as well as quantitative appraisal. This is an area of cardinal importance in design, in which a creative, often intuitive, approach can be the most successful. Unfortunately it is just not possible to rationalise the process and present it in textbook form, although attempts have been made to analyse the methodology of design. In general, however, we have to rely on the inate skill and experience of the designer to produce a good solution. This is a vital ingredient in achieving high mechanical reliability, true as much for the wear phenomena we shall discuss in the next subsection as for stress rupture. To ignore this aspect, concentrating solely on the quantifiable, is discounting one of the most potent means of achieving our objective. On the other hand, once the design has been formulated it is possible to make some quasi-quantifiable assessment of its reliability by methods we shall discuss later in this chapter. These methods only ensure, however, that an acceptable level of reliability may be expected. They do not ensure that there is not a better or cheaper solution 'just round the corner'. For that we have to rely on the designer.

### *10.3.2 Design for fatigue*

Before discussing design for fatigue it would be as well to clarify those aspects which will and will not be covered in this book. We shall not be concerned with stress analysis. It will be assumed that the maximum stress in an item is known, either by one of the more exact modern techniques such as finite element analysis, by one of the older less-exact estimates corrected by well-tried empirical factors, or from experimental investigation. Furthermore we shall assume that the known stress includes the effects of all stress raisers and concentrations, together with any dynamic effects, that is, stress variations arising from the way the load is applied. We shall likewise not be concerned with the geometry of the component which minimises these stresses. We shall also not be concerned with any of the material-production processes or finishing techniques which raise the fatigue limit. All these matters have been dealt with at some length in the literature and need not concern us here, though we may note that ignoring all or some of these factors is often the cause of premature failures in mechanical items. However, such failures must be attributed to ignorance or negligence, and will not be pursued here. What we shall be concerned with is variability, which occurs both in load and strength, and how the designer is to cope with this unavoidable aspect of design.

Given a reasonably accurate knowledge of the loads and strengths involved and of their variation, actual design may be undertaken. If the objective is to avoid fatigue failures in any circumstances, the procedure is fairly straightforward. The factors of safety advocated for stress rupture in section 10.3.1 may be used in conjunction with the fatigue limit strength. This will ensure that the item is intrinsically reliable with respect to fatigue. In other words there is virtually no chance of any load causing fatigue damage. This is considered fully justified if safety is involved, and may be necessary if a very large number of load cycles will be encountered in the lifetime of the item. For general purposes, however, this must be considered over-safe. Damage in fatigue has been equated with failure in stress rupture, which cannot be justified. In such circumstances only the extremely high loads could possibly cause fatigue damage and would have to be applied approaching a million times (not just once) before failure is likely. This is a totally excessive ratio for the probability of failure by fatigue *vis-à-vis* stress rupture. Expressed in a different way: an item subject to 1500 reversals per minute, if run continuously for 8 hours per day for 5 days per week for 52 weeks per year for 10 years would encounter $1.87 \times 10^9$ reversals *in toto*, for which figure 5.13 indicates a safety margin of 4.5 is required at rough loading for failure-free operation over the designated life span. To provide intrinsic reliability with respect to the fatigue limit requires a safety margin of 5.5. The lower factor of safety must be considered. Difficulties then arise because of the sensitivity of the failure-free life and subsequent failure rate to the precise load and strength distributions when only the tails of the distributions are intersecting. This is the same problem encountered with stress rupture, but there appears to be no satisfactory criterion, corresponding to intrinsic reliability for stress rupture, which can be used to define an acceptable fatigue working strength. In its absence, the equivalent of figure 5.13 for the relevant load and strength distributions and *s–N* curve is probably the best guide, but it is only a guide. Regrettably, experience has still to be heavily drawn upon. The situation is complicated by the fact that the designer has only one parameter at his disposal – the factor of safety or safety margin – and this controls both the failure-free life and the subsequent failure rate. The designer is not therefore able to select and design for these independently. Attempts to control the failure-free life by means of the factor of safety also modifies the subsequent failure rate and characteristic life. Very high characteristic lives, that is, wide spreads in the life of the population, represent an extensive useful life which is thrown away when the specified (design) life is equated with the failure-free life. A different criterion is required. It is necessary for the designer to take into account every factor to reach a workable compromise. Compromise here means using part, if not all, the fatigue wear-out period as well as the failure-free period. For example, the designer should consider whether any failures starting in the later phases of the planned life are acceptable (psychologically at least, failures which set in later in life are more acceptable than random failures which can occur when the product is still new); he should consider whether condition monitoring can be used to prevent the actual occurrence of failures (this is a powerful technique when full use is made of modern fracture mechanics); he should consider whether any warranty scheme can be introduced to mitigate the effect of disparate lives (this should avoid any bad reputation from those that fail early but be paid for by those that have an extended life); and so on. He might also consider whether the planned life is commensurate with the rest of the design. Many components have a comparatively short required life expressed in terms of cycles, and it is then that lower safety margins are most effective. The reduced initial cost or better performance associated with lower factors of safety must be recognised. Figure 5.13 shows that safety margins well below that for intrinsic reliability are possible. There is no inconsistency in rejecting the sensitive region for stress-rupture design but accepting it for fatigue: stress rupture is unpredictable and final – fatigue is a slow wear-out process which can be monitored and avoided. This would involve higher maintenance costs, of course, which must be set against the advantages. However, although the optimisation of initial and maintenance costs was discussed in chapter 9, it is very doubtful if the methods presented there are going to be of much help in these circumstances because of lack of data on which to base an optimisation and the sensitivity of the result to that data. With improved data a more realistic approach may be possible, but the optimum involves many factors. Specification of some of these is a management problem as well as a design one, though many managers fail to realise the design implications of some of their decisions.

Exercise 5.3 was formulated to introduce the problems associated with fatigue design for long life. They were accentuated by the safety aspects involved in a disc failure. If safety is not involved, there is room for more flexibility. The author believes that the problems associated with the extended wear-out period are rarely faced up to. Designers seek the easy way out by ensuring their designs are intrinsically reliable with respect to the fatigue or endurance limit. Failures can then never occur and the designer feels he has fulfilled his function – a course only too easy to take as a means of avoiding possible recrimination. Faulty design has itself in a large measure given fatigue the evil reputation that it often enjoys. Some of that reputation is totally unwarranted and there can be no justification for using high factors of safety to guard against the bad design. It is proper component geometry specification not excessive factors of safety, for example, that ensures no critical (that is, failure governing) stress raisers are present. Equally, stress raisers that do not raise the local stress up to the failure limit need not be eliminated – as they often are as a matter of principle – if any disadvantage is incurred. There is also no excuse for excessive margins to safeguard against poor material properties resulting from non-existent or inadequate quality control, though a variation in material properties commensurate with the planned manufacturing technique must be allowed. The real trouble is that levels of variation in both basic material properties as well as production-induced variation, and in the variation of the applied load, are not assessed. As theory has demonstrated, and practice confirmed on all too many occasions, the consequences of these variations can be catastrophic. Deterministic design just cannot take them into account, so the caution with which designers grounded in deterministic techniques approach fatigue is fully justified, but their solution is often not. The solution is a proper stochastic methodology. It requires a full knowledge of all variations to be expected and the acceptable failure level, derived from a clear understanding of the function of the component in the complete system.

It may seem strange in a book on mechanical reliability to find design for fatigue being criticised for being too reliable, but the author really believes it to be true more often than not. Most unacceptable fatigue failures are due to rank bad basic design. This 'buck' stops with management, who must employ properly qualified design staff. It is the first step. The way ahead is then clear: as stated, design must be based on stochastic methods, though it would be naive to think that the methods presented in this book offer a complete solution. A great deal of empiricism, experience and know-how are necessary to produce a satisfactory practical design. Nevertheless the methods presented here do offer a first order approximation, and draw attention to factors which must be taken into account. Essentially, effective design for fatigue often implies the possibility of failures at some level setting in at some point in the rated life. Success is achieved by acknowledging and assessing these failures and providing methods – maintenance, warranties etc. – to make them cost effective and acceptable. There is a very long way to go, and designers need to stand back and look at the problem anew, and, of importance, force research workers to study those problems that are relevant to design. Coming into this field from a totally different sphere (aerodynamics and thermodynamics) the author was appalled by the lack of answers to the simplest basic questions associated with stochastic fatigue design.

As the failure-free life is decreased the design problem becomes easier, partly because the sensitivity of the reliability to the design parameters is reduced as the load and strength distributions interact more strongly, but more particularly because at low failure-free lives (low cycle fatigue) the subsequent failure rate becomes unacceptably high. No long life of low failure rate exists, and withdrawal from service is essential before the failures set in. The choice of factor of safety can then be based solely on the desired life. It should perhaps be emphasised that though there is no fundamental difficulty in doing this, the practical difficulty that design has to be based on non-existent or very rudimentary data often remains. The data can be obtained if the effort is economically justified, as it is in a number of important applications. In spite of this, however, design for satisfactory fatigue reliability is very much an uncertain process at present, and designs must always be verified by prototype trials unless considerable experience is available in the relevant field.

*Exercise 10.6*

A part, whose failure-controlling mechanism is fatigue, is made from a material whose basic fatigue limit variation is represented by a normal distribution having a standard deviation of 7 per cent of the mean. It is subject to a finishing process which is claimed to raise the fatigue limit by a factor of 1.2. However, we can expect that this process is itself not perfect, and it is reasonable to assume that the percentage strength increase is normally distributed with a standard deviation of 7 per cent of the mean or quoted value. What would you assume for the strength variation of the finished product?

*Exercise 10.7*

You are considering the design of a machine part which is subject to a repeatedly applied tensile load which is normally distributed with a standard deviation of one-third of its mean value. It is expected that it will have to sustain $10^{11}$ applications of the load during its rated life. Four possible design criteria are to be considered

(i) the item is to be intrinsically reliable with respect to the fatigue limit
(ii) it is to have a failure-free period equal to the rated life
(iii) a safety margin of 4.5 based on the fatigue limit will be used
(iv) a safety margin of 3.5 will be used.

For the intended application the cost of a scheduled replacement can be taken equal to the initial cost, but an unscheduled replacement will cost 2.5 times the initial cost. As a good first approximation you can take the initial cost as proportional to the cross-sectional area of the piece.

The fatigue limit of the material is normally distributed. Strictly, the loading roughness should be assessed separately for each design criterion, but to simplify this exercise it may be taken as equal to 0.9 in all cases, enabling you to use figures 5.11 and 5.12 instead of having to estimate the fatigue failure pattern in each case. For the same reason you can neglect any second generation failures by using the figures directly where necessary.

Which of the design criteria would you use?

### *10.3.3 Some general notes on strength assessment*

The mechanisms of mechanical failure are almost infinite. In section 10.3.1 we were able to discuss design for stress rupture mechanisms in a fairly comprehensive manner because it is governed by a single failure criterion – load exceeds strength.

With wear, there is no unique criterion since loads degrade the strength in many diverse ways before actual failure. It is the degradation which mainly controls the age to wear failure rather than the initial load/strength distribution interaction, though the initial distributions do affect the degradation process. The dominance of the degradation is a consequence of the sensitivity of the failure rate to the safety margin, so that as soon as the safety margin reaches a critical value in the degradation process, failures certainly, rather than probabilistically, arise. In most cases the distribution of life is the distribution of the strength degradation mechanism. In section 10.3.2 we have dealt with just one wear process – fatigue – albeit probably the most important, but we have offered no treatment for the other wear mechanisms. If the same load gives rise to both wear and failure then the treatment of section 5.4 could be valid, and design would follow the same general lines adopted for fatigue in section 10.3.2. The treatment would be more positive because the very extended life would be curtailed in the absence of a 'limit', though the age range of expected failures could still be significant. Failing such a treatment, and certainly in those cases where the same load does not control wear and failure, an empirical treatment, based on section 5.2, is the only one possible at present. A great deal needs to be done before a comprehensive body of knowledge is available to deal with the many forms of wear.

Even in those areas treated in detail in this chapter, a cautious approach is essential. There is no agreed methodology for stochastic design, and so it must be introduced in parallel with existing conventional techniques. In well-established areas there is no need for change. For example, the design of an actual wooden ladder would be based on BS 1129, which incorporates endless experience, and is totally adequate for the purpose for which it was drawn up. The stochastic method used in the solution to exercise 10.4 is just not necessary. This would be true of most existing mechanical items. Experience which has a sound foundation should never be abandoned, and in such circumstances it would be quite irresponsible to introduce new design techniques without a valid reason or objective. One could imagine, however, that the stochastic approach could be used as an exploratory measure in an established field to gain experience and confidence in the method (as, for example, in exercise 10.4). Where practical experience is limited or does not exist, then stochastic methods may be used in parallel with the best deterministic methods available. The solution adopted would depend on engineering judgement and itself introduces an interesting decision-making problem. Stochastic methods really come into their own when the deterministic methods break down and result in an unsatisfactory reliability, or when the deterministic solution has to be modified to obtain a better performance at the controlled expense of reliability. More often than not, however, they can at present only offer indications. These may be very valuable to those knowledgeable in the field, but conclusions so reached must be verified experimentally. Eventually, of course, stochastic design must replace all deterministic methods if the optimum combination of performance and quality is to be achieved. In no way can deterministic methods represent what is essentially a stochastic situation. Hence even if there is no intention of making practical use of such design techniques at present, there are good and valid reasons for building up the necessary background knowledge for their future use.

### *10.3.4 Checklists*

Engineers will be only too well aware that there is a lot more to design, even detailed design, than the sizing of parts to meet a known load, important though that topic is. Perhaps the most difficult part of conventional design is recognising that some loads exist at all. Once a load is recognised it can usually be sustained, but we first need to be sure that every form of loading has been considered. Particularly difficult is the identification of internally generated loads, especially those that have no connection with the operational load. In this category, for example, are inertia loads, vibrationary loads, loads due to thermal expansion and other effects, loads due to clearances which may change with time as a result of wear, loads due to wear debris accumulation, loads arising from material incompatibility, and so on. The list is endless, yet the internally generated loads are often the ones responsible for failure, chiefly because they have not received design attention.

An experienced designer working in a familiar field will assess the difficulties almost subconsciously and make proper provision for all eventualities. But the less experienced designer, or even the experienced designer working in an unfamiliar field, can easily overlook possible failure modes. With the steady, and in some areas dramatic, advances in technology, and with the ever-increasing tendency for design staff to move on, either to widen their experience or seek higher salaries, such situations are becoming more and more common. Although it is axiomatic that each designer must anticipate all the modes in which his design can fail, positively identifying these modes becomes elusive. A random appraisal of failure mechanisms is not good enough: a systematic planned approach is essential. For this reason all well-established and well-run design offices have checklists to ensure that all possibilities have been considered, and that the mistakes of the past are not going to be repeated. When kept up to date by constant feedback of field experience, these checklists are a form of experience retention.

Checklists or questionnaires will obviously differ from organisation to organisation, and indeed from product to product. The following checklist has been drawn up in general terms to illustrate the approach. It is not intended to be used implicitly, though it may be suitable for general purposes and could form the basis of a more specific questionnaire. It has been based on the systematic fault analysis questionnaire proposed by Green in ref. 62. It has been divided into a number of categories. Categories 1-12 are fairly general and would apply to most mechanical designs. Categories 13-18 apply to specific but commonly occurring components. This section is by no means complete and any organisation would have to augment it by questions directed specifically at the particular components appearing in their designs. Finally categories 19-23 revert to a more general nature, but refer to aspects which are best dealt with after the behaviour of components has been resolved. This section also is not complete and would need adjusting for particular applications. As just one illustration, no attempt has been made to deal with equipment involving explosives.

The general categories adopted in this questionnaire are

1. Design concept
2. Environment
3. Operation
4. Material(s) selection
5. Stressing
6. Thermal effects
7. Wear
8. Ageing
9. Surface finishes
10. Dimensional factors
11. Manufacture
12. Assembly
13. Prime movers
14. Bearings
15. Seals
16. Rotating parts
17. Pressure vessels
18. Hydraulic, fuel and similar systems
19. Connectors
20. Maintenance
21. Instrumentation
22. Testing
23. Safety considerations.

The rules for completing the questionnaire have been postulated as follows

(1) All questions must be answered fully. If a question or section does not apply to the device under review, the reasons for its non-applicability must be set out.
(2) Indented questions relate directly to a preceding question and may require a number of answers, depending on the number of answers to the original question.
(3) Depending on the progress of the design, it may not be possible to answer all the questions during the early design stages. However, the earlier in the design process that the questions are completed, the greater will be their value.
(4) For each design, the design team will formulate additional questions specifically related to the components under review.

*1. Design concept*

What are the desired functions of the device under review?
Where are these functions specified?

Is this specification comprehensive?
Does the design meet all the desired functions?
In what areas will it perform better than required?
Will complexity or cost be reduced if these areas are eliminated?
Is the design as simple as possible?
What alternative methods of achieving the desired function were considered before adopting the present design?
Why were these alternatives not used?
If it were possible to start again, what changes would be desirable?
What does this device have in common with existing successful designs? How does it differ?
Are any departures from existing designs justified?
How have schedular considerations affected the design choice?
Have all interfaces with other design areas been identified?
On what basis were safety margins/factors of safety established?

*2. Environment*

Describe the normal operating load cycle and environments associated with the device: external, internal, etc.
To what other environments will the device be subject during shop testing, inspection, shipping, storage, handling, installation, start-up, maintenance?
What areas are subject to corrosion in any of the above environments?
How is this corrosion controlled?
What areas are affected by dust, grit, etc.?
How is dust excluded from these areas?
What areas are affected by high humidity or condensation of moisture?
What is done to protect or prevent this?
How is the device lifted?
What special tools are required for installation?
Who supplies these?
What is the effect of freezing during site storage?
What is the effect of excessive temperature?
Will the device be handled by skilled, semi-skilled or unskilled operators?

*3. Operation*

What special starting or stopping procedures are required?
What document sets out the operating procedures?
Are any special skills required for operation?
Has due consideration been given to safety aspects of operation?

*4. Material(s) selection*

Identify the material(s) used for each part.
What three material properties were most important in the selection of this material for this application?

Is any manufacturing process (welding, forming, etc.) likely to destroy the required properties?
What is the required metallurgical state of the finished product?
Were any unusual material(s) used?
If so was adequate information on the relevant properties available from recognised sources?
What non-destructive tests have been called for to ensure material properties?
Have the material quality control procedures been specified?

*5. Stressing*

What methods of stress analysis were adopted?
What document presents the stress analysis for the device?
What areas of stress concentration are present?

*6. Thermal effects*

Identify areas where heat is produced.
Identify areas where heat is lost.
Prepare an energy balance for the device.
Identify any areas carrying large temperature gradients:
(a) during steady conditions, (b) during transients.
What are the maximum stresses induced by these gradients?
For what areas has a differential thermal expansion study been performed?
What provisions are there for degradation of heat transfer surfaces by fouling, etc.?
Has full allowance been made for any creep arising from the operating temperatures?

*7. Wear*

Identify any areas subject to wear (sliding, rubbing, rolling, impact surfaces).
How might this wear be avoided?
How might this wear be reduced?
What is the effect of the wear?
Can this effect be tolerated?
What indications are there in the operation of the device to alert the user to wear taking place?

*8. Ageing*

What is the design life of the device?
What component controls this design life?
Identify any electrical insulation.
At what temperature does this normally operate?
What is the maximum temperature to which it could be subjected?
What class of insulation is this?
For which components has a fatigue study been performed?
What were the results of this study?

For which surfaces has surface fatigue been considered?
What is the maximum level of nuclear radiation to which the device is subject?
What non-metallic materials are used in the device?
  What is their life expectancy under the above radiation level?
Are any ageing phenomena likely to occur if the device is left idle for any period?

*9. Surface finishes*
Identify any surfaces on which: (a) sliding, (b) rubbing, or (c) rolling takes place.
  What finish is specified for these surfaces?
Identify any plated surfaces.
  Why is this surface plated?
  To what specification is the plating done?
Identify any painted surfaces.
  Why is this surface painted?
  To what specification is the painting done?
What requirements are there for cleaning surfaces?
  How are these requirements met?
Can the appearance be improved by more attention to exterior surfaces?
What methods are used to prevent galling in screwed joints?
Identify any hard-surfaced areas.
  Why is this needed?
  What method of hard surfacing is used?
  What procedure controls the application of this hard surfacing?

*10. Dimensional factors*
For what areas has a tolerance stack-up study been performed?
What essential clearances must be maintained?
  How are these clearances maintained?
What are the effects of thermal expansion?
What are the effects of deflection due to loading?
What essential alignment must be maintained?
  How is this alignment maintained?
What interchangeability provisions are incorporated in the design?
What standard parts does the device employ?
Where can 'non-standard' parts be replaced by standard parts?

*11. Manufacturing*
What novel or unique manufacturing methods are required to achieve the design?
What non-destructive testing techniques are used?
If welding is used, what welding and weld repair procedures are used?
What are the largest dimensions requiring machining?
What heat treatment is involved?
What special cleanliness controls are required in the process?
What tolerances will prove most difficult to hold?
Have quality control procedures been adequately defined?

*12. Assembly*

How is the device supported? Does this allow essential, but only essential, movement?
What levelness or verticality requirements are there?
Where are these requirements specified?
What happens if they are exceeded?
How are these requirements met?
What is the direction of rotation of any moving parts?
Is it possible to reverse this rotation?
What is the effect of reverse rotation?
What external connections are required?
Where are these detailed?
Is it possible to incorrectly assemble any part?
Have all nuts, bolts, screws, studs and other connectors been adequately locked?
Have all fluids and their flow direction been identified on all pipelines, valves, etc.?
Have all moving parts been given adequate clearance?
What are the maximum overall dimensions for installation?
What is the smallest opening through which the device must pass for installation?
What services are required: electrical, air, fluids?
What are the weights of major components?
What pull-space access is required?
Are any special assembly techniques involved?

*13. Prime movers*

Identify any prime movers in the device.
Where do these get their power?
How is the drive transmitted to the driven members?
Are any prime movers bought-in components?
What evidence exists to substantiate their reliabilities?
Are the current environmental conditions the same?
If new designs, how has this reliability been established?
Are the prime movers new integral designs?
If so, apply whole questionnaire to unit.

*14. Bearings*

Identify any bearings in the device.
For each of these:
What type of bearing is it?
What alternatives were considered?
Why were the alternatives not used?
Identify the direction of loading.
What are the effects of misalignment?
How is the bearing lubricated?

How is the bearing cooled?

What is there to prevent debris, dust, grit, etc. from entering the bearing?

What is the effect of such dust, grit, etc.?

What self-aligning features are present in radial bearings?

What provisions are made to prevent brinelling of anti-friction bearings during shipping?

Are non-flammable lubricants used in hazardous environments?

*15. Seals*

Identify any rotating or sliding seals in the device.

What type of seal is this?

What testing has been performed on this actual seal?

What is the maximum permitted leak rate?

What happens if this leak rate is exceeded?

How is excessive leakage detected?

What back-up is provided for this seal?

Identify any static seals in the device.

What type of seal is this?

What happens if this seal leaks?

What testing has been performed on this seal?

How much does the application of the above seals represent an extrapolation of proven performance?

*16. Rotating parts*

At what percentage of its critical speed does each shaft rotate?

What provisions are made for balancing?

What thermal transients have been considered that might cause shaft distortion?

What features are likely to cause a reduction in the stiffness of the rotating assembly?

What features are likely to cause a reduction in the stiffness of the support structure?

How is axial thrust from the rotating parts resisted?

*17. Pressure vessels*

Identify any components carrying large pressure differentials.

What code requirements apply to these components?

What document(s) identify compliance with these code requirements?

What areas have nil ductility transition temperature requirements?

What areas require a crack propagation analysis?

What areas require a fatigue analysis?

For what emergency conditions have the pressure vessels been designed?

To what hydro-test conditions will the pressure vessels be subject?

*18. Hydraulic, fuel and similar systems*

Are all fluids and parts of the system compatible?

Does the system contain a pressure relief valve or similar device?
What is the relief pressure?
Can this pressure be adjusted: (a) during normal operation, (b) accidentally?
Can pressure be released in an emergency? or for servicing?
Are all parts of the system accessible for inspection/servicing?
Are all replenishment/refuelling points readily accessible?
Are all points clearly marked and replenishment fluid identified?
Is the full state readily identified?
What happens to spilled fluid?
How can the fluid become contaminated?
Can filters be serviced/replaced without draining the system?
How is filter blockage indicated to operator or service engineer?
Can air/vapour locks occur?
Has flow direction been identified on all pipe lines?

*19. Connectors*

Identify all electrical, gaseous and hydraulic connectors.
  Have all electrical connectors adequate mechanical strength?
  Are relevant electrical connectors weatherproof?
  What sealing arrangements have been made for hydraulic/gaseous connectors?
  Are all gaseous/hydraulic connectors compatible with working fluid?
  Have all connectors been checked against expected vibration level?
  What are the consequences of a mildly leaking connector?
  Can any connector be subject to a spurious mechanical load: (a) during normal operation, (b) during servicing?
  Are all connections readily broken/remade?
  Which type of connection has not been previously used in the environment?
  Can any connector itself apply a spurious load to any other part?
  Are adjacent connectors keyed, sized or marked to prevent cross connection?
Identify all screwed or bolted joints.
  What locking method is used on each?
Identify all torque-carrying joints.
  What component actually carries the torque?
  What is the effect of torque reversal?

*20. Maintenance*

How often does the device require attention?
What component necessitates this?
Would it be desirable to increase this interval? Why?
How does maintenance affect plant availability?
What other components require periodic attention?
  At what intervals?
  What attention is required?
  Describe how access may be gained to any parts requiring inspection or adjustment.

Describe how any parts requiring replacement may be removed.

What critical clearances are there that are upset in the above operations?
How might the design be changed to avoid upsetting critical clearances?
For how long will the device be out of service for each of the above operations?
Is the device accessible for inspection or maintenance?
What special tools are required for maintenance?
Who supplies these?
What is done to ensure that experience from in-service equipment is fed back to the designers?

*21. Instrumentation*

What is measured?
Why is this measured?
How is this measured?
What is the accuracy?
What is not measured that could be?

*22. Testing*

What tests are performed on components of the device?
What is the reason for these tests?
Who is responsible for ensuring that the results of these tests are considered by the design team?
How are these tests reported?
What tests are performed on the device as a whole:
(a) to ensure the device meets its performance requirements, (b) to explore margins above design requirements, (c) to investigate expected life?
How are these tests reported?
What production tests are performed?

*23. Safety considerations*

Identify any moving surfaces exposed to contact by persons close to the device.
What provisions are made to prevent contact?
Identify any hot surfaces exposed to contact by persons close to the device.
What provisions are made to prevent contact?
Identify any large moving masses.
What provisions are made to retain pieces if these masses should fracture?
What provisions are made for grounding and shielding electrical circuits?
Has a safety hazard analysis been carried out on the device?
Where is it reported?

*Exercise 10.8*

Figure 10.5 shows drawings of various complexity and detail of a number of common engineering components. Quite deliberately the designs have been produced incorporating features which could mitigate against high reliability. It is also possible that similar features have been introduced unintentionally! You are asked to examine these drawings and comment on the designs from the reliability point of view.

*Exercise 10.9*

Develop a Pareto diagram of the importance of the checklist items listed above for a product with which you are closely concerned. Add any other checkpoints which are special to the product you have chosen and are not included in the above list, placing them at their appropriate ranking on the Pareto diagram. (No solution is given to this problem as the possibilities are so large.)

Good designers have, of course, had checklists of their own for years, sometimes written down formally, and sometimes carried in their heads. It is believed that some form of checklist is essential. It is the only method known to the author that ensures a design is systematically examined for all foreseeable failure modes. It is the technique by which the easily overlooked failure, the ones that are so common in practice, can be averted. If the checklist is developed over a period of time by correlation of design and field data for a specified product, then it will incorporate the summated knowledge on its failure modes and, as stated earlier, will retain that knowledge for future designers and reliability assessment teams. If it is to be used in this way, it becomes a firm's or organisation's formal document which must be regularly updated. So much is learnt, only to be forgotten in a changing design team, that a systematic approach is essential. If the lessons of the past can be properly recorded there is a reasonable chance they will not have to be learnt again. The cost of relearning can be very high. Even if it is a formal document, the checklist is for use where the work is done – in the drawing office. It must be printed indelibly on every designer's mind, as well as on paper. It should also be the basis of the more widely used failure modes and effects analysis discussed in the next section.

### *10.3.5 Failure modes and effects analysis*

Failure modes and effects analysis (FMEA) is one of the major techniques, rating in some peoples' evaluation only after the inherent skill of the designer himself (though some would say an integral part of that skill), by which high reliability can be achieved. It therefore warrants close study. In many respects the title is only a modern description of a technique that has been employed by all competent chief designers over the last 100 years, though more recently attempts have been made to formalise the procedure and extend it in some respects. The lead has come from the electronics world, where the strictly limited range of components and lack of environmental complications has eliminated many of the variations on the theme. The recognisably structured product of the electronic

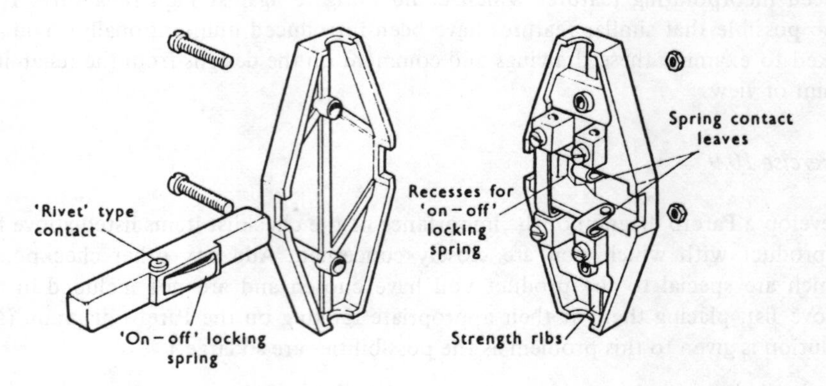


**Figure 10.5** (a)

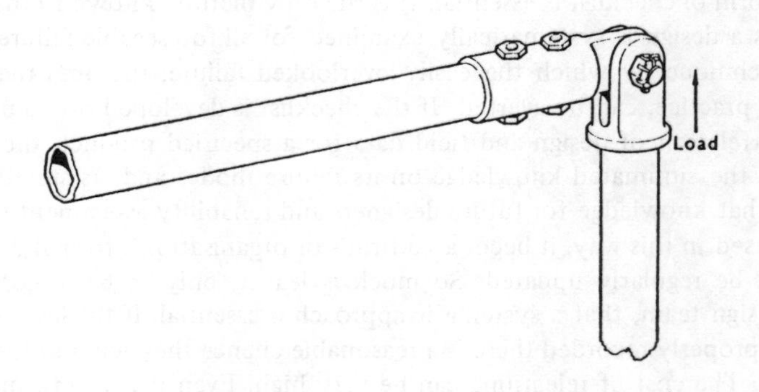


**Figure 10.5** (b)

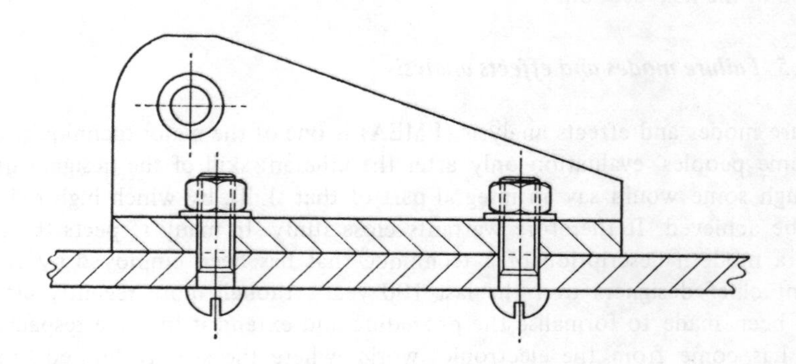

**Figure 10.5** (c)

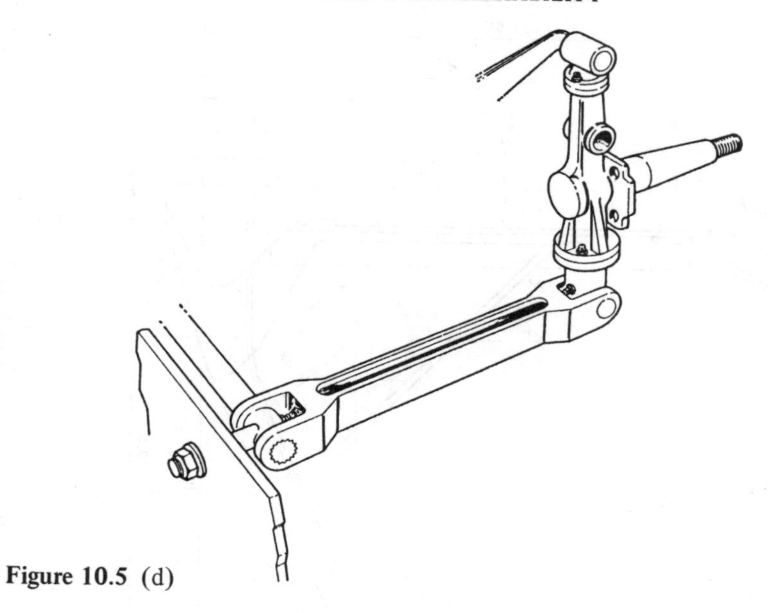

**Figure 10.5** (d)

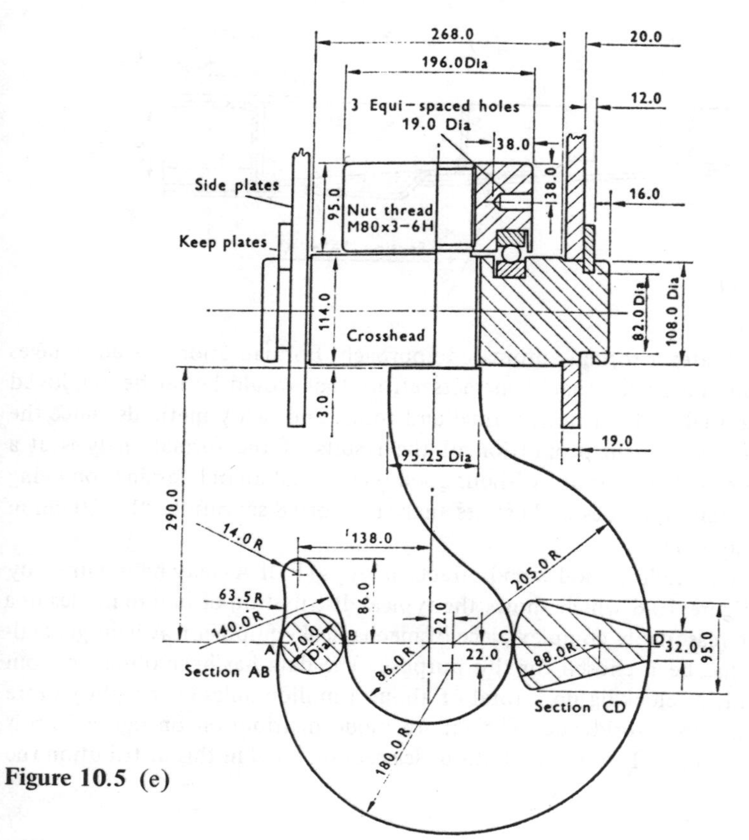


**Figure 10.5** (e)

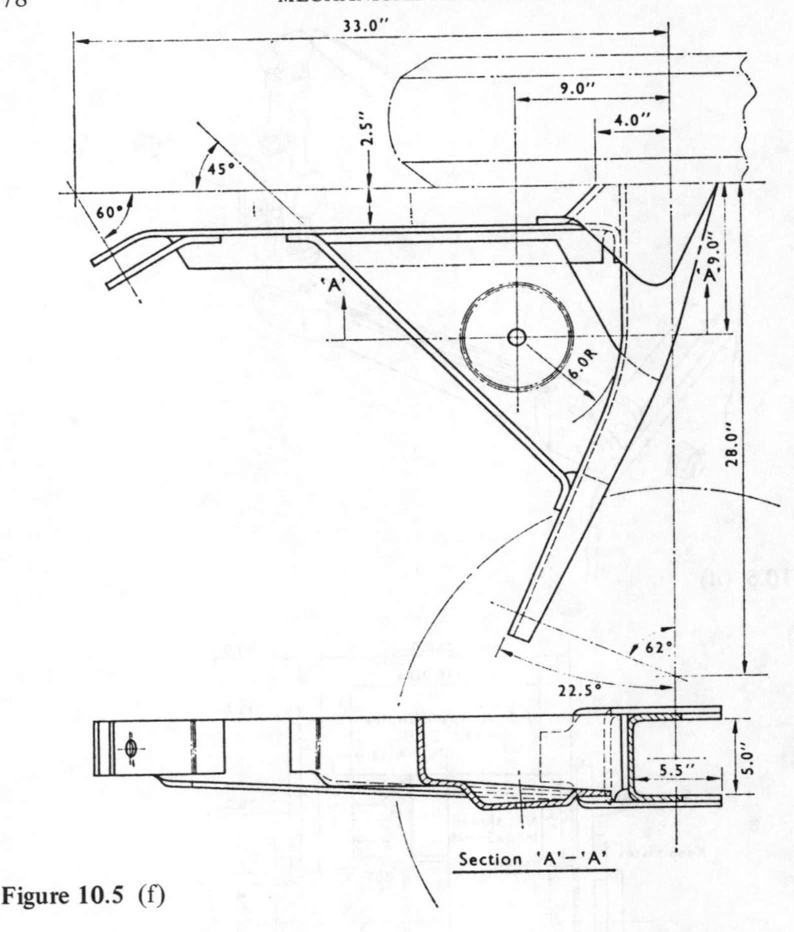


**Figure 10.5** (f)

industries facilitates a formal automated approach. Formalisation has advantages if it is not pushed so far that it attracts effort that would better be employed elsewhere. We shall look at both formal and rough and ready methods, since the latter may achieve a high proportion of the results of the formal analysis at a fraction of the cost. It was not without good reason that an old designer on being introduced to failure modes and effects analysis retorted scornfully "I do them in my head all day long."

The essence of failure modes and effects analysis (FMEA) may be obtained by referring to figure 10.6 which shows the typical distribution of failure modes in a representative reasonably complex piece of mechanical equipment. A light general-service truck has been chosen for this purpose. The data has been obtained from some 180 vehicles clocking up a total of about a million miles over slightly more than one year's use worldwide. The failure mode distribution on figure 10.6 is typical in following a Pareto distribution (see section 6.6). In this distribution the

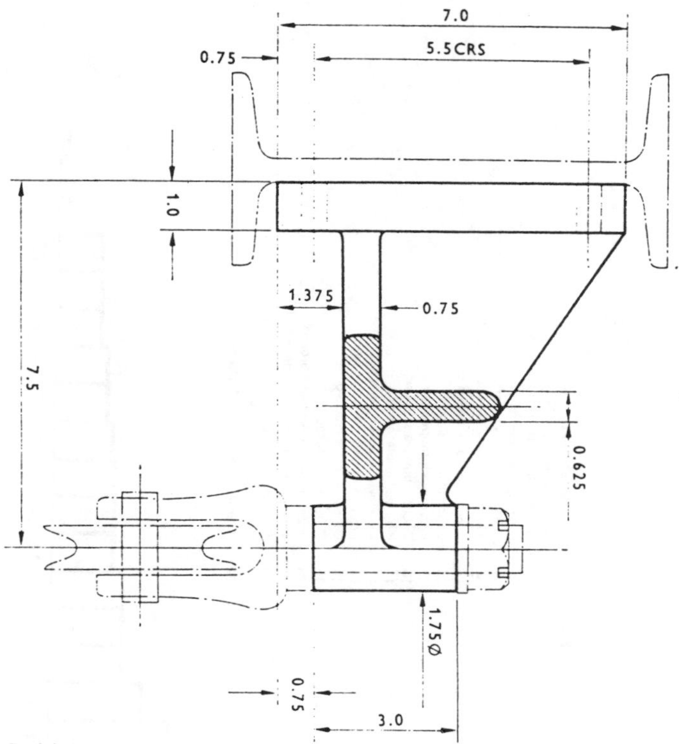


**Figure 10.5** (g)

**Figure 10.5** Design malpractices: (a) Switch – plastic; (b) helicopter control rod; (c) draw bar bracket; (d) torsion bar suspension; (e) crane hook, working load 20 000 kg, material – forged steel (all dimensions in mm); (f) plan view, rear suspension geometry, material – 6 swg mild steel plate; (g) swivel pulley bracket, material – cast iron (all dimensions in cm)

first few items dominate the overall behaviour of the system. Thus in this particular case the first three failure modes account for 40 per cent of the total failures: the first five account for just over 50 per cent and the first ten account for 70 per cent of all failures. All the failures are accounted for by 65 out of the thousands of components that make up the vehicle. This is typical of mechanical equipment. Even in as sophisticated an area as modern military aircraft, we find that 22 components accounted for 61 per cent of total failures, some 40 accounted for 75 per cent and approximately 200 accounted for 94 per cent. This is an item where 3500 modifications were introduced for basic system and role changes (ref. 63). We have, of course, been quoting actual service data, which would not be available at the design stage. The object of failure modes and effects analysis is to produce

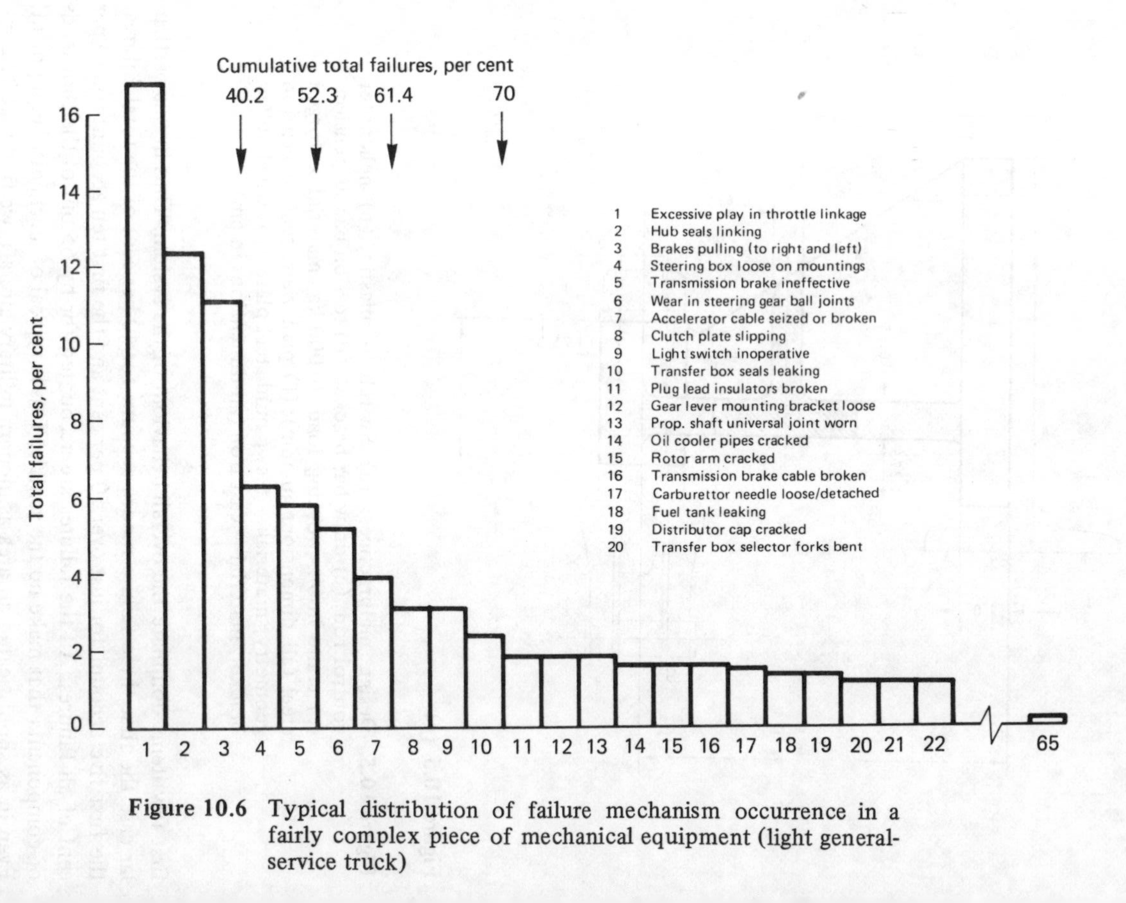


**Figure 10.6** Typical distribution of failure mechanism occurrence in a fairly complex piece of mechanical equipment (light general-service truck)

such distributions at the design stage and, by identifying Pareto's vital few, eliminate major failure modes from the production version. If, for example, we could have identified the first three failure modes, not even in the correct order, that were subsequently encountered on the general-service truck and have eliminated them by redesign, we should have reduced the failure rate by some 40 per cent. If we could have identified the 'top ten' – again not necessarily in the correct order of 'merit' – then we could have hoped to reduce the failures by some 70 per cent – no mean achievement. It should be noted that we have only to identify the failure modes which detract most from the operation of the equipment. It follows that extreme accuracy is not necessary. Of course, we will not identify each and all of the vital few failure modes, but there are good reasons for expecting a high success rate. Writing in the *1978 Conference on Reliability of Aircraft Mechanical Systems and Equipment* (ref. 64), G. W. Beasdate stated "Looking back, in retrospect, some of the defects may seem extraordinarily naive in what we like to think of as a high technology industry." At the same conference G. Jones wrote "Many operational problems are due to relatively minor details . . . . . . it must never be forgotten that the end product is an engineering assembly of sheet metal, pipes, wires and a variety of other components held together by nuts, bolts and rivets. The attention to detail design cannot be over-emphasised." After a review of major failures in a wide range of equipment, Wearne concluded in ref. 65 that the larger the project the greater the risk of lack of attention to detail. Yet again, Joseph Levine, chief of the Reliability Division of NASA (Houston) writing (ref. 66) of the Space Shuttle concluded: "The attention-to-detail lessons learned from the previous programs and the early emphasis placed on preventative disciplines cannot be overstated when dealing with high technology programs." These are just four of many other similar quotations: it is general experience. That most failures arise from faulty detail design rather than sophisticated error leads us to believe most could be detected by a straightforward, but rigorous, appraisal of the design, conducted on a systematic basis. A general assessment suggests that at least 50 per cent of the vital few could be identified by a FMEA when dealing with high-class design, but a more average figure would lie between 70 and 80 per cent.

The first step in the systematic approach of the formal modes and effects analysis is to define the limits of the system to which the analysis is to be applied, and then to specify the lowest level (that is, integral part, component, sub-sub-assembly or sub-assembly level) at which the analysis is going to begin. We shall return to this point when we know more about the technique, but for the moment we shall assume a decision has been made. In the informal approach this step is nearly always omitted, and the next step taken on a common sense basis as necessary: usually not very consistently!

The second step is to take each of the lowest level items in turn and list each, and ideally every, mode of failure. In practice the work involved can be excessive, and more often than not only a limited number of the most likely modes of failure of each item is listed. Selection must be by engineering judgement. The best aid to the identification of failures of mechanical parts for a FMEA is a good checklist. It is, however, at this stage that the first of the difficulties arises. This centres around the interpretation of the word 'failure' in this context. We have already stated that the environment, or loading, can have a far greater effect on the reliability of an item than any lack of strength, and we have already noted that the internal environment is as important as the external; indeed, for this analysis it is all-important because we conduct the analysis on the basis that the external environment has already been specified and is known. Hence amongst the failure modes we must include any unplanned contribution to the environment which an item under consideration may itself make. This may be clarified by an illustration. Suppose we have a rotor in which some of the parts are not positively located. Because of spurious movement this can lead to out of balance. The out of balance may be insufficient to interfere with the prime purpose of the rotor in any way, but it could well increase the general vibration level of the equipment as a whole and so lead to failure of what may be a completely dissociated part. It is immaterial whether the optimum solution is redesign of the rotor or the strengthening or isolation of the vibrating part. FMEA must record a potential out of balance of the rotor as a possible failure mode (in the example we have taken) and then consider the consequences in the subsequent part of the analysis. Vibration is one of the obvious mechanisms by which an item can affect the environment, but direct loads may also be involved. Thermal effects are another common area, to which can be added compatibility with surrounding materials, noise and other fluid effects, gaseous and fluid emissions from a component and so on. Note that at this stage we are identifying any aspect in which the item fails to meet its design role, even though it may not cause any physical failure of the item itself, or failure to carry out its prime function. The inclusion of such failure modes is absolutely essential, for it is only by such an approach that we can hope to identify all the loads that each item has to sustain.

There is another respect in which electrical/electronic FMEAs and mechanical FMEAs differ. Electricity may 'flow' with different currents at different voltages as, for example, air may flow at different rates and pressures, but electricity is always electricity whereas air is not always air! Air can carry differing amounts of dust, have different humidities, be at different temperatures, have different impurities, and so on. The same applies to all fluids, liquids or gases. It is therefore necessary to treat all working fluids, lubricants, hydraulic fluids etc. as components in their own right and include them in the FMEA. If this is not done a number of vital failure modes can be missed. Some may refer to failure of the fluid itself, as for example when a vapour lock can be formed, but more often the failure mode is the creation of an additional load on some other component, leading to the situation we have just discussed.

At the end of the second step we should have a list of all the failure modes that could occur at the basic level chosen for the start of the analysis. It is essential that considerable care is taken to complete the second step accurately and comprehensively, because if any failure-inducing feature is missed at this stage it could invalidate the whole of the subsequent analysis.

It is now possible to tackle the third step, either qualitatively, following old style practice, or quantitatively following more modern practice. In the qualitative approach failure modes are ranked in order of most likely. In this way we are able to identify the failure modes which constitute Pareto's vital few, though we make no quantitative assessment of their occurrence. In the quantitative approach each failure mode is assessed and the expected frequency of failure in a significant time span estimated, thus making a quantitative assessment of their relative occurrence. In this way the equivalent of figure 10.6 with an arbitrary scale is produced. It is essential at this stage that any 'risks' taken in the design be allowed for and the frequency of failure be the 'highest likely' rather than the 'hoped for' value. In practice, very few mechanical designers will be able to make a realistic estimate of the failure frequency directly, and it may well be more appropriate to estimate the lives expected of the various items. If the life of 5 per cent of the items ($B_5$) is estimated together with say the $B_{10}$ or maybe the median life ($B_{50}$), it is then possible to derive the Weibull distribution of the failures, as in exercise 6.13, since the value of the shaping parameter, $\beta$, will be largely defined by the failure mechanism. The frequency of failures at any age can then be calculated, taking into account any scheduled maintenance it is envisaged that the item should receive. It is, of course, essential that any proposed scheduled maintenance be included in the assessment. Whether the frequency of each failure mode is calculated from a life estimate as described above or estimated directly must depend on the preference of the analyist in each particular case and, of course, different methods can be used for different items in the same equipment. Before we go on to the next step of the analysis, ideas will be consolidated if you tackle the exercises below.

*Exercise 10.10*

Figure 10.7 shows a conventional stopcock which is to be used both as the main stopcock and to isolate sections of the domestic water system of a medium/large sized house. What do you think would be the three or four most likely forms of failure that could be encountered in the first ten years' use of this stopcock? This may be described as the rough and ready approach to the second step of a FMEA. Now proceed to carry out the second step quantitatively (the first step has already been taken for you) by listing in detail all the likely modes of failure of each component designated in the drawing.

*Exercise 10.11*

Carry out the third step of the FMEA on the stopcock started in exercise 10.10 by estimating the frequency with which each of the failure modes identified in that exercise will arise in the first ten years' use. Draw up a Pareto diagram of the expected occurrence of each failure mode.

The fourth step in the FMEA is to assess the effect of each failure mode, ultimately on the system as a whole. We start by assessing if each failure mode

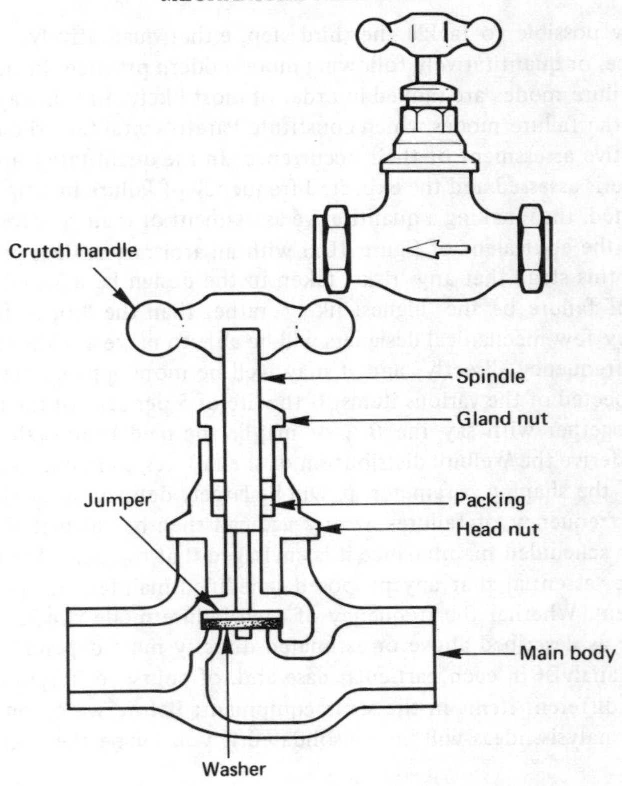


**Figure 10.7** Typical stopcock used in domestic water supply systems

identified in the second step could induce further secondary failure modes at the first level of the analysis. These are added to the first level failure mode list and the process repeated until all possible first level failures have been identified. Then the effect of each of these first level failure modes on the next higher level is assessed. If this is to be done in a comprehensive manner it is necessary to consider not only the effect of each of the primary level failure modes on the next level, but also the effect of every combination of primary level failure modes on the next level. In the case of a component the next higher level could be sub-assembly, sub-sub-assembly, or perhaps just a more complicated component. At the same time any additional failure modes associated specifically with this higher level must be introduced into the analysis. Strictly, the last action is not necessary, but because of interaction, which will be discussed later, it should be included. Once again it is necessary to consider if any of the effects identified at the second level can induce secondary or tertiary failures at that level. Unfortunately it is also necessary to examine if any second or subsequent level failure can feed back into the first level and set off an hitherto unaccounted failure sequence. The process is repeated for each level of the analysis, such as sub-assembly, followed in turn by assembly, equipment and ultimately the mission to be accomplished. The effect of each basic failure mode or combination of basic failure modes is thus traced through each level to its ultimate physical effect on the complete system. Very often this is a straightforward job, but there are occasions when a failure has to be followed through in detail. At the end of the fourth step a list can be produced of all the effects experienced at the ultimate level, which is the one that influences our use of the equipment and the one that matters, arising from failures within the equipment, and each effect can be traced back and allocated to its prime cause.

In the fith step of the FMEA we give a rating to each of the effects at the ultimate level determined in the fourth step. The rating describes in some way the importance of each failure on the operation of the equipment as a whole. In the rough and ready analysis each effect could be put into one of the broad qualitative groupings introduced in section 7.1: over-critical, critical, major, minor, and so on. In the quantitative approach each effect is given a numerical rating ($\alpha$), as also outlined in section 7.1. Most analysts arrange that $\alpha$ can take a value lying between 10 and 0, where 10 represents complete disaster, probably involving safety, and 0 represents no effect at all. Some effects, the repair cost of a failure is a good example, may be assessed fairly accurately (for this purpose) and quantitatively. Other effects, those involving safety for example, may have to be assessed more subjectively.

The sixth step in the FMEA is to derive a criticality for each of the failure modes. This depends on both the frequency and the effects rating derived in the second and fifth steps respectively. Thus the consequences of a major failure which occurs only a few times could be much more serious than those of a minor failure which occurs more frequently, so that it might assume a higher priority. As an example, suppose we are only interested in the repair costs of the truck whose failure modes were illustrated on figure 10.6. If the repair cost for each of the first 20 items listed on figure 10.6 were in the ratio 1, 2, 10, 0.5, 4, 6, 3, 15, 4, 10, 2, 3, 15, 6, 3, 4, 3, 3, 2, 7, then the relative cost of repairing each mode, which is given by the product of frequency and cost, would yield the Pareto distribution shown on figure 10.8. Our priority for remedial action, if based on cost, is therefore changed. On the other hand we may be more concerned with safety, in which case a different factor could be involved. In general we can define a criticality, $c$, for each failure mode such that

$$c = f\alpha \tag{10.9}$$

if only one effect arises from the failure, or

$$c = f\Sigma\alpha \tag{10.10}$$

if there are several effects.

Having calculated $c$, a new Pareto distribution can now be formulated using criticality instead of frequency, and hence the vital few critical to the success of

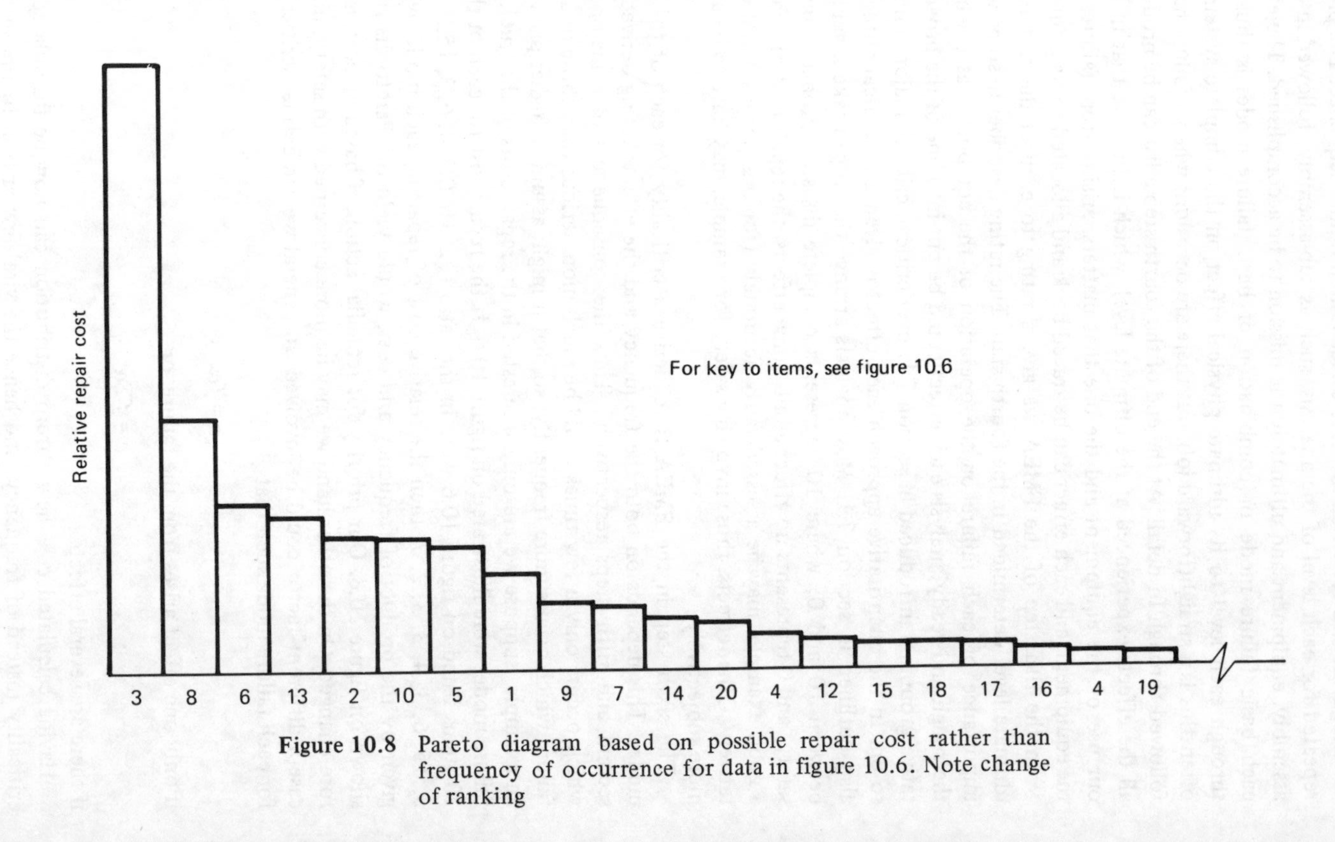


**Figure 10.8** Pareto diagram based on possible repair cost rather than frequency of occurrence for data in figure 10.6. Note change of ranking

the design, by whatever criterion has been adopted, can be identified. Old style chief designers drew up lists in ranked order of criticality almost subconsciously, based on engineering judgement and common sense, and indeed this can often be done for modern designs without resorting to the numerical technique described above. However, noting that it need not be conducted at anything more than a medium level of accuracy, the quantitative numerical approach can have advantages if many items are involved, or if different assessors covering different technological areas are making a contribution to the analysis.

With the Pareto distribution of criticality at hand it is possible to identify the failure modes which will mostly prevent the successful operation of the equipment in the defined role. The final step is one of design, for it is now necessary to propose feasible methods of eliminating those basic faults that were responsible for the critical effects. This will usually be done in close collaboration with the designer. In some cases, as an alternative to design modification, some critical items will be selected for special attention during development. This is most likely to arise if no clear design improvements are possible – for example, a feature might have been known as doubtful when designed – or when opinions differ regarding the best step to take.

*Exercise 10.12*

Pursue the FMEA started in exercises 10.10 and 10.11 by examining the effect of each failure mode on the higher levels of the system. In this case the next level can be taken as the stopcock itself. The level after that is the complete domestic water supply (it could be sections of it, but treat the whole system), and the ultimate level is the whole household. At the level of the stopcock you should consider whether there are any additional failure modes associated with this item as a whole, though you will not be able to do this for the level of the system itself in the absence of detail design data.

*Exercise 10.13*

Place the items you noted in the first section of exercise 10.10 – the common sense approach – in an order of importance, again based on common sense and experience. Then as an alternative approach, give an effects rating on a scale from 10 to 0 to each of the final effects you listed in exercise 10.12. From this and the frequencies determined in exercise 10.11 calculate the criticality of each failure effect. Present criticality as a Pareto diagram. On what aspects would you wish to see attention concentrated if you were seeking a higher than average level of reliability? Do these aspects differ in any way from those you listed above on the basis of common sense?

*Exercise 10.14*

Using the experience gained in carrying out exercises 10.10, 10.11, 10.12 and 10.13, comment on the relative value of a detailed numerical FMEA and the more rough and ready subjective one.

We must now return to the problem we introduced at the beginning of our discussion of FMEAs, but postponed: that is, the level at which the analysis should start. As stated, FMEA is a formalised approach to a fairly ancient practice. The formalisation is largely the result of electrical/electronic practice which took over the old style mechanical practice and rationalised it for that purpose. In the electronic application it is easy to leave the starting level as an arbitrary choice. It makes little difference, for example, whether this is at component or sub-assembly level. The reason, not often stated, is that there is comparatively little interaction between the items. When we come to deal with mechanical systems, however, there is considerable interaction between the components – indeed it may be all important – and in these circumstances it matters a great deal how we choose the level at which to start the FMEA. This may be illustrated by reference to an internal combustion engine. One cannot consider pistons, piston rings, cylinder block, combustion chamber, cylinder head and valve assembly as separate items: the interactive effects dominate this situation. What is satisfactory in one design may be totally inadequate in another. For example, we cannot say in abstract that a piston ring is a good or bad design – it depends on the application. We may not be able to eliminate interactions entirely, but we must choose the basic level at which to start the FMEA so that interactions are minimised. This often implies that we should begin the analysis of mechanical systems well above the basic component level: a conclusion which is reinforced by the large number of items that are bought in on a commercial basis and which often have a fair number of parts. There is little point in breaking down such items, which should essentially be treated as entities in themselves. It follows that the lowest level of a FMEA of mechanical equipment should be much higher than the corresponding analysis of electronic equipment. But such a conclusion is at variance with the observation that it is the detail design of small parts which so often accounts for the unreliability of mechanical equipment. Since both these aspects have to be catered for, it is necessary that no unique assembly level be chosen as the starting point for a mechanical FMEA. The first stage could well – indeed should be expected to – have mixed assembly levels. This may well include complete sub-assemblies on the one hand, and individual nuts and bolts on the other. Certainly all connectors and fasteners should be included individually at the first stage. Great skill lies in choosing the items that form the first stage of the analysis, minimising interactions without eliminating areas of potential failure. The description 'not very consistently', describing the old style designers' approach, could be accurate, but if so it is probably more of a compliment than the adverse criticism implied. This fundamental approach to a mechanical FMEA contrasted with an electronics FMEA must be properly understood if success is to be achieved.

The steps which make up the complete FMEA for a mechanical system can now be summarised as follows:

Step 1 defines the limits of the system to be analysed and the lowest levels at which the analysis is to begin
Step 2 lists all the primary modes of failure at the lowest levels set in step 1
Step 3 determines the frequency with which the failure modes listed in step 2 will occur in a defined accounting period
Step 4 (a) examines the effects of all failure modes listed in step 2 on all lowest level items, that is, determines all lowest level secondary, tertiary etc. effects
(b) examines the effects of all failure modes listed in steps 2 and 4(a) on the next higher (that is, second) level, including any secondary etc. failure modes at that level and any recursive effects on first or basic level
(c)
•
• repeats (b) for each of the higher levels in turn
•
( )
Step 5 allocates an 'effects rating' to each of the effects recorded at the highest level in step 4
Step 6 derives a criticality for each primary mode of failure listed in step 2 using its frequency from step 3 and its ultimate effect rating from step 5
Step 7 takes corrective action either by redesign, special development attention etc. for all the critical (Pareto's vital few) failure modes identified in step 6.

Some organisations like to formalise the procedure even more by requiring that the analysis be carried out on special pro-forma of the type shown on table 10.1. It can have the advantage that the analysis is carried out thoroughly and, perhaps of more importance, provide a reference against which to check subsequent experience. Equally important, it can form the basis of the development programme, which is to be discussed in section 10.4, and of the formulation of the maintenance strategy. However, as stated at the outset, it is undesirable that the FMEA be conducted with such formality that the process distracts from more productive activities.

**Table 10.1** Typical pro-forma for FMEP

FMEA of ______________________

Level of analysis
Next higher level

| *Item* | *Identification code, Ref. No., etc.* | *Failure modes* | *Frequency* | *Effects* | *Effects rating* | *Criticality* | *Repair action* | *Remarks* |
|---|---|---|---|---|---|---|---|---|
| | | | | | | | | |

Failure modes and effects analysis is a study in cause and effect. An alternative is a study in effect and cause, known as Fault Tree Analysis (FTA). In this approach the final effect, often called the 'top event', is taken as the starting point, and each is traced back to the primary cause of failure mode. It may be looked upon as a reversed FMEA. The end product is a fault tree of the kind commonly, for example, published for finding faults in car engines, though much more sophisticated.

An example of such a fault tree is shown in figure 10.9 – taken from ref. 67. The top event is taken to be the presence of sufficient energy to ignite a sugar dust concentration in a sugar refinery. This is shown in a rectangular box – this type of box indicating a failure which results from a combination of other failures. The standard symbols usually adopted for fault tree analysis are shown on figure 10.10. The logic symbol below the top event on figure 10.9 shows that sufficient energy may be present if the temperature exceeds the maximum ignition temperature (M.I.T.) *OR* the energy exceeds the maximum ignition energy (M.I.E.). Excessive temperature may be due to a heated surface *OR* a naked flame, in turn possibly due to flame release from unsafe work *OR* to external flame transfer. However, for external flame transfer it is necessary to have both an external flame *AND* the lack of flame break – see 'AND' symbol on figure 10.10. Both these events are considered to be basic failures which are statistically independent of other events, and are hence shown on circular boxes. Quite often the logic symbol for an OR gate contains a plus (+) symbol – for set union – while AND gates often contain a product (.) symbol – for set intersection. The gates on figure 10.9 are shown thus. Sometimes a less mathematically, but more linguistically, expressive notation is used in which the AND gate contains an ampersand (&) symbol while the OR gate is left empty. Once the logic tree is constructed, all logically possible failure scenarios (called minimal cuts) can be obtained. There are many algorithms and computer programs for finding minimal cuts. The mathematical analysis of fault trees is based on Boolean switching theory, but is not pursued here, and the reader who is interested is referred to the Reading list.

Although often treated as separate techniques in the literature, the author regards fault tree analysis and failure modes and effects analysis as essentially the same. In some cases it is easier to work from cause to effect, while in others the reverse is more amenable. The analyst should clearly choose the easier, and there is no reason why both should not be used on the same project. It is often claimed that Fault Tree Analysis is particularly suitable for use in the early stages of a project because it can be used to identify those items in a design that are likely to contribute most to a particular system failure, and therefore merit closest attention. This may be true if particular effects require special attention – safety aspects spring immediately to mind in this respect – but the generality of the claim is doubtful. The essential difference between FMEA and FTA is that the former starts with the failure mode while the latter has to find it. In an FTA, therefore, every mode *has* to be investigated, since it must be assumed each mode can contribute to a top event until proved otherwise. To the extent that FTA

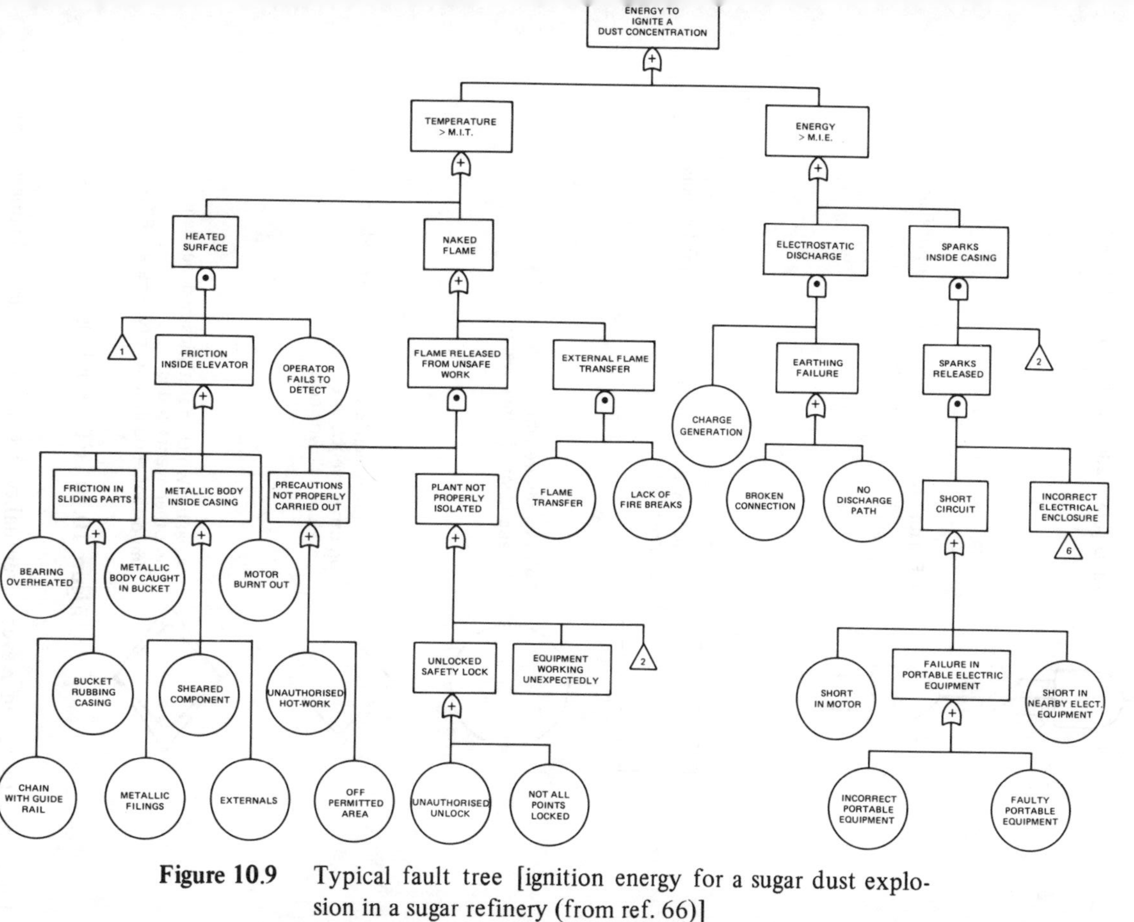


**Figure 10.9** Typical fault tree [ignition energy for a sugar dust explosion in a sugar refinery (from ref. 66)]

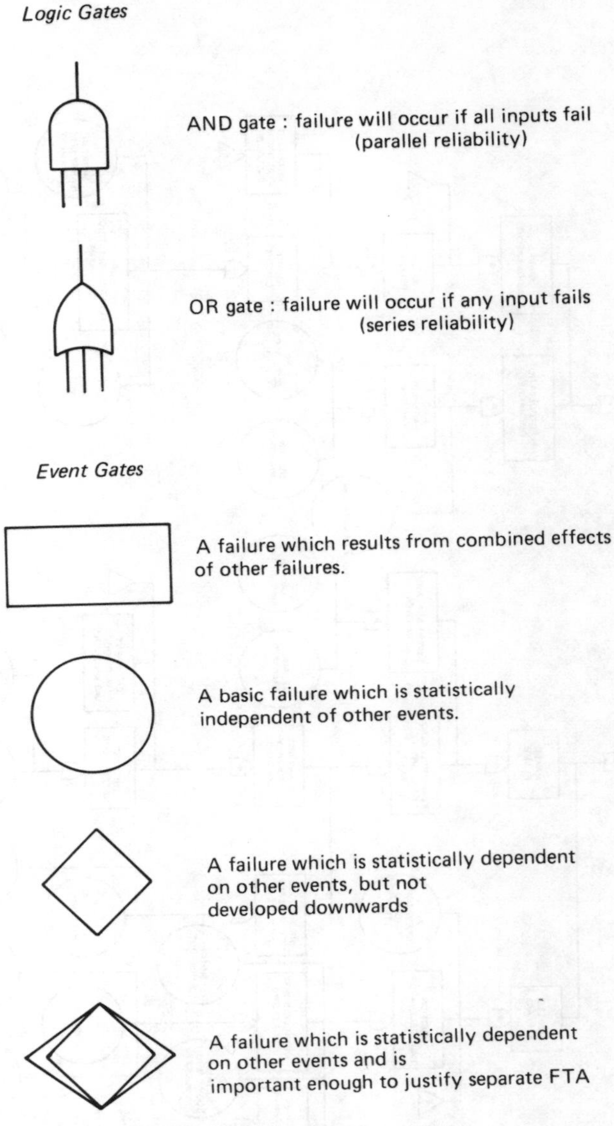


**Figure 10.10** FTA symbols

forces one to consider every eventuality it may be considered superior, but more often than not time and effort are not available. In an FMEA every mode should be investigated, but does not have to be, so that effort can be concentrated if it is insufficient for a full investigation. This is particularly relevant to mechanical systems, in which the number of failure mechanisms can be very large indeed, as contrasted with electronic systems where the number of failure modes is comparatively limited. Indeed, the effect of the number of modes is often overlooked. This is brought out by figure 10.9, which shows, for example, 'bearing overheated' as one prime event that may be responsible for the top event. But this tells us very little: there are a thousand and one reasons why a bearing may overheat. This FTA follows the general pattern in being cut off just when it is becoming interesting. To follow up this example: a bearing may overheat because it is not properly lubricated, so we should have to follow up the whole lubricating system, and indeed the interactive effects of every other component on that system. It may overheat because of tolerance stack-up, which leads to other components and perhaps to thermal effects on those components and any other components which contribute to heat addition or subtraction. At a less likely but still very possible level, a locating nut may loosen as a result of vibration, which in turn may originate in any (which means 'every' for FTA purposes) of the other components of the complete assembly. To pursue the 'tree' further than shown in figure 10.9, a great many more branches – some with pretty nasty thorns – must be drawn. This is necessary because of interaction between mechanical components, already referred to. If carried to its logical conclusion the mechanical part of the fault tree soon takes the form of a huge, tangled, inpenetrable undergrowth, rather than an elegant cultivated tree such as grace the pages of so many reliability texts. Perhaps the stage reached on figure 10.9 *is* the right place to stop the analysis. But if so, it has not tackled the real mechanical problem. There clearly are difficulties. The

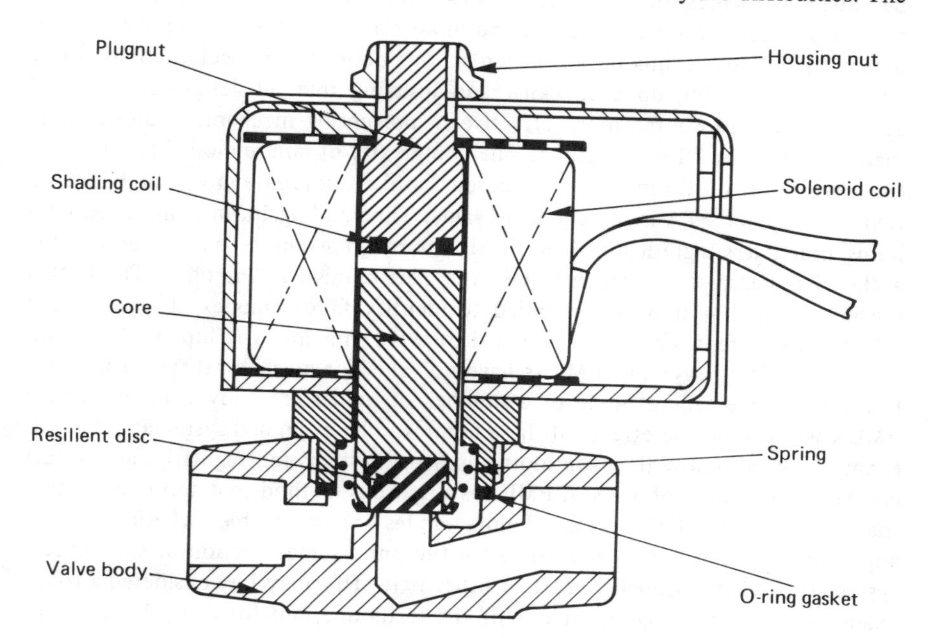


**Figure 10.11** Typical two-way solenoid valve

author would thus prefer not to use either FEMAs or FTAs for mechanical investigations, but to treat them as complementary, using the one which proves most amenable to any investigation.

*Exercise 10.15*

Figure 10.11 illustrates a simple, two-way, direct acting, normally closed, packless, solenoid-operated valve for the control of fluid flow. To operate the valve, current is passed through the solenoid so that the core is moved against the spring. List the top events, that is, the failure modes that can be sensed, for this valve and then the possible factors which could give rise to each of the top events.

Rather than comparing FMEAs and FTAs, we should perhaps be querying the value of either for mechanical systems. It must be understood that neither technique reveals a failure mode. That can only be done by an engineering appraisal. As a classical example involving safety, consider the Flixborough accident. Neither a FMEA nor a FTA, if carried out, could have predicted that accident if the inadequacy of the mechanical design of the faulty by-pass duct had not been appreciated – as it evidently was not. Without a correct engineering input both FMEAs and FTAs are meaningless. In the case of electronic systems, for which the techniques have been developed and proved so satisfactory, the number of failure mechanisms are comparatively limited, and hence a standard analysis may be expected to pick them all up. For mechanical systems the number of failure mechanisms are practically infinite, and more elaborate methods are necessary to detect them. The technique advocated in this book is the checklist. It is firmly believed that setting up a checklist and insisting that all designs are checked against it will prove far more effective in preventing mechanical failures than many a FMEA or FTA without it. The complexity of failure mode identification (may we call it FMI?) must be recognised. Until every failure mode can be identified, a FMEA or FTA is of very little value. Time and again, in connection with items much less significant than the Flixborough example quoted above, the author has heard the statement "Oh, the FMEA missed that one." The FMEA missed nothing! – it is not intended to identify failure modes, it only studies effects. A different technique is needed to avoid 'missing' a failure mode – FMI.

This is not to say that FMEAs have no value for mechanical systems. Sometimes a possible failure mode or a number of modes are known. In that case a FMEA will enable the effects of these failures to be examined systematically. For example, we all know that automotive exhaust systems fail regularly, and in fact can fail in a number of ways. A FMEA is a rigorous method to determine whether the consequences of the various failure modes are acceptable: whether, for example, if a bracket fails it is possible for the unrestrained section of the exhaust system to foul the propeller shaft or other part. This may be considered a trivial example not warranting the rigour of the technique, but for many large plants, such as nuclear power stations, the rigour of the method is essential. This is particularly so if safety is involved, though it applies equally to many process systems and control systems where the cost of failure is high. The methods are extremely powerful for studying the consequences of a *given* failure. They ensure an adequate back-up is provided or fail-safe measures taken. This must sometimes be done even when the primary failure is considered extremely remote. The procedure is very costly, but can be justified in special cases. But even then its success depends on the recognition of all primary failure modes: if a critical primary failure mode is not included the findings of an analysis could be very misleading. FMEAs and FTAs are generally most potent in studying the reliability of a system in which the component reliabilities are known, when the components of that system can be individually identified as separate entities, when the functional relationship between them is complex, and when the physical interaction between the components is minimal. In addition to these particular cases, there is little doubt that FMEAs and FTAs can contribute to the reliability of more everyday equipment; but there is also little doubt that they are very time-consuming and expensive if the full formal procedure is followed. A judgement has to be reached on economic grounds in each case. The author believes that the main advantages of a full FMEA can be achieved by encouraging all designers to perform 'mini-FMEAs' on the limited sections of the whole design for which they are responsible. Without the critical appraisal which FMEAs inculcate, proper design of even the smallest item is impossible. Furthermore, many mini-FMEAs greatly expedite a full FMEA if required. This proposition may be queried, as it is often contended that the main merit of a FMEA is a compilation by a body *outside the design team* of all failure modes together with an assessment of their consequences. However, it may equally be queried whether anyone completely outside the design team can isolate and assess every problem involved in mechanical design – without, that is, undertaking a complete redesign! The advantage of a full FMEA or FTA lies in its rigour – the 'leaving no stone unturned' and so on, and it seems this advantage can be achieved if the FMEA is carried out honestly by the design team, or at least its senior members. However, the extreme rigour of an independent body cannot be ignored, and the advantages of the latter should always be considered. Much will depend on the faith one has in the designer or design team, and how far one is prepared to adjudicate on irreconcilable differences of opinion at a later stage! Independent assessments can sometimes be counter-productive in the long term. The degree to which a FMEA or FTA is formalised becomes a management decision in which many factors prevailing at the time will contribute. However, no matter what decision is reached some formal documentation is necessary for three reasons

1. It provides the basis for the planning of a development programme (this is described in the next section). Without such documentation the programme could be very much a 'hit and miss' affair.
2. It provides the basis on which maintenance plans can be made. Although essentially provisional at this stage, the time scale is such that some planning must go ahead.
3. It provides a basis against which design progress can be assessed when field data becomes available.

In conclusion, then, the limitations of these methods, the circumstances when they are of value, and the format they would take must be recognised. In spite of some advocacy they are no universal panacea, but have a role to play. The real problem is the identification of the failure-inducing loads – particularly those that are internally generated – which have not been properly taken into account by the designer. (Much the same problem, though it centres around external loads, bedevils software reliability, for which existing techniques have proved so inadequate.) The author firmly considers setting up a relevant and comprehensive checklist (or any other FMI) for the product in question, and ensuring it has been used, to be the first priority.

### *10.3.6 Design or milestone reviews*

It is inevitable that during the actual design process the chief designer, or possibly the project leader, will hold informal meetings with those actually involved in the design to assess the manner and extent in which the design is achieving the requirements. However, it is most desirable that on all except the smallest project there should be some formal occasions when progress can be assessed by a representative cross-section of those involved. This constitutes the Design Review or Milestone Review. Its object is the formal documented and systematic study of a design by management, by the production team, by the user if he is involved, and (when appropriate) by specialists not directly concerned with the design. In the larger organisations to which we are becoming more and more accustomed, it is difficult to establish that old personal contact between customer, designer and producer. In the quotation (at the beginning of this chapter) from the author's survey of reliability conferences, good communications were identified as one of the most important factors in achieving high reliability. In many ways, Design Reviews are a modern formal replacement of the old informal contacts, and should be regarded as a method of communication. If they fail in this, they fail in their prime purpose. It is the duty of the chairman, and to some extent the secretary, to ensure that all matters relevant to the stage the design has reached are discussed in a design ambience. Since the intention is to prevent undesirable aspects finding their way into the ultimate design, the reviews should start as early as possible, even though many aspects cannot be resolved until the later stages. For most designs a Preliminary, two Intermediate, and a Final review are adequate, though for large projects additional Intermediate reviews may be necessary as soon as, and whenever, important design features are defined. The Preliminary review should precede the conventional design process, and ideally take place towards the end of step (2) of section 10.1. It should consist of a complete evaluation of existing knowledge in order to select the optimum design from the alternatives available. Any areas of high risk should be singled out at this stage. These may be pin-pointed by an early FMEA. The review should also ensure that there is a comprehensive plan to achieve the reliability and maintainability targets at least up to the prototype trial stage. The first (and any additional) Intermediate Design Review should essentially be concerned with component and equipment design. It will probably include any revised estimates of reliability of life arising from the more detailed design, and hence will bring in maintainability. The findings of any FMEA, which can now be based on actual drawings, should be considered at this stage. The second Design Review will probably take place between steps (6) and (7). Appraisal of any component testing should be made at this stage, leading to the plans for the prototype trials, the accompanying defect-reporting system, and the evaluation and assessment of the trials. The Final Design Review, like the Preliminary Design Review, really falls outside the province of conventional design but is best taken here for continuity in treatment of Design Reviews. The Final Review sets out to confirm (or possibly otherwise!) that the design meets its specification, so far as that is possible, that it is fully defined by its drawings and associated specifications, and that it can be produced by the resources available. The design drawings and specifications should then be frozen, and on approval by the sponsor, manufacture can begin.

A Design Review must cover much more ground than reliability, which alone has been outlined above. One of the pitfalls is that reliability will be given second place to more readily and immediately assessed qualities. It is the duty of the chairman, supported by the secretary, of the Design Review Meeting to ensure that reliability is not lost at this stage.

*Exercise 10.16*

Assume you are the secretary of a Design Review Board. Make a list of the subject heads you would deem necessary to cover the broad description of design reviews given above. Place the subject heads in order of importance with respect to achieving high reliability and indicate at which of the Design Reviews (Preliminary, Early intermediate, Late intermediate or Final) they should be taken.

### *10.3.7 Some concluding remarks on conventional design*

We close this section with some general remarks on conventional design. There can be little doubt that design is the key to reliability. The foundation is laid at that stage. If the foundation is unsound it *may* be possible to correct it, but often only at a very high price. Clearly, the best way to good design is via good designers or design teams. If this can be achieved, much that is written about design is superfluous. It must be recognised, however, that more recently the normal responsiblities of the old style chief designer have to a greater or lesser extent been taken out of the hands of individuals who were directly and intimately concerned with the design process. The technical scale of many projects has unfortunately promoted this trend. Many of the so-called new techniques are an attempt to ensure that all the thought processes of the old designer are in fact taken by those without design experience. Whether this can be done is for debate. It must imply that all concerned are capable of the necessary thought processes. Before setting up a Design

Review Committee, it might be advisable to ask how many of the proposed members have actually carried out any design work themselves, for example. If the result is less than 50 per cent (excluding chairman and secretary for which the figure must be 100 per cent) urgent management action is required. Over-emphasis on meetings and monitoring techniques is to be avoided – it is the equivalent of the amateur gardener who digs up his potatoes every weekend to see how they are growing! It has already been stated that design is an intellectual creative art. All monitoring processes must operate in this ambience, and not destroy the whole object of the enterprise of which they are but a small part.

If any changes are required to the design system that has served many major and minor companies so well in the last 100 years, they should be concerned with updating the current technical methods rather than with unnecessary changes to administrative procedures. The latter step is so often taken by a management which does not understand the design process but feels it ought to be doing something (or seen to be doing something!), whereas the prime need is so clearly for the elimination, so far as is possible, of empiricism. Some empiricism is a necessary evil which will always be with us, but as time goes by empirical factors, which often originally had some physical significance, incorporate more and more secondary functions so that the tail is soon wagging the dog. It then becomes unclear what role is played by the individual constituents of the factors and how the empirical factors themselves should be changed to meet changing circumstances. Fortunately, great strides are being taken in technical understanding in a number of fields, so now is the time to prune empiricism. The biggest single step would be the abandonment of the deterministic treatment upon which design is almost universally based. At least the empirical 'reserve' factors correlating experience (essentially stochastic) with design (deterministic) will no longer be required. The way will then be open to the use of modern technology, principally stress analysis, but more particularly many long-established techniques which require computer power to make them feasible design tools. Variability can then be introduced in its own right. This is what computer aided design should be all about. Thus we have reached a critical stage where design has properly to encompass reliability or stagnate. If we really want to produce high-quality goods the way ahead is quite clear: abandon deterministic design techniques and develop stochastic methods.

### 10.4 Development

It must be recognised that although a lot can be done to specify loading conditions in general terms, they will hardly ever be specified with an accuracy that allows actual design to follow the mathematical statistical model with any certainty. Furthermore, in his search for excellence the designer must necessarily step outside the bounds of established knowledge. Applied mechanical technology has always had a significant lead over theoretical quantitative understanding. With both loads and strengths subject to unknown factors, any advanced or novel design must be expected to lack precision. In particular, reliability, which depends essentially on variations in these quantities, becomes elusive. The writer would go so far as to say that numerical values of reliability at the conventional design stage are only rough guides. There is only one way in which the reliability of mechanical equipment can be determined and that is by testing under actual conditions. It is perhaps significant that in the aircraft industry, where high reliability is so obviously a necessity, research and development account for about 90 per cent of the total cost of military aircraft. The corresponding figure for land-based equipment is nearer 10 per cent.

The objective of development testing has been specified by Lovesey in ref. 68 as firstly taking an embryo equipment from the drawing board up to maturity as a production job and caring for its initial service years by correcting defects; and secondly, in modifying the basic established design to produce higher performance, to give longer life, or to suit new requirements.

In essence the development procedure is simple. A limited number of the product in question (the prototypes) is made to the agreed design. Sometimes products to slightly differing designs are made for comparison. The prototypes are then tested to demonstrate that the specification has been achieved. This would apply to both performance and reliability. If it is not achieved, and there will inevitably be some areas in which it is not, it is necessary to establish why the specification was not achieved, and from that knowledge the design is modified to rectify the fault. Since the modification itself may not meet the specification it is necessary to retest, and if necessary to repeat the process, until the specification is achieved. The process is one of test, assess, redesign ('fix') if necessary, retest, re-assess and so on: repeated until the specification, be it performance or reliability, is achieved. To isolate those areas needing attention and to establish the nature of any problem, development work is conveniently divided into a number of categories.

(1) *Full-scale testing under real conditions*. This is clearly the ultimate and necessary objective. Unfortunately it is expensive, but more particularly 'real conditions' may not be available for sufficient time. They often cannot be controlled, so testing can be capricious and repeat tests are often impossible. Furthermore, it is often difficult or expensive to include the detailed instrumentation which is an essential of some development work. Testing under real conditions has therefore often to be augmented and sometimes replaced by category (2).

(2) *Full-scale testing under simulated conditions*. This is probably the most powerful development technique. Conditions can be closely controlled and a detailed planned investigation can be carried out to a time schedule. Its weakness lies in the unforeseen real operating condition that is not simulated, so that it can never wholly replace (1). It can be very expensive, and sometimes cannot be adopted on account of the scale involved. In other cases it can offer the best approach.

(3) *Full-scale component testing.* Components can with advantage be isolated and tested separately when greater information can be obtained from a more elaborate instrumentation, and perhaps the component can be run under a wider range of conditions than is possible when contained within the full unit. This form of test is invaluable in establishing the merits of a component before the design is finalised, so that major hold-ups and expensive modifications are avoided at a later stage. It is obviously less expensive than testing the final product.

(4) *Detailed rig testing.* The fourth weapon is a very detailed investigation of a relatively small part in a special test rig. This may be carried out on an actual bit of the equipment or on a model, and is intended to provide a fundamental understanding of the manner in which the item is operating – the exact distribution of stress in a loaded member, for example. This type of work is closely akin to research, except in as much as it has a very definite objective in view.

In the minds of many people, all development is akin to research – often linked as 'R & D'. Development *has* been defined as 'research done quickly'! The parallel objectives were brought out by Lovesey, who shows that development is essentially a process of understanding the mechanisms responsible for unexpected behaviour so that corrective action can be taken. The art is to minimise effort. Accumulating knowledge for its own sake is bad development (it may be good basic research!) though a full knowledge of all relevant factors is essential for a quick and effective solution to the problem. Development is both a costly and time-consuming process. Herein lies the rub. Very often neither the requisite time nor money are included in the initial assessment, and the price paid is lack of reliability. Lack of time is particularly insidious, and methods of assessing the rate of reliability growth during development will be discussed later. Although the basic principles of development are simple enough, the implementation of an actual programme is often far from simple. This arises from the large variety of incidents that may require attention. They cover a very wide range of technology, and make heavy demands on development engineers. Furthermore, interaction between incidents can mask the fundamental nature of the problem.

In drawing up a development programme it is first necessary to establish how an item will be used and, in the light of this knowledge, list all the modes of failure that are likely to occur. For this purpose 'likely' must be interpreted as 'even remotely likely'. The reader may well observe that this action will have already been taken as part of a failure modes and effects analysis (FMEA). That this information is subsequently necessary for development purposes is, in the author's opinion, the most valid reason for following a more formal procedure for an FMEA at the design stage, and recording the analysis for subsequent use. Firstly it opens up a line of communication between the designer and the development engineer which is essential for a satisfactory product, and secondly it provides the basic data for a development programme. If no FMEA has been formally recorded, either as part of design itself or as part of Design Reviews, then the development engineer has to do the job. Such a procedure does waste effort, however, in as much as no advantage can then be taken of the FMEA to improve the design itself before prototype manufacture and development effort (that is, cost) is expended. The use of FMEA information for design and development does differ in one respect: for design we are usually only concerned with the vital few, for development we are concerned with all failure modes. With this information, tests must be planned to create circumstances in which these failures could occur. In most instances this will only involve normal operating conditions, though from the very nature of the investigation one would imagine that the upper end of the operating spectrum would receive the most attention. These tests must be carried out under controlled conditions, so that the exact circumstances which give rise to failure can be established. Simultaneously, the impact of any failure on the operation of the item can be assessed. At the same time, steps must be taken to promote the unlikely (or unexpected) failure by testing under realistic but less-controlled conditions.

An actual development programme may well start before the complete design has been finalised: major components or sub-assemblies and certainly any critical items should be tested under category (3) above to ensure that they are acceptable before they are incorporated in a design. Although such tests do not have the rigour of tests under categories (2) and (1), in that unexpected interactions may not be simulated, nevertheless they do provide some assurance that major defects have been eliminated and that the whole design is not going to finally fall down on account of some over-optimistic assumption in the feasibility or project stages. Such tests are really aimed at gross deficiencies in the design but, even so, attention must be drawn to the necessity of simulating the load cycle as accurately as possible and of representing a full life. Category (4) tests may be called up if the category (3) tests reveal any deficiency which cannot be obviously corrected.

Testing under categories (1) and (2) can only proceed when the initial design has been completed and prototypes are available. The difficulties at this stage are mainly economic: it is necessary to ensure that sufficient prototypes are available to make a representative population, and also to ensure that adequate test time is available to cover fully the proposed life of the equipment. In this connection it should be realised that a development programme will not only be concerned with reliability – many see the objective as the achievement of a specified performance. So often within a development programme conflict can arise. Adequate performance is also achieved by test, assessment, modification, retest and so on, but proceeds fastest when mechanical failures are reduced to a minimum and components can be modified quickly and at little cost for retest and re-evaluation. Performance evaluation rarely involves prolonged testing, and as soon as adequate data is available there is every incentive to proceed with the next step. By contrast, reliability can only be properly assessed when failures have occurred. A number of failures must be induced at normal operating conditions. If the design is at all successful this requires prolonged testing. A development programme which is devoid of failures tells us very little. With no failures recorded on, say, ten prototypes, one is 95 per cent confident (in the statistical sense) that the MTTF is 3.3 × the cumulative test life under the operating conditions of the test. This is far from ensuring intrinsic reliability under general operating conditions, and a wear-out mode could arise immediately after the longest individual test life. If the wear is measurable (growth due to creep, for example), and a correlation between wear and life exists, some assessment can be made, but it is an extrapolation. To make any firm deductions a representative number of failures must be induced. This holds up performance evaluation. Realistically, every failure mode cannot be assessed – indeed most eventualities will not be covered; but it is totally uneconomic to carry out development work which purports to establish reliability yet provides no failure data. That can be positively misleading. To get some idea of the information that can be obtained from data and the amount of information that is required to make firm deductions it is suggested you try the following exercises. Detailed solutions and comments are given in chapter 15.

*Exercise 10.17*

As part of a development programme to increase the scheduled maintenance interval of an existing type of vehicle from ten thousand miles to fifteen thousand miles, a manufacturer arranged for the testing of 20 engine oil filters. Each filter was tested in a vehicle run under representative conditions. Every one thousand miles, each filter was removed and evaluated; if satisfactory, it was returned to the vehicle and testing continued for a further one thousand miles before re-evaluation. A failure was deemed to occur when the measured pressure drop exceeded a specified level or the filtration efficiency fell below the minimum required. Once a failure occurred, the filter was removed from the test. The results showed that failures had occurred at the following thousand mile intervals: 23 (one failure), 25 (one failure), 27 (one failure), 28 (one failure), 29 (two failures), 31 (two failures), 32 (one failure), 33 (one failure), 34 (two failures), 35 (one failure), 36 (two failures), 37 (one failure), 38 (one failure), 39 (one failure), 41 (one failure) and 42 (one failure). You are asked to make an appraisal of this data and hence make recommendations for the course of action to be followed by the manufacturer. You must justify all decisions you make.

*Exercise 10.18*

A new material has been proposed for boots intended for use in arctic conditions. Tests have proved the functional claims of the material. To test its wear properties, nine pairs of boots were issued at random to representative users and worn until failure occurred, when they were returned for examination. The test terminated after 18 months. The time in use before return is listed below. What comments and observations can you make on the basis of this data alone?

1st pair returned after 371 days
2nd pair returned after 236 days
3rd pair not returned at end of test
4th pair returned after 300 days
5th pair returned after 470 days
6th pair not returned at end of test
7th pair returned after 530 days
8th pair returned after 148 days
9th pair returned after 88 days.

The previous two exercises are artificial ones specially designed specifically to illustrate how valid test data should be interpreted. They contain no spurious data or red herrings. You are now asked to tackle exercise 10.19, which is a very small extract from a much bigger trial. You will now be dealing with real data in the way it was first reported.

*Exercise 10.19*

You are asked to deduce as much as possible from the following data extracted from a trial report on 25 pre-production prototype general-purpose load-carrying vehicles. Of the 25 prototypes, six were retained and tested by the trial unit. The first three were subject to somewhat harsher treatment than normal to promote failures, but the second three were subject to normal usage. The remaining prototypes were distributed among representative users who operated them as part of their standard fleet. In order to keep this exercise within bounds, data is only given for the main engine unit. Data on auxiliaries and installation failures are excluded as is that on all other aspects of the complete vehicle. The engine is an orthodox spark ignition petrol engine as is found in most privately owned vehicles, so all readers should be familiar with its operation and the significance of any failure. In the table below, following a description of the failure, it is noted whether the onset of the failure was sudden or gradual and whether the failure could be graded as complete, partial or trivial so far as the operation of the vehicle is concerned. Finally, the conditions under which the prototype was operating at the time of failure are recorded. Occasionally, additional information on the failure is added. The data refers to failures and unscheduled maintenance activity only, but in addition the engine unit was serviced normally according to the maker's recommendations. This is not recorded. You should note that a certain amount of trouble had been experienced with spark plug leads on earlier versions of this engine and the makers have a modification which they claim eliminates any failures of this nature.

Specifically, you are asked what recommendations you would make on the basis of this data alone, and what additional work you would put in hand if you were in charge of these trials.

| | |
|---|---|
| Prototype No. 1 | Total usage 21 918 km |
| 000 478 km | Throttle arm clamping bolt loose. Gradual. Partial. Noted on pre-trial checks |
| 000 800 km | Choke cable insecure. Gradual and partial failure. General road work |
| 002 480 km | Distributor points pitted. Gradual and trivial failure. Noted on routine inspection. Cause not established |
| 003 975 km | Engine misfiring: plug leads off. Gradual and partial failure. Tracks and rough roads |
| 006 600 km | Throttle cable broken. Gradual and partial failure. Cross country |
| 007 288 km | Oil filler cap seal ineffective. Gradual and partial failure. Tracks and rough roads |
| 009 000 km | No. 4 spark lead adrift. Gradual and partial failure. Noted on routine inspection |
| 009 000 km | Throttle pedal bush worn. Gradual and trivial failure. Noted on routine inspection |
| 009 956 km | Engine misfiring. Gradual and trivial failure. Cause not established. Noted on routine inspection |
| 011 031 km | Modified plug leads fitted. Gradual and partial failure. Result of routine inspection |
| 011 981 km | Throttle linkage clamping bolt loose. Gradual and partial failure. General road work |
| 012 352 km | Engine misfiring: modified spark plugs fitted. Gradual and partial. General road work |
| 012 554 km | Rear exhaust pipe to chassis clamping bracket worn. Sudden and trivial failure. Tracks and rough roads. Considered to be reasonable wear-out. Tracks and rough roads |
| 013 515 km | Throttle arm clamping bolt loose. Gradual and partial failure. Noted on routine inspection |
| 014 459 km | Throttle clamping bolt loose. Gradual and partial failure. Noted on routine inspection |
| 015 062 km | Fuel leak filter/carb. pipe. Gradual and partial failure. Tracks and rough roads. Attributed to poor manufacture |
| 015 218 km | Throttle arm clamping bolt loose. Gradual and partial failure. Tracks and rough roads |
| 015 218 km | Throttle cable frayed at cable end. Gradual and partial failure. Tracks and rough roads |
| 019 122 km | Throttle arm clamping bolt loose. Gradual and trivial degree of failure. Noted on routine inspection |
| 020 071 km | Throttle arm clamping bolt loose. Gradual and trivial degree of failure. Noted on routine inspection |
| Prototype No. 2 | Total usage 25 001 km |
| 003 091 km | Plug leads off No. 3. Gradual and partial failure. Tracks and |

| | |
|---|---|
| | rough roads |
| 003 556 km | Plug leads off. Gradual and partial failure. Tracks and rough roads |
| 003 843 km | HT lead foul exhaust. Gradual and partial failure. Tracks and rough roads |
| 004 829 km | Throttle pedal arm loose. Gradual and partial failure. Tracks and rough roads |
| 007 773 km | Throttle pedal coupling bolt loose. Gradual and partial failure. Tracks and rough roads. |
| 008 258 km | Engine misfiring: ignition system damp. Gradual and partial failure. Tracks and rough roads. Adverse weather conditions reported at time of failure |
| 008 258 km | Oil filler cap seal ineffective. Gradual and partial failure. Tracks and rough roads |
| 008 716 km | Oil filter union leaking. Gradual and trivial failure. Tracks and rough roads |
| 011 165 km | Throttle linkage maladjusted. Gradual and partial failure. Noted on servicing check |
| 012 387 km | Modified plug leads fitted |
| 013 600 km | Choke cable stiff in operation. Gradual and partial failure. Considered due to wear. Noted during workshop inspection |
| 014 830 km | Misfiring ignition: modified spark plugs fitted. Gradual and partial failure. Noted during routine inspection |
| 018 230 km | Excessive movement in throttle linkage. Gradual and trivial failure. Noted during routine inspection |
| 023 580 km | Distributor full of water after washdown. Gradual and believed trivial failure due to human error |
| 024 151 km | Throttle cable broken. Sudden and complete failure. Tracks and rough roads |
| Prototype No. 3 | Total usage 14 810 km |
| 004 315 km | Coolant leak rear hose. Sudden and complete failure. Tracks and rough roads |
| 004 832 km | Temperature gauge inoperative. Gradual and trivial. Noted on routine inspection |
| 011 019 km | Throttle cable frayed at pedal end. Gradual and partial. Noted on routine check |
| 012 009 km | Slight leak from water pump. Gradual and trivial. Noted during routine inspection |
| Prototype No. 4 | Usage 23 245 km |
| 002 847 km | Misfiring: new sparking plugs fitted. Gradual and partial. Tracks and rough roads |
| Prototype No. 5 | Usage 2568 km |
| No failures recorded | |

| | |
|---|---|
| Prototype No. 6 | Usage 4443 km |
| 002 660 km | Ignition retarded. Gradual and partial |
| 002 660 km | Choke connecting balance pipe defective. Gradual and trivial |
| Prototype No. 7 | Usage 14 480 km |
| 001 377 km | Throttle linkage sticking. Gradual and partial. General road work |
| 002 756 km | Misfiring: ignition dwell angle reset. Gradual and partial. General road work |
| 002 756 km | Oil filter seal leaking. Gradual and partial. General road work |
| Prototype No. 8 | Usage 14 091 km |
| 003 281 km | No. 1 plug lead loose. Gradual and partial. Noted on pre-trial checks |
| 006 331 km | Misfiring: ignition contact breaker points renewed. Gradual and partial. Tracks and rough roads |
| 006 331 km | Throttle pedal clamping bolt loose. Gradual and partial. Tracks and rough roads |
| 008 057 km | Fit modified plug leads. Routine inspection |
| 008 970 km | Throttle cable broken. Gradual and partial. Tracks and rough roads |
| 012 362 km | Throttle linkage out of adjustment. Gradual and trivial. Noted on routine inspection |
| 012 362 km | Choke cable outer unserviceable. Gradual and trivial. Noted on routine inspection |
| Prototype No. 9 | Total usage 13 814 km |
| 005 562 km | Throttle arm clamping bolt loose. Gradual and partial. General road work |
| 006 145 km | Throttle arm clamping bolt loose. Gradual and partial. General road work |
| 006 979 km | Throttle arm clamping bolt loose. Gradual and partial. General road work |
| Prototype No. 10 | Total usage 1071 km |
| No failure recorded | |
| Prototype No. 11 | Total usage 11 228 km |
| 005 328 km | Throttle arm clamping bolt loose. Gradual and partial. General road work |
| Prototype No. 12 | Total usage 6947 km |
| 003 636 km | Fit modified high tension leads |
| 004 279 km | Accelerator movement stiff. Gradual and trivial. Noted in workshops |
| Prototype No. 13 | Total usage 6226 km |
| 003 654 km | Fit modified high tension leads |

| | |
|---|---|
| Prototype No. 14 | Total usage 2550 km |
| No failure recorded | |
| Prototype No. 15 | Total usage 484 km |
| No failure recorded | |
| Prototype No. 16 | Total usage 1601 km |
| No failure recorded | |
| Prototype No. 17 | Total usage 6711 km |
| 005 390 km | Slight leak water pump. Gradual and trivial. Tracks and rough roads |
| 006 451 km | Contact breaker points closed to 0.006″ and 0.007″. Gradual and partial. Tracks and rough roads. Cause unestablished. |
| 006 451 km | Spark plugs renewed. Gradual and trivial failure. Tracks and rough roads |
| Prototype No. 18 | Total usage 626 km |
| No failure recorded | |
| Prototype No. 19 | Total usage 10 263 km |
| No failure recorded | |
| Prototype No. 20 | Total usage 1579 km |
| No failure recorded | |
| Prototype No. 21 | Total usage 1472 km |
| No failure recorded | |
| Prototype No. 22 | Total usage 1792 km |
| No failure recorded | |
| Prototype No. 23 | Total usage 7849 km |
| 006 531 km | Throttle arm clamping bolt loose. Gradual and partial failure. General road work |
| Prototype No. 24 | Total usage 8679 km |
| 000 801 km | Throttle arm clamping bolt loose. Gradual and partial failure. General road work |
| Prototype No. 25 | Total usage 7884 km |
| 006 520 km | Accelerator pedal sticking. Gradual and partial failure. General road work |

The preceding exercises were intended to introduce you to some simple prototype trials data assessment, and at the same time it is hoped that exercise 10.19 will have drawn your attention to some of the difficulties involved in interpreting the actual raw reported data. To some extent the difficulties can be minimised by good trials planning, but you will probably have noted that a little more information on what at first appears to be a minor fault would have made the intepretation much more positive. However, trials cannot be held up, and it is inevitable that a lot of information will not be recorded because every fault cannot be followed through in detail. On the other hand, there will be a lot of useless information that must be discarded when the analysis is carried out. The biggest mistake, which occurs frequently, is the discarding of apparently trivial information which is subsequently seen to be most relevant when taken in conjunction with other data. The best safeguard is a well-planned programme with defined objectives (though one has always to be careful not to plan out the unforeseen failure-inducing circumstance that is later going to prove all important). It is therefore suggested that before we go on to discuss some special development techniques adopted by mechanical engineers, you prepare a development programme for a relatively simple item by tackling the exercise below.

*Exercise 10.20*

High power/weight ratio is the prime factor controlling the performance of any vehicle, and in the case of the bicycle depicted in figure 10.12 this must be interpreted as low weight (since the designer has no control of the power pack!). Unfortunately, special light-weight contemporary machines cost a lot more than the average. However, a manufacturer of these bicycles believes he can equal the lowest current weights at mass market prices by the extensive use of special materials now becoming more generally available, coupled with advanced production processes. He realises that he may run into difficulties; in particular, the reliability of the product may not be up to the standard of his existing machines, which by gradual refinement over a long period of time is very good and has given his firm a good reputation. This he cannot afford to lose. Since he is departing from the evolutionary process previously adopted by his firm to gain a step advantage over his competitors, he plans a significant development programme to prove the production techniques and the reliability; but he wants to know what this would entail before going ahead. You are asked to specify that part of the programme which would ensure that the reliability of the proposed product is at least as good as the existing model, using figure 10.12 as a guide to the type of machine he has in mind. You should assume that the manufacturer will make all the parts, except the tyres which will definitely be bought in, even though he may eventually contract out the manufacture of some parts.

It must have become apparent that considerable effort has to be put into development of mechanical equipment if we are to achieve worthwhile results. Part of this is due to the very large number of failure modes that are possible for even the simplest component, and even more so for the final product which may have a multitude of components with different, and different combinations of, failure modes. It is also due to the many different load conditions that may usually be expected in practice, and which have to be covered in any effective development programme. All this is compounded by the fact that a great deal of testing is necessary to establish the failure pattern, particularly if this is a wear

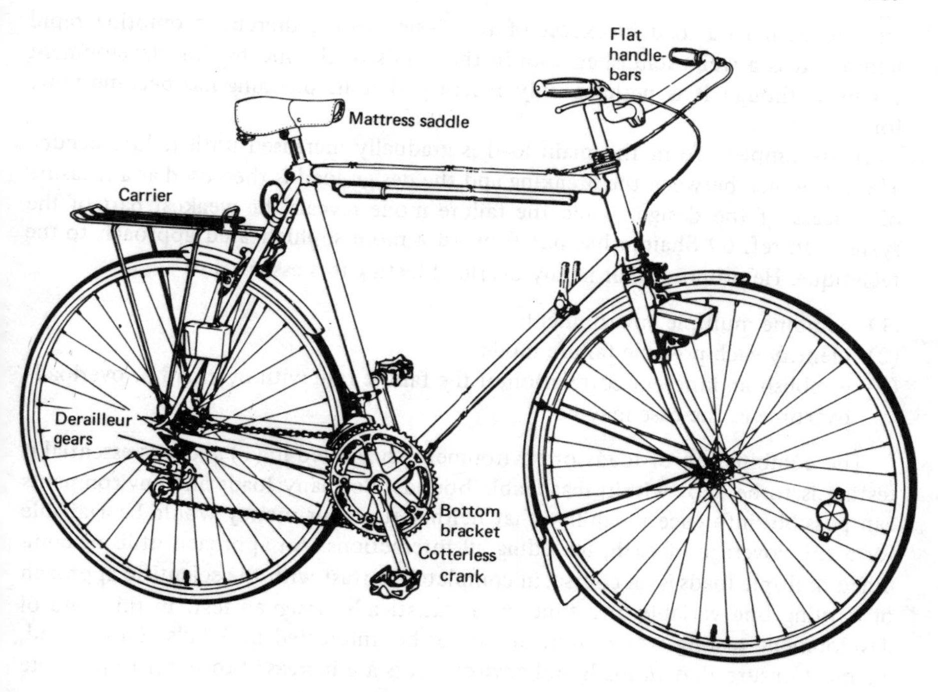


**Figure 10.12** A typical bicycle

mechanism which has a substantial failure-free period followed by an unknown distribution of failures. The simple constant failure rate which can be established over a relatively short period of testing is just not an adequate representation of most mechanical failure mechanisms. The amount of testing needed is illustrated by the following example.

Suppose that the tests on car exhausts, used to illustrate Weibull analysis in section 6.4.2, were part of a development programme to demonstrate a target reliability of 99 per cent (1 per cent failure) at 20 000 miles. Extrapolating the straight line on Weibull paper (shown dotted on figure 6.8), we see that we can expect 3 per cent of failures at 20 000 miles (that is, for $t - t_0 = 20\,000 - 9\,700 = 10\,300$ miles). Clearly, these exhaust systems do not meet the requirement, even though the minimum life of the sample of ten was 24 791 miles, with one lasting nearly 50 000 miles. This illustrates how time-consuming development and trials testing can be: many tests lasting well in excess of the specified life are necessary to establish the reliability with any degree of confidence and to reveal all significant failure modes. Exercises 10.17, 10.18 and 10.19 should confirm this statement. Lengthy tests can be expensive, the more so if several different environmental circumstances have to be covered. A way out of this difficulty, favoured by many mechanical engineers, is overload testing. As its name implies, this involves subjecting the item to a load in excess of the design value, thereby promoting rapid failures. It is a well-established tool in the hands of the mechanical development engineer, though it is perhaps only recently that its planning has become more formalised.

In its simplest form the main load is gradually increased until failure occurs. The difference between the breaking and the design load is then used as a measure of success of the design, while the failure mode reveals the weakest part of the system. In ref. 69 Shainin has put forward a more sophisticated approach to the technique. He proposes that in any overload testing it is essential to

(1) combine multiple environments
(2) identify each possible failure mode
(3) establish an inherent correlation of the failure rate with overload or overload–overtime as the case may be

The combination of loads or environments suggested in (1) above is essential if testing is to be kept within reasonable bounds. So many loads and environments can possibly influence reliability that neither time nor money would be available for a full coverage of each, including all interactions. An approach utilising combined multiple loads is, of course, in complete contrast with the scientific approach of varying one variable at a time, or a statistically designed test. In this kind of development testing we are not, or cannot be, interested in details. To this end, we must ensure that *all* loads and environments are increased together to promote every possible failure mode.

The possible failure modes referred to in (2) above comprise random failures (a constant $\lambda$ mode), fatigue-type wear-out failures, and the type of wear-out failures associated with a more gradual strength deterioration. The common distribution and location of these failure modes is illustrated on the load–time map in figures 10.13a, b and c for each of the modes respectively, and as a combined contour plot on figure 10.14.

It will be appreciated that the location of the failure zones shown in figure 10.14 will vary from application to application. For example, the life of some components is dictated solely by fatigue and the gradual strength deterioration type of wear-out may be of little importance, so this zone would be moved well to the right. Other components are subject to little alternating stress but suffer gradual strength deterioration, so that the fatigue zone is lifted to a higher stress level. The contour plots corresponding to figure 10.14 for these regimes are shown on figures 10.15 and 10.16. For simplicity, the diagrams have been drawn for a single mode, whereas in reality a very large number of modes will be present. The situation overload that testing is designed to reveal is any overlap of the load and strength zones which would give rise to failures. This is illustrated in figure 10.17. Now the closest proximity of the load to the instantaneous strength occurs at the end of the planned life. Hence it is necessary to concentrate the testing on this aspect. To do so the item must be conditioned by normal operation for its design life before overload testing can begin. Pre-conditioning is essential because the

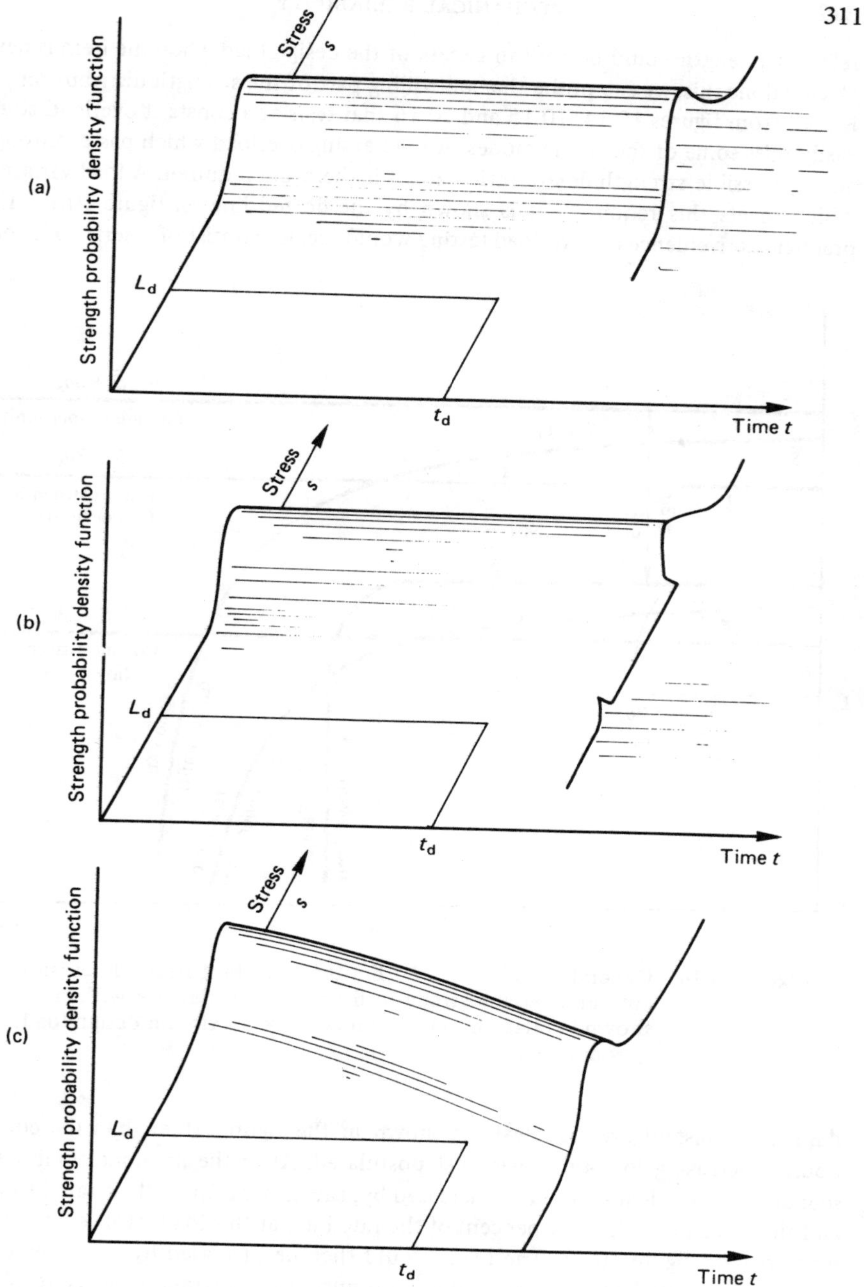


**Figure 10.13** Location of strength probability density functions in relation to load at various ages. $L_d$ = maximum design load. $t_d$ = design life. (a) Invariate strength; (b) strength variation due to fatigue; (c) strength variation due to erosion etc.

relevant strength could be well in excess of the design load when the item is new. Having done this, we can probe the controlling part of the strength distributions. It is seen from figures 10.14, 10.15 and 10.16 that tests at a constant overload could easily miss some of the failure modes. An increasing overload which passes through all the possible strength deterioration zones is therefore required. A load variation which meets this requirement is shown by the dotted line on figure 10.18. For practical convenience the overload testing would ideally consist of a series of short-

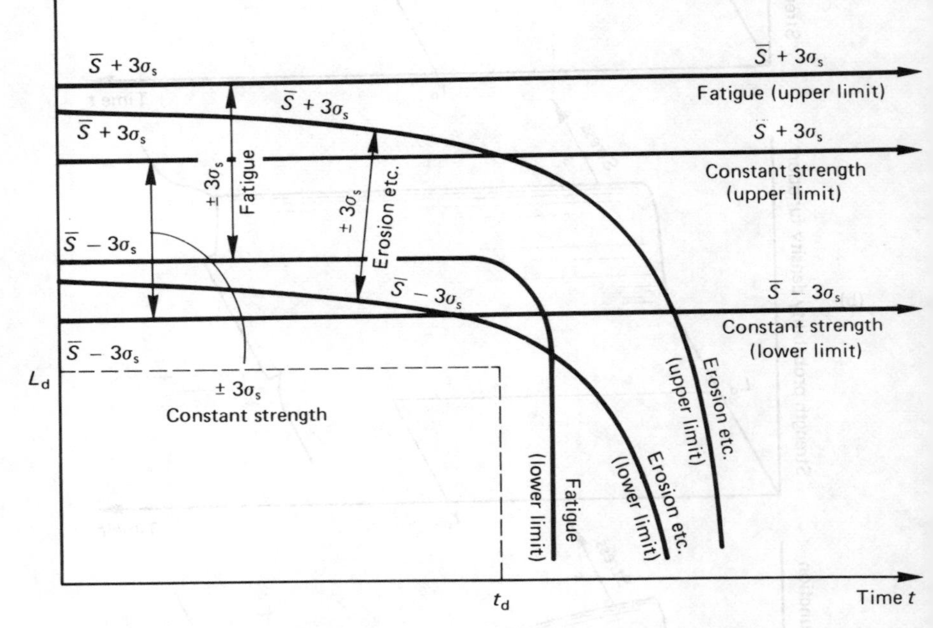


**Figure 10.14** Generalised plan of $\pm 3\sigma$ limits in constant strength, erosion etc., and fatigue strength distribution variations with time, shown in relation to load limits. $L_d$ = maximum design load. $t_d$ = design life

duration, constant-overload tests as shown in the figure rather than the continuously increasing load-time map just postulated. After the pre-conditioning test sequence, ***all*** the loads would be increased by, say, $x$ per cent of their rated values, and the test continued for $x$ per cent of the rated life at this load level as illustrated by step 1 on figure 10.18. The loads would then be increased by a further $x$ per cent of their rated values and testing continued for a further $x$ per cent of the rated life – step 2 on figure 10.18 – and so on until failure occurred, or the load increase was so great that the mode of probable failure would have changed. This latter point may need illustration. Suppose we are investigating failures associated

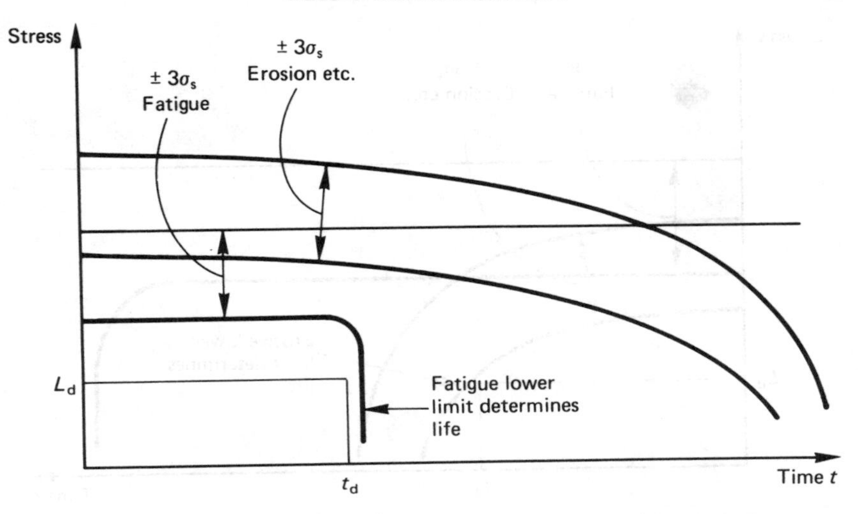


**Figure 10.15** Location of fatigue and erosion strength zones in relation to load when life is limited by fatigue and erosion is of little significance except for long-life items. Constant strength zone not shown

with cylinder bore wear in a reciprocating engine. This process is slow and we may wish to accelerate possible failures by increasing the load. However, any degree of overload testing which promotes failure by seizure of the pistons in the cylinder tells us nothing of the failure mode under investigation. This would be unacceptable and cause the test to be terminated. (It may still, of course, be valid for the study of this alternative failure mode.) Within this limitation, then, multiple loads and environments can be combined and the test conducted to reveal all failure modes, thus achieving objectives (1) and (2).

The essential objective (3) is best achieved by plotting the cumulative percentage failures against the percentage increase in 'stress-time', that is, against $x$, $2x$, $3x$, . . . etc. It is usual to use Weibull paper for this purpose on account of the flexibility of the Weibull function to represent any distribution. A typical plot is illustrated in figure 10.19. If the curve is concave downwards as illustrated, then $t_0$ will be greater than zero and we can conclude that no failures will occur when $x = 0$, that is, over the rated life and load conditions covered in the pre-test, so the item is intrinsically reliable for this duty. If the curve is concave upwards then we conclude $t_0$ is negative and hence there will be a finite failure rate at $x = 0$. This can be evaluated by straightening the curve, and extrapolating back to $x = 0$, which will give the failures at the end of the rated life under the specified load conditions of the pre-test. (If the raw data plot lies on a straight line, direct extrapolation is possible to obtain the same information.) The correlation between failures with stress-time established above is purely empirical and cannot be used to ascertain failure mechanisms. This must be done by the physical examination of the failure.

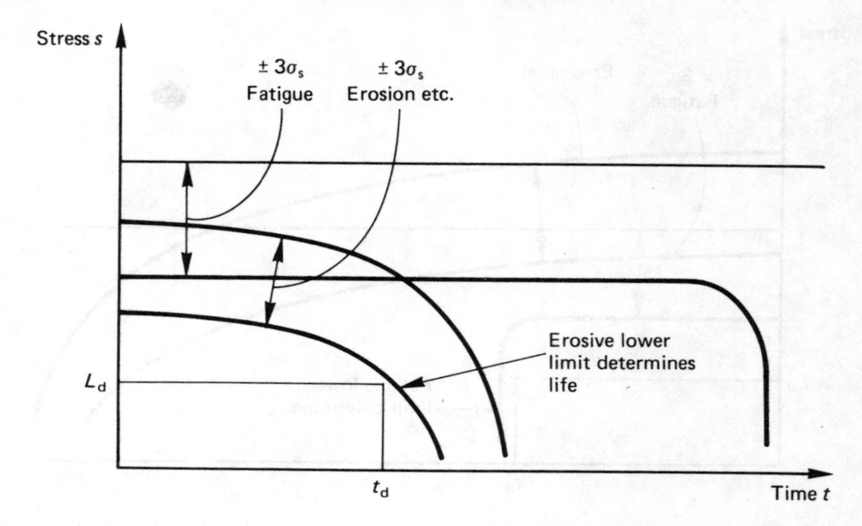


**Figure 10.16** Location of fatigue and erosion strength zones in relation to load when life is limited by erosion or similar phenomena and fatigue is of no significance. $L_d$ = maximum design load. $t_d$ = design life

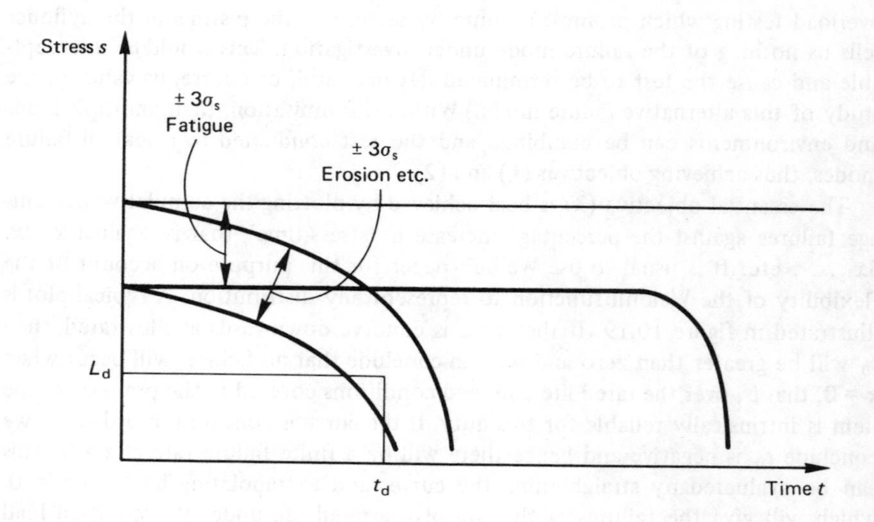


**Figure 10.17** An unacceptable situation in which a strength zone enters the design load–time domain, resulting in excessive failures. These failures may not be apparent until later in the item's life. Note that as drawn the erosion zone enters the load–time domain, but any strength zone may do this in practice

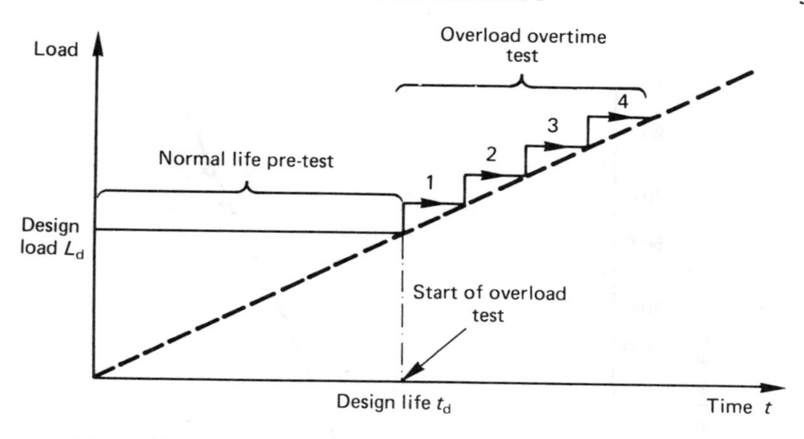


**Figure 10.18** The load–time programme for overload testing

Although many regard the quantitative determination of the reliability as the ultimate objective, the writer believes that the identification of the weak components and the failure mechanisms, so facilitating remedial design action, is the real pay-off from overload testing. The strategy has to be carefully worked out to promote all possible failures. Indeed, if this first step is not successful the validity of the whole test is in question.

*Exercise 10.21*

(a) Do you consider the pre-test sequence in which rated loads are applied for the rated life as an essential step in overload testing?
(b) You are carrying out an overload test on a small number of an item of equipment which is of a very complex nature. Soon after commencing the overload tests, a component in one test specimen fails. Would you terminate the test on this specimen or replace the component with a new one and continue testing?

**Exercise 10.22*

An automatic rifle is specified to fire a total of 50 bursts, each burst having a specified pattern of 20 rounds, with a reliability of 98 per cent. Tests are carried out on 12 prototypes as follows

(1) Each rifle fires 50 bursts of the 20 specified rounds, allowing the rifle to cool between each burst. All rifles pass this test.
(2) Each rifle fires an incremental 10 per cent of bursts, each having a 10 per cent incremental number of rounds, that is, an additional five bursts of 22 rounds are fired.

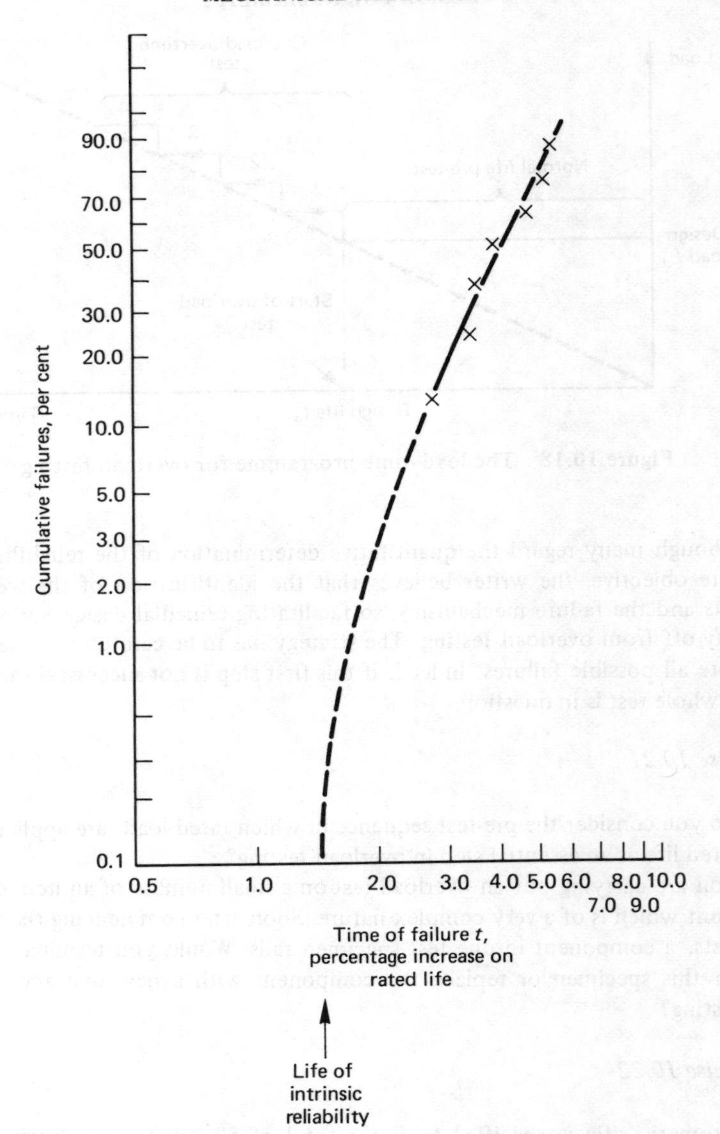


**Figure 10.19** Overload test data in which seven failures were recorded, plotted on Weibull paper

(3) All surviving rifles fire a further incremental 10 per cent of bursts, each having a 10 per cent incremental number of rounds, that is, an additional five bursts of 24 rounds are fired.

(4) The incremental (1) to (2) equal to (2) to (3) is repeated, added and the guns fired after each addition until all fail.

The total number of rounds fired by each rifle (that is, including the initial 1000 rounds of the pre-test sequence) is as follows

| | | | | | |
|---|---|---|---|---|---|
| 1401 | 1533 | 1649 | 1764 | 1859 | 1918 |
| 2003 | 2025 | 2121 | 2171 | 2222 | 2376 |

Does this rifle meet the reliability requirement? Suppose the reliability of this rifle had to be established by testing the 12 prototypes to failure without any overload. How many rounds would you expect the longest surviving rifle to fire before failure?

The success of overload testing techniques depends very largely on the selection of the loads which are to be increased, and the subsequent identification of the resulting failures with those experienced under normal conditions in the field. This will only stem from a sound appraisal of the design and operating environment (internal as well as external). In order to save time and cost, one is deliberately embarking on a test that will not provide full and complete understanding of all the phenomena involved. It could well be that additional tests of the type described in categories (3) and (4) on page 300 will be necessary to establish the real cause of any failure and the necessary remedial action. On the other hand, if a numerical reliability is all that is sought and it is within specification, such work would be unnecessary – other than as part of a general programme aimed at a future generation of equipment. As stated, the author places more value on the nature of failures revealed by overload tests than on any numerical data, but their incidence has to be related to possible occurrence in the field (that is, are they important?), either through experience or the numerical methods described above.

A detailed knowledge of the technique underlying overload testing, whether it is adopted or not, provides a sound understanding of the relationship between all the various time-dependent strength distributions and the loading cycle over the whole life span. The technique is worth study from that point of view alone. The reader may well draw up more precise diagrams corresponding to figure 10.14 for a number of specific failure mechanisms of real products.

It is appropriate at this point to draw a distinction between accelerated testing and overload testing. The author would prefer to restrict the description 'accelerated' to those tests in which the time scale alone is increased, by, say, testing 24 hours a day when the expected usage is 8 hours a day, but in which the loads are kept at the expected operational level. This nomenclature is not universal, and sometimes tests in which an adverse environmental feature is increased in concentration, or the temperature is increased, to promote rapid corrosion for example, are referred to as 'accelerated'. 'Overload' would seem more accurate and consistent, but ingrained practice cannot easily be changed and care is needed in reading the literature.

Another technique to reduce testing time is the 'sudden death' test. This is more appropriate to the cheaper product, in which a reasonably large number of items can be made available for test. As usual, the object is to reduce the cost of the test. As described in the literature, the total test population is divided, at random, into, say, five or six equal sized groups. Testing is carried out on each group until the first failure in the group is recorded, upon which all the other members of that group are discarded (sudden death). Thus one failure is obtained from each group, and the test data can be then interpreted on the basis that the remaining members of the group are drop-outs at the group first failure life. This would seem a good way of reducing test time if the failure pattern is close to random, since the test failures would be distributed more or less uniformly over the test life span. However, for mechanical items the onset of wear-out failures is probably the most important part of the curve. It would then seem better, in the author's opinion, to use a different strategy which allowed the early failure distribution to be more accurately defined; for example, by testing all items until, say, one-fifth had failed, followed by sudden death on the remaining four-fifths divided, at random, into four groups. In this way, the early part of the wear life (the curved part of the Weibull plot) would be accurately defined and the straight, less important, part of the curve is where the economies would be made. This is just one alternative strategy. Many variations are possible, but the underlying principle is the same: artificially create drop-outs to reduce testing time. An essential requirement is that their creation should be unbiassed. A practical consideration is that they should not suppress valuable information.

Although the various techniques described above allow the testing time to be reduced, they are essentially test techniques to eliminate the grosser failure modes, and the most relevant information must come from field trials using as near normal operating conditions as possible. There is no way of avoiding this, and the final reliability assessment must be based mainly on such trials.

In theory, the interpretation of the data from a prototype development programme should be straightforward and logical. One assumes that some components or sub-assemblies of an equipment on trial do not meet expectations. The failures that arise are examined on an engineering basis and statistically for cause, and a remedy is proposed. At the same time, the failure rate is assessed and from such data it can be established if the higher failure rate can be accepted or perhaps countered by a revised maintenance schedule. An optimum course of action is evolved, and if this requires redesign then the new item has to be evaluated in the same way as the original. In practice, the process can be very different, and is much more of an art than a science. In the first place, the prototypes will be continually modified, to obtain additional information on performance if for no other reason. The environment of any component will consequently be a changing one, which of itself casts doubt on the representative nature of any failure data. Many failures will be secondary ones, consequential upon other failures. Finally, the duty cycle of the equipment is more likely to be controlled by performance evaluation rather than by the duty cycle of the developed equipment. The result is a whole crop of 'unrepresentative' failures which are usually eliminated from any assessment on little more than a subjective basis. Recall how in exercise 10.19 several of the throttle arm failures were designated as trivial even though this subsequently proved the most significantly persisting and possibly dangerous main engine failure mode. Any subsequent analysis must recognise that many valid failures will have been discarded and many invalid ones retained. Never does the difficulty of separating baby and bathwater seem so great! Finally, what is left may be too sparse for proper analysis. The distributions of exercise 10.19 are examples. Yet a careful analysis of the in-service failures of a piece of equipment showed that no failure mode was encountered in the first year of operation that had not been previously recorded in prototypes trials. Most had been discarded as 'irrelevant'! Many other workers have confirmed this experience. None are so blind as those who do not wish to see. It is essential that an effective data-collecting and reporting organisation be set up. An integrated engineering/statistical analysis of the data must then be followed in a ***determined*** fashion if worthwhile results are to be obtained. Carried out in this way, prototype trial can be invaluable – essential indeed – in achieving high quality.

Reference has already been made to the serious consequences of a curtailed development programme, so that it becomes essential to make adequate allowance, in both time and cost, for this work during the initial planning of the procurement cycle. In the author's opinion, nothing can replace a detailed estimate by a development engineer who is experienced in the relevant field. In spite of this, 'management' is continually seeking some formula which will give 'instant' estimates, and it is essential that we should assess the value of this approach before leaving the subject of development. We recall that the nature of the work is such that one starts with an immature product whose reliability can be assumed to be significantly below specification. By operating the item in its intended manner and environment, the causes of any unacceptable unreliability are established, one by one, over a period of time. As each is determined, corrective actions are taken and incorporated for further proving. In a good organisation most actions are successful, though some may be less so and require a second bite at the cherry. As a consequence of such action the reliability increases steadily, but often intermittently, towards the specification value. A typical plot of reliability against development activity is shown on figure 10.20. Discontinuities in plots of this nature are not unusual in the development of mechanical items. Once a source of failure is ascertained it usually takes some time to effect a satisfactory redesign, remanufacture the defective piece, and install it in the parent equipment. Furthermore, to expedite the programme several modifications are often incorporated in the prototype at the same time, in the same way as several maintenance activites are scheduled simultaneously. The result is a sudden jump in reliability. Such step changes are particularly likely during the initial stages of a programme when the 'vital few' which exert the major control on reliability are being remedied. Sudden falls are also a feature of this work, not because of any lack of success, but because a sudden crop of failures can be encountered at any time as a result of wear-out of an intrinsically reliable component or sub-assembly. A certain ***sang-froid*** is necessary in any development engineer because, owing to the above reason, the reliability may fall continually during some stages of a programme in spite of successful action. These are factors that must be lived with. In spite of such factors, attempts

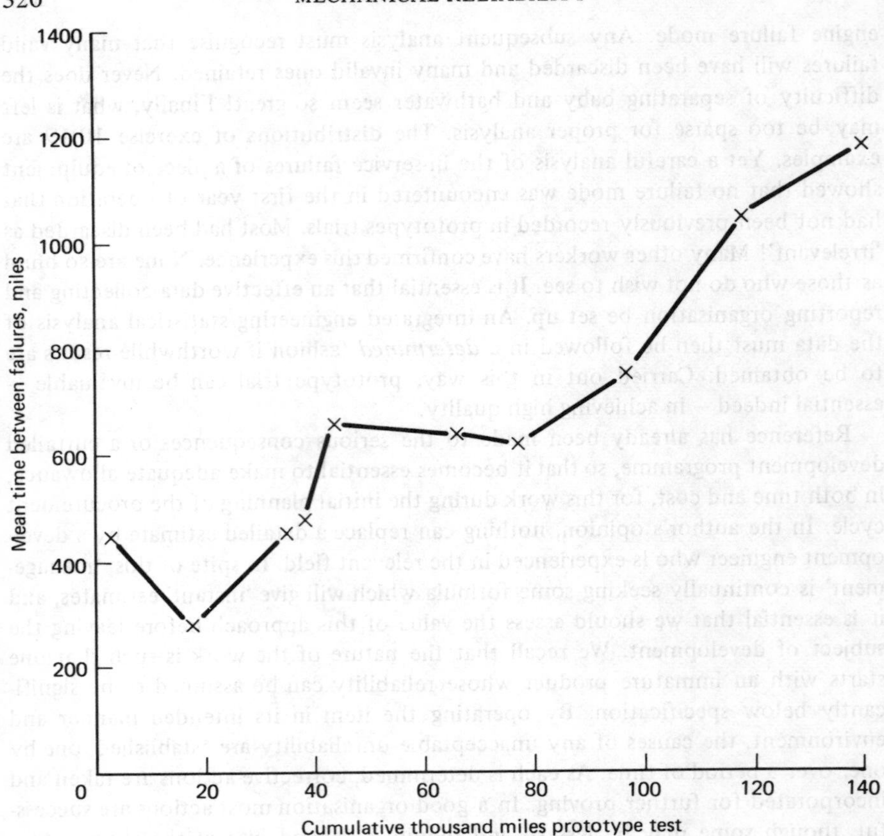


**Figure 10.20** Typical plot of cumulative failure rate against cumulative distance covered in prototype testing of a vehicle

have been made to replace the discontinuous reliability–time curve with a continuous one which can then be represented mathematically. By generalising, it is believed that reliability growth can be planned, costed and monitored.

The most favoured assessment of reliability growth is due to Duane (ref. 70). From an examination of a number (limited) of mechanical and electrical devices, Duane observed a linear relationship between the logarithm of the cumulative failure rate and the logarithm of the total or cumulative operating time. Expressed mathematically

$$\log \lambda_c(T) = -\alpha \log T + \log K \qquad (10.11)$$

where

$\lambda_c(T)$ = cumulative failure rate = $N/T$

$T$ = total or cumulative operating time

$\alpha$, log $K$ = constants

$N$ = total number of failures.

Equation (10.11) may also be written as

$$\lambda_c(T) = \frac{K}{T^\alpha} \tag{10.12}$$

We now encounter a difficulty in that equations (10.11) and (10.12) indicate an infinite failure rate at the commencement of testing – when $T = 0$. To overcome this difficulty it is usual to postulate a start point at a finite value of $T$ from which reliability growth is measured. Let

$T_s$ = the cumulative time from which growth is to be measured

$\lambda_s$ = the cumulative failure rate at $T_s$

then from figure 10.21

$$\alpha = \frac{\log \lambda_s - \log \lambda_c(T)}{\log T - \log T_s} \tag{10.13}$$

$$\log \lambda_c(T) = -\alpha(\log T - \log T_s) + \log \lambda_s \tag{10.14}$$

or

$$\lambda_c(T) = \lambda_s \left(\frac{T_s}{T}\right)^\alpha \tag{10.15}$$

Hence from equation (10.12)

$$K = \lambda_s T_s^\alpha \tag{10.16}$$

The instantaneous failure rate, $\lambda(t)$, will be given by $(\mathrm{d}N/\mathrm{d}T)$ where $N$ is the total number of failures. But

$$N = \lambda_c(T) \times T = \lambda_s \left(\frac{T_s}{T}\right)^\alpha T = \lambda_s T_s^\alpha T^{1-\alpha} \tag{10.17}$$

therefore

$$\lambda(t) = \frac{\mathrm{d}N}{\mathrm{d}T} = (1-\alpha)\,\lambda_s \left(\frac{T_s}{T}\right)^\alpha \tag{10.18}$$

or

$$\lambda(t) = (1-\alpha)\,\lambda_c(T) \tag{10.19}$$

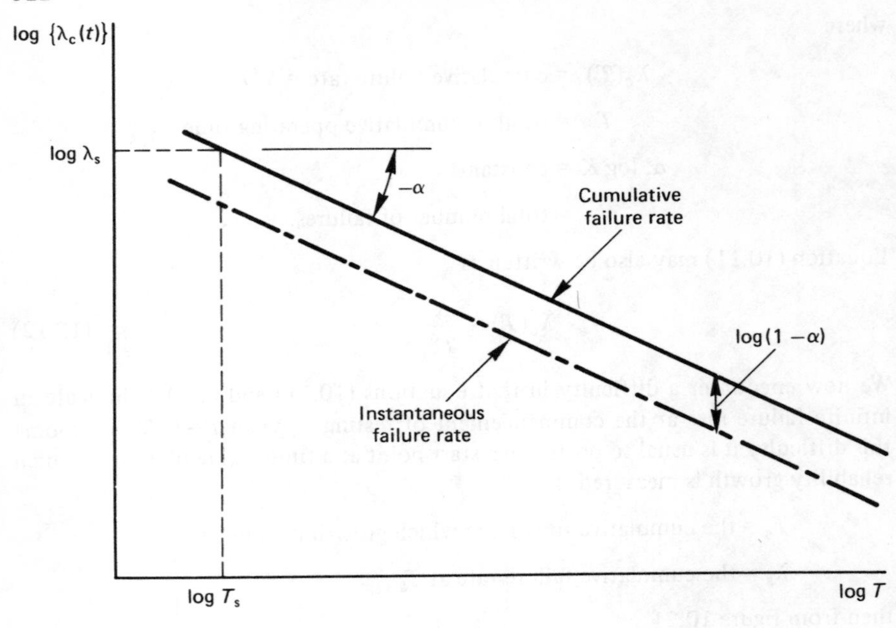


**Figure 10.21** Duane plot

Equation (10.19) can be written

$$\log \lambda(t) = \log \lambda_c(T) + \log (1 - \alpha) \tag{10.20}$$

When plotted on the log-log scale recommended by Duane, the instantaneous failure rate will therefore follow a straight line parallel to the line for the cumulative failure rate and displaced from it by log $(1 - \alpha)$.

The Duane model is very straightforward and easy to apply either numerically or graphically. Straight lines in the latter case can easily be extrapolated to yield predictions of on-going development programmes. Some workers use the quantity $M_c$, which is the cumulative mean time between failures, rather than the cumulative failure rate, but this leads to exactly the same results since $M_c = 1/\lambda_c(t)$. In either case the parameter $\alpha$ is known as the growth rate, and is generally considered to be determined by the intensity and efficiency of the development programme. A value of 0.1 is believed to represent a development programme which concentrates on performance and pays little specific attention to reliability. A value of 0.6 represents a programme in which reliability is the main objective. A value of 0.4 is often quoted as an average value. However, this contention should be treated with some reserve, because the value is affected by the choice of the starting time.

Figure 10.22 shows some field data plotted on log-log scales, taken from Duane's original paper, illustrating the growth in reliability or decrease in failure rate for some mechanical devices. Correlation in these cases is seen to be good over the main part of the development programme. However, the reader should

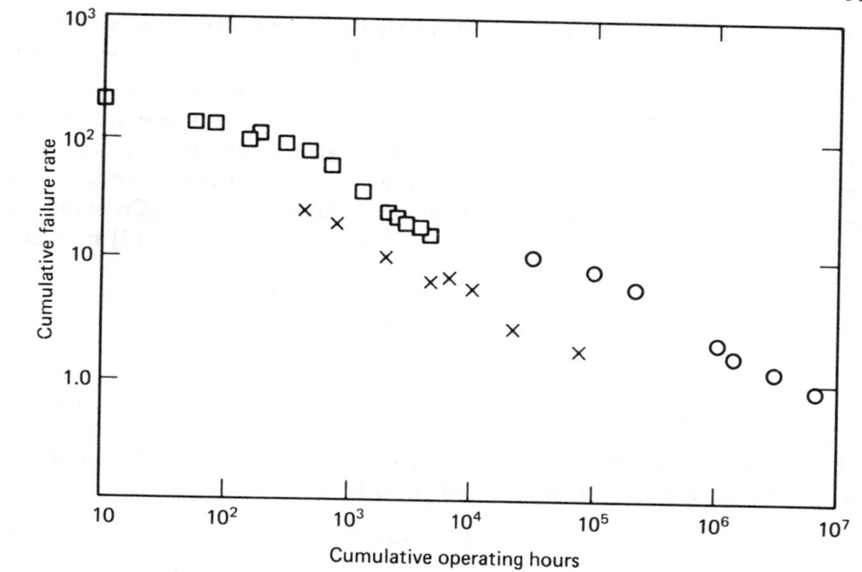


**Figure 10.22** Some of the data quoted by Duane to support a linear log–log relationship during development.
⊡ Aircraft generators.
× Complete aircraft jet engine (during early stages of introduction into service).
⊙ Complex hydromechanical device

notice the indications in two of the examples shown of asymptotic values at the beginning of development, and more particularly at the end, suggesting that extrapolation of the Duane curve is not always justified.

Crow has proposed an extension to the Duane model which allows a more rigorous statistical representation of the pattern of reliability growth. In this model the average number of failures accumulated by time test, $T$, is represented by the Duane equation as at (10.12), but the actual number of failures observed to that time is taken as a random variable. Thus the total number of failures, $N(T)$, accumulated on all test items in cumulative time $T$ is described by the Poisson distribution. The probability that exactly $N$ failures occur between the initiation of testing and the total test time, $T$, is

$$P[N(T) = N] = \frac{m(T)^N \, e^{-m(T)}}{N!} \qquad (10.21)$$

In which $m(T)$ is the mean value function – that is, the expected number of failures expressed as a function of test time. To describe the reliability growth process, this function is given the Duane relationship

$$m(T) = \lambda T^{\beta} \qquad (10.22)$$

in which $\lambda$ and $\beta$ are positive parameters.

It will be noted from equation (10.21) that the mean value, $m$, equal to $\lambda T$ in the normal or homogeneous Poisson process, has been replaced by the term $m(T)$, allowing the mean to be an arbitrary, rather than linear, function of $T$. It can thus be made to represent reliability growth. It results in the non-homogeneous Poisson process. The essential difference between the homogeneous and non-homogeneous Poisson processes is that under the latter, successive times-between-failures are not independent and identically distributed random variables. Otherwise Crow follows the Duane model. Comparing equation (10.22) with equation (10.12) it is seen that

$$K(\text{Duane}) = \lambda(\text{Crow}) \tag{10.23}$$

and

$$\alpha(\text{Duane}) = 1 - \beta(\text{Crow}) \tag{10.24}$$

The number of failures occurring in the test times interval $T_1$ to $T_2$ is a random variable having the Poisson distribution with mean

$$m(T_1) - m(T_2) = \lambda(T_1{}^{\beta} - T_2{}^{\beta}) \qquad (10.24)$$

The number of failures occurring in any specified interval is statistically independent of the number of failures occurring in any other interval that does not overlap the specified interval. Only one failure can occur at any instant.

The rate of change of the mean value function is called the intensity function of the process. For the reliability growth model under discussion, the intensity function is obtained by differentiating (10.22) to give

$$\rho(T) = \beta T^{\beta-1} \tag{10.25}$$

The probability of the occurrence of a failure between time $T$ and time $T + h$ is given approximately as $\rho(T)h$ where $h$ is a small interval of time. The intensity function is very similar to the failure rate [compare the expression just quoted with equation (2.7)], but should not be confused with it. For this reason the time of development has been represented by $T$, reserving $t$ for the life distribution. However, the end of a development programme may be identified theoretically with zero further reliability growth, so that the anticipated failure rate for the product in service should be equal to the intensity function at the end of the development.

In the above equations $\lambda$ is a scaling parameter, determined by the units chosen for $T$. The parameter $\beta$ is the practically significant one, for it determines the shape of the intensity function curve. If $\beta$ is equal to unity, the intensity function is constant and the reliability of the item is not changing through the development programme. The failures are exponentially distributed with a failure rate of $\lambda$. If $\beta$ is less than unity the expected number of failures in an interval of fixed length decreases as the starting time of that interval is increased, that is, the development programme is improving the reliability. For a good development programme one can expect $\beta$ to have a value of about 0.5. In a badly organised development programme – one that concentrated on performance at the expense of reliability – $\beta$ could be greater than unity. But note that the prime objective of a development programme *could* be improved performance. Ideally, $\beta$ should be equal to unity for such a programme, but a value greater than unity would indicate that reliability is being lost in order to achieve that increased performance, and that corrective action is necessary unless it were intended.

Using the above model, which is usually referred to as the AMSAA model (it was formulated by the U.S. Army Material Systems Analysis Activity), field data may be plotted on log–log paper to obtain a straight line, since equation (10.22) can be written as

$$\log m(T) = \log \lambda + \beta \log T \tag{10.26}$$

in which $T$ is the time at which each discrete failure occurs and $m(T)$ is the corresponding cumulative failures. From the straight line the constants and $\beta$ can be evaluated as in the Duane representation. Equation (10.25) then gives the failure rate of the production version – if identical with the prototype!

For a more detailed treatment, including a statistical estimation of the parameters by maximum likelihood techniques, goodness of fit estimation and so on, the reader is referred to MIL–HDBK–XXX (see Reading list), which also contains a description of many other models of reliability growth that have been devised. The AMSAA model described above is the one generally used for practical reliability growth monitoring during development. Some workers use the SIL model which has asymptotic values for the failure rate at the beginning and end of the test time. This is a closer approximation to reality – see figure 10.22, for example – but requires extra arbitrary parameters which make the curve fitting more difficult. Indeed, the more exact models often cannot be determined until the test is very advanced, and thus it is too late to take corrective action.

For our purpose, an assessment of the value of the models for estimating reliability growth during development of general mechanical items is more pertinent than a deep statistical treatment of the many models that have been proposed. We first note that all the models are based on a constant failure rate when no remedial action is taken. They therefore do not represent wear-out processes. It may be noted, though, that some unacceptable components may fall into this category by exhibiting true random failures. We must then ask ourselves how closely this interpretation of the models represents the actual circumstances.

In describing the basis and the justification for a FMEA, it was shown that the majority of failures in a mechanical system were due to few causes – the Pareto principle. In the initial stages of a development programme, corrective action must be concentrated on the observed failure mechanisms of this small number of failure modes. Early action is very much a process that is taken in discrete steps, and it is difficult to see how it can be represented by the smooth mathematical function which represents the summation of a large number of events. This criticism has been met to some extent by Crow (see ref. 71), who has extended the AMSAA model to cover this aspect. However, if any major design change is involved, as may well be the case, the resulting jump in reliability is unrelated to prior history. No growth model could possibly span the process both before and after a major design change. The reliability after a major design change must be estimated by the same appraisal techniques that were used in the original design, and the time scale for its achievement estimated from a knowledge of the resources that can be brought to bear to effect a redesign, manufacture, and introduce it into the prototypes. In ref. 71, Crow has allowed for the efficiency of the operation by an empirical 'fix effectiveness' factor. He also recognises that there are some failure modes for which no remedial action will be taken for economic reasons. The treatment then becomes a lot more complicated than the simple AMSAA model, and its validity when a major design change is involved is still questionable. As the development programme advances and major problems are resolved, however, it is likely that more and more attention will be paid to the far larger number of miscellaneous minor failure modes which make up Pareto's trivial many. Reliability growth will then be much smoother, and it may be expected that the statistical reliability growth models will be more valid. Likewise one would expect these reliability growth models to be valid and valuable for monitoring the reliability behaviour during subsequent development – the Mk IIs, Mk IIIs etc., unless drastic design changes had been introduced. A word of warning may be necessary, however. A development programme may be settling down into the steady growth phase which can be mathematically modelled, only to run into a major snag later in the programme. The bird ingestion responsible for failures of the initial design of the first stage fan blades of the RB 211 turbo fan engine is an example. Arising late, in the development programme, they caused the bankruptcy of Rolls Royce. It is difficult to see how any statistical model can predict the future occurrence of a so far unexperienced failure mode, even it it is a random one. Equally, the inability to predict a catastrophe of that magnitude may be regarded as more than a little local difficulty. However, it is when equipment first enters service that we can expect closest correlation. It is general experience that on first entering service, equipment reliability is well below that achieved by the later prototypes. There are two main causes. In the first place production techniques will undoubtedly differ from the 'one-off' methods used to produce prototypes. Serious isolated problems can arise at this stage, but in general there will be a lot of minor differences that will give rise to a fair crop of trivial failures before they are ironed out. Secondly, the equipment will be in the hands of inexperienced operators (in the new equipment), who will be unlikely to treat it with the care and consideration that is given to prototypes. This gives rise to multitudinous minor failures. The phenomenon is well known (see figure 10.23). As experience is gained by producer and operator, the reliability grows to its prototype value. The process is a steady continuous one – even with complex mechanical equipment – and there is a wealth of evidence to support it. Such situations are well represented by statistical growth models. It is worth noting that several of the curves used by Duane to support his original thesis in fact referred to the period of introduction into service, rather than the embryonic period

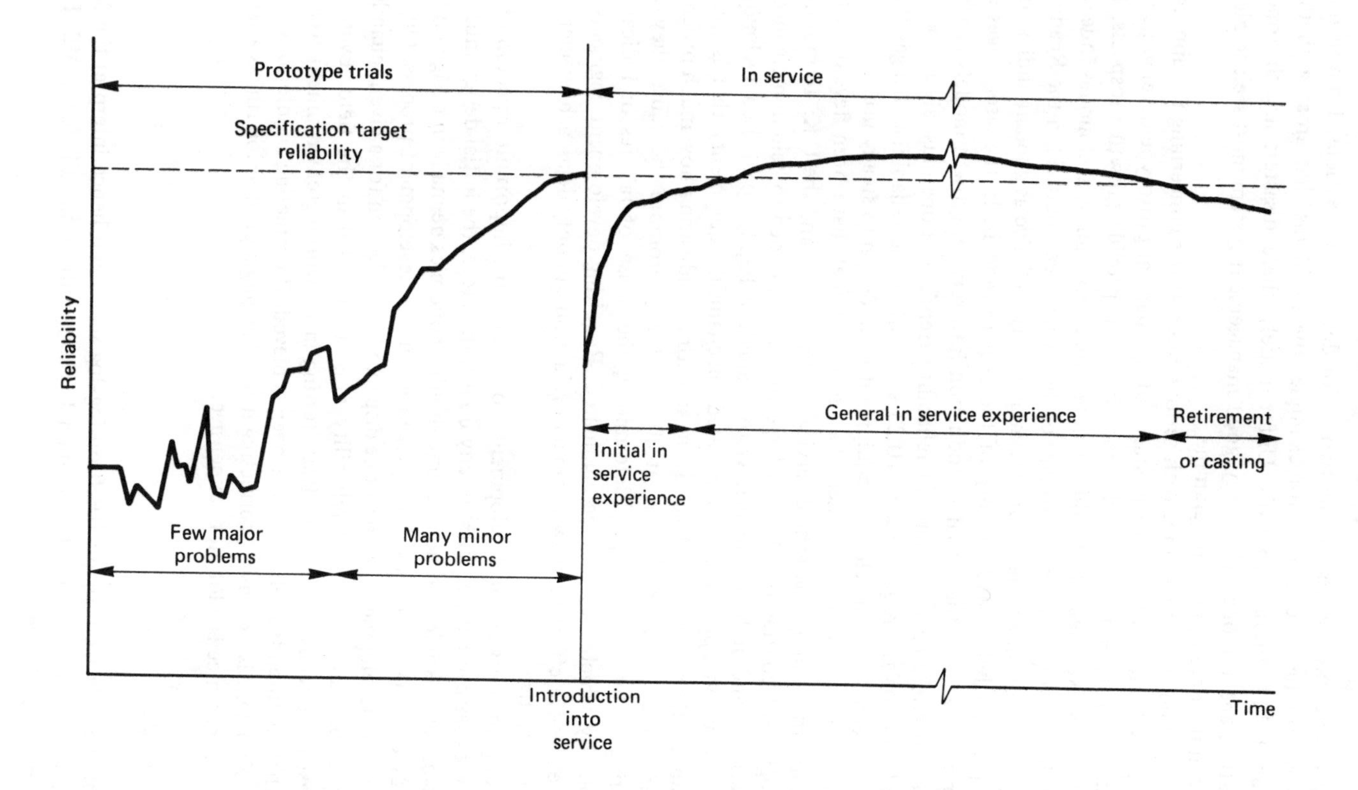


**Figure 10.23** Typical lifetime variation of reliability

seen by Lovesey as the first of the two major development actions. Even in this case there can be exceptions – for example, the problems that arose with the American F.111 aircraft controls. Unfortunately these disasters are the ones that matter, and the ones which any good management system must seek to pick up early in the development programme.

We must conclude, therefore, with serious reservations concerning the application of growth models to the early stages of mechanical prototype development. An engineering appraisal is much more likely to provide growth prospects. If 'prediction' is excluded, it should be noted that very simple techniques indeed, such as the plotting of a moving average or the cusum (see Chatfield in the Reading list for a description of the cusum technique) may well give as good an indication of the progress being made as any of the growth models. In the later stages of development, and in the initial introduction into service, growth models can be expected to give a close estimate of reliability trends, but one must always be on the look-out for the rogue. Figure 10.23 summarises the reliability changes that may be expected from mechanical equipment from design to steady use.

Finally, it should be recognised that growth models have been developed by large concerns, mainly the armed services of the U.S.A. and the U.K., to monitor the progress of contractors. By and large they are intended to help non-engineers (civil service and military managers) to evaluate a highly specialised technique (especially when applied to complicated mechanical equipment) that is much more an art than a science. Apart from sometimes indicating how much progress has been made, they do nothing at all to actually promote the cause they are measuring – by, for example, determining the cause of failures or indicating methods by which they may be eliminated. Requiring development engineers to divert excessive effort to growth model evaluation can only delay achievement of the real objective.

Taking an overview of development work, the most important aspect must be its very expensive nature. Without any doubt, therefore, the initial design must be as reliable as possible. On no account should economies be made in design on the grounds that inadequacies will be picked up in the development process. On the other hand, development can never be dispensed with. Design for a given reliability is impossible, and the actual reliability must be verified in the field. Even the simplest item needs, at the very least, testing in its working environment to confirm that its intended reliability has been achieved. If correcting an inferior design by development is expensive, correcting it by changing a production run as a result of field experience is a lot more expensive.

*Exercise 10.23*

Replot the data from figure 10.20 on log-log scales and hence interpret the data so far as possible using the Duane model. The specification calls for an MBTF of 6000 miles for this vehicle.

*Exercise 10.24*

Make an engineering assessment of those failures recorded in the prototype trials of exercise 10.19 which one could reasonably expect to be removed by development. Hence assess the failure rate to be expected of the main engine in service.

*Exercise 10.25*

Discuss the validity of using reliability growth models based on a constant failure rate for items known to have wear-out modes.

## 10.5 Some final design considerations

The treatment in this chapter has followed the chronological sequence of events in the production of equipment, and the next step of manufacture is usually regarded as quite outside the design process. However, the designer's job is by no means at an end. He has a firm obligation to find out how far his design actually achieves its objective. This knowledge can only be assembled when the equipment is in the hands of the user. The feedback of this information to the benefit of subsequent designs is an essential feature of planning for reliability. In general, the small-scale user of equipment will not be able to collect the necessary quantity of data for significant assessment. In such cases the designer must do the work if he wishes to achieve a reliable product, a satisfied customer, and keep himself in business. However, large-scale users of equipment are often in a much better position to collect and process the data than the designer. In this book we shall assume that the latter condition prevails, and the whole process is discussed under the responsibilities of the user in chapter 12. If this is not so, the actions outlined in that chapter are additional functions to fall on the designer. Except in extreme cases, however, this is a joint designer/user or maybe designer/maintainer function. The danger is that being a divided responsibility it becomes nobody's responsibility. The feedback of information is essential. Without it the whole design process must stagnate. It is a management function to decide where responsibility primarily lies; and in this connection it is worth recording that satisfactory operation of this activity could do as much as anything else to improve design for reliability.

In conclusion, we should note that while the basis for any design achieving its target is set in the drawing office, the complete act of design brings in many more activities. For example, whoever draws up the initial specification is taking an integral part in the design process. So is the development engineer. At the centre is creative design; and all the supporting functions must contribute to, and be in harmony with, that act. There can be no opting out of any step if high quality and reliability are to be achieved, but on the other hand none of the supporting activities must be allowed to dominate the main activity. It is essential that those in control have a clear idea of the nature of design and of its requirements, from concept through the early character-forming stages and later adolescence, to a final maturity that makes it fit for its job.

# 11
# System or equipment reliability

## 11.1 Introduction

We have so far in this book really been concerned with the component, the building brick on which everything else depends as it were, though we have found it necessary to relate the component to the system in connection with maintenance and to some extent in connection with design. We must now tackle the vexed question of system or equipment reliability. The problem is the assessment of the multi-component system reliability given that of the individual maintained components. This is usually required for the selection of the most cost-effective system, or for redesigning the system for higher reliability. It is generally accepted that this is feasible for electronic systems, and even if the accuracy of the assessment is not very high, it is believed that alternative competing systems can be placed in the correct order of merit. This greatly assists decision making at the feasibility and early design stages. It is sometimes claimed that similar techniques can be applied to mechanical systems but, since it has been observed on numerous occasions in this book that there are significant differences between mechanical and electronic behaviour, it would be unwise to assume that electronic practice can be carried over to mechanical systems. We shall therefore examine mechanical system reliability assessment from first principles, but pay attention to those techniques which have proved successful in the electronic world.

## 11.2 The reliability block diagram (RBD)

The first step, when following accepted electronic reliability practice, is to draw up a Reliability Block Diagram (RBD) which depicts the logical operational or functional connections between the components of the equipment, as contrasted with the physical connections. The RBD shows identifiable items of hardware in a functional relationship, and thus indicates which ones must operate successfully for the equipment to perform its intended function. To illustrate the way in which reliability block diagrams are drawn up, figure 11.1a shows the RBD for the

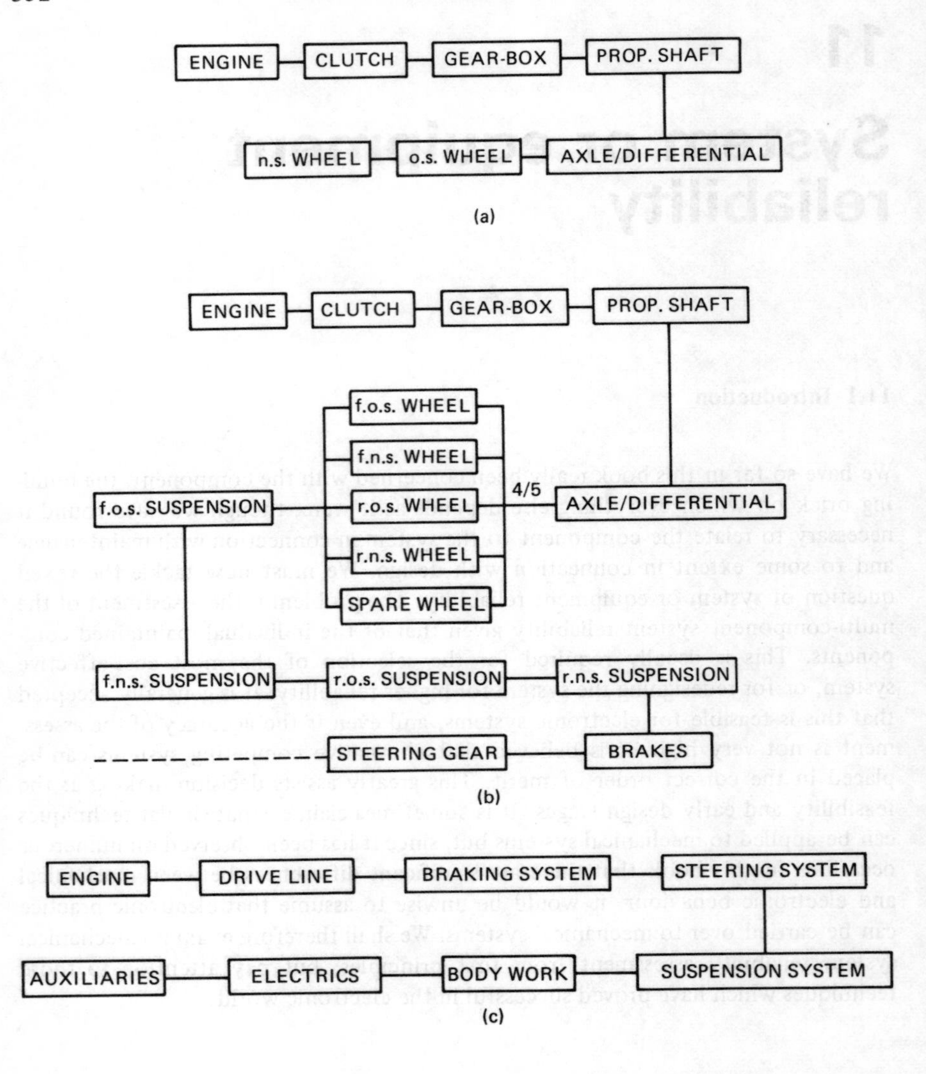


**Figure 11.1** Reliability Block Diagrams (RBD): (a) drive line, (b) part vehicle, (c) simplified vehicle

drive line of an automobile. It will be seen that the familiar components appear as blocks which are connected in this particular example as a simple series system, in which the failure of any one item implies the failure of the whole system – in this case the drive line. Of some importance, it will be noted that for reliability purposes the near-side and off-side wheels are in series even though physically they are in parallel. This follows from the necessity for both wheels to function if the system is to function. RBDs represent a functional and not necessarily a physical relationship. Although it is usually convenient to arrange components in the sequence in which their functions are often thought to be performed, there is no necessity to do this. Any order would lead to the same conclusion. In the same manner in which we drew the RBD for the drive line, we can build up the RBD for the whole vehicle. A rudimentary diagram is shown in figure 11.1b. This shows, in particular, how the parallel redundancy introduced by the spare wheel is represented functionally. In the ordinary motor car the spare is in a state of standby redundancy, that is, it is in a passive state, and is only brought into active use when one of the four active wheels fails. This is usually denoted by '4/5 redundancy' in which only 4 of the 5 elements are required to be in an active state. Any number may, of course, theoretically be in standby. By comparison with passive redundancy, in active redundancy all components are continuously operating, but a reduced number is capable of sustaining the load on failure of active unit(s). The aero engines of exercise 3.8 are a good example of active redundancy. The advantages and disadvantages of redundancy are discussed in more detail in section 11.4. Here we are concerned with the functional relationship.

Unfortunately, we find that there is no unique reliability block diagram, and different diagrams apply to different loads and loading conditions. The diagrams to represent any secondary failures may also differ. The difference in the required diagrams may be illustrated by further reference to that shown in figure 11.1b. The placing of the four wheels in a 4/5 redundancy is only true of a 'slow' failure in any one of the four active members. A sudden failure, such as a blow-out, could result in an utter disaster which the spare would be useless to control. For this failure mode the four active wheels must be shown in series, as in figure 11.1a. Incidentally, the author regards the carrying of spares as a maintenance aid rather than redundancy, but clearly there are different views on this subject. In whatever way we approach RBDs, alternative versions are necessary to meet differing circumstances, and furthermore all alternatives must be drafted. This is a feature of electronic as well as mechanical RBDs and can lead to a considerable amount of work.

The reader may have observed that the RBD of figure 11.1b is a very curtailed one. Many other components must be included in the full diagram: for example, windscreen, windscreen wipers, doors, catches, switches, battery, lights, bodywork, seats and so on should appear. Furthermore, it will be appreciated that all these items must be placed in series with those on figure 11.1b. This leads to the problem invariably encountered with the reliability block diagrams of mechanical systems. Whereas electronic equipment makes great use of redundancy to achieve high reliability, it is not possible to do this to anything like the same extent with mechanical equipment. Consequently, reliability block diagrams for mechanical equipment become little more than a component list drawn in series. The sophisticated techniques available for reducing a complicated series/parallel diagram to its essential logic – so valuable to the electronics engineer – are of little use to the mechanical engineer dealing almost exclusively with series diagrams. There may be some instances for which a RBD offers worthwhile insight – a complicated plant could be a case – but in general they have little to offer the mechanical engineer as a technique.

It may be worth recording at this point that many reliability engineers approaching mechanical equipment like to bulk components into fairly substantial sub-systems. The reason for doing this is often a desire to replicate the electronic systems approach. It is contended that multi-component sub-systems exhibit the constant failure rate of the pseudo-random condition and can then be treated in the same manner as electronic components. The result is a very simplified reliability block diagram as shown for example in figure 11.1c for the automotive illustrated in figures 11.1a and 11.1b.

## 11.3 Reliability prediction

The prediction of the system reliability, given the component reliability, is fairly straightforward for electronic equipment. Since the loading roughness is very low, equations (3.3) and (3.6) can be applied to series and parallel sections of the RBD. Exercises 3.7 and 3.8 are examples of the technique which has to be employed. There is an extensive literature dealing with this aspect of reliability and the Reading list provides a good start for those who wish to pursue it. For mechanical systems, the loading roughness is most often sufficiently high to militate against such a basically simple approach. Thus it would be quite improper to evaluate the reliability of the drive line represented in figure 11.1a as

$$R = R_E \times R_C \times R_G \times P_P \times R_A \times R_{WN} + R_{WO} \qquad (11.1)$$

Most likely, in the rough loading we can anticipate, putting $R$ equal to the lowest of the terms on the right-hand side of the above equation would be a closer approximation, but could still involve a significant error. It has been demonstrated both by physical and mathematical reasoning that loading roughness is a critical factor not only in determining the reliability of a component in a given environment, but also in determining the manner in which the individual component reliabilities should be combined to yield the system reliability. It should not be forgotten that in the presence of wear the strength distribution, and hence the loading roughness, will be a function of time, so that not only are the components' reliabilities a function of time but the precise manner of their combination is also a function of time. It is recognised that in the development of the thesis of chapter 3 only one load was considered, whereas most mechanical equipment is subject to a multi-load environment in which some load combinations are totally independent. However, the alleviation offered by this feature is limited in as much as the independent loads for one set of components are often the dependent loads for a different set of components. Very rarely indeed can complete independence be assumed. We are therefore forced to the conclusion that with the information that is likely to be available to us in the foreseeable future, there is no possible way in which we can calculate the reliability of a mechanical system from the information on its basic components. This conclusion stems directly from the effects of loading roughness. It is a distressing conclusion, but one that must be firmly faced if we are to remain practical engineers.

The practice of lumping components together into recognisable major components may be superficially attractive. If the major components are chosen to be sufficiently independent, the product rule could be used. As an illustration, we may refer to the RBD for an automotive at figure 11.1c. The failure modes of the braking system, steering system, suspension system, body work, electrics and auxiliaries may be seen to be so significantly different as to warrant the assumption of independence. The engine and drive line may be thought of as dependent, though together independent of the other items. It would then be possible to write

$$R = (\text{Min. of } R_E/R_D) \times R_B \times R_{ST} \times R_{SUS} \times R_{BY} \times R_{ELECT} \times R_{AUX} \quad (11.2)$$

This is still an approximation since extreme environments have been assumed in allocating the operators. Furthermore, some variations in the environment would be common and affect the numerical values of several of the quantities on the right-hand side of the above equation. For example, this would arise if the car were used for rallying as compared with ordinary private use. More seriously, what values are we going to use for the individual quantities appearing on the right-hand side of equation (11.2)? To use general values based on experience begs the question, because this would only lead to a general value for the equipment as a whole, which could equally well be obtained directly from experience. It could not lead to the value for the particular design, and to different values applicable to competing schemes. To assess that value it does seem to the author that it is essential to take into account the reliabilities of the individual components arising from their design and manufacture in relation to the assembly as a whole. So, attractive as it may appear, and in spite of the strong advocacy for this method the author has experienced from competent reliability workers, he rejects this technique also.

The fact remains that on many occasions the relative merits of competing equipments have to be resolved, often at an early stage in the procurement process, and one of the major qualities that has to be taken into account is reliability. Some method of assessment which is not wholly subjective is necessary. The author here believes that a more realistic assessment, that is acknowledged to be approximate, is better than exact calculations based on an imperfect mathematical model, or perfect mathematical models for which there is no numerical data. We start from the knowledge that in spite of much criticism, particularly of the so-called consumer items, the reliability of mechanical items is very high indeed. Many writers have pointed out that Pareto's principle applies to mechanical reliability and that it is in practice only a small number of items – Pareto's vital few, which cause all the trouble. To assess the reliability of any equipment, therefore, we first identify the troublemakers – the vital few failure modes. This is a matter of engineering design, and the method by which the vital few are identified has been fully described in section 10.3. In the present chapter we shall assume that the vital few failure modes and mechanisms, together with the components or integral sub-assemblies involved, have been identified and the expected number of failures evaluated, taking into account all maintenance. With this information available it can be determined which, if any, of these few failure modes are dependent. This is a straightforward matter of engineering assessment. The reliability of the system may then be evaluated from the data on each of the vital few failure modes in the same way as we deduced the reliability of the limited system of figure 11.1c. Mathematically

$$\lambda(t) = \sum_{i=1}^{i=m} \{\text{value of } \lambda(t) \text{ for dependent failure modes}\}_i \qquad (11.3)$$

The value of $\lambda(t)$ for each grouping of dependent failure modes could be taken as the highest individual value if the loading roughness is very high, or the sum of the individual values if the loading roughness is very low (see section 3.2). A compromise value can be taken for medium loading roughness if a firm rough/smooth allocation cannot be made. If desired, the overall value finally calculated can be modified by a small amount to take into account the influence of the trivial many, though this would be irrelevant for ranking purposes. It can with justification be based on experience. The number of failures to be expected in any small time, $\Delta t$, is then given by

$$F = \lambda(t) \times \Delta t \qquad (11.4)$$

and hence the reliability of a mission of duration $\Delta t$ by

$$R = 1 - F = 1 - \lambda(t) \times \Delta t \qquad (11.5)$$

It will be noted that the above calculation is very similar to the method based on a limited reliability block diagram. Of prime importance, however, the limited RBD has been replaced by a rational and systematic engineering assessment of the items to be included. These form no functional or other block diagram grouping. They are a 'vital few grouping'. The only additional information required is an appraisal of the dependence or true independence of the failure modes which constitute Pareto's vital few. This should not be difficult. It is therefore considered that the derived value is every way superior to that deduced from concepts based on reliability block diagrams. Even so, we must beware of placing too much reliance on values so calculated – they are guides only.

The main source of error in the above estimation of reliability is, of course, the environment or loading roughness, which can significantly affect the numerical input values. However, the maintenance procedure can also introduce major unknowns, particularly if the target is availability or cost rather than reliability itself. It was shown in chapter 9 how decisive a role is played by maintenance, and yet it is often only when an item has actually entered service that the maintenance can be finalised. Even when the item is in service, the maintenance plan can be changed. Attention must also be drawn to marked variations of reliability with time in some of the wear-out modes. A change of the age at which reliability is assessed could therefore result in significantly different reliabilities. The best equipment for one time scale may not be the best for another. When all these factors are taken into account the accuracy of any prediction is extremely doubtful, and the author would not care to base any important decision on such estimates.

The above pessimism needs some qualification. One of the first observations of this book was that one man's system was another man's component. It should be apparent from the treatment in this sub-section that the author's system is an identifiable integral piece of machinery, quite likely fairly complex in construction and functioning, and probably requiring regular servicing or maintenance to keep in full working order. An example often used in this book is a vehicle, be it a private motor car or a main battle tank. But this 'system' could be part of a much wider system. Thus a locomotive would be a system in the author's interpretation, but to others the complete railway network or system is the entity that should be considered. The locomotive is but a sub-system of this wider system, which comprises not only locomotives but other rolling stock, track (offering some scope for parallel operation), tunnels, bridges, signalling equipment, maintenance services, fuel storage and supply facilities, goods-handling equipment, etc., plus many other non-engineering sub-systems. Gas, water and electricity supply systems are further examples of larger systems. A feature of these larger systems is that they nearly always consist of discrete sub-systems, albeit connected by track, pipe, cable, etc. Many process plants also fall into this category. The division between the integral systems discussed earlier in this sub-section and these discrete systems is sometimes blurred, but more often than not quite apparent. The treatment of discrete systems *can* differ from that already described. In the first place a discrete system is more readily represented by a reliability block diagram, in which the blocks define the discrete elements and the connecting lines represent the track, cable, pipe or whatever. The mathematical modelling of such a system then follows more closely that used for an electronic system; though a warning must be made against ignoring any dependency between the 'discrete' sub-systems. Such modelling has reached a fairly advanced stage and space precludes an adequate treatment in this book. Readers who wish to follow up this specialist aspect are referred to well-established books – such as that by Green and Bourne in the Reading list. The modelling itself is probably fairly accurate in the hands of experienced workers. However, the input data to such models is the occurrence of failures in the integral sub-systems (or systems in the connotation of this book) which make up the system. We have already expressed strong reservations about its estimation. On the principle of 'garbage in – garbage out' one must transfer the reservations to the output of such models, unless greater confidence can somehow be attributed to the source of the input data.

As already emphasised, it would be naive to calculate the input data for these models from first principles, and confidence is achieved by using established known data for each discrete sub-system. Its source is usually a databank. A big question mark must be put against this procedure but, with care, workable values can be derived in certain circumstances. In the first place it should be noted that it is quite useless obtaining generic data from databanks for application to a given sub-system. For example, it should be quite impossible to obtain a value for the occurrence of failures that would be expected in a particular pump by referring to 'pumps', but it might be possible to obtain a figure for pumps made by 'Bloggs & Co.' used in a known environment. This is possible because firms rarely design *ab initio* but incorporate proven parts wherever they can save design and development costs, if for no other reason. Constituent items of a discrete system may therefore be very similar to previously produced products. The occurrence of failures in closely similar items supplied by the same manufacturer and used in the same circumstances can therefore sometimes be used as raw data. This figure must be adjusted to take into account any modifications that have been made. FMEAs and FTAs are not too difficult in these circumstances, because the number of new parts is limited and the original bits and pieces can be treated as an entity. Using these techniques it should be possible to detect factors which would cause departures from the values experienced in the field for the basic design, though putting figures to the extent of the departure can be a lot more difficult. From what has been stated over and over again in this book, it is obviously essential that the basic data refers to the same load conditions. Fortunately, the load cycle in many industries using discrete systems is known and does not vary very much from one application to another, so that within a given industry it is often possible to obtain representative figures. It *may* be possible to make adjustments as a result of experience if the basic data does not conform to the expected load cycle, but the sensitivity is so great that considerable inaccuracy must generally be anticipated unless a good deal of experience is at hand. This is the greatest source of error. Finally, the basic data must be adjusted for any change in maintenance policies and more particularly for any change in the quality of maintenance. The success of using databank figures depends on having to make only the minimum of modifications to the basic data, that is, on having data which is directly applicable. Unfortunately this is not always possible, least of all when the overall design is at all innovative, that is, when most progress is being made and reliability assessments are most in demand.

Given the input data, the actual modelling follows established lines. The accuracy of the output must be commensurate with that of the input, but should be adequate to allow broad planning to go ahead. Decisions of the type 'Are two or three generators in a power station necessary to achieve a good availability?' are possible. The sensitivity of the conclusions to the input data should always be assessed, however, and the confidence limits evaluted. This means true engineering confidence limits, not narrow statistical ones. Since the success of these methods must depend heavily on the quality of the input data, the author would contend that true success can only be attained by a fuller understanding of those factors which control the mechanical reliability of the elemental parts. That is the key to the quality of data. There is just too much uncertainty in the basic data for any complacency at present.

## 11.4 Reliability apportionment

Closely allied with reliability assessment is reliability apportionment, by which is meant the allocation of the reliability to be achieved by various sub-assemblies of a system in order to achieve an overall system reliability. This is usually considered a necessary action at the early design stage when several sub-contractors are involved in a project, so each can be allocated a target figure. Apportionment in the electronics field is usually achieved via the reliability block diagram and the product rule. From the discussion in sections 11.1 and 11.2 it will be seen that such techniques are not available for mechanical reliability. Indeed the whole question of reliability apportionment needs to be treated with some reserve in that application. As brought out in chapter 9, maintenance is the key to mechanical reliability, so that the targets set to individual sub-contractors should aim at a coordinated maintenance programme rather than an inherent sub-assembly reliability which may be achieved by a diversity of reliability/maintainability combinations. An uncoordinated maintenance programme is almost certain to involve high total costs. With the object of achieving a coordinated programme, the ages of scheduled first and second line maintenance should be specified at the feasibility stage. A target can then be set to sub-contractors in terms of the life of intrinsic reliability, $B_0$, or alternatively in terms of $B_5$ or $B_{10}$ if these are thought to be more appropriate and relevant to the expected failure mechanisms. Clearly, these should be integer multiples of the scheduled maintenance intervals. By this means the rudiments of an integrated maintenance plan can be established. Doubtless it will be changed in the light of development experience, but initial compatibility will have been achieved. The danger in specifying target reliabilities which may be achieved by incompatible maintenance schedules will have been averted. Of equal importance, a life target is much more tractable to the designer than reliability expressed in terms of a probability.

The specification of sub-assembly reliability then follows along identical lines to equipment specification and has been fully discussed in section 10.2. The real difficulty is the usual one of specifying the load cycle and environmental conditions, the more so since these will be controlled more significantly by the equipment internal circumstances than by the user requirement, and load identification is complex.

## 11.5 Parallel redundancy

The fundamental advantages that may accrue from redundancy were examined in section 3.1, and practical forms of redundancy specified in section 11.2. It is now necessary to pull these together in an appraisal of redundancy as a means of achieving very high reliabilities. It would be as well to observe at the outset that this technique has been used with great success in the electronic world, and in view of the very high reliabilities theoretically possible it must be considered as a means of achieving such reliabilities in all systems. It will be recalled, however, from section 3.1 that the big gains from parallel channels were only theoretically available at low loading roughnesses. Also, electronic applications are associated with low loading roughnesses, so the gains theoretically possible can be realised in practice in that application. At a loading roughness of unity, the high reliabilities will not even be realised theoretically: reliability of a parallel system will be only that of the most reliable channel (see exercise 3.9). Clearly, we can expect intermediate gains at intermediate roughnesses. A practical curve for an aeronautical application is shown on figure 3.11 (taken from ref. 11). It corresponds to medium loading roughness, and lines up with theoretical expectations. Hence parallel redundancy is only likely to be of value for those mechanical applications where the loading roughness is on the low side – a severe restriction since most mechanical loading roughnesses tend towards unity rather than zero. The physical reason for this varying response to redundancy is that at low loading roughness it is the dispersion of the items' strength which causes failure, so that the duplication or triplication at random from a population increases the chance that one item is the strong one which can withstand the load on behalf of the system. At high loading roughness it is the distribution of load which gives rise to failures, so that if a high load from a wide distribution is encountered it is likely to cause failure of all items – whether they be strong or not, in parallel or not – and the system fails. This is the source of the common load, common mode or common cause as it is now sometimes called, failure.

It is noticeable that a very large literature has grown up amongst those using redundancy techniques on the subject of common mode failures. One cannot help suspecting that the designers of electronic systems are running into the consequences of loading roughness that have beset mechanical engineers all along. The inadequacy of their techniques has only become apparent when dealing with parallel rather than series systems. (As pointed out in section 3.1, the error is on the credit side when dealing with series systems but on the debit side when dealing with parallel systems.) The objective of most of the literature is to make allowance for these common modes in calculations which have been based on perfectly smooth loading; that is, calculations which are fundamentally unrepresentative of the actual circumstances and therefore unsound. The author has doubts concerning the efficacy of such methods, and has been pleased to note a more rational approach in recent literature (see refs 72 and 73, for example). In this approach the probability of failure is split into dependent and independent parts. This requires a 'black or white' allocation in place of the real 'grey' scale, but is a reasonable approximation. Then we can write

$$\lambda = \lambda_i + \lambda_c$$

where $\lambda$ = overall failure rate of channel
$\lambda_i$ = independent failure rate
$\lambda_c$ = failure rate associated with common load.

Since the probability of a channel being in the failed state is proportional to the overall failure rate

$$F = F_i + F_c$$

Suppose, now, that we have two channels. If there were no common mode failures the probability of failure of the two channel equipment would be given by

$$F_i = (F_i)_1 \times (F_i)_2$$

where the suffixes 1 and 2 refer to the values applicable to the two channels respectively. Now this must be the lowest possible value for the equipment failure probability, so that

$$F_L = F_i$$

where $F_L$ = lower limit of failure probability.

If all the failures were due to common causes (that is, the loading roughness = 1) the probability of failure, $F_c$, would be the greater of $(F_c)_1$ and $(F_c)_2$. In refs 72 and 73 this is taken as the highest possible value of the equipment failure probability, so that

$$F_U = F_c$$

In the above references it is considered that the true probability of failure must lie between the upper and lower bounds just defined. In ref. 73 a log-normal distribution of possible values is assumed, from which the best estimate of the probability of failure of the two channel system is

$$F = \sqrt{(F_L \times F_U)}$$

However, the author would have thought that if the independent modes were truly independent then the probability of equipment failure is the product of the two independent modes and the greatest of the dependent modes, that is

$$F = (F_i)_1 \times (F_i)_2 \times F_c = F_L \times F_U$$

In ref. 72 an empirical formula is given which grades the overall failure probability between the median value quoted above and the lower bound. This would seem very reasonable for the rough descriptions of the channel attributes given in the reference. It could well be argued that it is unnecessary if the individual failure mechanisms are fully identified and treated in their own right.

It will be noted that the above treatment assumes constant failure rate, that is, it applies to stress-rupture common failure modes. It may be thought to apply to pseudo-random failures, but this does need closer examination. It seems to the author that a lot of confusion over common mode failures arises because writers do not state clearly whether they are dealing with stress-rupture or wear-out failures, or both. At least the author gets confused if others do not. To illustrate the problem, let us return to the solution to exercise 10.1A dealing with parallel redundant pens. Let us assume that the failure mode accounting for the lower single pen reliability quoted by the supplier is due to the wear mechanism in which the ink runs out. Furthermore, assume that all pens are carried in the same jacket pocket and that the selection for use is made at random. Then the loads to which the pens are subjected are the same and the failure mechanism is the same; so one would conclude that any failures would be in a common mode and hence the product rule is invalid and the solution to exercise 10.1A is not correct. (It assumes independence and that the product rule is applicable.) According to the treatment in refs 72 and 73 (which are implicitly if not explicitly based on stress rupture) the reliability of the three pens is 95 per cent. But is such a conclusion valid for wear? Although the ink requirement per unit usage may be very variable, that is, the loading is rough, this is not the failure-inducing load. The failure-inducing load for wear mechanisms is the cumulative load. For wear-out phenomena the strength distribution is transformed into a life distribution [for example, by equation (5.11) in the case of fatigue] by integrating the load distribution, so individual loads have no role of their own. Since these particular pens are chosen for use at random, one can expect the cumulative load on each pen to be much the same, that is, the failure-inducing load is narrowly distributed so that the effective loading roughness is low (smooth) and hence (by the solution to exercise 3.9) the product rule is valid, even though a common mode is involved. For wear phenomena, common mode failures (as generally understood) are the consequences of a common failure mechanism and a common *cumulative* rough loading which gives rise to that failure mechanism. It is these circumstances which invalidate the product rule. The validity or otherwise of the product rule needs very careful examination when wear phenomena are involved.

The difficulty in dealing with exercise 10.1A, and often with practical problems, is that a reliability for the required 'mission' has been quoted, that is, a single value, when in fact that value is a function of time. Suppose I had not used the pens at random initially, but had used only one continuously for half its life of intrinsic reliability (assuming intrinsic reliability was the design intention), thereafter using the two pens at random as before. Then very high reliability indeed would be achieved, as calculated by the product rule, because one pen would be intrinsically reliable when the other was losing reliability as a result of wear-out. That pen would be replaced/refilled on failure, giving an intrinsically reliable pen which would safeguard against wear-out in the other. High reliability is continuously achieved. It is the technique the author adopts with his pens – though it does not obviate the common mode failure associated with leaving both pens in the other jacket pocket! With wear-out phenomena the reliability of parallel channels is a function of time, and that function depends on the age mix of the individual channels. It is difficult to see how any theory which ignores this can give the right answer, and real systems are bound to appear capricious when compared against theories which do not truly represent the wear. This difficulty is the equivalent of 'good as new' compared with 'bad as old' for a multi-component series system. With a limited number of mixed-age parallel channels, the state could well be described as 'better than old', though not 'good as new', after repair.

The lesson to be learnt from this example is that it is the cumulative loading roughness which indicates the validity of the product rule for common wear-out mode failures. Even so the system reliability can only be assessed when the variations of channel reliability with time are known. Since these variations are so disparate, it is essential to treat each failure mode separately. Bulking of failure data for different mechanisms is only likely to obscure the basic processes involved – a conclusion we have reached before on other grounds.

The above observations regarding the product rule follow from the theory of section 3.1, which assumes implicitly that all the parallel channels are substantially the same. If totally independent channels can be used, the system is decoupled, at least in the functional sense employed in reliability block diagrams. The product rule would then be valid with all loading. The words 'totally independent' have to be interpreted strictly for this to be true, but in spite of this there are circumstances when substantial independence can be achieved. It is a matter of conceptual design to realise this theoretical state in a practical form. The gains to be achieved are, of course, considerable. It is probably the royal road to very high reliability via parallel channels in mechanical systems.

The merits or demerits of redundancy have to include rather more than just reliability. As indicated in earlier chapters, the whole system quality must be taken into account. Thus on a pure weight for weight basis, an extra channel adds weight that could alternatively be deployed to increase the safety margin of a single channel. For the same weight, an increase by as much as a factor of 2 is possible. Reference to figure 4.13 shows that would reduce the chances of a random failure by a realistically unknown but very large factor indeed. It could offer a higher reliability from a single channel than from two parallel channels of the same overall weight in roughish loading conditions. Single channel mechanical systems can be very reliable indeed, so there is no need to rush into parallel redundancy. There is probably no single mechanical system on which so much safety depends than the steering gear of motor cars. Millions of cars rush at each other all hours of the day all over the world, relying on their steering gears to avoid collision. Collisions rarely happen through failure of the gear, so in one case at least very high reliability must be achieved. Steering gears of cars are never duplicated! High reliability commensurate with safety can be achieved in a single channel, even by such a maligned industry as motor manufacture. It can surely be achieved elsewhere. A thorough appraisal is essential in each particular case.

The circumstances in which parallel redundancy is most valuable are those where a high degree of decoupling, and hence statistical independence, is possible. A good example is the provision of standby generators in hospitals and similar institutions. The system meets all the requirements: the standby channel is a completely different system from the main channel (usually a diesel-driven generator as compared with steam or gas turbines at the grid generator stations); the standby is by way of separate lines; and the standby system is installed and operated by entirely different personnel. There are few such good examples. In most other instances the operators and/or maintainers constitute some common element in the parallel channels even if the channels are physically totally different. Since mechanical reliability is so sensitive to the operating conditions and dependent on maintenance, true independence, which is essential to success in this technique, is difficult to achieve. Even in this case 'time' constitutes a common element, and must be decoupled by running the standby unit from time to time although it may not be an operational requirement. This is necessary to safeguard against that form of mechanical wear that can take place by 'not use'. Another example of effective parallel channels is manual standbys for powered systems. They can often be made to meet the requirements, incorporating significant independence. Systematic testing, or maintenance, is again needed to ensure that the standby system is fully functional at all times. It must be clear that the reliability of even a smoothly loaded parallel redundant system will deteriorate rapidly once any wear-out processes set in, and strict maintenance is essential to prevent such deterioration.

There are, of course, numerous examples of duplication, or excess holdings, of parts, sub-assemblies or even complete equipments, which may be regarded as a form of redundancy, but this is not a safeguard to the equipment. It provides higher maintainability, and hence higher availability, but the equipment reliability remains unchanged.

Clearly, redundancy has a part to play, and should be considered and exploited to the full whenever very high reliability is the prime requirement and cannot be achieved by a single channel, but only then. The grave danger is that excessively high reliabilities that can be erroneously calculated lead to a fools' paradise, and hence away from the true path to high reliability via high-class design, adequate development and rigorous maintenance. There is certainly no excuse whatsoever for using parallel redundancy which employs channels which are not themselves the most reliable that technology can offer.

## 11.6 An overview of system reliability

The author believes it would be wrong to attach too much importance to system reliability estimation for mechanical equipment. The problem does not lie there. It lies in achieving acceptable component reliabilities under the multitude of loading conditions to which they will be subjected in real use, and in planning maintenance schedules to keep the reliability at that level. It *must* be true that one can only determine whether an item is strong enough if both the loads and strengths are fully known, and if all the failure mechanisms (both stress rupture and wear-out) are substantially understood. Without basic information on all aspects of component behaviour, system reliability estimation can be little more than conjecture. Realistically, much of the above prerequisite information is not available, and worthwhile estimates cannot be made for components and hence for systems. It is better to face the situation than create a make-believe one. The best estimate of equipment reliability at present could well be based on the probability of the designer or development engineer having thought of all eventualities. In colloquial terms, the track record of the designer, design team or manufacturer for the particular job under review is probably the best guide.

Most progress towards optimum reliability will be achieved by trying to understand why strength is distributed, how failure mechanisms cause fracture or other malfunction, and what loads are actually applied to mechanical parts. Appropriate action – specification, design, development, maintenance – can then be assessed. It is to that theme which we now return. When all that is known, the calculation of component or system reliability should be comparatively easy.

# 12
# The role of the user in achieving reliability

## 12.1 The user/designer relationship

It has been more than adequately demonstrated that the distribution of load is far more influential on the reliability that is achieved by an item than is the distribution of its strength. The reader could well refer back to sections 4.1 and 4.5 for confirmation. Nothing dismays the author more than to become involved in an argument between a designer or manufacturer and a user regarding the responsibility for reliability: often a futile exercise in blaming the other guy without regard to realities. Abuse can clearly cause failure. While there may be some circumstances when intentional abuse is justified, these are essentially cases of calculated risks and outside the scope of this book. Unintentional abuse is a real cause of failure. There is only one remedy, and that is adequate training of the operator. The operator must be fully aware of the limitations designed into the equipment and the conditions of use for which it was intended. However, these should be agreed between designer and user. Only too often what is abuse to a designer is common practice to the user. The user and designer must come to terms. How this is to be achieved will obviously depend on the circumstances. If the purchaser is calling for a special product to meet his needs, he has a duty to define, in its specification, the exact conditions under which he intends to operate the equipment. Furthermore, he has a duty to see that the equipment is subsequently operated within those conditions, though perhaps less of a duty than an act of self-interest. Success in this connection is primarily achieved by adequate operator training and rigorous discipline. It is amazing how much money and effort is put into conventional quality control, which is control of the product, and how little into operator quality control. Very real gains are available to those who invest in a sound operator training programme and back it up with an assessment of actual usage. Some hostile environments and some adverse conditions cannot be avoided. These should be specified so that the design makes due allowance for them, or the possibility of failure is accepted. Otherwise reliability is in the hands of the user.

The purchaser of the isolated product – typically the consumer items – can do little to affect the design. He must necessarily take the product 'as it comes' and try to operate accordingly. Here the onus falls on the designer to state clearly the circumstances and environment in which he expects his product to be operated. Handbooks are woefully inadequate, and often seem to be written by people who have never seen, let alone handled, the equipment. The operator must be given every opportunity to match his use of the equipment to the designer's intentions. It is, of course, in the manufacturer's interest to ensure that the design meets the requirements of his potential customers.

We see therefore that the onus for correlating the design with the use may lay primarily with either the designer or the operator. It may sometimes be more equitably distributed. However, we can be certain of one thing: if the design is not prepared in a true knowledge of the operating conditions, the result is – inevitably – unreliability. The essential link by which information is fed back from the field into design was illustrated in the first figure of this book, and our subsequent study of reliability had underlined that requirement and emphasised its importance. It is more than a formal link. The designer must be able to place himself in the position of the user if he is to achieve the objective. To do this he must receive an adequate feedback of the in-service behaviour of his design. Feedback of field experience has two aims

(1) to let the designer know of defects in existing equipment, both so that remedial action can be taken by means of 'mods' and so that the designer can build up a background knowledge of the behaviour of his products in the user's hands
(2) to let the designer know how it is intended to operate and maintain future equipment so that new designs can best meet the requirement.

The small-scale user can do little to promote these objectives but the large-scale user is in a very strong position to initiate action. It is with the large-scale user in mind that the rest of this chapter has been written.

## 12.2 Collection and collation of user experience

Before he can advise the designer, the user must collect and collate all the reliability information available to him. Several methods are open to him

(1) special service engineers can be deployed in the field to note and assess all failures in their areas
(2) the user in the field can report to a central source any failure which he encounters by means of 'defect reports'
(3) the jobs undertaken by repair workshops can be examined to provide failure data.

The above methods have been given in order of increasing amenability to automation but decreasing built-in professional evaluation. The first method has the advantage of on-the-spot investigation, evaluation and assessment by a professionally capable engineer. There is, however, no guarantee that the engineer is notified and investigates all failures in his area. Furthermore, when making any investigation he could be unaware whether he is tackling a one-off type of failure or a recurring fault. The second system dispenses with specially trained personnel and relies on the user to report any failure. Any on-the-spot assessment is eliminated from this system, but theoretically all failures are reported. A statistical appreciation should therefore be possible and, by using specially prepared forms which require only simple questions to be answered, it is possible to obtain a significant amount of technical data even though the user who completes the form is completely non-technical. This is a relatively cheap system since it dispenses with special personnel and is adaptable to a wide scale and variety of operations. The weakness of the system is the reluctance of a user to fill up 'another – – – form'! In particular, users do not raise defect reports on minor recurring failures so that the data recorded at the central point can be in error. It would seem that this system does not reveal the vital few that create the many 'well known' failures, simply because the failures become so 'well known' as to be not worth reporting! This defect is overcome by the third method, which analyses all the information taken from standard (or only slightly enlarged) workshop information such as job cards. The advantage of this system is that it is comprehensive and can be carried out for various levels of activity. Its disadvantage is that a very large amount of detailed data has to be sifted, but if the scale of operations is sufficiently large to justify an automatic data-processing facility it can become extremely flexible and provide a great deal of information not only on reliability but on maintenance and other management aspects. As an example of such systems, the Army's FORWARD (Feedback Of Repair Workshop And Reliability Data) is described by Haywood and Metcalf (ref. 73). This is one of several systems which have come into operation. The paper referred to describes the kind of information that is sought on a wide range of equipment, and how it is collected – in this case from a flimsy copy of a standard job card. The information on the report includes the holding and repair units, equipment type, serial number and utilisation, reason for repair, job progress information, the work done (repair or modification), tradesmen used, man hours taken, spare parts used and faulty parts information, on three categories of equipment. The paper gives examples of the coding employed and describes the data output. FORWARD outputs are: (a) standard print-out at regular intervals, (b) interrogation of the repair data for specific purposes, and (c) referring reliability assessments to control data and printing-out exceptions. Lastly, and perhaps of most importance, the paper describes the flow of reliability data from source to areas of direct use. The weakness of all such systems could lie in the coding necessary for the ADPS, and that being 'automatic' it cannot provide the technical appreciation one would expect from an on-the-spot engineer. The temptation to make such a system provide so much information that 'in-office' diagnosis is possible, must be firmly resisted. This would merely clog the fruitful communication channels. It is also essential that the output from the system can be readily interpreted by those responsible for taking action, and indeed that the information does get to those directly concerned.

### *Exercise 12.1*

The manufacturer of the bicycle described in exercise 10.20 is proposing an export drive with his new model. To ensure its reputation he is setting up a number of service stations at various centres, and in order to keep a record of the performance is arranging for each of these centres to report to his headquarters all the jobs they carry out. Each centre will report under the following code. Examine this code and comment on its suitability.

*Category of failures*

I(i) Failure: The bicycle cannot be used.
I(ii) Fault: The bicycle can be used but either,
(a) the adjustments for rider comfort are unsatisfactory,
(b) undue effort and/or skill would be required, or
(c) wear or maladjustments exist which will lead to eventual failure or fault as (a) or (b).

*Order*

2(i) Primary: Failure/fault not due to another failure/fault.
2(ii) Secondary: Failure/fault due to another failure/fault.

*Cause*

3(i) Random: Failure/fault arising from chance condition other than accident damage.
3(ii) Fair wear and tear: Failure/fault arising from the effect of normal wear or deterioration.
3(iii) Neglect: Failure/fault arising from lack of maintenance (lubrication, routine adjustments, etc.) or from improper storage.
3(iv) Damage: Failure/fault caused by careless handling or misuse or accident.
3(v) Manufacture: Incorrect manufacture.
3(vi) Design: Inability of any part to withstand design loading for design life span.

Reports will give the following information in the order:

I. Category: whether a failure as in 1(i)
a fault as in 1(ii) (a) or (b)
or a fault as in 1(ii) (c).
II. Order/cause.
III Part number(s) of item(s) replaced from necessity.
IV Part number(s) of item(s) replaced as a precaution.
V. Frame number.

Details of code as below.

| | Code |
|---|---|
| Failure as in 1(i) | 01 |
| Fault as in 1(ii) (a) or (b) | 02 |
| Fault as in 1(ii) (c) | 03 |

| *Order of failure/fault* | *Cause* | *Code* |
|---|---|---|
| | Random | 04 |
| | Fair wear and tear | 05 |
| Primary | Neglect | 06 |
| | Damage | 07 |
| | Manufacture | 08 |
| | Design | 09 |
| Secondary | | 10 |
| Routine maintenance | | 11 |
| Parts replaced from necessity | | 12 |
| Parts replaced as a precaution | | 13 |

*Example*: 01/07/11/12–P4, P12/13–P6, P6/2001 months indicates that: the bicycle was unusable as a result of damage sustained. Some routine maintenance was carried out with the repair. Parts P4 and P12 had to be replaced; P6, P6 were replaced as some wear was evident. The frame number was 2001.

It is now some ten years since the U.K. Army's FORWARD was put into service, and somewhat longer since it was first planned. Many other reporting systems in both the U.K. and the U.S. and elsewhere are contemporary with FORWARD so that we are in a position to judge how effective they have been. The author must be among the first to acknowledge the great deal of factual information on failures he has obtained from such systems, particularly FORWARD. They are indeed the source of virtually all data. Most of the concepts upon which this book has been based arise from a study of field data. It should be noted, however, that it is possible to search the records for the appropriate data to support or refute a proposition without confining oneself to a particular equipment, but this is not the way the system is supposed to work. The object of any reporting system is to deal with specific failures in specific equipment. The situation is then far from satisfactory. The main difficulties have been illustrated in exercise 12.1 but it would be as well to list the unsatisfactory aspects of most ADP systems.

(1) The coding in any ADP system cannot be fully comprehensive, so that a good deal of the original information is 'lost'. It is very often not possible to ascertain the exact nature of any failure – for example, the difficulty of distinguishing between a fatigue and stress-rupture failure has already been

noted. Such systems are incapable of recording much significant engineering data and information: often the most that can be deduced is that a failure (or a suspected failure!) of an item has occurred.

(2) As just noted, many 'failures' are recorded when no actual failure has occurred. This may arise from an incorrect diagnosis. Thus a part may be declared faulty and replaced even though it is perfectly satisfactory. It is impossible to place any figure on this action, but experience of parts removed as faulty and returned to the base workshop or factory for repair, only for no fault to be found, suggests the proportion may be very high. Certainly it can often be high enough to invalidate any statistical analysis. Very few ADP systems incorporate any means of correcting such false data. In addition to faulty diagnosis, repair action may be taken in advance of any actual failure as a precautionary measure, particularly if the operative is aware of a commonly recurring failure. You should re-examine the clamping bolt failures of exercise 10.16 to see how 'anticipation' may invalidate the data. Then again there can be coding errors – some genuine mistakes, but others due to laziness in checking a coding. It is much easier to jot down any frequently used code. The number of sparking plugs that have been changed on Diesel engines according to ADP print-outs is legion! Finally, there is malicious false records entered solely to account for wasted time when no real work has been done. Corrupt data of one kind or another is a very serious source of trouble and error in analysis and interpretation. It can be particularly serious when only limited data is available, as is so often the case, so that the misrecording of even a single failure can influence the interpretation.

(3) Very few ADP systems have the facility for tracing the past history of any part, or perhaps more importantly, any sub-assembly.

(4) Finally, the most serious omission is the absence of any data on the loading. It may seem boring and unnecessary to point out once again that the loading is all important but, without a knowledge of the total loading conditions to which a failed component has been subjected throughout its life, very little can be done to remedy any serious fault. To the author this is the most serious criticism of all the systems he has examined. Furthermore, it is a criticism which it is difficult to overcome, for it means continually recording the operation of every unit. This may be justified in some cases: for example, many airlines keep very comprehensive data on each of the aero engines in their fleet of aircraft, though this is done more as an essential part of on condition maintenance than as data collection for subsequent analysis. It is claimed that the cost is justified on a total cost basis, which could well be true in this instance. Even in this highly sophisticated field, however, the RAF find it difficult to follow the same procedure because of the widely varying duty cycle of various missions. Except in very special circumstances, the cost of a comprehensive survey of loading is prohibitively expensive.

How can we sum this up? The first thing one must do before setting up any ADP system for mechanical reliability is to be absolutely sure of the purpose it is intended that it should fulfil. When most of the existing systems were designed, the importance of the loading had not been recognised. The collection of failure data and hence the formation of a 'databank' was often seen as an end in itself. It was believed, very largely on the claimed experience of the electronic industry, that the failure rates of components could be held in such banks, referred to when new designs were under consideration, and hence the reliability of the system could be calculated, and its maintenance policy established. Such a view is now naive in the extreme. However, it still may be possible to achieve part of this goal on a very narrow front. Thus a manufacturer could establish data peculiar to his product. Even so he would have to be very careful what data he collected and be sure he was able to process that data for his desired purpose before embarking on an advanced data-collection and analysis system. Furthermore, he would have to keep in mind that future change of circumstances could invalidate both the data and the analysis.

In spite of all the above criticism, we must be realists and recognise that no progress can be made without data of some kind and, also, as a generalisation, the more data the better. The obstacle is cost, for the cost escalates rapidly with the comprehensiveness of the data. Any well-organised company or institution must have data available for management purposes, and it would seem best to make as much use of this data as possible for reliability purposes as well, at least in the first instance. This is how FORWARD started. Special processing of the data is, of course, necessary, but the object should be to keep the cost of data collection as low as possible. The initial approach must be via a Pareto analysis, which will draw attention to those areas most in need of detailed investigation. The next and very important step would be to 'home in' on these areas only for a full investigation. This may well involve the compilation of more comprehensive data, but it will be over a limited area with associated cost limits. The exact nature of the data needs to be well thought out in conjunction with the method of analysis. For example, a firm exporting moderately delicate equipment told the author that they loaded one of their packing cases with some accelerometers and recording kit in addition to their product. The data recorded told them more about the causes of some unreliabilities in their product, and hence the measures necessary to effect a worthwhile improvement, than the most comprehensive record of failure data in the destination country could ever hope to have done. The process is essentially an engineering one in which both the data collection and the analysis is matched to the problem on hand. The writer can only observe that load cycles are often the trouble spot, and that while Weibull representation of the failures can help greatly in planning maintenance, it does little to reveal causes of failure. We must therefore be certain what we are aiming to do. It is all too easy for the user to adopt a somewhat complacent attitude aimed solely at optimising his maintenance policy for products he has already bought. Necessary though this is, if he really wishes to make progress he must assess the actual cause of failure, and remedy that. This may well involve closer liaison with the designer. So be it. In 1974, the author wrote: "the significant theme which has recurred over and over again is the matching of the design to the requirements of the user in its intended environment, and the necessity of good communications to achieve it." He sees no reason whatsoever, some ten years later, to change that statement in any way. In this chapter it is assumed that the user takes the initiative. The isolated user cannot take the initiative, and then the onus falls on the designer – but dialogue there must be. Filing cabinets full of computer print-outs achieve nothing: direct communication between designer and user is undoubtedly the key to success, and should motivate and shape all failure data compilation and assessment.

# 13
# Management aspects of reliability

The first step any management must take in connection with reliability is to convince itself that a high, or at least a defined, level of reliability in its equipment or its product is a worthwhile economic objective. The management of a manufacturing concern could well argue that money spent on achieving high reliability represents an equivalent loss in profits and the loss of the sometimes high profits associated with spares provisioning: reliability is something for the user to concern himself about, they would argue. Before contesting this point in principle it would be as well to observe that no man, or firm, is an island to himself. Although calling itself a manufacturing company, such a firm is very much a consumer too – a consumer of machine tools, presses, forges, furnaces maybe, a consumer of bought-in material of all kinds, of power, of heating and lighting, of packaging, of office equipment, of transport and of other items too numerous to mention for even the simplest and smallest concern. The consequences of unreliability in any of these items is scrap or lost work, and this should be very much the concern of the management of even a 'purely manufacturing' company.

It is obvious that the users of equipment will be in the best position to assess the effects of unreliability. It has been stated, for example, that Royal Air Force (RAF) expenditure on maintenance represents 50 per cent of the total expenditure. It is, of course, total maintenance and no figures are given for the cost of unavoidable maintenance. However, we see that a substantial proportion of the running costs is directly attributable to maintenance of one form or another, amounting to about twice the general running cost (that is, total cost less expenditure on new equipment and on maintenance). On the commercial side, in a similar field of activity it has been reported that the cost of unreliability (that is, unavoidable maintenance and hence loss of revenue) to British Airways amounts to 7 per cent of the total revenue. It is possible that in areas where reliability is not so closely associated with safety as to preclude any but the highest standards, the cost of unreliability could be very much higher indeed. There is clearly every incentive, therefore, for the management of a predominantly user organisation to purchase equipment of the high reliability necessary to minimise the whole life cost.

Turning now to the manufacturing aspects, we may assume that the cost of quality assurance is about 5 per cent of the gross turnover. Nixon in ref. 74 has claimed that by doubling the quality engineering effort (and hence cost of this activity) the total cost of quality assurance can be reduced by 35 per cent, achieved largely by reduced scrap. If such an activity is applied only to 60 per cent of the total scrap (this is about the total of Pareto's vital few) Nixon shows that it will result in a saving of 1.5 to 5.5 per cent of gross turnover. This compares with a profit margin of 7.5 per cent of gross turnover. There is no reason to believe that Nixon's figures are in any way an exaggeration for a firm that has not already introduced modern quality assurance techniques, and they could be on the low side for less advanced concerns. To this must be added the not inconsiderable savings which would be obtained from a reduced warranty budget, where such a scheme exists. The above figures are, of course, mean values which would not apply to all concerns, but they do show that substantial savings may be possible for manufacturing as well as operating organisations.

Thus we see that whether we are concerned with manufacturing or user or consumer organisations – and most actual organisations are a mixture of all – it is highly desirable that management should at least closely examine their *total* quality assurance costs and consider whether they have been truly minimised.

We have left to the last that factor usually designated as 'customer satisfaction'. While the very large purchaser may be able to estimate the whole life cost using the very latest data and methods, by far the greater number of purchasers rely on reputation. Shortages which enable a manufacturer to sell anything he could make have long gone and, once a reputation for producing an unreliable product is acquired, sales of perfectly adequate products by the same manufacturer begin to decline. It is impossible to put figures to this, but whoever disbelieves it should study the inroads made into the British motor car market by overseas competitors during 1970 following the publicity given by the British Consumers Association publication *Which* and the Automobile Association to the low reliability of some home-produced models. (Note that the validity of the claim is not the issue – the image is. Thus so far as the author is aware, the reliability of British cars was and is much the same as, say, Continental cars bought at random on the Continent. The image was somewhat different.) The situation in the United States is very similar indeed, and has been fully discussed by Miaoulis and O'Brian (ref. 75). With the worldwide growth of consumer associations and more recently consumer protection legislation, one can see no reversal to this trend. Reliability has now become a saleable product and the wise manufacturer will recognise it. He will also recognise that reputation is all-important in a field where facts are extremely difficult to establish.

And what if these appraisals show that high reliability is essential? It will have become apparent from the earlier chapters of this book that a great deal of collaboration and coordination between all concerned is necessary to achieve this quality. It is not without good reason that it has been said that "reliability is everybody's business." The essential links between what might otherwise be considered autonomous functions were presented at the outset (see figure 1.1) and have been emphasised again subsequently. Since it is essentially the responsibility of management to create and maintain these links, it becomes clear that the processes which produce the desired reliability are a close concern of management. This has led to a great deal of literature and much discussion on the role of management in reliability; but the management's responsibility towards reliability is no different from its responsibility towards any other quality. What applies to the management in general applies also to the management of reliability in particular.

As long ago as 1931 the American pioneer of statistical quality control W. A. Shewhart (ref. 76) laid down as the three essential steps for any quality control programme

(1) the *specification* of what is wanted
(2) the *production* of things to satisfy the specification
(3) the *inspection* of the things produced to see if they satisfy the specification.

More recently, in 1963, Nixon expanded these and set out the following seven essential requirements for the achievement of a reliable product (ref. 77)

(1) a satisfactory design of product, thoroughly proved by adequate testing in order to establish its reliability under the conditions to which it will be subjected in use
(2) a full specification of the requirements of the design which must be clearly understood by everyone concerned with the production of the constituent parts and of the complete end-product
(3) confirmation that the manufacturing processes are capable of meeting the design requirements
(4) full acceptance, by all those concerned with production, of the responsibility for meeting the standards set by the specification
(5) a check that the products conform with the specification; this check is required to protect the customer, to safeguard the reputation of the company, and to provide essential information regarding failure to conform – it is the responsibility of inspection, quality assurance or quality audit
(6) instruction in the use, application and limitations of the product
(7) a study of user experience, feedback to the department concerned and rapid remedial action.

Once listed, as above, it is difficult to see how these requirements can be ignored. Yet ignored they are: as is often too self-evident. The stumbling block is surely demarcation, be it the pursuit of sectional interest – such as the reduction of costs or speeding up delivery at the expense of reliability – or perhaps the non-pursuit of an interface activity. High reliability can only be achieved when all concerned pull together for a united purpose. In these circumstances the role of management is clear. They must let it be known throughout their organisation that one of their main objectives is high reliability, and must be seen to be acting on this. They must then create an atmosphere and company structure in which the necessary cooperation is possible and welcome. The detailed structure by which management facilitates cooperation and ensures that all the above requirements are met depends on too many factors to allow generalisation. Such features as the size of company or organisation involved, its existing structure, the nature of its products, the type of contracts involved and so on must all be taken into account. On one aspect there does, however, appear to be agreement. It is mandatory that one man – be he called the quality manager, reliability controller, reliability coordinator or any other title – be solely responsible for ensuring the effective operation of the quality management system, which should be specified in writing and available to all concerned. The appointee must have direct access to Board level to resolve any disputes. He must play a major part in formulating policy on the specification of the reliability in a project, the reliability aspects of the design reviews, the conduct of prototype trials, including the collecting and analysis of reliability data, the feedback of that data to appropriate departments, and the control of corrective action to ensure the specification is achieved; that is, he must be, and be seen to be, in charge of the reliability programme. The major difficulty in any organisation concerned with reliability is one of responsibility without power, which arises from the all-embracing nature of the subject. It is an essential management requirement to ensure that all persons in the reliability team who have responsibility also have the authority and resources to meet that responsibility. Any reliability team will itself probably be quite small because most steps will be taken in conjunction with other departments, which should be appropriately staffed to meet their role in achieving reliability. There can be no contracting out in this connection, and in particular management should make it clear to the chief designer and the production manager that reliability is every bit as important as any other performance parameter. However, the major difficulty facing management is not seen as one of devising an organisation, but more one of staffing it. For example, the reliability manager should have had some design experience, if not he must have had development experience involving redesign, and the preference is for both, essentially in the type of work relevant to the project. He should have experience of the production methods involved. He must have a good statistical background, though will not be a statistician. He must have a sound grasp of economics and must be able to work with all kinds of people. Such staff are not easily found. Not only will managers be in short supply, the same difficulty will be experienced in recruiting good designers on whom the success of the whole project so largely depends. Any good designer tries to get away from basic design to some more elevated post as soon as possible – he would certainly not wish to 'waste' time on the minor parts and components which we have seen are most often responsible for low reliability. This situation is aggravated by the complexity of modern equipment which necessitates a design team rather than a designer. Experience retention in a large changing team is, in those circumstances, a serious difficulty. Perhaps the introduction of computers into design will provide the necessary impetus; computer aided design is currently 'in'. Computing has status – design does not. The source of trouble then becomes design carried out by computer scientists! The onus falls on management to attract the right people to the necessary jobs, or to train competent staff. In many instances the only solution is 'to grow one's own'. It would often appear that low reliability is the result of ignorance at all levels rather than any malevolence, and the proper training of all personnel in the operative departments where day to day decisions are taken could be the most effective solution. Staff would then be able to deploy all modern techniques themselves and, of equal importance, they would appreciate what their colleagues in other departments were trying to do, and so cooperate effectively. Good education will provide the tools. Then there is a reasonable chance that management policy can readily be appreciated, assimilated and implemented.

*Exercise 13.1*

How far do you think that the quality manager should be entitled to impose his ideas on a design team?

Ensuring that appropriately qualified and experienced staff are appointed is undoubtedly management's first responsibility, because if that fails all other actions are likely to be little more than cosmetic. The second responsibility must be to establish good communications throughout the organisation. The necessity for good communications has been emphasised over and over again at every reliability conference attended by the author. One can only surmise that actual progress is limited, otherwise we should not witness repeated calls for this essential feature. We are not, in this book, going to draw the numerous line management diagrams which grace so many reliability manuals, though we shall look at their background logic. Such diagrams must exist but, as already pointed out, have essentially to be peculiar to each organisation. However, it must be appreciated that lack of communication goes much deeper than management structure. Paperwork and committee work is not communication. The lack of communication with which we are concerned is epitomised by the inability of many good practising engineers and many good statisticians to talk together: they speak a different language. They are often intolerant of each other's approach. Senior management must bridge this gap: they must establish the common goal towards which all should be working – they must be able to talk all 'languages'! Then they can turn to organisation.

The basic management logic which should underlie any organisation is set out in figure 13.1, (adapted from ref. 78). The central figure in this connection is the Quality Manager, who must oversee all the various activities indicated in the diagram. How this is to be achieved in detail is special to each organisation, but the essential areas indicated in figure 13.1 must be part of his remit. Before he can tackle any of the other activities his first priority is to formulate the Quality Policy. It is generally agreed that management must take a firm hold of the complete quality process by issuing and committing itself to a policy statement. This must state clearly what priority is to be attributed to quality and reliability: for example, whether the ultimate in performance is to be sacrificed for reliability. The policy statement must also define the strategy to be adopted in sufficient detail to enable each member of the organisation to see the part he is expected to

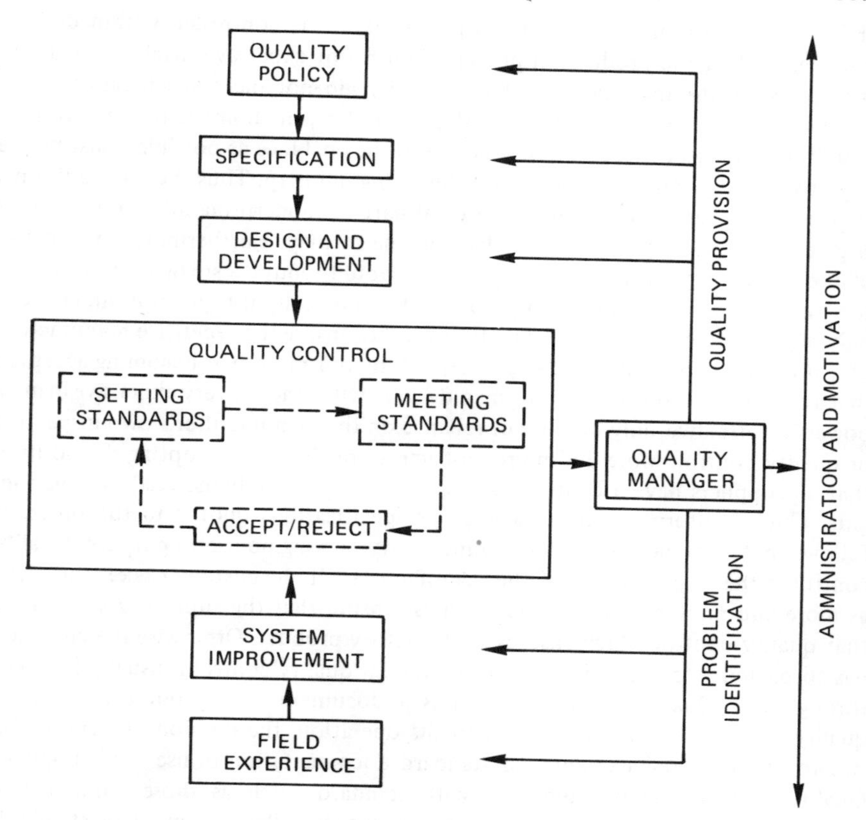


**Figure 13.1** Quality management logic

play and be clear about his role and responsibilities. The requirements under Specification, Design and Development, Field Experience and System Improvement have already been covered by previous chapters of this book and need no further comment at this point. However, the activities under the block headed Quality Control in figure 13.1 have not been treated because this is a well-established subject and adequate texts already exist. The subject is usually covered under workshop management. At the heart of any traditional quality control system was the inspector. He undoubtedly had a part to play, ensuring that the materials (a limited range) met specification, and that items were made to drawings and so on. But nowadays we are faced with a very much larger range of specialised materials, both metals and non-metals, which often have to be prepared and processed under scientifically controlled conditions. The specification is so tight that it is much easier to control the manufacturing process than to inspect the final product. Under these circumstances the inspection would amount to little more than a duplication of the manufacturing process. Likewise basic parts are made on automatic machines whose variation is controlled within defined limits as part of the production process. Control of expensive machines in a way that prevents the manufacture of parts which do not meet specification is far more effective than a post mortem inspect and reject. Immediate correction of any failure to meet specification 'as close to the coal face as possible' must be the objective of any capital-intensive manufacturing industry. Thus the old traditional inspection process has become an integral part of modern manufacture and to a large extent the classical inspector has no role to play. Furthermore, on account of the complexity of modern equipment it would require a small army of inspectors to carry out this function rigorously. Economically it is just not on. One can only rely on the manufacturer's quality control to ensure the repetitive manufacture of a proven component or sub-assembly. More and more it is becoming accepted that a customer's inspection during manufacture will either be very expensive in manpower, or effective only because of reliance on the manufacturer's own inspectors or quality control. More and more customers are therefore adopting the position that all suppliers have a contractual responsibility to supply materials or components which conform to the drawings or specifications quoted in the order. It follows that customer inspection is unnecessary, serving no useful purpose. Quality control is the responsibility of the manufacturer. If the customer takes this view, as more and more are doing, then he must ensure that the supplier *does* exercise that quality control which ensures a satisfactory product. Otherwise the customer has abrogated all control. This check on product quality control is usually achieved through the 'Quality Manual', which is a document setting out the supplier's quality system, the rules laid down for its operation, the division of responsibilities for quality, and examples of standard documentation in use. The supplier's Quality Manual can be compared with standards such as those contained in refs 79, 80 and 81. The minimum requirements called for in these standards ensure that

1. The management structure is clearly defined with lines of communication and responsibilities properly documented.
2. Every level of supervision below management fully understands the management policy and, in particular, what is expected of local control – and is technically qualified to implement that policy.
3. A self-audit procedure is established to ensure that the organisation is acting as described.
4. A quality head is appointed with clearly defined responsibilities.
5. Documentation, issued for control of work and disciplines, is itself under formal issue control.
6. Sub-contractors and suppliers are from an approved list and subject to similar control as that of the prime contractor.
7. Instructions to personnel are in written unambiguous terms.
8. Cost-effective corrective action is formally undertaken in all situations which could affect the quality of the delivered product.

9. All inspection and test equipment used to measure conformance to a customer requirement is calibrated against standards traceable to National Standards.
10. Adequate documented control of hardware and raw materials is instituted so that no doubt exists at any time as to the status of such hardware or material.
11. Adequate records are kept of all transactions affecting the delivered product.
12. Systematic Design Reviews take place during the design phase of contracts, with documented control of ensuing actions.
13. Proper records of development programme planning, execution, evaluation and remedial action are kept.
14. The procedure for agreeing specification changes with the customer is defined.

The implementation and efficient working of the total system as set out in the Quality Manual is the responsibility of the Quality Manager and his staff.

It would be wise to recognise that supplier-based or manufacturer-based quality control has been pioneered by big concerns: the Armed Forces, public utilities, etc. But what is sauce for the goose is sauce for the gander, and in the fullness of time the system could percolate through the whole of industry. Indeed it must do if most large concerns follow this plan, for the prime contractor must ensure that all sub-contractors are following the same procedure and so on. The value of the Quality Manual is based on the belief that well-documented control procedures and adherence to them will automatically build in the consistency which is a hallmark of quality and reliability. We may cast first doubts here by noting that such a system will also ensure the consistency of an ill-advised specification or design. The more effective the quality control, the more effectively are misdeeds perpetuated! Old Joe on the shop floor can no longer put things right. Furthermore, it is a bureaucratic solution emanating from bureaucracies. The preparation of a Quality Manual is no small task and involves high-level technical and managerial staff. It has to be followed by audit and validation. A medium sized firm may be audited by a few dozen customers and may in turn have to audit even more suppliers, although attempts are being made to set up national standards in many countries. There is great scope for bureaucracy here. The small army of inspectors referred to earlier, who at least had some technical background and contact with the product, could be replaced by an even larger army of bureaucrats, who have no technical 'nous' at all. Since most managers are bureaucrats one can expect the system to thrive, but one doubts if over-expansion will contribute anything to reliability. A very firm hand is necessary to ensure that the total quality assurance programme is an initially well-balanced one and continues to stay so. The system is a technical necessity and must serve that end with a minimum of administrative effort and interference. In setting the balance we should recall that a critical appraisal of actual product failures of a general mechanical engineering nature in countries ranging from the U.S.A. through Europe to Japan, which between them practise almost every imaginable variation of management, labour relations and productivity incentives, shows one consistent figure (see, for example, Nixon's book detailed in the Reading list). Only about 15 per cent of these in-plant faults stem from errors on the factory floor or from haphazard inspection. So manufacture and workshop quality control are not the villains of the piece. The Quality Audit should look elsewhere.

From a reliability standpoint it should be noted that the newer forms of quality control are more diffuse than the older ones. Exercise 10.6 shows how variability can escalate in multiple processes. One must therefore expect that the strength distribution will have a more substantial tail than that resulting from old style techniques, which were applied at the end point and therefore tended to produce a distribution with a cut-off in the tail. From our earlier discussion of tails, it seems that modern techniques are likely to make the achievement of a specified reliability or the estimation of the reliability of an existing product that much more difficult.

A less well-defined responsibility of the Quality Manager is motivation, and the promotion of a quality conscious attitude throughout the organisation. The author believes that motivation is something which is peculiar to the special circumstances and environment which surround any organisation. The Quality Manager has to select the most appropriate techniques. What is good for one organisation is ineffective in another. How often has this lesson to be learned in reliability work? Thus Quality Circles, for example, may fit the Japanese mentality, but do they fit the British? or the American? What happened to 'Zero Defects' and 'Right First Time'? No doubt some ideas to promote motivation which are generated both in this country and abroad are most valuable, but the general principles underlying their formation and implementation require a study in industrial psychology, which is not a subject for treatment in this book. What is not in doubt, but is relevant to the subject matter of this book, is that everyone from Board Room to Shop Floor must be fully motivated, in particular they must 'communicate', if high levels of reliability are to be achieved.

While questions of organisation appear paramount in many managerial discussions, the question of warranties often appears to be swept aside. There is, perhaps, a good reason for this in that nearly all the user organisations contributing to reliability studies and techniques are in fact large-scale users. Provided they can be satisfied with the average quality of a product, its variation is of little concern because, if some items fail dismally to reach the average, others do much better. Both customer and producer rely on 'swings and roundabouts' to see justice is done. Little cognisance need be taken of the wide disparity of life that may exist between products to the same nominal design. As we have seen, that disparity can be very wide and is no reflection on the competence of the producer. It is a fundamental feature of the wear process arising from distributed loads and strengths. However, to the small-scale user, and more particularly to the one-off user (that is, for the general run of consumer items) it matters a great deal if the one product he/she has purchased is a very good or a very bad one – the average is irrelevant if that is not the one purchased. Some time ago warranties were thought to be the solution to this problem. However, little progress appears to have been made. The reason seems to be that although there is a known inherent disparity in product lives (though the full extent seems not to be recognised), there is a superimposed substantial variation because of the way the product is used (that is, the loading roughness). Producers are afraid, with some justification, that disparity in user standards swamps all other variations. It does appear that in those cases where a warranty scheme has been tried to any appreciable extent, it has often come to grief over a definition of the loading environment. The 'your fault not mine' syndrome comes into operation. Note that it is the long-term warranty, guaranteeing maintenance costs over an extended period, that is being discussed: not the more usual short-term guarantee which the manufacturer makes sure only covers the pre-wear-out period. Some try to argue that most products consists of a reasonably large number of parts so that 'swings and roundabouts' does operate to some extent for the one-off purchaser, but this would seem akin to the argument that it is the failure rate and not the nature of the failure that matters. Another view is that with reasonable luck the purchaser who received a very short-lived product the first time will receive one of the longer-lived ones as replacement. It does not always work out that way! Manufacturers have 'got away with it' so far because few consumer groups realise the very wide disparity that can exist between the shortest and longest lived products, and the consequent injustice, through no fault of either the manufacturer or customer, that may be incurred. Warranties do seem the solution in spite of the difficulties, and management could well give thought to the method in which they must be framed while time is still available.

The management of mechanical reliability in many organisations does not impress one with its competence. The reason is not far to seek and has been stated on numerous occasions. Too much management is non-technical and cannot understand the factors involved. The trouble spots are easy to identify: management based on legal and financial backgrounds who care little for the product; the concentration on the short-term at the expense of the long-term development, and an inability to deal with the consequences; concentration on the 'broad view' when the product is being let down in the detail; lack of stability as a result of adopting the latest management fashions; and so on. During the author's life time, managers have pursued their goals by refusing to specialise with ever-increasing vigour: understanding less and less about more and more until the modern criterion of a good manager seems to be one who knows virtually nothing about everything! Staff at all levels soon comprehend. Disillusionment and frustration follow throughout the whole organisation, and quality and reliability 'fly out of the window'.

In concluding this chapter, then, we must revert to the prime role of management. While management may decide overall policies, those policies can only be implemented if its staff are capable and willing to implement them, and if management has created those conditions in which staff can work effectively towards their goals. This is particularly true of modern quality assurance which places emphasis on the individual. The old 'long stop' of a final accept or reject inspection is fast disappearing if not gone. Lord Hilton laid particular emphasis on the working atmosphere when he characteristically wrote, "The art of management is to create conditions in which staff can work with the assurance that they are not wasting their time. When a member of staff is able to say that, if the Board or his boss or someone else had given him better direction, he would have been saved time and effort, then that management is bad." To accompany the right environment we need the right individuals, for it is upon individuals that reliability ultimately depends. Summing up a very early conference on reliability, Hayes stated: ' human failing is the greatest source of unreliability, whether the failing is in the board room, the office, or the shop floor. Personal integrity and responsibility are vital." Management's prime duty is to provide the necessary leadership and inspiration, and to support staff who are capable of achieving the objective.

It is not possible to set concise exercises on 'Management' as has been done for the more technical aspects of this book. Because the subject is so wide ranging, it would be necessary to provide the complete background against which each problem would have to be solved. This cannot be done in the confines of a small book, since each exercise would become a case study – the conventional way of teaching management. Instead a number of discussion topics are given below, which you should examine against the background of your own organisation. No 'solutions' are given: there are no unique ones.

1. What is the status of designers in your organisation? Why isn't it higher?
2. Examine the accountancy system in your organisation. Is it possible to isolate the cost of unreliability and the cost of maintenance to your concern? If not, why not?
3. Have Quality Circles any role to play in your industry? If so, how do you think they should be adapted to fit the British/American temperament in general, and your own organisation in particular?
4. What are the outward and visible signs of QA that you would expect to find in any reliability conscious organisation?
5. Find out how the lower management levels of your organisation regard the achievement of reliability and customer satisfaction. Do the statements of those concerned agree with the actual practices under their control?
6. Discuss the difficulties involved in guaranteeing a whole life cost.
7. How is the feedback of 'in-service' data achieved in your organisation? Is it effective?

# 14 Concluding remarks

The study of 'reliability' in this book has been aimed at general mechanical equipment. The intention has been to present a theoretical background to the subject and build on that to more general aspects. The reliability of electronic and electrical equipment has been specifically excluded, but because that field has been so fully covered and because so much progress has been made, and because the achievements in that field have often been held up as an example to mechanical engineers, comparisons between the two topics have been made whenever appropriate. Thus it is hoped that the reader will have become aware of substantial differences in application, though the underlying theory is exactly the same. There can be little doubt that mechanical reliability is a much more complicated subject. This arises partly from the much greater diversity of mechanical products, and from the far greater range of operating conditions they can be expected to encounter: but it is also in a large way due to the preponderance of wear-out phenomena. Wear-out is a complicated subject, markedly non-linear, and totally inadequately understood and therefore modelled. Although a lot of good work has been done in this field, it has concentrated on the initial stages of the process, to reduce wear and hence prolong life: the life of intrinsic reliability in the phraseology of this book. For firm believers in intrinsic reliability this is the right priority. It does mean, however, that very little is known of the final stages and what controls the wear-out failure pattern. This field is one of almost total empiricism. Yet it is surprising how often estimates based on empirical information and sound practical experience can be used to great practical purpose and advantage, particularly when those making the estimates fully appreciate all the basic mechanisms and factors involved and how far to push empiricism. As a consequence we find that mechanical reliability is very high indeed – far higher than many of its detractors would have us believe, especially when not deliberately sacrificed for performance and economy. Much equipment, properly used, goes on for ever and ever, and is still durable when out-moded. Maybe we have not made great advances in mechanical reliability during the last decade or so – but then, we did not start from the thermionic valve!

Difficulties in putting a firm numerical value on mechanical reliabilities are compounded by maintenance procedures. In essence they control wear and retain it within acceptable bounds; but the processes themselves are so varied that any unified body of knowledge is difficult to establish. It is in such circumstances as these that a full understanding of the fundamentals is so important. It enables the practitioner to appreciate the major factors controlling any particular case. That has been my aim in writing this book. It is not a trouble-shooting handbook, neither does it set out standard procedures. For that purpose one must refer to specific manuals.

What I have set out to do is develop, in the first place, a simple mathematical model to represent the major physical factors which play a part in determining the reliability of mechanical equipment. It is a simple model. It would be totally wrong to introduce more complicated models unless they more closely represented reality. The problem is that we do not know what reality is: we do not know the shapes of the tails of load and strength distributions – only that reliability is critically dependent on them; most often we do not fully understand the mechanism of wear-out and the factors which control it – we certainly cannot quantify the complete wear–life failure pattern; we seem to know even less about fatigue; we do not know why there are such large discrepancies in repair times; and so the list grows. Furthermore there are so many physically different wear mechanisms that we need many models. Then perhaps we could start to think about interactive wear which is so often the real problem. Until such knowledge is available, simplicity must offer the best approach. Consequently it would be idle to pretend that the model followed in this book is all embracing, though there appears to be substantial field evidence to support it in general terms. But there will be many perturbations, some large perturbations, on the general conclusions reached in the preceding chapters when particular cases are examined. Some may be explainable via emphasis on special aspects which distort the general picture in particular circumstances. Some may be inexplicable. Whatever the situation may be, however, a full understanding of the basic factors at the lowest level *must* increase the prospects of solving a particular problem. The book has thus concentrated on the primary causes of mechanical failure as the ultimate factor controlling mechanical reliability, and on defining the measures we have to take to safeguard against any adverse effects. Although many current approaches attempt to deal with this subject without understanding the fundamental processes of mechanical failure, it is believed that in the end the optimum solution can only be reached when the nature of the failure–life relationship is fully understood.

No attempt has been made to study the field of functional reliability, which can be very important in some applications. By functional reliability is meant, for example, the failure of a control mechanism to take into account a particular set of circumstances, even though no mechanical failure was involved. As such, this aspect is rightly outside the scope of this book. But the picture does become a bit blurred when the control mechanism is itself required to safeguard against a primary mechanical failure. The number of possibilities is vast, and to incorporate a treatment of this subject would have unacceptably increased the size of the book, and introduced a second distracting theme.

For the same reason no attempt has been made to incorporate the effects of human factors on reliability. They do have a big part to play – there is such a thing as the 'Friday afternoon car'. On the other hand, the shop floor is not the villain of the piece; and with increasing automation, which locks up considerable capital sums, one may expect it to decrease in relative importance in future.

One could not, however, study reliability without reference to economics, although this is another vast ancillary subject. Economics has therefore been introduced, but only as essentially required and at the simplest level. Those wishing to pursue the matter further are referred to the specialised books (and the further references they contain) given in the Reading list. Cynics may well wonder whether values based on projected mechanical failure rates and projected inflation rates are worth any consideration! But some broad trends may be discerned.

One cannot leave the subject of mechanical reliability without some comment on the respective roles of engineers and statisticians. The author firmly believes that in the first place mechanical engineers must recognise that virtually every quantity that they deal with is distributed. It is no longer good enough to use deterministic methods with empirical reserve factors. Statistical methods are available to deal with distributed quantities and they should be used. This would eliminate much of the empirical morass that has grown up around many topics and allow more rational progress to be made. The lead must come from the universities, where still far too much basic engineering is taught deterministically. If the engineer must leave his deterministic make-believe world, the statistician must also leave his make-believe world. The constant failure rate regime has little part to play in mechanical reliability. The continued reliance on constant failure rate mathematics has done more to impede progress in mechanical reliability than anything else. Wear-out and non-constant rates are dominant. Glib representations that are devised more for mathematical tractability than physical representation must give way to realism. This implies that statisticians must understand the physical processes, and this often means collaborating with those who have often spent most of their lives studying some wear mechanism, and have a strong 'feel' for what is involved. The limitations of statistics in these circumstances must be recognised, as must also the fact that experience can often lead to the optimum solution without mathematical representation. As Dorfman has stated "if an operation is repetitive, well recorded, intended to serve a well defined goal, and is of the kind in which the influence of decisions on the attainment of the goal do not transcend technical knowledge (that is, the ideal situation for an operational researcher) then it is not unlikely that current practice will be far short of the most efficient practice attainable." (Is the level of unscheduled fatigue failures an example?) One must beware of concentration on the 'trivial many', which are more amenable to mathematical treatment because they behave as a large random population and therefore attract attention. It is the 'vital few' that are the source of trouble.

If I were asked what factors contributed most to a high-quality product, that is, one which satisfied the demands of the customer and which achieved the lowest whole life cost without sacrificing any specific performance objective, then I would say that the first requirement must be a management which at the highest levels really wanted that quality, and created the ambience in which it could be achieved. Without the will to achieve success, little is possible. Given that will, the concensus of informed opinion turns to design as the key activity. The foundation of high quality is laid by the initial design and if that is far off optimum there is very little that can be done subsequently, however well directed and intentioned – except at very high cost. With a bascially good design team the problem is most often an inadequate knowledge of the loading conditions, and the method of filling that gap is good communications between the designer/manufacturer and the user. However, there is an awful lot of bad design about, which may not be surprising in view of the understanding and experience necessary to apply successfully the current empiricism, compared with the lack of status and financial reward associated with this activity. I believe the role of maintenance in achieving mechanical reliability is grossly under-estimated. Even so, the maintenance policy and the means for carrying out maintenance (to minimise inefficiency and excessive effort) must be set at the design stage. Once the design is finalised it is too late and too expensive to seek other palliatives. My key technological area for attention in order to achieve high quality via optimum reliability is, then, design backed up by adequate development.

Finally I must refute, because I believe this is inherent in achieving our goal, the widespread belief that work in the quality and reliability fields is intellectually of second class status. It is a belief which drives many good people away from a fascinating and demanding subject, in which there is still a lot to learn, much of a truly fundamental nature such as the physics of the wear processes, and in which the rewards could be very rich indeed. It requires a very broad approach and an integrated knowledge of all the subjects of mechanical engineering, a great deal of common sense, and an appreciation of real life, plus a sound command of statistics. A working knowledge of economics also helps! Unfortunately, the necessary breadth cannot be an excuse for lack of depth. As we have seen so often it is, in the last resort, the detail in relation to the whole system that can bring disaster. The lines beginning "for want of a nail . . ." have become hackneyed beyond all measure, but even that stands as evidence, well over three hundred years after they were first written, to the value of the sentiments they express.

# 15 Solutions to exercises

*Exercise 3.1*

The suggested distributions are given in figure 15.1, (a), (b), (c), (d) and (e), but the following notes give the basis behind the choice.

(a), (b), and (c) form a related series in which the item is made of a natural material. Most natural materials have a fairly wide strength range. The loads applied may in all cases also vary over a wide range. Tennis and squash racquets are reasonably heavily stressed and a chance of breakage is accepted, but this is kept as low as possible, commensurate with other properties. The strength of a lead pencil depends to some extent on the sharpening, but since failure is not serious and only means simple resharpening, a fairly high risk of failure can be accepted; a reduced value of the safety margin is therefore permissible. In the case of a ladder we have again a natural material with a wide distribution of strength which is required to carry a wide range of loads. Failure in this case is, however, totally unacceptable and reliability is achieved by using a wide safety margin. Note the contrast between (a), (b) and (c).

The grandfather clock of (d) is driven by gravity and has the most certain load of all the examples quoted. The strength is distributed over a smaller range than (a), (b) or (c) since it is made of manufactured material, but a finite range does exist. In comparison the load on a car braking system may vary from light to heavy and is therefore widely distributed as in (e). Although the mechanical strength of the braking system may be carefully controlled, road adhesion is a varying quantity. A wide distribution of 'strength' may therefore be expected. This results in an only moderate reliability, although safety is involved (compare with (c)). In this case some reliance is placed on the operator to choose matching conditions (for example, not to brake heavily on icy roads).

All the above distributions have been sketched as roughly normal. What would you expect?

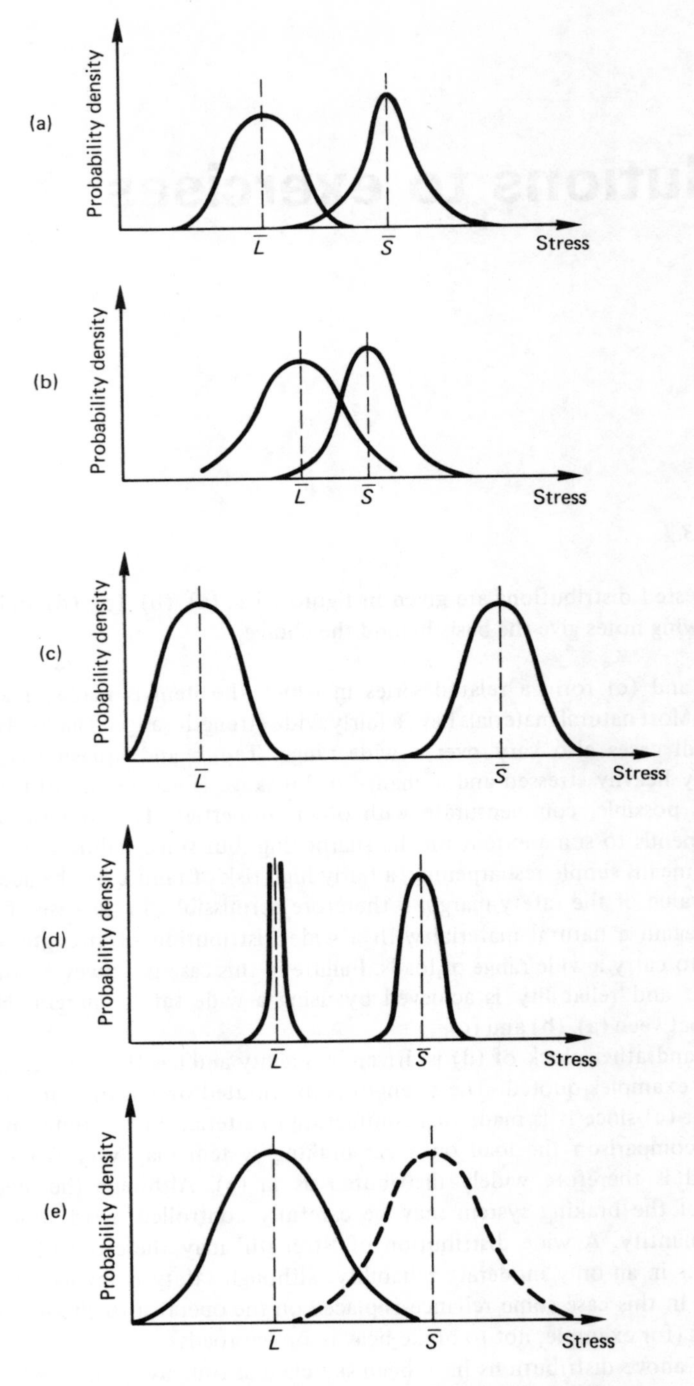


**Figure 15.1** Distributions of loads and strengths

*Supplementary exercise 3.1A*

What distribution of ultimate strength would you expect of a carbon steel whose nominal strength is quoted as 600 MN $m^2$? How would you expect the distribution of strength of a component made from this material and subject to simple tensile loading to compare with the basic material distribution?
(Some comments on this exercise are given at the end of this chapter.)

*Exercise 3.2*

Most statistical textbooks contain some comment to the effect that real values of a variate cannot extend from $-\infty$ to $+\infty$ as so many continuous mathematical distributions do, and in particular a negative value of a variate may be meaningless, for example, if the variate were the height of men. This contradiction is usually explained by stating that the theoretical distributions are approximate, and that the mathematical equation is only valid over defined limits. It is then claimed that the approximation is very good for the bulk of the population. It would appear that the authors of statistical textbooks consider the tails of the distributions to be of little significance since they contain so small a proportion of the total population. However, it will be appreciated from exercise 3.1 that reliability, or lack of reliability, arises from the interaction of extremely high loads and extremely low strengths, that is, those regions represented by the tails of the loads and strength distributions. One would imagine that the correct representation of the tails is all important in this application. It can be of little practical value for the purpose in hand to represent the bulk of a population by, say, a normal distribution, justifying the representation by standard goodness of fit tests, if it leaves representation of the tails unresolved. Not all distributions have infinite tails, of course, but they are all abstract mathematical equations having no physical significance. Extreme value distributions have been introduced by statisticians to deal with tails, but they can in no way infer the shape of the tail from the bulk distribution. They too are mathematical functions having no physical significance.

It must be recognised that there is no reason whatsoever why the items in the tail of a real population should follow any theoretical distribution. Statistical distributions are not derived from any physical base. Hence it is not possible to fit a distribution to the bulk of the population and expect it to represent the tail. The only way to represent a tail is to fit the distribution to that tail. Tails vary so much that no assumptions as to their nature can be made. This arises from the multitude of factors that can influence them. Regarding the strength distribution: this can arise from quality control, which can be applied at any of the numerous stages of manufacture, and which itself is not perfect and is distributed; it can arise from the mixing of batches, from peculiarities in the manufacturing process, as well as from inherent departures from standard and so on. Reference to figure 3.3 will show the kind of peculiarity which may be encountered and which it is so difficult to represent. One may expect load distributions to yield an even wider range of variations, being often subject to human influence yet sometimes subject to mechanical control (for example governing). All such distributions can only be determined physically by measurement.

Unfortunately, the experimental determination of the tail distribution is prohibitively expensive. We are looking essentially for the 'one in ten thousand' chance, and to determine that at any worthwhile level of confidence needs a lot of testing and measurement. It has to be done for all strength attributes and for all forms of loading. Furthermore, generalisation is not possible since the physical factors controlling the extreme values are largely unknown. Each item and each load has to be investigated separately. There is not the slightest chance of such work being carried out on even a remotely comprehensive basis in the foreseeable future: hence it must be recognised that standard distributions are unlikely to fit tails of real distributions and that real tails are effectively indeterminate. Any practically useful theory of reliability must acknowledge this fact.

*Exercise 3.3*

(a) The couplings of a goods truck will be subject to reasonably rough loading, both as a result of the steady load itself, which will depend on the train load, and adventitious shock loads. It is reasonable to assume that the conditions approximate to theoretically very rough conditions in which the reliability of the series system is equal to the reliability of the component.
Hence, reliability of train at 100 000 miles = 99.9 per cent.
It should be noted that real conditions cannot be ***infinitely*** rough and so a slightly lower reliability may in fact be experienced.

(b) Being part of a base load generator, the loading conditions on the turbine stages will be reasonably smooth. It follows that the product rule may be considered valid for this application.
Hence reliability of turbine = $(0.9999)^{10}$ = 0.999 = 99.9 per cent.
One can hope that since the conditions will not correspond to the ideally smooth situation a slightly higher reliability will in fact be achieved, but it is impossible to estimate this without being given the detailed load and strength distributions.

*Exercise 3.4*

The contradiction is based on the indiscriminate use of the product rule. If the load is widely distributed the reliability of the system will be that of its 'weakest link', that is, the reliability will be 90 per cent in the first example quoted, and 92 per cent in the second. Concentrating on these components, and increasing their reliability substantially, will increase the system reliability of the first example to 92 per cent and the second to 93 per cent.

If the load is narrowly distributed the product rule holds, as in the equations quoted. Improving one of the 99 per cent reliable elements to 100 per cent will only raise the system reliability to 30.4 per cent, that is, 0.4 per cent: improving the weakest element to 100 per cent will raise the reliability to 32.6 per cent, that is, by 2.6 per cent. If, however, all the 97 elements are identical, giving rise to the same failure, then eliminating this will raise the system reliability to 80 per cent, but this is a very special case of 97 identical components subject to smooth loading. It is not representative of mechanical systems. There is in fact no contradiction if the loading for each system is specified, and the system itself is adequately specified. The appropriate system reliability can then be calculated. As a general practical rule the best result is obtained by eliminating the identical failures which lead to the largest reliability deficit, whatever the loading conditions.

*Exercise 3.5*

Reliability can be written as

$$R = \left\{\begin{array}{c}\text{Probability that}\\ \text{component has}\\ \text{certain strength}\end{array}\right\} \times \left\{\begin{array}{c}\text{Probability that}\\ \text{load has}\\ \text{smaller value}\end{array}\right\}$$

or as

$$R = \left\{\begin{array}{c}\text{Probability that}\\ \text{load has}\\ \text{certain value}\end{array}\right\} \times \left\{\begin{array}{c}\text{Probability that}\\ \text{component has}\\ \text{greater strength}\end{array}\right\}$$

Equation (3.10) was derived from the first expression. Using the second and proceeding as before

$$R_{s_0} = \Big(L(s_0)\,\mathrm{d}s\Big) \times \left(\int_{s_0}^{\infty} S(s)\,\mathrm{d}s\right)$$

$$R = \int_0^{\infty} \left(L(s) \int_s^{\infty} S(s)\,\mathrm{d}s\right) \mathrm{d}s$$

*Exercise 3.6*

If the modes of failure and their causes are truly independent we can use the product rule to write

$$R_o = R_1 \times R_2 \times R_3 \times \ldots \times R_n$$

where

$R_o$ = overall reliability

$R_1$ to $R_n$ = reliability with respect to each of the $n$ nodes of failure.

Taking logarithms

$$\log R_o = \sum_{i=1}^{i=n} \log R_i$$

Differentiating with respect to $t$ gives

$$\frac{1}{R_o}\frac{dR_o}{dt} = \sum_{i=1}^{i=n} \frac{1}{R_i}\frac{dR_i}{dt}$$

or

$$\lambda_o = \sum_{i=1}^{i=n} \lambda_i$$

It should be noted that the conditions postulated are rarely satisfied in practice, but for failure modes which are grossly dissimilar it is often possible to assume the conditions as a reasonable approximation. If the conditions are not satisfied the product rule does not apply, so that the application of the theoretical equations is a matter of physically identifying dependent and independent modes.

*Exercise 3.7*

Probability of a critical dimension being incorrect $= \alpha$
Probability that such dimension is accepted $= P_1$
Probability of acceptance of one incorrect dimension $= \alpha P_1$
Probability that either of two critical incorrect dimensions is accepted $= 2\alpha P_1$

Probability that both critical dimensions are correct $= (1 - 2\alpha)$
Probability of passing gauge 1 $= (1 - P_2)$
Probability of passing gauge 2 $= (1 - P_2)$
Probability of passing gauge 1 and gauge 2 $= (1 - P_2)^2$
$= 1 - 2P_2 + P_2^2$

Probability of not passing gauge 1 and gauge 2 $= 2P_2 - P_2^2$

Probability of correct part not passing gauge 1 and gauge 2 $= (1 - 2\alpha)(2P_2 - P_2^2)$
Hence probability of failure $= 2\alpha P_1 + (1 - 2\alpha) P_2 (2 - P_2)$
Reliability $= 1 - 2\alpha P_1 - (1 - 2\alpha) P_2 (2 - P_2)$

*Exercise 3.8*

Applying the product rule the probability of system success where
(1) engines 1, 2, 3 and 4 survive $= R^4$
(2) engines 1, 2 and 3 survive and 4 fails $= R^3(1 - R)$

(3) engines 1, 2 and 4 survive and 3 fails $= R^3(1 - R)$
(4) engines 1, 3 and 4 survive and 2 fails $= R^3(1 - R)$
(5) engines 2, 3 and 4 survive and 1 fails $= R^3(1 - R)$
In all other circumstances the system fails. Hence

$$\text{Reliability} = \text{Probability of (1), (2), (3), (4) or (5)}$$
$$= R^4 + 4R^3(1 - R)$$

The validity of the expression depends, of course, entirely on the validity of the product rule. It is extremely unlikely that the four engines can be completely independent, although designers attempt as much independence as possible in order to achieve a higher reliability of the propulsive system as a whole; for example, 99 per cent engine reliability leads to 99.94 per cent system reliability in the above example. It is very unlikely, however, that the theoretical figure could be achieved, owing to lack of independence – even if it is only a common maintenance team. It is fascinating to the author that when A. A. Griffith proposed replacing large engines by a greater number of smaller engines for performance reasons (they have a higher overall thrust/weight ratio by virtue of the square-cube law) one of the major criticisms raised by practical engineers was the loss of reliability. Either the practitioners or the theoreticians were wrong! The author knows where he would put his money.

*Exercise 3.9*

The equivalent of equation (3.16) for a parallel system of $i$ components is

$$P_n(s > s_0) = 1 - \prod_{i=1}^{i=n} \left[1 - \int_{s_0}^{\infty} S_i(s)\,\mathrm{d}s\right]$$

whence the equivalent of equation (3.18) becomes

$$R = \int_0^{\infty} L(s) \left\{1 - \prod_{i=1}^{i=n} \left[1 - \int_s^{\infty} S_i(s)\right]\right\} \mathrm{d}s$$

For ideally smooth loading in which $L(s)$ is zero for all values of $s$ except those lying between $L$ and $(L + \mathrm{d}s)$

$$R = 1 - \prod_{i=1}^{i=n} \left[1 - \int_L^{\infty} S(s)\,\mathrm{d}s\right]$$

that is, the probability of a system failure is equal to the product of the probability of failure of the individual components.

For infinitely rough loading

$$\left[1 - \int_s^{\infty} S_i(s)\,\mathrm{d}s\right] = \begin{matrix} 0 \text{ for } s < S_i \\ 1 \text{ for } s > S_i \end{matrix}$$

Hence

$$\prod_{i=1}^{i=n}\left[1 - \int_s^\infty S_i(s)\,ds\right] = \begin{matrix} 0 \text{ for } s < S_{i\,max} \\ 1 \text{ for } s > S_{i\,max} \end{matrix}$$

where $S_{i\,max}$ is the maximum value of the unique strength. Hence

$$R = \int_0^{S_{i\,max}} L(s)\,ds$$

$$= \text{reliability of the most reliable component}$$

Thus for a parallel system in which all the channels are active, the probability of system failure is equal to the product of the probability of failure of the individual channels with ideally smooth loading, but when the loading is infinitely rough the reliability of the system only is equal to the reliability of the most reliable channel.

*Exercise 4.1*

(a) For rough loading $R = e^{-\lambda t}$ and this formula may be applied to the train couplings. Hence

$$0.999 = e^{-\lambda 100\,000}$$

Therefore

$$\lambda = \frac{-\log_e 0.999}{100\,000}$$

$$= 1.000\,5003 \quad 10^{-3}$$

Reliability at 50 000 miles $= e^{-\lambda 50\,000}$

$$= 99.95 \text{ per cent}$$

(b) For the smooth loading conditions of the turbine the reliability is invariant with time. Therefore

Reliability at 25 000 hours = 99.9 per cent

*Exercise 4.2*

We can write the general equation (3.8) as

$$R = \int_0^\infty \phi\,ds$$

where

$$\phi = S(s)\left(\int_0^s L(s)\,\mathrm{d}s\right)^n$$

Using the nomenclature illustrated in figure 4.3 and the values of the probability densities noted on that figure

$$R = \int_0^{s_1} \phi\,\mathrm{d}s + \int_{s_1}^{s_2} \phi\,\mathrm{d}s + \int_{s_2}^{s_3} \phi\,\mathrm{d}s + \int_{s_3}^{s_4} \phi\,\mathrm{d}s + \int_{s_4}^{\infty} \phi\,\mathrm{d}s$$

where

$$\int_0^{s_1} \phi\,\mathrm{d}s = 0$$

$$\int_{s_1}^{s_2} \phi\,\mathrm{d}s = 0$$

$$\begin{aligned}
\int_{s_2}^{s_3} \phi\,\mathrm{d}s &= \int_{s_2}^{s_3} h_\mathrm{s}\left(\int_{s_1}^{s} h_\mathrm{L}\,\mathrm{d}s\right)^n \mathrm{d}s \\
&= \int_{s_2}^{s_3} h_\mathrm{s}\left(h_\mathrm{L}s - h_\mathrm{L}s_1\right)^n \mathrm{d}s \\
&= \frac{h_\mathrm{s}}{n+1}\,\frac{\{(h_\mathrm{L}s - h_\mathrm{L}s)^{n+1} - (h_\mathrm{L}s - h_\mathrm{L}s)^{n+1}\}}{h_\mathrm{L}} \\
&= \frac{h_\mathrm{s}}{n+1}\,\frac{\{1 - (h_\mathrm{L}s - h_\mathrm{L}s)^{n+1}\}}{h_\mathrm{L}}
\end{aligned}$$

$$\begin{aligned}
\int_{s_3}^{s_4} \phi\,\mathrm{d}s &= \int_{s_3}^{s_4} h_\mathrm{s}\{1\}^n\,\mathrm{d}s \\
&= h_\mathrm{s}(s_4 - s_3)
\end{aligned}$$

$$\int_{s_4}^{\infty} \phi\,\mathrm{d}s = 0$$

whence

$$R = \frac{h_\mathrm{s}}{(n+1)}\,\frac{\{1 - (h_\mathrm{L}s_2 - h_\mathrm{L}s_1)^{n+1}\}}{h_\mathrm{L}} + h_\mathrm{s}(s_4 - s_3)$$

Since $(h_\mathrm{L}s_2 - h_\mathrm{L}s_1)$ must be less than unity this term raised to the power $(n + 1)$ decreases rapidly with increasing values of $n$, becoming zero when $n$ is reasonably large. The first term in this equation then decreases by virtue of the term $(n + 1)$

in the denominator, becoming zero when $n$ is very large. The reliability therefore falls steadily, from its initial value of unity, in the form of an early life characteristic. The ultimate reliability is given by the second term in the last equation as

$$R(\infty) = h_s(s_4 - s_3)$$

*Exercise 4.3*

The relevant load and strength distributions are illustrated diagrammatically in figure 15.2. The load distribution is cut off at 10 MN, so that all items whose strength is greater than 10 MN survive and all whose strength is less will ultimately fail. Failures are denoted by the shaded area in figure 15.2. The 10 MN load is $(15 - 10)/2.5 = 2$ standard deviations of strength from the mean strength, and since the strength is normally distributed the area representing failures may be obtained from standard tables as 2.38 per cent. Hence the ultimate reliability = $100 - 2.38 = 97.72$ per cent.

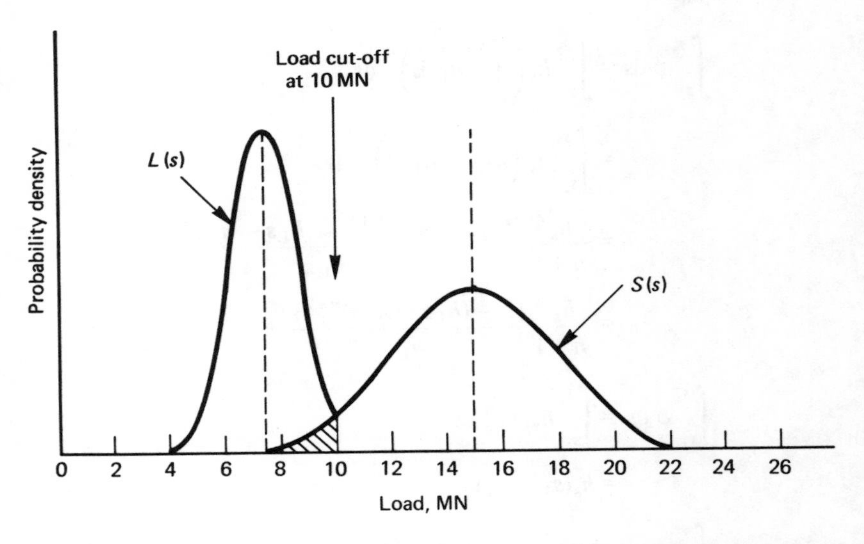


**Figure 15.2** Load and strength distributions for hydraulic installation

*Exercise 4.4*

Quality control action produces a cut-off in the low strength tail of the strength distribution so that no items have a strength below a given value. We may note that the single application of a load of unique value will have the same effect since all items whose strength is less than the value of the load must fail and those whose strength is greater must survive. Mathematically we can therefore represent quality control action by the application of a single defined load (a hypothetical load representing the quality control action) followed by the application of the real load. Equation (4.7) can therefore be rewritten as

$$R(m) = \int_0^\infty \left( S(s)\, \phi \left[ \int_0^s L(s)\,\mathrm{d}s \right]^n \right) \mathrm{d}s$$

where $\phi$ has the value zero for all values of $s$ less than the quality control cut-off value and the value unity of all $s$ in excess of this. Since no quality control can be perfect, a more accurate expression can be obtained by distributing $\phi$ about the cut-off value, that is

$$\phi = \int_0^s \phi(s)\,\mathrm{d}s$$

where $\phi(s)$ is the probability density function of an appropriate distribution representing the quality control process. Its mean value corresponds with the nominal cut-off value.

*Exercise 4.5*

The general solution giving the reliability of an item subject to a single application of load is given by equation (3.8) as

$$R = \int_0^\infty \left( S(s) \left[ \int_0^s L(s)\,\mathrm{d}s \right] \right) \mathrm{d}s$$

The solution giving the reliability of the item subject to $n$ repeated applications of load from the distribution $L(s)$ is given by equation (4.7) as

$$R = \int_0^\infty \left( S(s) \left[ \int_0^s L(s)\,\mathrm{d}s \right]^n \right) \mathrm{d}s$$

This may be rewritten as

$$R = \int_0^\infty \left( S(s) \left[ \int_0^s L(s)\,\mathrm{d}s \right]^{n-1} \left[ \int_0^s L(s)\,\mathrm{d}s \right] \right) \mathrm{d}s$$

$$= \int_0^\infty \left( S'_n(s) \left[ \int_0^s L(s)\,\mathrm{d}s \right] \right) \mathrm{d}s$$

By comparison with equation (3.8) $S'_n(s)$ is seen to be the probability distribution of the strength at the $n$th application of load. Hence

$$S'_n(s) = S(s) \left[ \int_0^s L(s)\,\mathrm{d}s \right]^{n-1}$$

*Exercise 4.6*

The probability of obtaining a component of strength lying between $s_0$ and $(s_0 + ds)$ from the initial population is given by

$$S(s_0)\,ds$$

and the strength of this component will deteriorate with successive applications so that it takes the following values

between $(s_0 - \Delta s_1)$ and $(s_0 + ds - \Delta s_1)$ at application of 2nd load
between $(s_0 - \Delta s_2)$ and $(s_0 + ds - \Delta s_2)$ at application of 3rd load

.        .
.        .
.        .

between $(s_0 - \Delta s_{i-1})$ and $(s_0 + ds - \Delta s_{i-1})$ at application of $i$th load

The probability of encountering a load which is less than the strength, $s_0$, at the initial application of load is

$$\int_0^{s_0} L(s)\,ds$$

and of encountering a load which is less than the strength, $(s_0 - \Delta s_{i-1})$, at the $i$th application of load is

$$\int_0^{(s_0 - \Delta s_{i-1})} L(s)\,ds$$

By the product rule, the probability of encountering a load which is less than the strength of a component in any of $n$ applications is

$$\prod_{i=1}^{i=n} \int_0^{(s_0 - \Delta s_{i-1})} L(s)\,ds$$

Hence the probability of obtaining a component of original strength, $s_0$, and a load which is always less than its deteriorating strength is

$$S(s_0)\,ds \prod_{i=1}^{i=n} \int_0^{s_0 - \Delta s_{i-1}} L(s)\,ds$$

If $s_0$ is now allowed to take any value from 0 to $\infty$ the reliability of a component taken at random from a large initial population, this expression becomes

$$\int_0^{\infty} \left( S(s) \left[ \prod_{i=1}^{i=n} \int_0^{s - \Delta s_{i-1}} L(s)\,ds \right] \right) ds$$

*Exercise 4.7*

The load on the rung of a ladder consists of the person plus any load he is carrying. The mass of the average person is known to be more or less normally distributed with an average value of about 70 kg (11 stone) and a standard deviation of about 16 kg (2.5 stone). The mass of the load is more difficult to assess. One imagines the distribution is skewed since the average payload is very light (paint pots and the like) but occasionally a very heavy load may be lifted. We shall assume that the upper tail of this load can be represented by a normal distribution of mean value 2 kg (4.5 lb) and standard deviation 10 kg (22 lb). Hence the mean mass carried by the ladder will be 72 kg and its standard deviation $\surd(16^2 + 10^2) = 18.9$ kg; that is

$$\left.\begin{aligned} \bar{L} &= 72 \times 9.8 = 705.6 \text{ N} \\ \sigma_L &= 18.9 \times 9.8 = 185.2 \text{ N} \end{aligned}\right\}$$

Also we have

$$\begin{aligned} &\text{Mean stress allowable} &&= 80 \text{ N/mm}^2 \\ &\text{Standard deviation of allowable stress} &&= 15 \text{ N/mm}^2 \end{aligned}$$

It should be noted that the standard deviation of the stress is not the standard deviation of the strength any more than the mean stress is the mean strength. However

$$\frac{\text{Standard deviation of strength}}{\text{Mean strength}} = \frac{\text{Standard deviation of stress}}{\text{Mean stress}}$$

that is

$$\frac{\sigma_s}{\bar{S}} = \frac{15}{80} = 0.188$$

In an application when safety is involved there can be no doubt that the safety margin should be at least equal to that providing intrinsic reliability at the relevant loading roughness. This is therefore taken as the design basis. In that case there are two unknowns ($\sigma_s$ and $\bar{S}$) which are related by the above equation and the empirical relationship is represented by the graph in figure 4.18. We can thus solve for these unknowns but it will be necessary to employ successive approximations on account of the empirical relationship in figure 4.18. To do this write

$$\frac{\bar{S} - \bar{L}}{\surd(\sigma_s^2 + \sigma_L^2)} = x = \text{unknown safety margin}$$

Squaring

$$\begin{aligned} \bar{S}^2 - 2\bar{L}\,\bar{S} + \bar{L}^2 &= x^2 \sigma_s^2 + x^2 \sigma_L^2 \\ &= x^2 \left(\frac{\sigma_s}{\bar{S}}\right)^2 \bar{S}^2 + x^2 \sigma_L^2 \end{aligned}$$

or

$$\left\{1 - x^2 \left(\frac{\sigma_s}{\overline{S}}\right)^2\right\} \overline{S}^2 - 2\overline{L}\,\overline{S} + \{\overline{L}^2 - x^2 \sigma_L{}^2\} = 0$$

Solving for $\overline{S}$ gives

$$\overline{S} = \frac{\overline{L} + \sqrt{\{\overline{L}^2 - (1 - x^2\gamma^2)(\overline{L}^2 - x^2 \sigma_L{}^2)\}}}{(1 - x^2\gamma^2)}$$

where the positive sign of the square root has been taken as that relevant to this problem. We have evaluated $\gamma = \sigma_s/\overline{S} = 0.188$, $\overline{L} = 705.6$ and $\sigma_L = 185.2$; hence to solve we give $x$ some rounded mid-range value, say 4, from which $\overline{S}$ is calculated as 3283.6 and $\sigma_s$ as 617.3. Then

$$\text{loading roughness} = \frac{185.2}{\sqrt{(185.2^2 + 617.3^2)}} = 0.287$$

From figure 4.18 the corresponding value of $x$ is 4.2, which differs from the assumed value of 4.0. We therefore guess a new value for $x$ (= 4.1 say) and repeat the calculation to obtain $x = 4.15$ and again to obtain $x = 4.13$. The corresponding value of $\overline{S}$ is 3614 and the loading roughness is 0.263 which is a close enough approximation to the value corresponding with $x = 4.13$ on figure 4.18. The rung must therefore be designed for a mean strength of 3614 Newtons. The remainder of the exercise is a straightforward problem in strength of materials and is given in outline only.

It is assumed that the rung is *encastré* at each end in the stiles of the ladder and that the rung is of constant circular cross-section as illustrated in figure 15.3. The

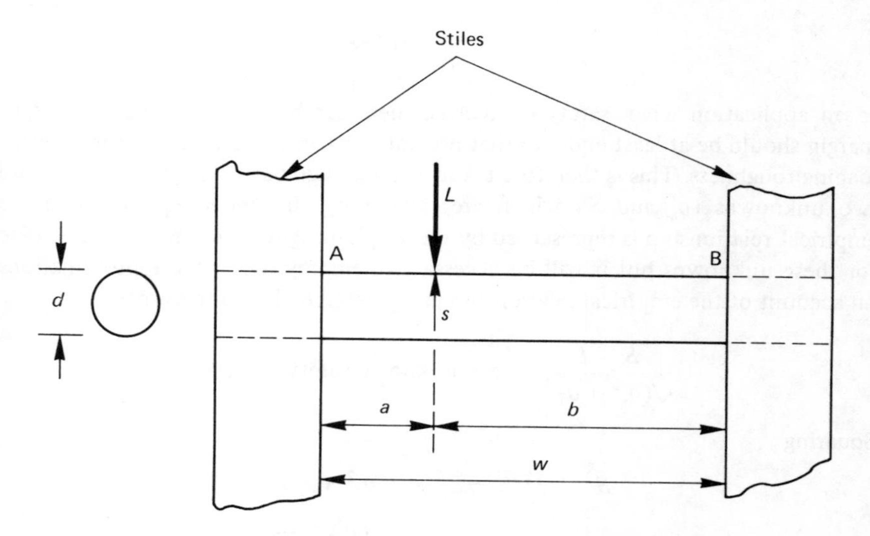


**Figure 15.3** Notation for rung of ladder

maximum bending stress in the rung occurs at A when $a < b$ and at B when $b < a$. Treating the case shown, the bending stress at A to provide a mean strength $\bar{S}$ is given by

$$\frac{\bar{S}\, a\, b^2}{w^2 Z}$$

where $Z$ is the modulus of the rung section = $\dfrac{\pi d^3}{32}$.

The stress at A is

$$\frac{\bar{S}\, a(w - a)^2}{w^2 Z}$$

Now

$$\frac{\mathrm{d}}{\mathrm{d}a}\{a(w - a)^2\} = w^2 - 4wa + 3a^2$$

$$= 0 \text{ when stress at A is maximum}$$

It follows that the maximum stress at A occurs when $a = w/3$. This is just about the point we may expect the foot to be placed in climbing the ladder, the foot applying the full load at some instant. The actual location will be statistically distributed, but since the maximum value that the stress can take is categorically determinate, the design is ba..ed on that value. Then the mean working stress to provide a mean strength, $\bar{S}$, is given by

$$\text{Mean stress} = \frac{\bar{S}\,(w/s)\,(2w/3)^2}{Zw^2}$$

$$= \frac{4}{27}\frac{\bar{S}w}{Z} = \frac{4}{27}\frac{32}{\pi}\frac{\bar{S}w}{d^3}$$

$$= 1.509\,\frac{\bar{S}w}{d^3}$$

But mean allowable stress = 80 N/mm$^2$. Therefore

$$d = \sqrt[3]{\left\{\frac{1.509\,\bar{S}w}{80}\right\}}$$

$$= \sqrt[3]{\left\{\frac{1.509 \times 3614 \times 240}{80}\right\}}$$

$$= 25.4 \text{ mm (1 inch).}$$

It is possible that you obtained a different diameter, which is most likely due to different assumptions for either the mean load or its dispersion. If the quantities are significantly different you should consider the implications of the assumptions regarding the load, for this is the all-important factor in achieving high reliability.

*Exercise 4.8*

The problem is best approached by rewriting equation (4.7). The term in square brackets in that equation is the load cumulative distribution. Hence equation (4.7) can also be written as

$$R(t) = \int_0^\infty S(s)\, G(s)^n \mathrm{d}s$$

where $G(s)$ is the load cumulative distribution. The distributions have been plotted on figure 15.4. It will be noted that $G(s)$ is always less than unity, and for some ranges of $s$ is considerably less than unity. Hence, while for the first application of load the whole of the $G(s)$ function contributes to the evaluation of $R(t)$ in the above equation, at higher values of $n$ the quantity $[G(s)]^n$ will be a second order one, and so negligible for some values of $s$. The integrand of the above equation will then be zero for these values of $s$. Thus the parts of $S(s)$ corresponding to these values of $s$ will play no part in determining the value of $R(t)$ after the first application of load, that is, parts of the $S(s)$ distribution have been effectively eliminated from the evaluation of $R(t)$ and hence from $\lambda(t)$ at the first applica-

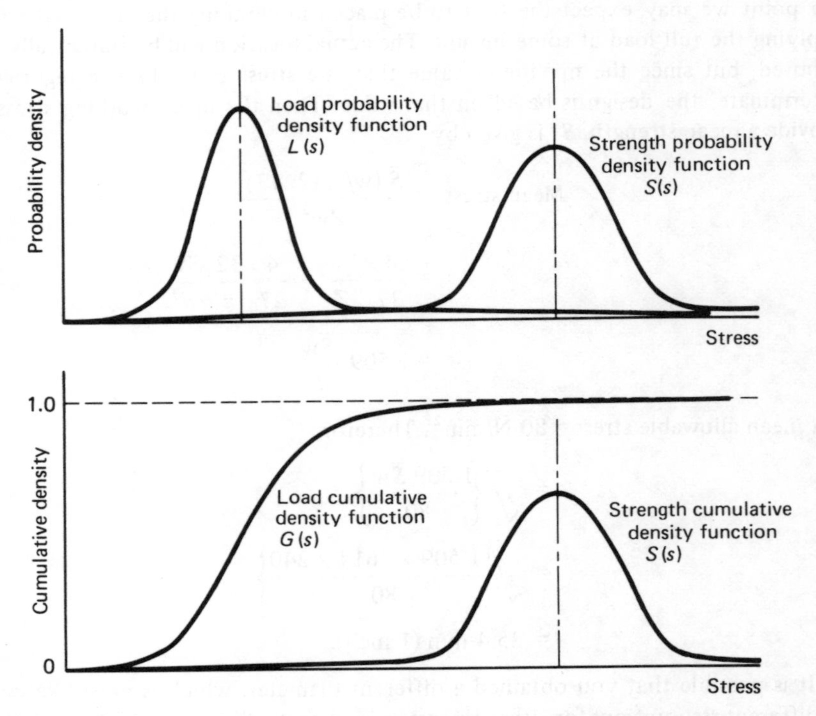


**Figure 15.4** Interference of probability density and cumulative functions (tails shown exaggerated)

tion of load. Reference to figure 15.4 shows that the relevant part of the distribution is its extended lower tail. Physically, the weak items which form that tail of the distribution are eliminated (by failure) at the first application of load and play no further part in the process. It is this group of weak items which gives rise to the higher failure rate at the application of the first load. The magnitude of the first load failure rate is therefore strongly dependent on the tail of the strength distribution.

*Exercise 4.9*

In the solution to exercise 4.9 we saw that the application of the first load was a special case, and that the concept of intrinsic reliability could not apply in that instance since the tail of the strength distribution played an important role which disappeared for the subsequent applications of load. However, for the one shot device we are only concerned with the application of one (that is, the first) load. The difficulties arising from the indeterminacy of the tails of the distributions reassert themselves. A procedure involving only a consideration of the steady constant failure rate is no longer available to us. Intrinsic reliability has little meaning. It would appear that there are four alternative means of surmounting the problem

1. A single-shot device should be designed in the same way as a multi-shot device using intrinsic reliability as the criterion. This implies that a very small proportion – virtually negligible – will in fact fail. Even this may not happen if the tail of the strength distribution is practically not as extensive as the one theoretically assumed.
2. More rigorous quality control can be adopted to ensure that the tail does not exist, thus positively achieving the 'hoped for' condition 1 above. A cut-off at the notorious $3\sigma$ point of a normal distribution, which is often used to encompass the whole population, would reduce the failures at the first application of load to zero for an intrinsically reliable item (for normal distributions). This does not seem an impossible target.
3. Many so-called one shot devices can in fact be operated several times. Certainly components of one shot devices may often be operated continuously for a short period of time. Thus a satisfactory quality control procedure is often available through limited operation or testing by the manufacturer.
4. It is of course theoretically possible to design for a given level of reliability via equation (3.8). This will not achieve intrinsic reliability because the variation of failure rate with safety margin is not steep enough at one load application for us to identify an asymptotic value. Of more importance, however, this technique demands a thorough knowledge of both load and strength distributions, which is both difficult and costly, if not impossible, to achieve.

The true one shot device does present major difficulties, but we should recognise that such devices are much less common than may be supposed.

Of some general importance, an examination of one shot devices highlights that intrinsic reliability is not a valid concept for the first application of load in the case of all multi-load or continuously operable devices. However, failure at first load does mean that these devices would not operate at all when new. Although strictly a reliability problem, few people would regard it as such, because the failure is immediately obvious: "It never worked." Both user and manufacturer would regard such a failure as the manufacturer's responsibility, and he would undoubtedly correct any fault involved. One might reasonably suppose it would have been revealed by the manufacturer himself before the user was involved. Nevertheless it is a form of concealed quality control which is available for the continuously operated device to ensure intrinsic reliability in actual use, which is not available for the one shot device.

*Exercise 5.1*

Many solutions are possible to this exercise. As an illustration the author would give

(1) A *drill* (Indeed, any tool or device which has a cutting edge or whose performance depends on certain dimensions could be quoted.)
(2) A *journal bearing*. (Most machines with mating moving surfaces tend to 'wear in' before they start to wear out. This category is a lot more common than may at first be expected.)
(3) A *round of ammunition*. (Anything which 'ages' comes into this category. The characteristic is a finite 'shelf life'.)

*Exercise 5.1A*

If you have not already included it, in which of the categories, if any, do you consider fatigue should be placed?
(The solution to this exercise is given at the end of this chapter.)

*Exercise 5.2*

We can expect the failures of the roof tiles to be normally distributed about a mean of 150 years with a standard deviation that would give effectively zero failures at 30 years. Since the variate of the normal distribution ranges from $-\infty$ to $+\infty$, the value corresponding to effectively zero failures has to be fixed arbitrarily. We shall take this as four standard deviations from the mean, giving it 1 in 31 544 chance of a failure. This corresponds to about 0.2 of a tile failure out of 5000 which we can safely treat as zero. Then

$$4\sigma = 150 - 30 = 120$$
$$\sigma = 30$$

An age of 60 years corresponds to $(150 - 60)/30 = 3$ standard deviations from the mean. Standard tables indicate that 0.001 35 of the population fall outside this limit as indicated on figure 15.5 by the shaded area. Therefore number of tiles failing and needing replacement in 60 years = 0.001 35 x 5000 = 6.75 = 7 tiles. An age of 90 years corresponds to (150 – 90)/30 = 2 standard deviations from the mean. Standard tables indicate that 0.0228 of the population fall outside this

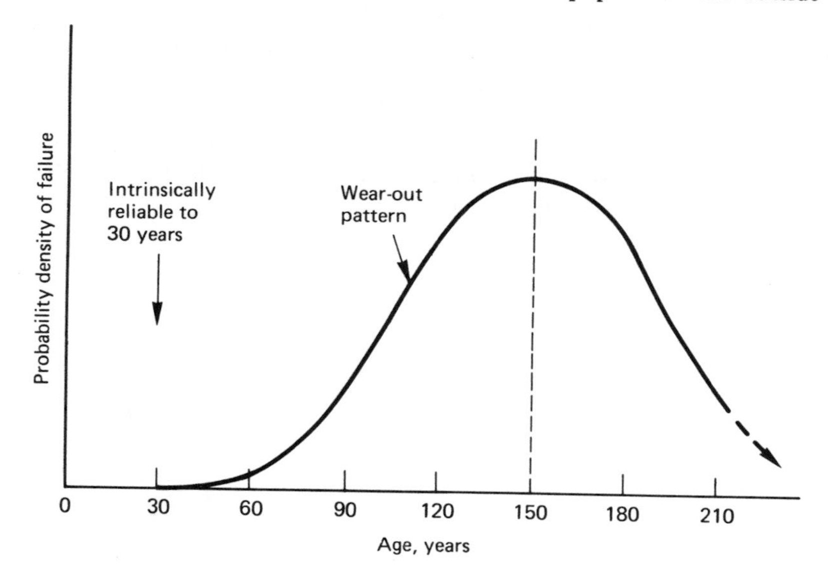


**Figure 15.5** Failure–age pattern

limit. Hence the smallest number of tiles failing and needing replacement in 90 years = 0.0228 x 5000 = 114 tiles. This number could be slightly on the low side because some replacement tiles will be over 30 years old and could also have failed. The number is probably zero at 90 years, that is, 60 years of age for the second generation, but would increase significantly thereafter. The problem of calculating failures with replacement is discussed fully in chapter 9.

*Exercise 5.3*

For disc
- Mean elastic endurance limit ($\overline{F}$) = 1000 MN/m$^2$
- Standard deviation ($\sigma_F$) = 72 MN/m$^2$

In helicopter role
- Mean load ($\overline{L}$) = 1200 MN/m$^2$
- Standard deviation ($\sigma_L$) = 50 MN/m$^2$

In earth-moving role
- Mean load ($\overline{L}$) = 675 MN/m$^2$ ($1200 \times 0.75^2$)
- Standard deviation = 101 MN/m$^2$

Using equations (5.10) and (5.11) the failure rate-age curves can be calculated. These are plotted in figures 15.6a and b for the helicopter and earth-moving roles respectively. There are marked differences in the curves for the two applications. In its helicopter role this disc is intrinsically reliable for 7500 cycles, representing 7500 missions. Since disc integrity is absolute in this application, it would have to be removed before this (a factor of at least 2 is used in practice). It is to be noted, however, that once the life of intrinsic reliability has expired, the failure rate rises rapidly. This is reflected in the cumulative failures, plotted on figure 15.6, which indicates that even if operation could be continued beyond the life of intrinsic reliability, 85 per cent of the population would have failed by 10 000 cycles (33 per cent increase). Not a great deal of useful life is therefore lost by restricting the useful life to that for intrinsic reliability.

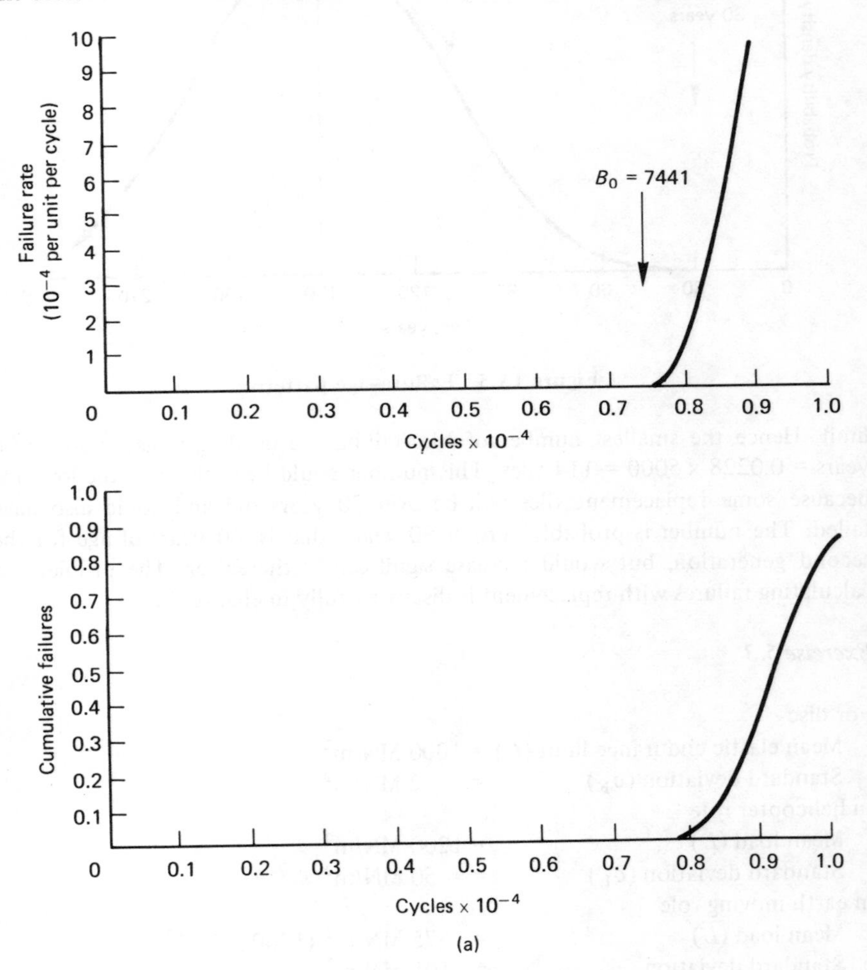


**Figure 15.6** (a) Failure rate and cumulative failures of turbine disc in helicopter role

The reduced rating in the earth-moving role has produced a marked increase in the life of intrinsic reliability to $2.5 \times 10^5$ cycles, though because of the greater number of cycles per mission this represents only 2000 missions. It is doubtful if this would be acceptable in this role, and an increased life has to be sought by operating outside intrinsic reliability. A decision to do so is difficult. On the one hand the failure rate is so low after intrinsic reliability has expired (after 8000 missions only 30 per cent of the population would have failed) that the proposition is very attractive. On the other hand, unless the mode of failure was particulaly innocuous it is unlikely that failures of turbine discs could be tolerated. Further

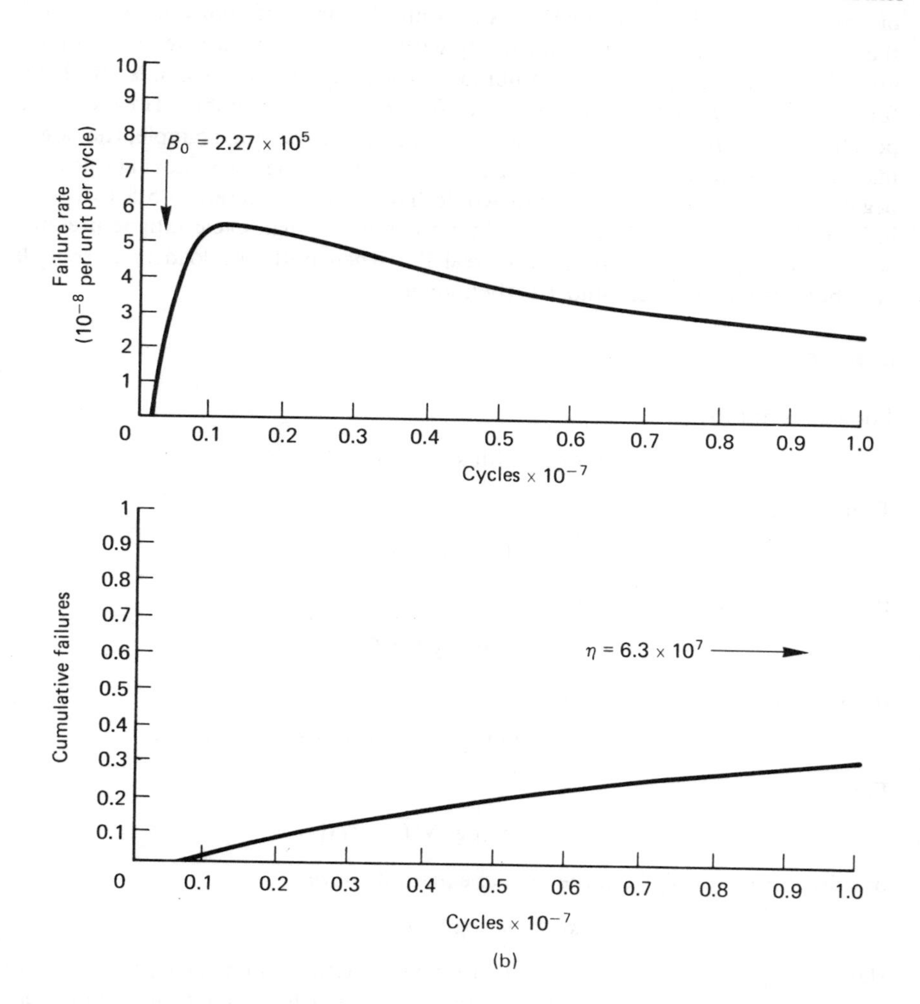


**Figure 15.6** (b) Failure rate and cumulative failures of turbine disc in earth-moving role

derating (unacceptable on performance criteria) or redesign of the disc (unacceptable on cost criteria) are the alternatives.

It should be noted that low lives of intrinsic reliability in low cycle fatigue (LCF) imply a high subsequent failure rate, so that limiting the planned life to the failure-free life is cost effective. On the other hand high lives of intrinsic reliability in high cycle fatigue (HCF) imply a low subsequent failure rate. If the application in the latter case is such that some failures are acceptable (that is, no safety or other serious consequences are involved) then it is possible to achieve even higher lives at a low, but not zero, failure rate. If failures are not acceptable a great deal of this life has to be thrown away. An infinite life may of course be achieved if the design is based on intrinsic reliability with respect to the fatigue or endurance limit. This may be essential if the number of load cycles per mission is very high, for example, reversed loading in a high-speed rotating member. However, the penalty can be quite high at only medium high reversals. For example, the safety margin expressed relative to the fatigue limit is 2.88 in the earth-moving role (it is negative in the helicopter role) but would have to be increased to 5.5 to ensure infinite life in fatigue. Thus the optimum design for high cycle fatigue presents considerable difficulty, especially in real life when both the load and strength distributions are only very imperfectly known.

*Exercise 5.4*

For a constant load

$$L(s) = 0 \text{ except for } s = \bar{L} \text{ when } L(s)\mathrm{d}s = 1$$

Therefore equation (5.11) reduces to

$$N_f = N(\bar{L} - s_f) \text{ for } \bar{L} > s_f$$

If the mean fatigue curve is given by

$$s = m \log N + C$$

the $s$–$N$ relationship is given by

$$s - s_f = m \log [N(s - s_f)] + C$$

For $s = \bar{L}$

$$\bar{L} - s_f = m \log [N(\bar{L} - s_f)] + C$$

or substituting for $N_f$ from the second equation above

$$s_f = -m \log N_f + (\bar{L} - C)$$

Hence, if $s_f$ is normally distributed the transformation will make $N_f$ log-normally distributed so long as $\bar{L}$ is always greater than $s_f$, which for practical purposes can be taken as about three standard deviations above the fatigue limit. It will be recalled that many workers consider that the log-normal distribution is about the best theoretical representation of the life pattern experienced in standard fatigue tests. This theory would agree with that experience, except to predict increasing departure as the test load approaches the fatigue limit stress. As we have seen, the departure can be significant in certain circumstances. Nevertheless, there would appear to be no marked conflict between experience and theoretical prediction.

*Exercise 5.5*

The $s$–$N$ curve is defined by

$$s - s_f = m \log(N/N_0)$$

where

$s_f$ = the fatigue limit strength
$N_0$ = cycles above which $s$ = constant = $s_f$.

Equation (5.13) can be written as

$$N_0 = 1/I$$

where for concentrated loads

$$I = \frac{q}{N(s - s_f)} = \frac{q}{\exp\{-(s - s_f)/m\}}$$

$$= \frac{q}{\exp\{-s/m\} \times \exp\{s_f/m\}}$$

where $q$ is non-dimensional frequency of the load application at stress $s$, that is, the actual frequency of an individual load application divided by the total number of load applications, and is thus equal to the probability density at $s$ for concentrated loads.

Since the transformation (5.13) does not change the cumulative failures associated with $s_f$ or $N_f$, the above expression can be rewritten as

$$I = \frac{1}{\exp\{[s_f]/m\}} \times \frac{q}{\exp\{-s/m\}}$$

where $[s_f]$ = value of $s_f$ associated with a specified cumulative number of failures. That is

$$I = \text{function of load}/\exp\{[s_f]/m\}$$

Hence

$$\frac{[N_f]_{B10}}{[N_f]_{B50}} = \frac{[I]_{B50}}{[I]_{B10}} = \frac{\exp\{[s_f]_{B10}/m\}}{\exp\{[s_f]_{B50}/m\}}$$

where $B10$ and $B50$ denote the value applies to 10 per cent and 50 per cent cumulative failures respectively. It follows that the ratio of cycles for 10 per cent cumulative failures to cycles for 50 per cent cumulative failures is independent of the loads, and is dependent only on the distribution of the fatigue limit strength and the slope of the linearised *s–N* curve.

This result is of some practical importance: the standard application of Miner's law to the mean/median *s–N* curve to estimate the life with multiple loading must give the life to 50 per cent cumulative failures. However, this is clearly unsatisfactory as a life estimator for virtually all design purposes. The life for a much lower, say 10 per cent, cumulative failures is required. Thus we should write

$$\frac{n}{N} = 1 \text{ for 50 per cent cumulative failures}$$

$$\frac{n}{N} = \frac{[N_\mathrm{f}]_{B10}}{[N_\mathrm{f}]_{B50}} \text{ for 10 per cent cumulative failures}$$

The value of the Miner cycle ratio summation for design purposes can therefore ideally be derived from the fatigue limit strength distribution and the slope of the linearised *s–N* curve by means of the above theory. For typical values corresponding to a roughly normally distributed fatigue or endurance limit with an *s–N* curve already described in the text, the value of the Miner cycle ratio summation is 0.62. An empirical value of 0.8 is often used though values as low as 0.4 have been quoted. For the accurate calculation of the value a much more exact representation of both the distribution and the shape of the *s–N* curve itself would be required. However, the roughly calculated value does seem to be 'in the right street'. Of more importance, the theory shows that the value is dependent on the material properties only, and once determined for a given material (in the finished state) can be used for all load distributions.

The reason for Miner cycle ratio summations less than unity are often said to arise from variations in the loading sequence. This analysis suggests it can be fully accounted for as a 'reserve factor', that is, to line up deterministic and stochastic values. It must be recognised that the theory is very simplistic, but if it is accepted that the fatigue limit is distributed – and that would seem indisputable – it follows that individual members of the population must have individual *s–N* curves, and hence some reserve factor which cannot be far from the theoretical value is necessary. On the other hand, it would be unwise to generalise too much from this exercise, which applies to a particular form of the loading and may not represent the general case.

*Exercise 5.6*

In order to obtain a solution it is necessary to solve equation (5.11), where $L(s)$ is chosen as the normal distribution having the parametric constants given in the question. The function $N(s - s_\mathrm{f})$ is a power relationship in this case. The solution of equation (5.11) must be carried out using numerical methods. The probability density of failure with respect to time is then the transform of the strength distribution using the solution to (5.11). From this, the cumulative failure function and hence the failure rate can be calculated. The calculation is very tedious if carried out by hand, but is easily computerised. The resulting failure patterns are given in figure 15.7, both for a constant load life which is inversely proportional to the load and also for a constant load life which is inversely proportional to the cube of the load. The failure pattern for a fifth power relationship is also shown. When the constant load life is inversely proportional to the load there is a failure-free life of about half an hour before the failure rate becomes finite and then continues to increase more and more rapidly with age. This pattern is very similar

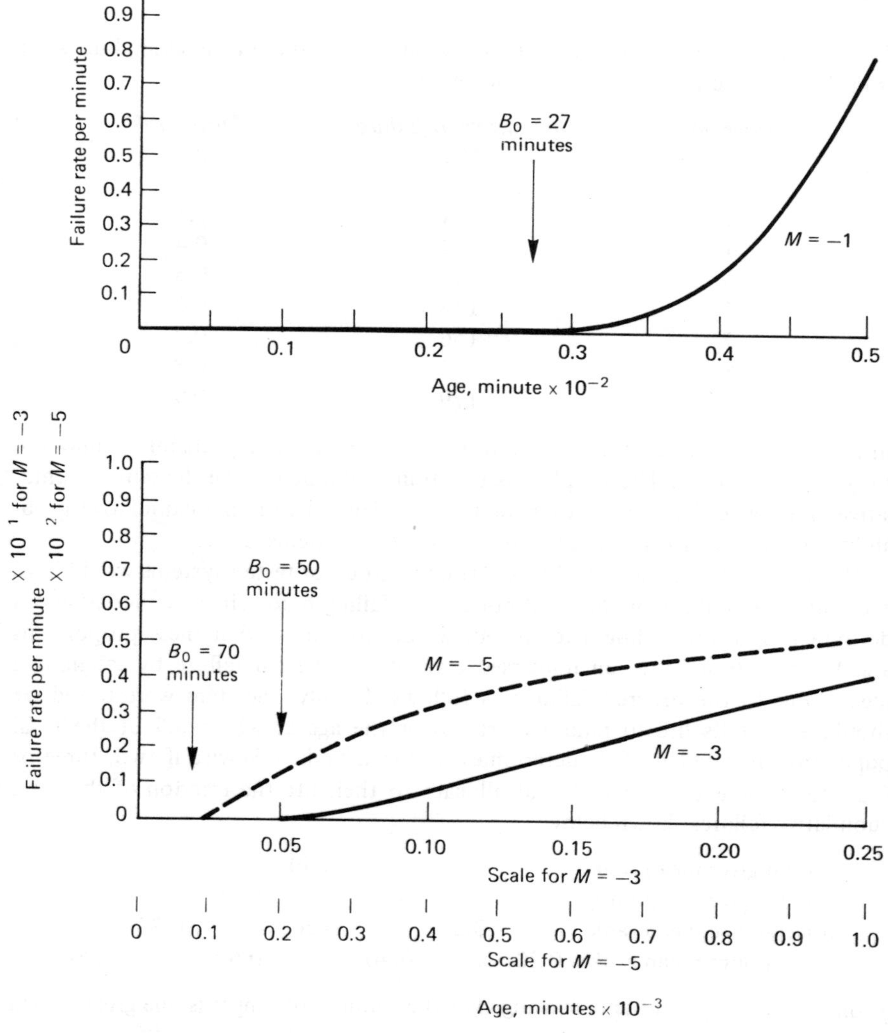


**Figure 15.7** Failure rate–age patterns for different tool wear rates

to that estimated theoretically for erosive or corrosive type wear in which the failure-inducing load is different from the wear load – see section 5.2. However, with the higher inverse powers for the load–life relationships the increase of failure rate with time is less pronounced. In no case examined by the author, however, has the failure rate decreased with age as it does with a fatigue mechanism. It seems that the decreasing failure rate is due to the strength cut-off at the fatigue limit. The failure patterns are consequently of a different nature. Even so, one cannot identify a strictly limited range of failure patterns for each mechanism. Not all *s–N* curves have a limit, for example.

*Exercise 6.1*

The first step is to rank the number of operations to failure in an ascending order as in the first two columns of the table below

| *Order no.* | *Operations to failure* | *Mean rank* |
|---|---|---|
| 1 | 102 | 0.1 |
| 2 | 115 | 0.2 |
| 3 | 124 | 0.3 |
| 4 | 132 | 0.4 |
| 5 | 141 | 0.5 |
| 6 | 148 | 0.6 |
| 7 | 156 | 0.7 |
| 8 | 165 | 0.8 |
| 9 | 179 | 0.9 |

In order to plot these data we need to know the cumulative probability function, $F(t)$, at each failure. The simplest assumption would be to take the sample cumulative fraction of failures at each failure age. Thus the sample cumulative probability function of the $i$th failure from a sample of $N$ items is $i/N$.

However, the first failure of this particular group of tested systems would have a definite percentage of the total population failing before it, and the last has a definite proportion failing after it, but we do not know what these proportions are. We therefore adopt an unbiassed estimate of the $i$th failure by assigning a mean rank to the ordered failures as follows. If only one item were tested we should expect its life to failure to represent the age at which 0.5 of the total population had failed. This determines its mean rank. Likewise if two, three or four items were tested we should allocate to their life the fraction of the total cumulative failures shown below

| | | | | | | | |
|---|---|---|---|---|---|---|---|
| 1 item gives mean rank | | | | 0.50 | | | |
| 2 items give mean rank | | | 0.33 | | 0.66 | | |
| 3 items give mean rank | | 0.25 | | 0.50 | | 0.75 | |
| 4 items give mean rank | 0.20 | | 0.40 | | 0.60 | | 0.80 |

From this display, and appreciating that the number of ranges is one greater than the number of items, a general expression for the mean rank of the $i$th failure from a total of $N$ items will be given by

$$\text{mean rank} = \frac{i}{N+1}$$

Clearly if $N$ is very large the difference between $i/N$ and $i/(N+1)$ can be ignored, but for small samples, as the one under discussion, the latter gives a better estimate. The mean rank has been calculated by means of the above equation and recorded in the right-hand column of the above table for the particular example under discussion.

This data has been plotted on standard probability paper in figure 15.8. Since the data plots as a straight line it follows that it can be represented by a normal distribution. For this distribution the mean is equal to the median and can be read off the graph as 140 operations. The area under the normal curve between the mean and a deviation equal to the standard deviation is equal to 34.13 per cent of the total area under the curve (see standard tables). Hence we can read off an estimate of the standard deviation from the graph by evaluating the difference in abscissae between the mean and a point whose ordinate is $50 - 34.13 \approx 16$ per cent. This is illustrated in figure 15.8 from which we see that the standard deviation is 29.5 operations.

To estimate the intrinsically reliable life we must postulate the number of standard deviations beyond which we would regard the possibility of a failure as negligible. We may regard 4 standard deviations as a suitable value, giving the chance of failure outside that range as less than 1 in 30 000 (see figure 6.1). Hence, intrinsically reliable life = $140 - 29.5 \times 4 = 22$ operations.

*Exercise 6.2*

A job report arises as a result of a failure, and we may therefore take the mean time between failures as 271 miles. Therefore

$$\lambda = \frac{1}{271} \text{ miles}$$

and

$$\text{Reliability} = e^{-x/271}$$

where

$$x = \text{distance (miles)}$$

Hence probability of success at 100 miles is the

$$\text{Reliability} = e^{-100/271}$$
$$= 69 \text{ per cent}$$

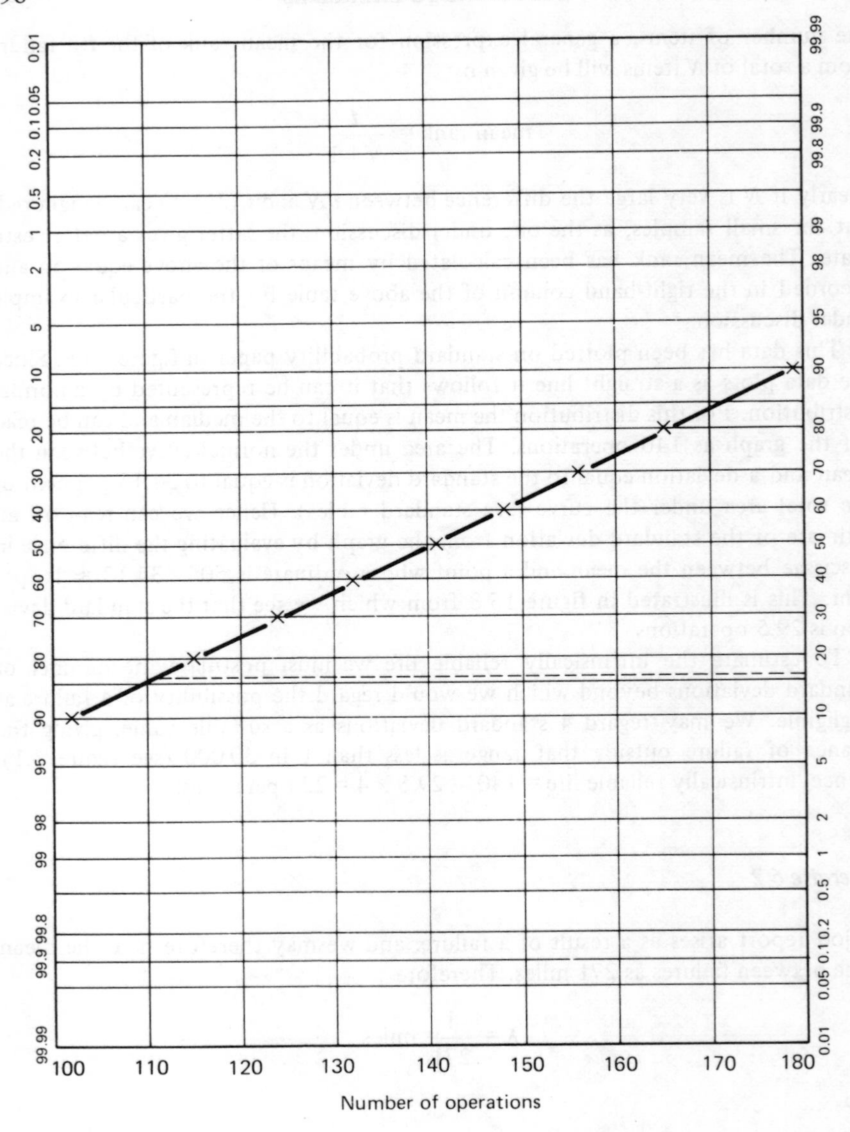


**Figure 15.8** Failure data plotted on standard probability paper

*Supplementary exercise 6.2A*

Suppose you were given exercise 6.2 as a practical question; is there any additional information you would seek before quoting the value of 69 per cent?

*Exercise 6.3*

The failure rate, $\lambda$, is $0.05 \times 10^{-3}$.
Hence we can write the reliability $R(t)$ for random failures as

$$R(t) = \exp\{-0.05t \times 10^{-3}\}$$

We must now state the minimum value of the reliability we can accept. The author would specify 99 per cent. Then

$$0.99 = \exp\{-0.5t \times 10^{-3}\}$$

or

$$0.01 = 0.05t \times 10^{-3}$$

giving

$$t = 200 \text{ h}$$

On the assumption that the reliability from random failures should not drop below 99 per cent, we could therefore use a stylus for 200 h, that is to play 400 sides of a long-playing record. However, we must consider how wear-out relates to this. Simple engineering common sense should raise doubts whether an unlubricated sapphire point 0.0012 to 0.0025 mm radius would survive 200 h rubbing. If you did not know, you should have ascertained from appropriate data that a sapphire stylus would wear out (in the sense that it would start to introduce distortion) after some 100 sides of a long-playing record, that is after about 50 hours playing. Thus replacement should follow after a further 50 sides of long-playing records have been played. This example is controlled by the wear-out process and not by random failure processes.

This exercise illustrates the importance of recognising the failure modes with which one is dealing, of confining the mathematical analysis to the mode for which it is applicable, and finally considering the importance of all failure modes to the question under consideration.

*Exercise 6.4*

$$\text{MTTF} = \frac{\int_0^\infty t\, f(t)\,\mathrm{d}t}{\int_0^\infty f(t)\,\mathrm{d}t} \quad \text{from equation (6.8)}$$

$$\int t\, f(t)\,\mathrm{d}t = t\int f(t)\,\mathrm{d}t - \int\left(\int f(t)\,\mathrm{d}t\right)\mathrm{d}t$$

$$= t\int f(t)\,\mathrm{d}t - \int F(t)\,\mathrm{d}t$$

$$= t\int f(t)\,\mathrm{d}t - \int\{1 - R(t)\}\,\mathrm{d}t$$

$$= t\left(\int f(t)\,\mathrm{d}t - 1\right) + \int R(t)\,\mathrm{d}t$$

Therefore

$$\mathrm{MTTF} = \frac{t\left(\int_0^\infty f(t)\,\mathrm{d}t - 1\right) + \int_0^\infty R(t)\,\mathrm{d}t}{\int_0^\infty f(t)\,\mathrm{d}t}$$

If all items fail $\int_0^\infty f(t)\,\mathrm{d}t = 1$

$$\mathrm{MTTF} = \int_0^\infty R(t)\,\mathrm{d}t$$

In some failure modes the whole population never fails; some assembly failure modes fall into this category. In that case

$$\int_0^\infty f(t)\,\mathrm{d}t \neq 1$$

The full equation has then to be used.

*Exercise 6.5*

The cumulative failures, estimated by the mean rank as described in the solution to exercise 6.1, have been plotted against number of cycles to failure on log–linear graph paper in figure 15.9. The points plot as a linear curve and hence we can conclude that they follow a negative exponential distribution. The mean time to failure, equal to the mean time between failure, is the number of cycles for which $F(t) = 0.632$, that is 460. The failure rate is constant and equal to $1/460 = 0.00217$.

A comparison between the two sets of test data on the extinguishers indicates that the rival product shows no signs of wear-out after 700 operations. By that number of operations the manufacturer's own product would all have worn out! So from a simple point of view his rival would seem to have a superior product. However, it is worth considering whether such a product would be operated that number of times and whether such a life without wear-out does not represent over-design in respect of the wear-out failures and under-design in respect of random failures – a not wholly desirable state of affairs in reality. The manufacturer may well observe that his rival's product has a fairly high random failure rate of some two in a thousand, which would apply to a new extinguisher as much

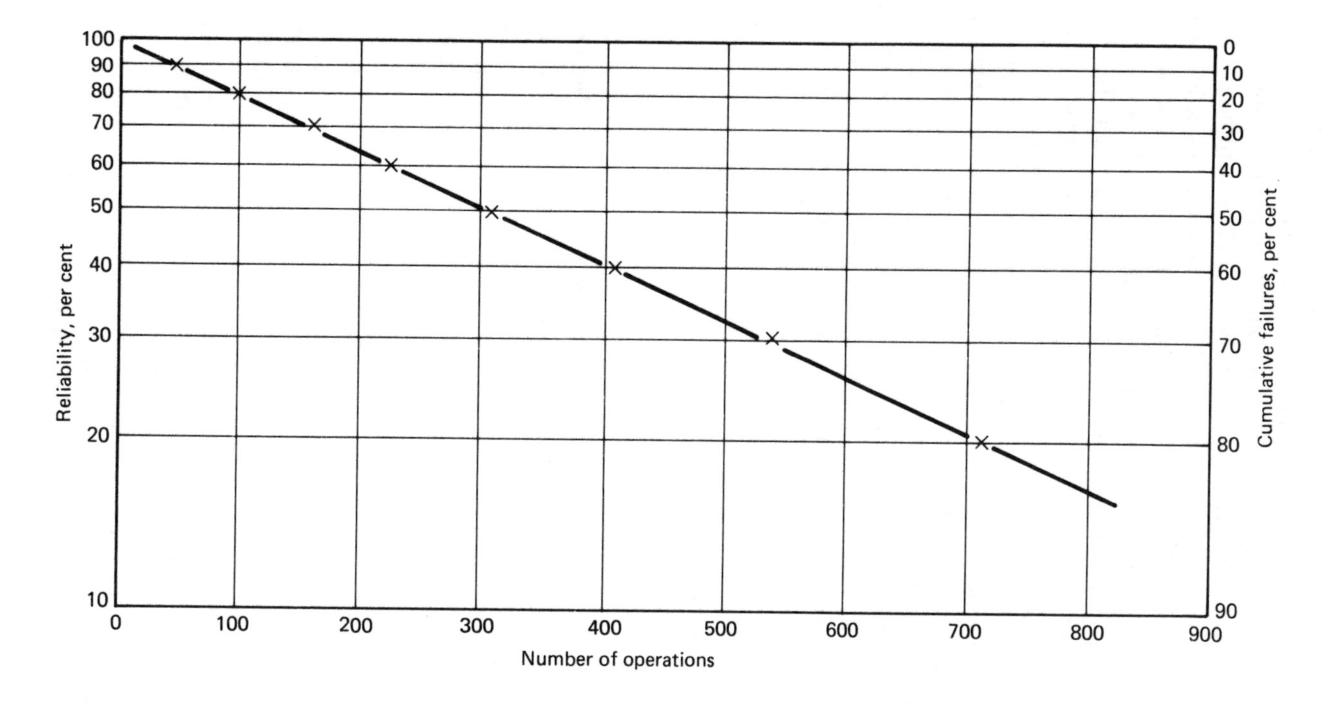


**Figure 15.9** Failure data plotted on log–linear graph paper

as to an old one. He may well consider, and of more importance his customers may well consider, that such a failure rate is too high in a product which is specifically bought to safeguard against what could be a disastrous emergency. At least his own product could be guaranteed to deal with 22 emergencies, and one would hope that unintentional fires are not all that prevalent in one location. Furthermore, the constant failure rate of the rival product could be sensitive to environment, and hence much higher in some circumstances. On the whole, therefore, the manufacturer may consider he has basically a better product. The danger is that the extinguishers could be operated either as part of training drill or for test purposes a significant number of times before being required in earnest. The life of intrinsic reliability could therefore be used up before any emergency occurred. His course of action should be to identify the source of wear by physical examination of his own products after test, with a view to either redesigning for longer intrinsically reliable life, or if that is not possible by replacement of the worn part after, say, twenty operations. By this means he could ensure continual intrinsic reliability – at a cost. This leads to a consideration of maintenance which is treated fully in chapter 9, but may offer a better solution than redesign.

*Exercise 6.6*

The distributions are plotted in figure 15.10.

*Exercise 6.7*

| *Order number* | *Hours to failure* × $10^{-3}$ | *Mean rank* |
|---|---|---|
| 1 | 1.8 | 0.09 |
| 2 | 2.4 | 0.18 |
| 3 | 2.8 | 0.27 |
| 4 | 3.2 | 0.36 |
| 5 | 3.6 | 0.45 |
| 6 | 3.9 | 0.55 |
| 7 | 4.4 | 0.64 |
| 8 | 4.7 | 0.73 |
| 9 | ? | 0.82 |
| 10 | ? | 0.91 |

These results are plotted on Weibull paper in figure 15.11. A linear curve is obtained from the raw data indicating $t_0 = 0$. We then deduce that

(a) $\beta = 2.70$, representing a wear-out type of failure

(b) $\eta = 4300$ h = characteristic life

(c) Failures at 2000 h = 11 per cent, or

Reliability = 89 per cent at 200 h

This is hardly satisfactory.

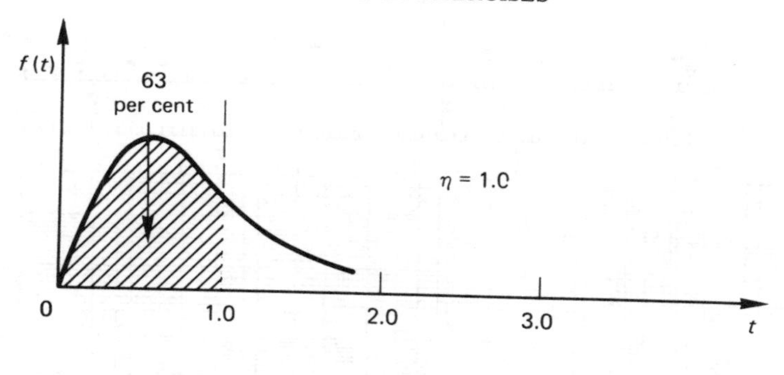


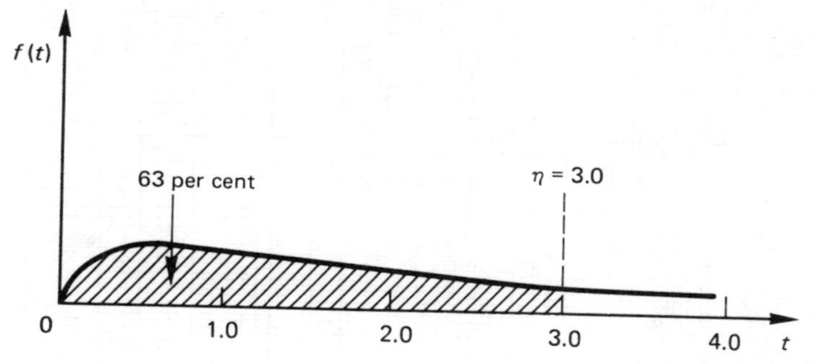


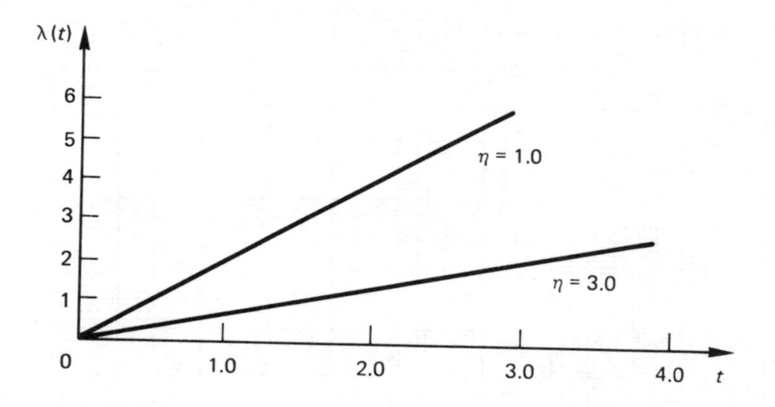


**Figure 15.10** Weibull distributions with different values for parameters

*Exercise 6.8*

After ranking the bearing cages in order of failure, calculating the mean ranks the raw data can be plotted on Weibull paper as in figure 15.12. This reveals a curved line which can be straightened by the technique described in section 6.4.2. The

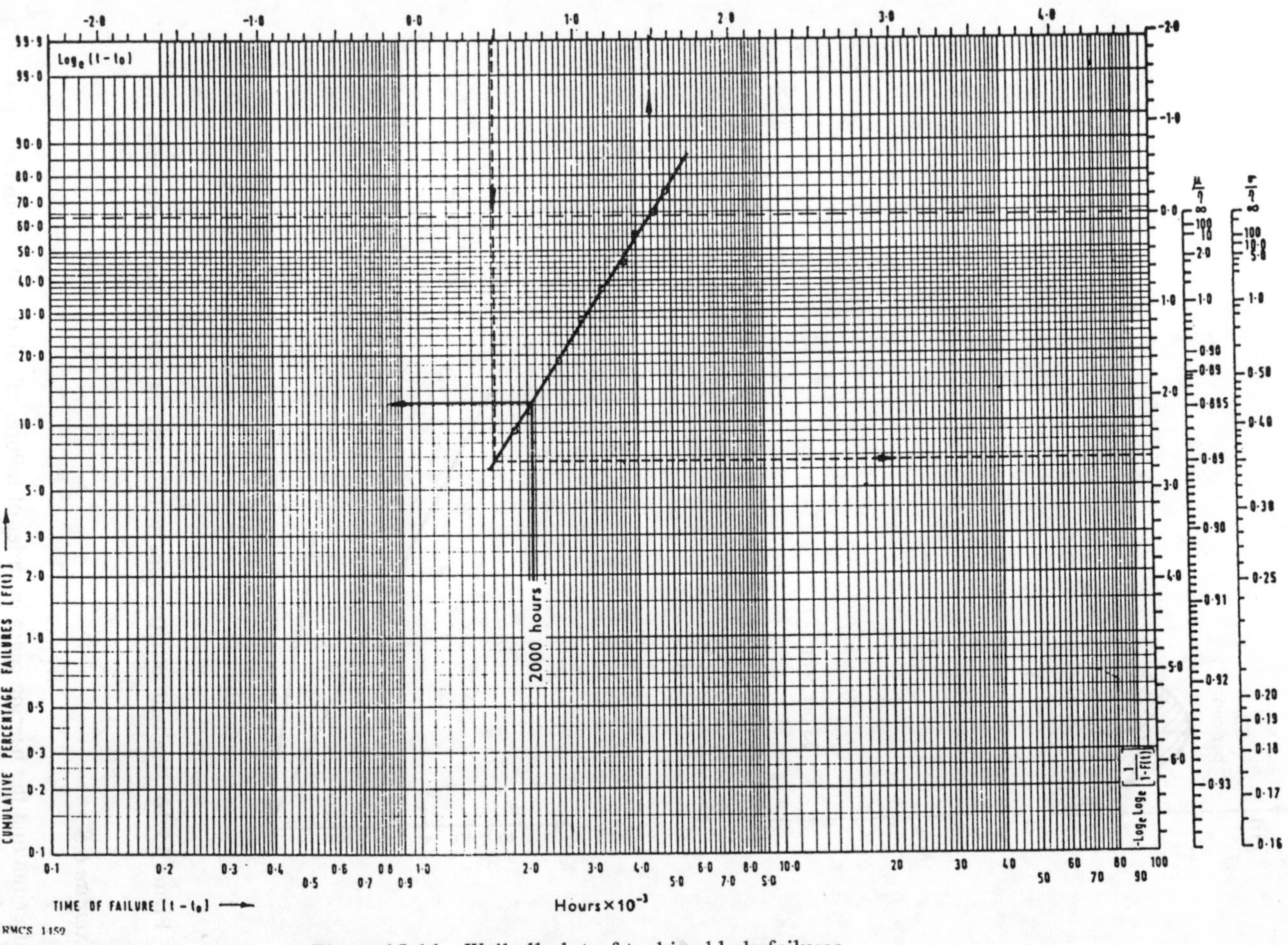


Figure 15.11 Weibull plot of turbine blade failures

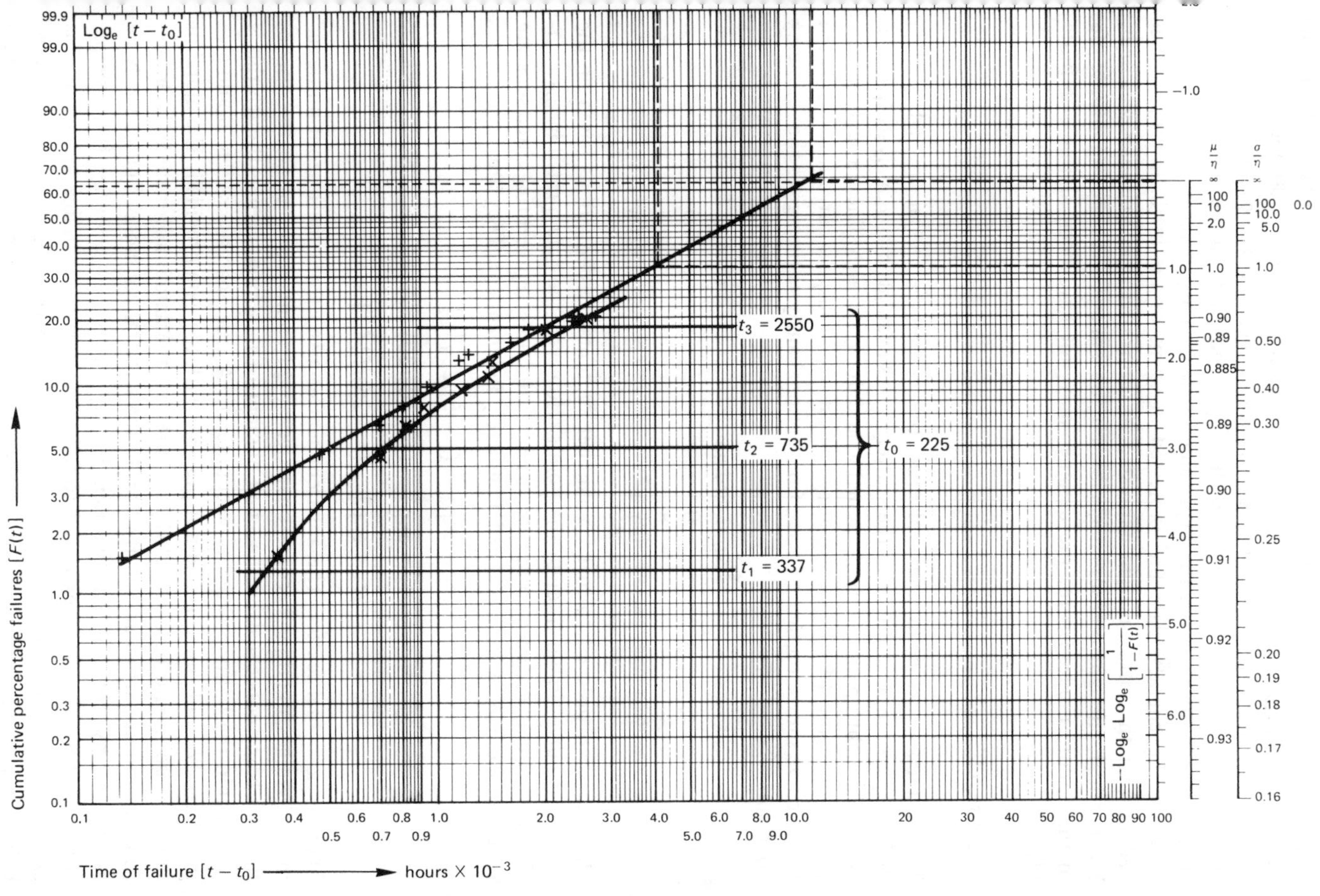


**Figure 15.12** Weibull plot of bearing failures

construction is shown on figure 15.12, giving a value of $t_0$ = 225 hours. From the derived straight line on figure 15.12 the characteristic life is 13 500 hours and the shaping parameter is 0.95. From this we may deduce that the cages are failing in fatigue, as we could well expect in this application, after a life of intrinsic reliability of 225 hours. From the high value of the shaping parameter and the high value of the characteristic life expressed in terms of the intrinsic life, we can deduce that the safety margin expressed in terms of the fatigue limit was high, but not high enough to ensure negligible fatigue damage and hence an infinite life in this failure mode. (*Note*: the designer may not have been aiming at infinite life.)

*Exercise 6.9*

This exercise was deliberately planned to deceive the reader, though with a definite purpose in mind. With the scatter in this example many analysts (particularly if brought up on electronic items) would draw a straight line through the raw data as on figure 15.13. A more wary analyst would suspect curvature and sketch in the curve shown on figure 15.13. Conventional analysis then gives the following constants.

| | *Straight line* | *Curved line* |
|---|---|---|
| $t_0$ | 0 | 4.5 |
| $\eta$ | 18 | 10.5 |
| $\beta$ | 1.4 | 0.8 |

One can then proceed to make a number of tentative deductions based on the particular curve drawn through the points. However, the data quoted in the exercise was actually deduced from three Weibull functions representing three separate failure modes which were assumed completely independent. They are

$$R_1 = \exp[-\{(t - 5)/20\}^{0.5}]$$

$$R_2 = \exp[-\{(t - 10)/20\}^{0.8}]$$

$$R_3 = \exp[-\{(t - 25)/5\}^{3.5}]$$

There is *no* actual scatter. The apparent scatter arises from the discontinuities as the individual functions become operative at different values of $t_0$, the discontinuities being concealed by the small number of data presented – though this is well within the limits practically encountered. A true analysis, which should give the above Weibull functions, is not possible with the data provided. The moral of this exercise should be clear: always ensure that the data is adequate and comprehensive, and in particular that it refers to the same failure mode. This can only be established by physical examination. It may not be so essential for a simple component which can be expected to fail in a single recognisable mode, but it is essential for a multi-component item, or a multi-part component. Some statisticians argue that the many failure modes of a complex item – the engine of an

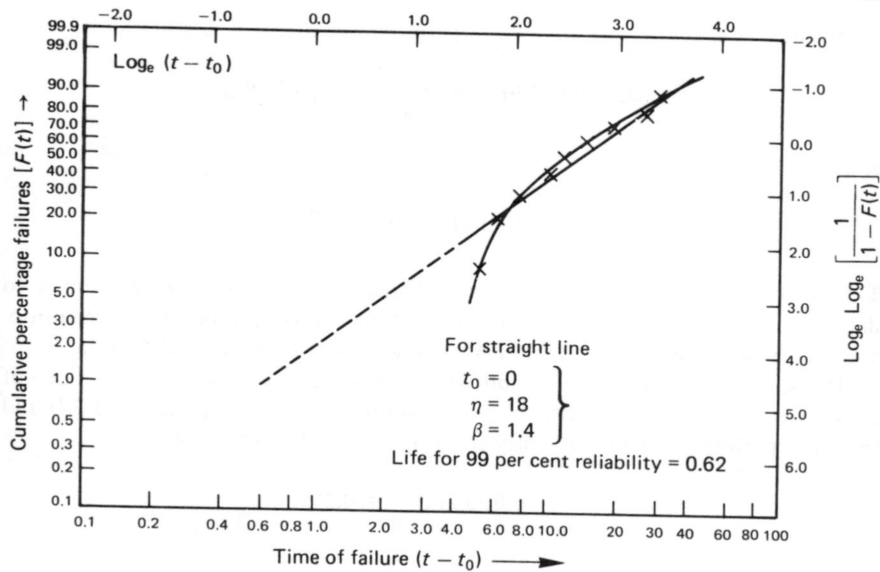


**Figure 15.13** Apparent scatter of data as a result of multiple failure modes

automotive for example – can be bulked and a distribution fitted to the behaviour of the whole item. The failure rate for an automotive engine is a plausible example, but figure 15.13 shows how dangerous this approach can be. Bulking of the failure patterns of a large number of components whose individual parameters are randomly distributed will lead to a random (that is, constant) overall failure rate, no matter what the individual failure age characteristics may be. Physical separation of the constituent failure modes and individual analysis is essential. Recombination may be practically convenient for many presentation purposes, of course, but it should only be done with a true knowledge of the constituent modes.

## *Exercise 6.10*

A typical *s–N* curve is plotted in figure 15.14. The full line represents the mean line usually plotted and the hatched area shows the region within which results may be expected to lie (the scatter). From this curve we observe that at a stress level of $\pm 200$ MN/m$^2$ the mean life is $10^3$ cycles. At this stress level one can expect the scatter to be represented by a standard deviation of 20 cycles. Now the number of cycles of bracket vibration per mean hour of machine operation is given by

$$1 \times 3600 \times \frac{0.01}{100} = 0.36$$

Therefore

$$\text{mean life of bracket} = \frac{10^3}{0.36} = 2777.8 \text{ h}$$

and

$$\text{standard deviation of scatter} = \frac{20}{0.36} = 55.6$$

The scatter will have a roughly normal distribution, and so our problem is to determine the Weibull distribution which corresponds to a normal distribution, and which has a mean value of 2777.8 h and a standard deviation of 55.6 h.

The shape of the Weibull distribution is controlled by $\beta$, which has a value of 3.44 for a roughly normal distribution. Additionally, from equations (6.37) and (6.38) or preferably reading directly from figure 15.14, for $\beta = 3.44$

$$\frac{\mu}{\eta} = 0.899 \text{ and } \frac{\sigma}{\eta} = 0.29$$

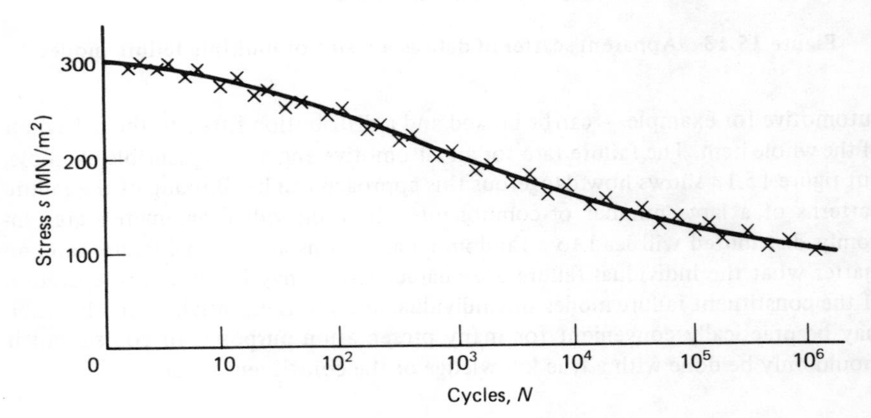


**Figure 15.14** Typical *s–N* curve for bracket

From the latter

$$\eta = \frac{\sigma}{0.29} = \frac{55.6}{0.29}$$

$$= 192 \text{ h}$$

$$\text{mean value} = 0.899 \times \eta = 0.899 \times 192$$

$$= 173 \text{ h}$$

Therefore

$$\bar{t} - t_0 = 173 \text{ h}$$

where

$$\bar{t} = 2777.8 \text{ h}$$

$$t_0 = (2777.8 - 173) \text{ h}$$

$$= 2605 \text{ h}$$

Hence the required Weibull distribution has

$$t_0 = 2605 \text{ h}$$

$$\beta = 3.44$$

$$\eta = 192 \text{ h}$$

Note the following simplifications made in this exercise which have enabled a solution to be calculated.

(1) The fatigue strength of the bracket has been taken equal to that of the material, and more particularly the scatter in the fatigue strength of the bracket has been taken equal to that of the material. The following factors could invalidate this.
  (a) The manufacturing process could affect the material properties and introduce scatter in its own right.
  (b) The manufacturing process will produce a dimensional scatter and hence an additional scatter in bracket strength.

(2) The vibration has been taken at either a constant level of stress or zero. In practice a stress spectrum would be encountered. It could vary from model to model. Furthermore, the simple addition of cycles would not be valid except in simple cases like the one quoted.

In spite of these criticisms, which must be overcome in any practical calculation, this exercise illustrates how a failure distribution arises from a variation in the material properties from which the piece is made.

*Exercise 6.11*

We can use an estimate which will be too high half the time and too low half the time, that is, the median rank. The median rank may be obtained from the binomial expansion

$$(R + F)^N = [(1 - F) + F]^N$$

The expression in square brackets can be expanded. The $i$th median rank is then obtained by equating the $i$th term to 0.5, and solving for $F$.

The median rank is often preferred by statisticians because it enables a consistent estimate to be made of the confidence limits. For example, the 5 per cent and 95 per cent ranks, calculated as above by equating the $i$th term to 0.05 and 0.95 respectively, can be used to give a 90 per cent confidence interval. The calculation of the ranks as described above is somewhat tedious and in practice the values are taken from tables. Alternatively, the median rank can be calculated from a commonly used approximation due to Benard, which gives

$$i\text{th median rank} = \frac{i - 0.3}{N + 0.4}$$

However, there is no corresponding formula for other percentage ranks, which rather limits the use of the above expression.

*Supplementary Exercise 6.11A*

Indicate how the confidence limits may be plotted on Weibull paper when the appropriate ranks are known.
(The solution to this exercise is given at the end of this chapter.)

*Exercise 6.12*

The Weibull distribution is

$$F(x) = 1 - \exp\left\{-\left(\frac{x - x_0}{\eta}\right)^\beta\right\}$$

where

$$F(x) = 0.5 \text{ for } x = 1.5$$
$$F(x) = 0.95 \text{ for } x = 2.5$$
$$F(x) = 0.99 \text{ for } x = 3.0$$

Substituting gives three equations which can be solved for the three unknowns ($x_0$, $\eta$ and $\beta$) by any of the standard numerical techniques to give

$$x_0 = 0.5, \quad \eta = 1.2, \quad \beta = 2$$

*Exercise 6.13*

In this case we have to make use of the known nature of the failure mode to put a value to the shaping parameter. For low cycle fatigue we should expect a low value for $\beta$ – say 0.5. Then substituting in the Weibull equation we obtain

$$0.05 = 1 - \exp\left\{-\left(\frac{2000 - t_0}{\eta}\right)^{0.5}\right\}$$

$$0.10 = 1 - \exp\left\{-\left(\frac{3500 - t_0}{\eta}\right)^{0.5}\right\}$$

This gives two simultaneous equations which can be solved for $t_0$ and $\eta$ to obtain

$$t_0 = 1500 \text{ hours}, \eta = 150\,000 \text{ hours}$$

Using these values of the Weibull parameters, the proportion failing in 2500 hours is calculated as 8 per cent.

*Exercise 6.14*

Follow conventional Weibull analysis as in the first three columns of the table below

| *Order no.* | *Time to failure* | *Mean rank* | $\left(\text{Mean rank}\right)_\infty$ | $1 - \left(\text{Mean rank}\right)_\infty$ |
|---|---|---|---|---|
| 1 | 13 | 0.0152 | 0.9394 | 0.0606 |
| 2 | 42.2 | 0.0303 | 0.9545 | 0.0454 |
| 3 | 50.4 | 0.0454 | 0.9697 | 0.0303 |
| 4 | 54.25 | 0.0606 | 0.9848 | 0.0152 |

Plotting the mean against time to failure on Weibull paper gives the curve indicated on figure 15.15. Conventional calculations give a negative value of $(t - t_0)$ and hence an imaginary Weibull distribution. We therefore try a modified distribution. The mean rank for the infinite distribution is calculated by means of equation (6.63), and the last two columns of the above table then completed. The reliability (= 1 − mean rank) of the infinite distribution is then plotted against $(t_c - t)$ using a guessed value of 60 for $t_c$, to obtain the curve on figure 15.15. This is straightened by the usual technique which gives $\Delta t_c = 5.2$ and hence $t_c = 54.8$. The data can now be replotted using this value of $t_c$ to obtain the straight line shown on figure 15.15. From this we find

$$t_c = 54.8 \text{ weeks}$$
$$\eta = 102\,000 \text{ weeks}$$
$$\beta = 0.35$$

Substituting these values in the modified Weibull distribution the failure rate can be calculated and plotted against time as in figure 15.16.

*Exercise 6.15*

The distributions are plotted on figure 15.17.

*Exercise 7.1*

No failure – the system is functioning exactly as designed.
*Notes*: You should note that the trade term for the item known colloquially as 'overflow pipe' is 'warning pipe'. This latter nomenclature illustrates its dual function: in addition to allowing excess water to escape, the overflow of water is intended as a warning that action is necessary before complete failure occurs. You will appreciate that the overflow pipe is rarely capable of dealing with the full

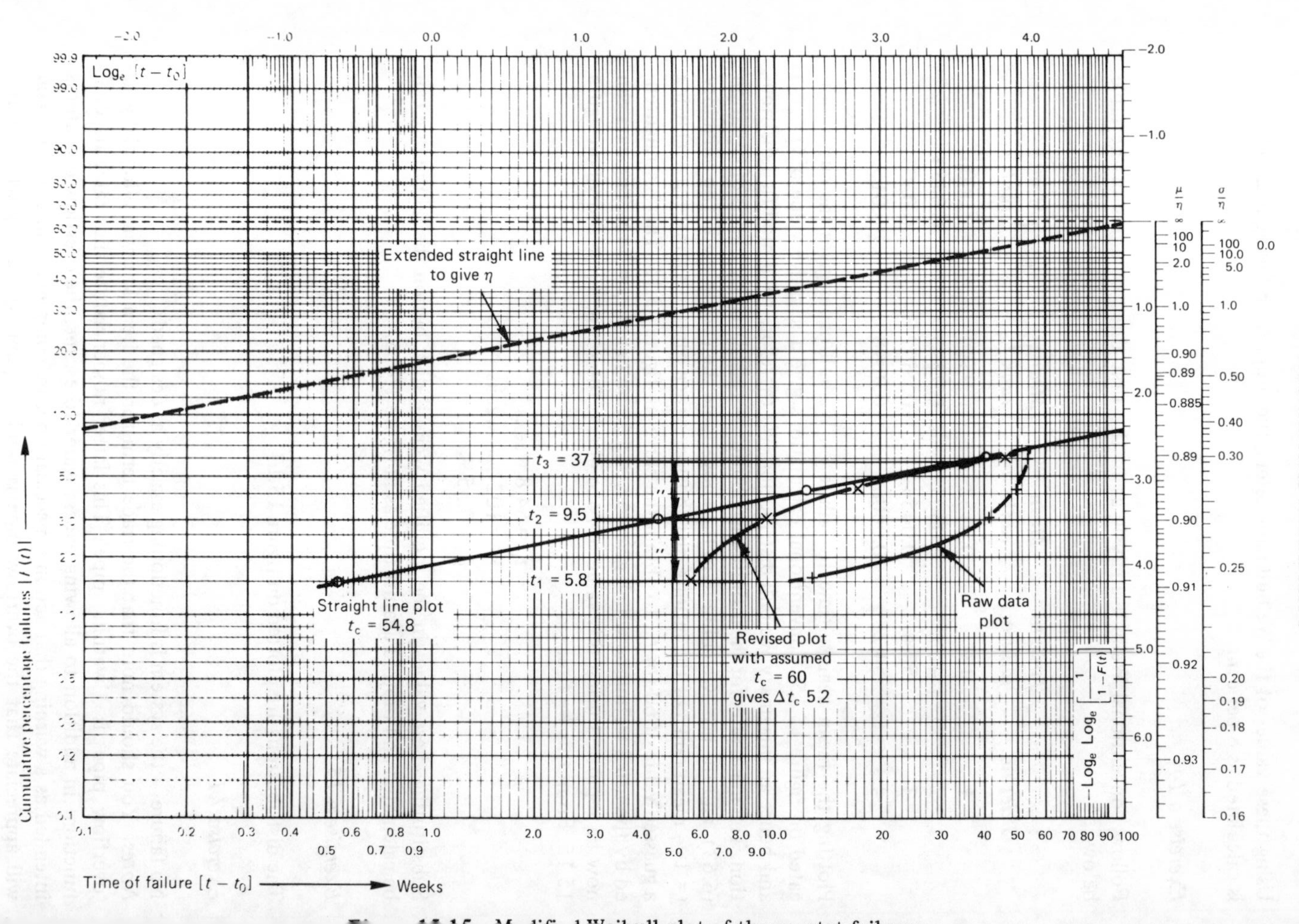


Figure 15.15 Modified Weibull plot of thermostat failures

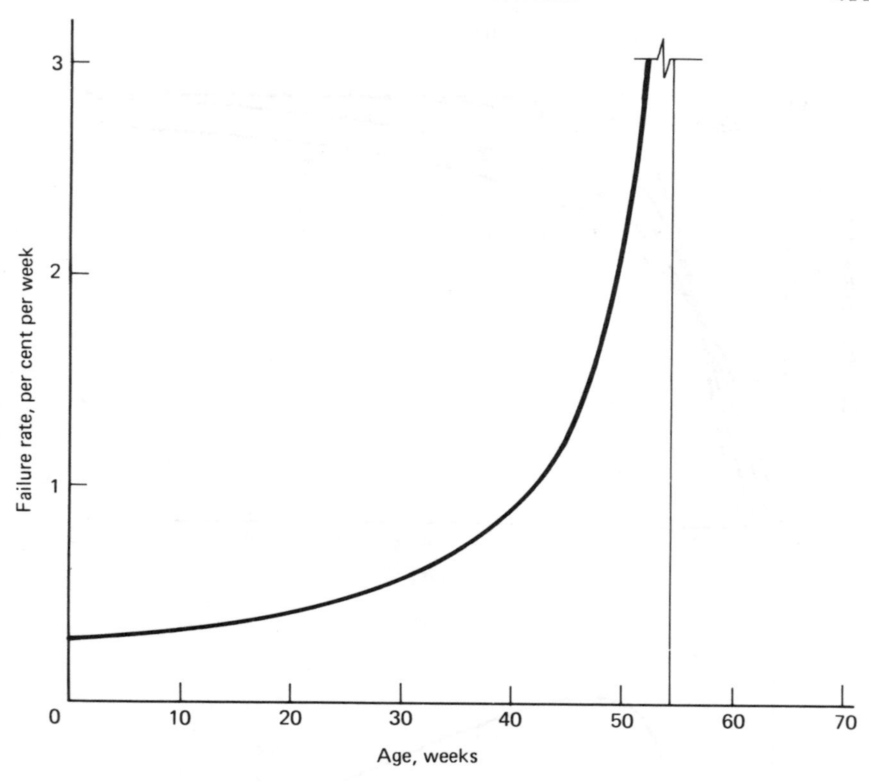


**Figure 15.16** Failure rate for thermostat valves

supply flow. Its main function is therefore warning. The so-called overflow is a well-established example of an early warning device. Early warning techniques in general are a developing topic and discussed more fully in section 9.4. The intention is that on receipt of an appropriate warning – in this case observation of water issuing from the warning pipe – remedial action is taken. In the case we are discussing this involves replacing the washer so that no actual failure of the system occurs.

You should also note that this old fashioned piece of domestic equipment contains a feature which has recently attracted much attention under the modern phraseology of 'fail safe'. According to this doctrine a critical component can fail, but the design of the system is such that the system continues to function in an acceptable manner until remedial action is possible, or if this is not possible the system adopts a secondary mode of operation which avoids secondary damage or shuts itself down without incurring secondary damage. You are reminded that the secondary damage may be several orders of magnitude more expensive than the

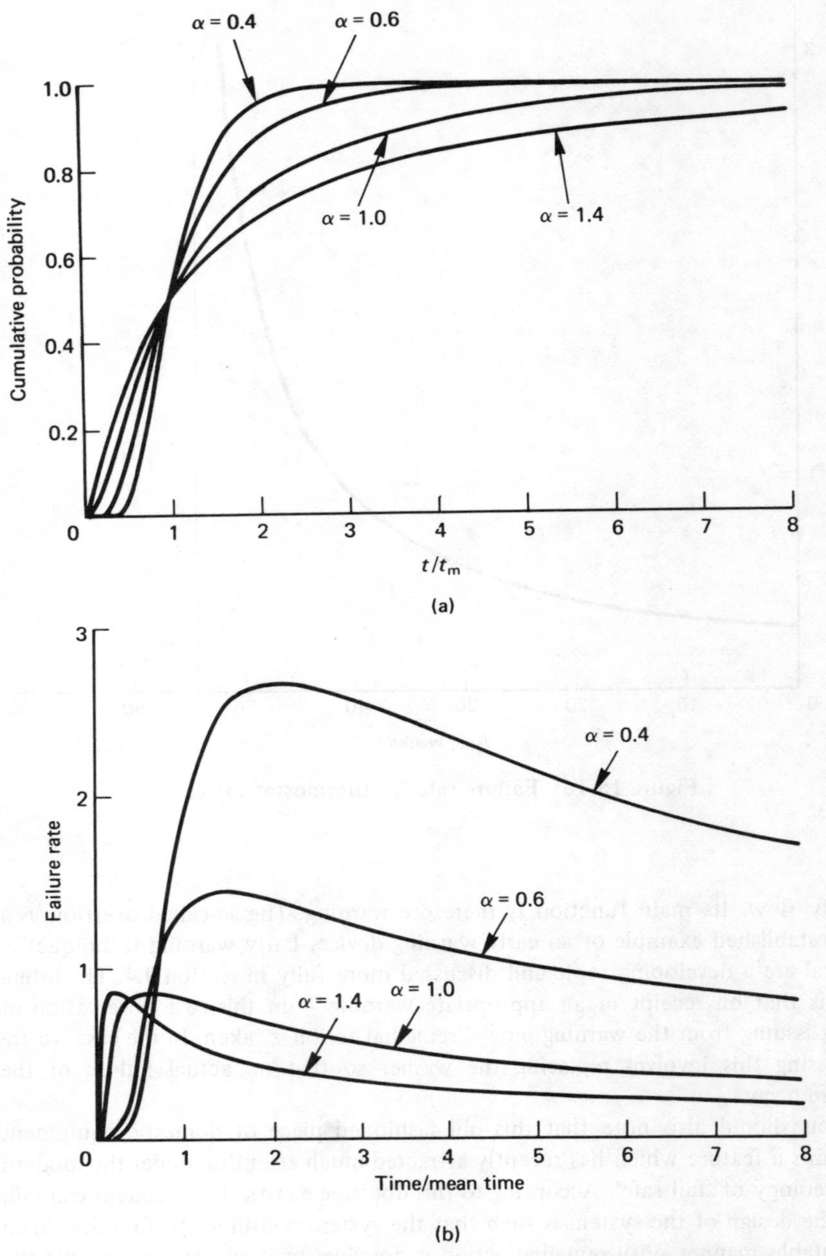


**Figure 15.17** (a) Cumulative probabilities for the log-normal distribution at various dispersions ($\alpha$). (b) Failure rate for log-normal distribution at various dispersions ($\alpha$)

initial failure and hence the importance of this technique. In the case under discussion the washer has failed, but the system has 'failed safe' and is able to operate in an acceptable manner until the washer can be replaced. Obviously it is essential that any fail-safe arrangements should give clear warning of the failure that has or is about to occur.

*Supplementary exercise 7.1A*

Suppose now the situation envisaged in exercise 7.1 occurred in winter so that the overflow froze up and the tank flooded. Is this a failure? If your solution to exercise 7.1 was 'no failure', how do you reconcile your views?
(The solution to this exercise is given at the end of this chapter.)

*Exercise 7.2*

As presented, this is a meaningless question: we really need to know the purpose for which the information is required. A number of simple cases spring to mind

(1) If the data were being assembled for use by a self-drive car hire firm, this would count as a failure. Vehicles owned by such a firm are operated by inexperienced people and by those who, although having some experience, are unfamiliar with all types. Yet success is essential for business reputation. Operator inexperience must then be accepted as a random variable.
(2) If the data were being collected for a taxi-cab firm, on the other hand, it would be reasonable to expect professional expertise from their drivers, and so to discount this particular failure mode from your statistics.
(3) If the data were being collected for engineering purposes, it would clearly be necessary to know the object of the investigation and to define the operating conditions and terms of failure more closely.

This example shows that an incident may or may not count as a failure, and illustrates the necessity of defining circumstances very carefully.

*Exercise 7.3*

The raw data is plotted on Weibull paper in figure 15.18. Although the engineering reports on these failures drew no distinction between any of the failure modes encountered, the plotting suggests very strongly that the early batch of failures follows a modified Weibull distribution, which must be associated with some inherent manufacturing or assembly defect, while the latter batch follows a more conventional Weibull wear-out distribution. The curves which most closely fit the points have been drawn in figure 15.18. Of some interest is the discontinuity between about 8500 and 10 500 miles. Detail analysis of both curves using the techniques already described gave the following results

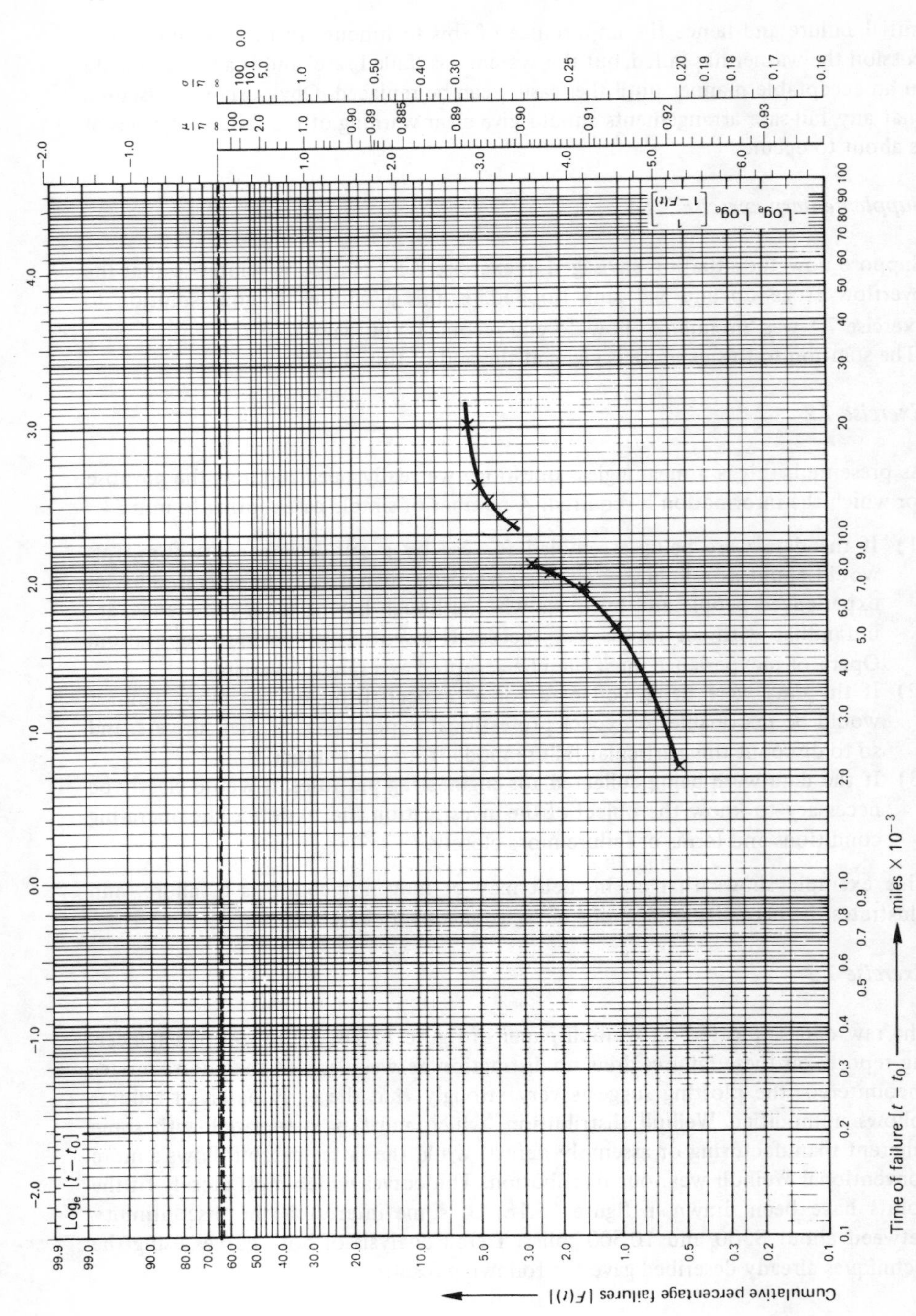


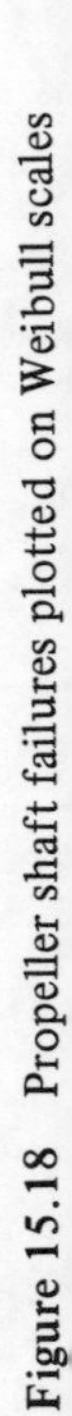

Figure 15.18 Propeller shaft failures plotted on Weibull scales

Early life curve – modified Weigull distribution for which
$t_c = 8455$ miles, $\eta = 1.1 \times 10^7$ miles, $\beta = 0.45$
Later life curve – standard Weibull distribution for which
$t_0 = 10\,513$ miles, $\eta = 3389$ miles, $\beta = 0.59$

It may be deduced from the analysis that the intial group of failures do follow a modified Weibull distribution in which failure is due to some wear-out process whose cause cannot be specified without a further examination of the shafts themselves. We may suspect it arises from some assembly fault which, however, only affects about 4 per cent of the shafts. Failures from this cause would not arise after 8455 miles. Even so it should merit attention: since 96 per cent of the shafts did not suffer this defect one imagines a remedy is at hand.

Following the cessation of the early life failures there is a period in which no failures arise until what appears to be a conventional fatigue mode sets in. This is indicated by a $\beta$ less than unity accompanied by a finite $t_0$. The estimated value of $t_0$ is probably not far out, but the value of $\beta$ should be treated with some suspicion because (1) all the drop-outs arising from the early life failures occur in the most sensitive (that is, the initial) part of the curve, even though they are a small proportion of the total, (2) there are only five analysable points, and (3) there is some suspicion that the last point may be a rogue. In spite of these doubts, the analysis suggests by virtue of the low $t_0$ and low $\beta$ that there is strong interaction between the strength and loading distributions. A higher safety margin does appear to be required, though the necessity should be assessed economically.

This is an interesting example in that it shows both early life and a fatigue wear-out as quite separate modes. It will be noted that there are no random failures and hence an apparent period of intrinsic reliability between 8455 miles and 10 513 miles. Physical assessment with a confirmatory separation of the modes would have added considerably to the strength of the conclusion, but is not evident in the initial reports. The importance of physical evidence would be even greater if the two failure modes merged, as often appears to happen, to give an S-shaped curve which is difficult to interpret using purely analytical techniques.

*Exercise 7.4*

The two sets of graphs make an interesting comparison. Taking a common feature first: all the graphs have a curvature which is concave downwards, indicating a finite positive value of the locating constant in the Weibull distribution. We may therefore conclude that in all cases there was a period of intrinsic reliability before failures were recorded. Here similarity ends. Although it is not possible to deduce accurate values, owing to the scale of the graphs, it is readily seen that the slopes of the lines on figure 7.11 are all considerably greater than those of figure 7.12. Reference 39 quotes values well in excess of 3, which may easily be confirmed from figure 7.11 in spite of the scale. Reference 40 quotes a shaping parameter of unity for the curves in figure 7.12, and again this may be confirmed from that figure. There is certainly no slope in excess of unity. These deductions are emphasised by the curves which have been drawn through the points on figures 15.19 and 15.20, though these are not those drawn by the authors of the references quoted.

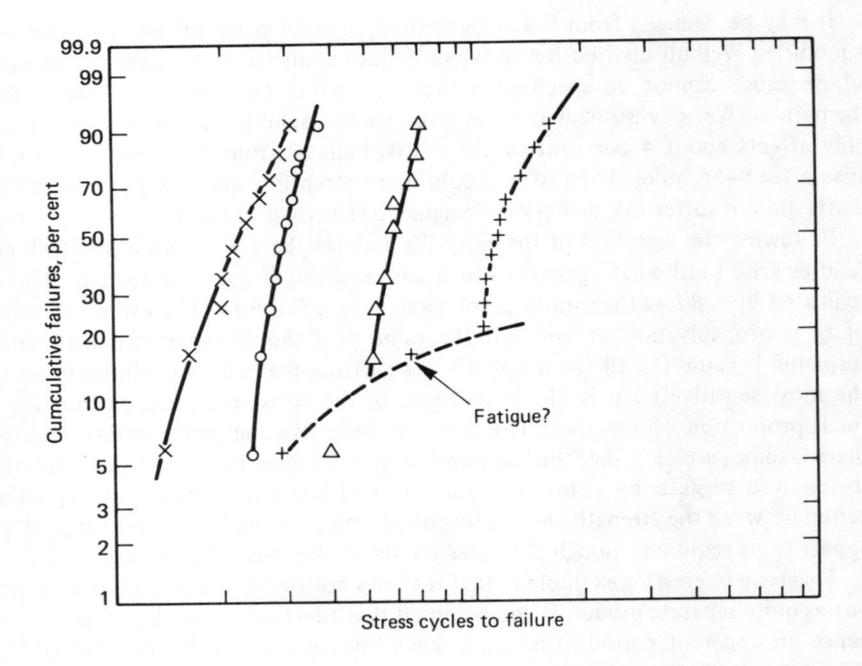


**Figure 15.19** Cumulative distributions of RHPN 1005 bearing failures for various levels of debris filtration plotted on Weibull scales (data from ref. 39).
x 40 $\mu$m filtration, ⊙ 25 $\mu$m filtration, △ 8 $\mu$m filtration, + 3 $\mu$m filtration

The importance of these results lies in the physical identification of the failure mechanism in refs 39 and 40 for the same, if not identical, components (that is, the balls of a ball bearing). In those tests for which the failure mechanism has been identified as fatigue, the shaping parameter is close to or slightly less than unity. In those tests for which the failure mechanism has been identified as surface damage or wear followed by disintegration, the shaping parameter is greater than three. Although this test data refers to laboratory conditions rather than true field conditions, it does provide important confirmatory evidence of the distinctive failure-age patterns for true erosive wear and fatigue wear. It is as near a direct comparison as may be obtained in this kind of work, and lends substantial support to the general concept of wear-out failure patterns deduced in chapter 5.

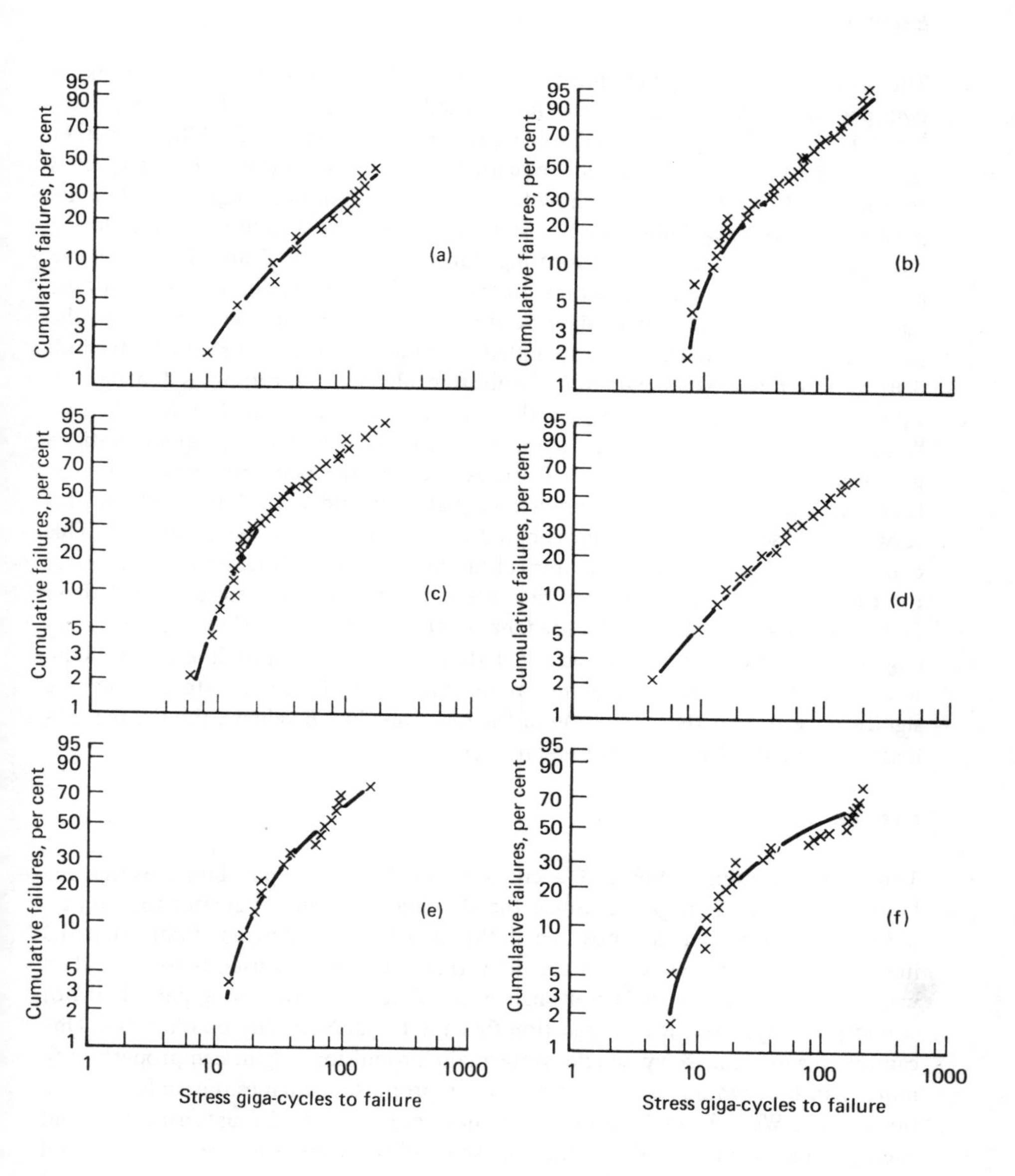


**Figure 15.20** Cumulative distribution of AISI 52100 ball failures with various lubricants, plotted on Weibull scales (data from ref. 40). (a) Paraffinic mineral oil. (b) Formulated tetraester oil. (c) Tetraester base. (d) Traction fluid 1. (e) Traction fluid 2. (f) Synthetic paraffinic oil

*Exercise 7.5*

The author regards the equation with some suspicion: it looks very much like a two parameter Weibull equation. Indeed, substituting $R = 0.9$, $t = B_{10}$, and $\beta = 1.17$ into the equation and solving for $\eta$ gives a value of $6.84B_{10}$! Whether this was the basis of the empirical equation or not, it can certainly be interpreted in that way. It is difficult to see how this can be justified. For fatigue, which is the predominant bearing failure mechanism, there must be a failure-free life during crack nucleation, that is, the locating constant must be finite. Indeed, some authorities recognise this by adding an 'excess life' to the value derived from the equation. It is not clear why they do not use a three parameter function in the first place. It should also be noted that representing a three parameter Weibull distribution by a two parameter distribution always over-estimates the derived value of the shaping parameter. So the value of 1.17 may be higher than the true Weibull value. This would then agree more closely with the concepts of fatigue presented in chapter 5. If the differences between the two and three parameter functions can be accepted, the empirical equation would seem simple and straightforward to use, but circumstances could arise when it could be misleading. This could be so when the further empirical factors for load, lubrication etc. are used; for the way they are introduced into the empirical equation suggests that these factors modify the characteristic life (that is, the standard deviation) only, whereas engineering considerations indicate that they modify the failure-free life as much, if not more, than the subsequent failure pattern. One can therefore anticipate significant errors in the estimation of the low lives (low $B$ values), though the estimation of median lives may not be far in error.

*Exercise 7.6*

This is an intriguing example. The curvature of the Weibull plot indicates that the locating constant is negative. Since the shaping parameter is greater than unity, only two explanations are possible if this is a true distribution: firstly that the item has a limited shelf life, or secondly that the item was used before the data were collected. Although Diesel engines can deteriorate over long periods if not stored properly, there is no suggestion that the Royal Navy buy up old stock. One cannot imagine the Navy would store main propulsion engines improperly any more than they would buy second-hand main propulsion engines. Feasible explanations of the Weibull distribution just do not hold water. One must suspect the old enemy – mixed modes. The author has seen other curves (some relating to Diesel engines) of the type indicated. They arise when only one or two failures in each of half a dozen or so different modes have been collectively plotted and analysed as an entity. The rapid rise of the cumulative failures with age shown on this plot is then due to an increasing number of modes participating in the pattern rather than to the peculiarity of any failure mechanism. Hence one would require more evidence before such data could be interpreted. All cases of such patterns which the author has been able to examine in detail showed that mixed modes had been used in the analysis; but in the majority of cases, of which this is one, no conclusion can be reached, because of inadequate data.

In a similar manner, one sometimes comes across equipment which apparently has a very high Weibull shaping parameter – values of 10 are not unknown. Again it is suspected that these are due to several components ceasing to be intrinsically reliable at the same time. This may be no bad thing, because it indicates that maintenance action can be taken on all items simultaneously, so reducing cost. It does demonstrate, however, that Weibull analysis is not the whole story. The author finds it frustrating that most exceptional failure patterns cannot be assessed properly because of inadequate data: much valuable information is lost.

The reader may feel it was unfair to set this exercise; but the author believes that the presentation of corrupt data without caveats is fully justified to inculcate a sceptical view of purely statistical analysis. Nearly all practical data appears to be corrupt in some way, and rarely comes with a Government warning on the packet!

As a result of trying to analyse much field data the author would state categorically that each mode must be analysed separately, and that only data up to first failure should be included since subsequent data can be corrupted by maintenance. This may seem a council of perfection, as indeed it is, and it is fully appreciated that it cannot often be followed in real life. Nevertheless this council of perfection should always be kept in mind and, if it cannot be followed, great care should be taken in interpreting analysed data. Purely statistical analysis should be regarded with suspicion and only accepted when it agrees with an engineering appraisal. Practical necessity may prevent this, but a knowledge of the dangers can avoid an unwarranted conclusion being reached.

*Exercise 9.1*

To conform with modern maintenance scheduling it will be assumed that the replacement interval for the pump could be set at 12, 24, 36 or 48 thousand miles with a total vehicle economic life of 95 thousand miles (that is, just before a scheduled maintenance which is thus not implemented). Using the Hastings' computer programme, the number of both unscheduled and scheduled replacements, for each of the scheduled maintenance intervals listed above, can be calculated and are given in the table below.

| *Maintenance interval* | *Unscheduled replacements* | *Scheduled replacements* |
|---|---|---|
| 12 | 0 | 700 |
| 24 | 45 | 292 |
| 36 | 51 | 179 |
| 48 | 54 | 95 |
| 95 | 55 | 0 |

One can expect a cost ratio, $K$, around 5 in this application, but to cover most practical cases the total cost of maintenance (expressed in terms of the cost of one scheduled replacement) have been calculated for $K = 1$, $K = 5$ and $K = 10$, and given in the table below.

| *Maintenance interval* | $K = 1$ | $K = 5$ | $K = 10$ |
|---|---|---|---|
| 12 000 | 700 | 700 | 700 |
| 24 000 | 337 | 517 | 742 |
| 36 000 | 230 | 434 | 689 |
| 48 000 | 149 | 365 | 635 |
| 95 000 | 55 | 275 | 550 |

In all cases it will be seen that the cheapest maintenance cost occurs at a replacement interval of 95 000 miles, that is, the rated life, indicating that a repair maintenance policy, the one adopted in practice, is the most economic one.

A scheduled maintenance policy will only be more economic if $K$ is greater than $700/55 = 12.7$ when the scheduled replacement interval should be 12 000 miles, that is, that retaining intrinsic reliability.

*Exercise 9.2*

$$\text{Availability} = \frac{\text{up time}}{\text{up time} + \text{down time}}$$

$$= \frac{(\text{up time} + \text{down time}) - \text{down time}}{(\text{up time} + \text{down time})}$$

Now (up time + down time) is the total time available and we may take this as a constant: in the limit it is 24 hours per day, though in other circumstances it may be less, that is, say a working day of 8 hours. However, in any application the total time is constant, and hence to maximise availability we must minimise down time.

Let $T_s$ = time for scheduled replacement
$T_u$ = time for an unscheduled replacement

Then down time = $N_s T_s + N_u T_u$

Where $N_s$ and $N_u$ have the same definitions as in the main body of the text. Then

$$\frac{\text{down time}}{\text{down time to retain intrinsic reliability}} = \frac{N_s T_s + N_u T_u}{N_i T_s}$$

$$= \frac{N_s + K N_u}{N_i}$$

$$\text{where } K = \frac{\text{time for unscheduled replacement}}{\text{time for scheduled replacement}}$$

This is the same as expression (9.14), both of which have to be minimised. It follows that with a different interpretation placed on $K$, which could, of course, lead to different numerical values in any particular case, the same general procedure regarding the choice of maintenance policy should be followed whether cost or availabily is used as a criterion.

*Exercise 9.3*

The following factors all tend to increase the value of the cost ratio $K$

1. Rarely has a scheduled repair to be performed at a precise time and generally can be planned at the most convenient time to suit the operational programme. Lost revenue is therefore minimised.
2. The provision of the spares for a scheduled repair can be planned in advance so that the capital value of spares held in store is minimised.
3. In the same way human resources can be minimised.
4. For an unscheduled replacement the equipment has either to be taken to the repair man/team, or the man/team has to be taken to the equipment. This involves transport costs and additional down time.
5. A repair man/team may not be available if several unscheduled failures coincide. Further time is lost. If more men/teams are employed they could be idle during slack time which would again involve additional cost.
6. Similar considerations apply to material resources. A comprehensive store locks up capital and itself costs money. On the other hand, a rudimentary store could well involve delay while spares were sought from a central store or supplier.
7. The administrative costs associated with an unscheduled replacement must be higher and take longer than those associated with a scheduled replacement. This implies both higher cost and lost revenue.
8. An unscheduled failure can incur secondary damage. The cost of the latter could even be higher than the cost of the primary failure. In any case it is an additional cost.

*Exercise 9.4*

Imperfect maintenance can be represented by a standard Weibull distribution with a shaping parameter less than unity or by a modified Weibull distribution, depending on the type of the imperfection. There is no difficulty in assimilating these into the calculation of the failures with replacement, as described in section 9.2. The real difficulty is attributing values to the various parameters.

A simpler approach which is more amenable to calculation assumes a maintenance efficiency ($\eta$), which is the proportion of the maintenance activities which is successful. A proportion $(1 - \eta)$ will have to be repeated. In this simple approach it is assumed that these would be detected very soon after re-introduction into service, so no 'life' need be introduced into the replacement theory calculation. It is also assumed that every second attempt is successful. In that case we can re-write equation (9.13) as

$$\frac{\text{total cost of maintenance}}{\text{cost of retaining intrinsic reliability}} = \frac{N_s + (1-\eta)KN_s + KN_u + (1-\eta)KN_u}{N_i + (1-\eta)KN_i}$$

in which any repeated action is assumed to be an unscheduled action. It should be noted that the cost of retaining intrinsic reliability in the left-hand side denominator of this equation is now the cost of retaining intrinsic reliability with imperfect maintenance.

Then

$$\frac{\text{total cost of maintenance}}{\text{cost of retaining intrinsic reliability}} = \frac{N_s[1 + (1-\eta)K] + N_u[K + (1-\eta)K]}{N_i[1 + (1-\eta)K]}$$

$$= \frac{N_s + K'N_u}{N_i}$$

$$\text{where } K' = K\,\frac{[1 + (1-\eta)]}{[1 + K(1-\eta)]}$$

Since $K$ is always greater than unity for any scheduled maintenance policy then

$$K' < K$$

It has already been shown that the lower the value of $K$ the longer should be the period between scheduled maintenance activities. It follows that with inefficient maintenance the time between scheduled maintenance activities should be ***greater*** than with perfect maintenance.

The effect of imperfect maintenance can be significant. If the maintenance efficiency is 90 per cent, the effective value of $K$ is reduced from 5 to 3.7 for example, and from 10 to 5.5. If the efficiency is only 60 per cent, the reductions are from 5 to 2.3 and from 10 to 2.8. Such figures are not impossible – the number of items that are removed from aircraft, only to discover at base that they are perfectly satisfactory, can be high. Reference 63 states that 60 per cent of removals were unjustified! One wonders what the figure is in less sophisticated industries. Tests by the Consumers' Association using 'planted' faults in typical consumer items confirm the above indications and suggest the figure could be a lot worse. On the basis of the above theory these figures could often imply a change from scheduled maintenance with ideal maintenance to a repair policy with inefficient maintenance. It may well account for the old adage of the cynical mechanical engineer – "if it's doing what you want to your satisfaction: leave it alone." Certainly it is necessary to take into account maintenance efficiency if optimum maintenance policies are to be evaluated. Unless this is done estimates could be widely adrift.

It may be worth recording that the above inefficiency derives from the failure to perform some maintenance action correctly; but a second form of inefficiency may arise from the disturbance of components, including those not directly involved in the maintainance, during the maintenance process. This aspect of mechanical maintenance also needs recognition, because it always occurs to some extent even with the best of maintenance. Moving mechanical parts wear *in* during their initial operating period. Replacement parts will most likely have different dimensions, owing to manufacturing tolerances, even if the wear is negligible. Hence the 'fit' after maintenance will not be the same, and tolerance stack-up could give rise to significant differences of fit. While parts can rewear in, this will take time and is often a different process since the mating surface finishes are no longer in the 'raw' state. Sometimes the final result is not as good as the original, particularly if a lot of wear has already taken place. It is a case of 'new wine in old bottles'. The condition after maintenance can then be not just 'bad as old' but 'worse than old'! How many of us have encountered this at some time or other. It is difficult to make any generalisation on this phenomenon, and each case has to be treated on its own merits.

*Exercise 9.5*

Items which could be monitored and the value of so doing can be enumerated.

1. *Delivery pressure*. This is dependent on circuit resistance (number of radiators in use etc.) and would show too great a variation for monitoring purposes.
2. *Load*. The load could be measured electrically, but is again expected to vary over too wide a range in normal practice to be of much use for monitoring purposes. An upper limit – say 10 per cent above normal maximum – is only likely to be exceeded during certain failures and would be of little use for monitoring. Note that the fuse does this job anyway, and only signifies a failure has already occurred. It is also worth noting that in this application a low load could also be indicative of impending failure. For example, excessive erosion of the impeller could reduce the load below the standard demand level. Since the lower limit is variable, it is difficult to see how this feature can be utilised, however.
3. *Temperature*. There could be some temperatures which are indicative of state, but in this application the temperature of the various parts would depend on the chosen water temperature and hence vary widely. Greater temperature variations would only be recorded when failure was actually occurring. Hence temperature is of little use for monitoring purposes.
4. *Vibration*. Vibration would give indication of several types of failure. Because of the influence of installation it would be difficult to establish a general level and the change of vibration in any particular installation would have to be used as a criterion.
5. *Noise*. Noise is an obvious quantity to monitor and can give first indication of impending failure to the acute ear without artificial aid. However, this is im-

precise. Automatic monitoring would have to be based on change of noise emitted since the quality would depend on the installation.
6. *Lubrication*. There is no oil system which can be used for monitoring in these simple pumps.

In general, 4 and 5 offer the most hopeful technical solutions, but both require rather expensive equipment so that any consideration of a continuous monitor must be discarded. Even an intermittent monitor by a service engineer could well be too expensive. It must be appreciated that cost is the deciding factor in all maintenance schemes. You should consider if any simple versions of 4 or 5 could be produced to do the job in question, that is provide a continuous monitor and an early warning to the user of any impending failure so that maintenance action can be taken. If possible, would they give adequate protection?

It is also worth noting that, on account of the low cost of pumps, this may be a case where a standby unit is the cheapest solution if high reliability is required. This is particularly so if some consideration is given to ease of replacement in the initial design.

You should also consider how far reliability is of itself important, or whether availability is not the better criterion. If so, what additional factors have to be taken into account before deciding on an optimum course of action?

*Exercise 9.6*

A fan is in many respects similar to a pump and hence the factors which could most easily be monitored are substantially the same as in exercise 9.5. Although the environmental conditions are different it is still considered that the normal variations in suction or delivery pressure, in load, and in temperature would be greater than that to be expected prior to failure. Vibration and noise are therefore the main contenders, with a preference for the former as a simpler technique. In this application the cost of failure could be high and hence could justify the expense involved.

To examine its efficacy we first list the major failure modes likely to be encountered

1. Failure of fan blades either by corrosion or because of the polluted atmosphere.
2. Mechanical failure of the fan blades.
3. Mechanical damage to the fan blades from any debris or solid matter carried in the air stream.
4. Bearing failure arising from wear.
5. Failure of the stator element: this would particularly be associated with loosening of fasteners, etc.
6. Loss of tip clearance, mainly associated with distortion or incipient failure of the stator element.
7. Rough running associated loss of 'line' with driving motor, or loss of balance of the rotor assembly.

Since a catastrophic failure is extremely unlikely in this application – all the failure modes, except 1 above, would be preceded by some vibrational effects, either because of out-of-balance forces arising from any damage to the rotor, rough running in the case of bearing failures, or general rattling and so forth associated with the pre-structural failure of the stator element. It is possible, though by no means certain, that some indication of corrosion would be given by a change of vibration characteristics.

Vibration would therefore appear to be an attractive monitor. It is suggested that an accelerometer mounted on the fan casing would provide the necessary signals, though two mounted to record vibration in two perpendicular planes are preferable. The output from the accelerometer could be integrated to provide the mean velocity of vibration as a single measure of the unit's condition. Although the absolute value would be useful in the initial setting up, it is considered that change of vibration level would be the most relevant parameter for condition monitoring. As a more expensive alternative which would provide more information, the output from the accelerometers could be analysed to give the level/frequency relationship at discrete intervals of time. An increase in the level of vibration at a known frequency is not only an indication of impending failure, but could give a lead to the source of failure.

Since vibration monitoring would not reveal any corrosive action, it is necessary to back up vibration recording by direct visual inspection at discrete intervals of time. Access for this must be provided at the design stage.

Hence visual inspection together with vibration monitoring are considered to be an adequate protection against the most likely failure modes.

*Exercise 9.7*

To calculate the reliability we have to take into account that a failed vehicle would be repaired, if possible within the time constraint, and then carry on with the mission. It may fail a second or third time, and on the occasion of each failure a repair would be attempted. From the Poisson distribution we obtain the probability that the vehicle will complete the mission

$$\text{without a failure} = e^{-\lambda x}$$

$$\text{with 1 failure} = \lambda x e^{-\lambda x}$$

$$\cdot \qquad \cdot$$

$$\cdot \qquad \cdot$$

$$\text{with } i \text{ failures} = \frac{(\lambda x)^i e^{-\lambda x}}{i!}$$

Now every time a vehicle has a failure the probability of successfully completing a repair is, for a time constraint $\tau$

$$(1 - e^{-\tau/\phi})$$

so that the probability of a vehicle having one successfully repaired failure is

$$\lambda x \mathrm{e}^{-\lambda x}\,(1-\mathrm{e}^{-\tau/\phi})$$

Similarly, for a vehicle with $i$ failures the probability of every failure being repaired in the time $\tau$ is

$$\frac{(\lambda x)^i \mathrm{e}^{-\lambda x}}{i!}\,(1-\mathrm{e}^{-\tau/\phi})$$

thus the probability of repairing all failures is

$$\sum_{t=0}^{t=\infty} \frac{(\lambda x)^i \mathrm{e}^{-\lambda x}(1-\mathrm{e}^{-\tau/\phi})}{i!}$$

$$= \mathrm{e}^{-\lambda x} \sum_{t=0}^{t=\infty} \frac{(\lambda x)^i (1-\mathrm{e}^{-\tau/\phi})}{i!}$$

$$= \mathrm{e}^{-\lambda x}\,\mathrm{e}^{\lambda x(1-\mathrm{e}^{-\tau/\phi})}$$

$$= \mathrm{e}^{-\lambda x}\,\mathrm{e}^{-\tau/\phi}$$

This completes the exercise as set, but it should be noted that the above analysis can be carried out, for example, using the Weibull distribution in place of the negative exponential, though the final solution is not so simple. In either case, when $\tau$ is very large, as may arise between missions, or $\phi$ is very small, or zero as in the case of ideal repair maintenance, then the reliability approaches 100 per cent for all values of $x$. This arises because all failures are being repaired; but the occurrence of failures ($\Lambda$) remains finite (equal to $\lambda$ for the negative exponential distribution). However, during a mission $\tau$ may be expected to be small and, unless $\phi$ is also very small, the failure rate approaches its value with no maintenance (equal to $\lambda$ for the negative exponential distribution), that is, in the limit when little or no maintenance is possible during the mission, the occurrence of failures during a mission is roughly equal to the failure rate at the appropriate age.

*Exercise 9.8*

The relevance of $M$ as a measure of maintainability depends on the relevance of the time constraint to the particular situation. If time itself is the limiting factor then $M$ would seem to be a suitable measure of maintainability. One would therefore expect it to be a measure of most first line maintenance which has often to be fitted into operational requirements. For example, with 10 minutes between lectures in a standard college timetable, 10 minutes is clearly the time constraint that could be allowed for any maintenance of a projector in a lecture room, although a longer period would apply at the end of the day and at weekends – and possibly during vacations. The quantity $M$ clearly has operational relevance. In particular, many tactical situations are time limited, and this has resulted in $M$

being adopted for many defence purposes. There are obviously civil counterparts, as in the example quoted above, but its general value should not be over-rated. Even in the defence field it is doubtful if it is relevant to second line maintenance, and certainly not to third line. Cost is the relevant factor in these circumstances. The majority of mechanical maintenance activities have *got* to be carried out at some time, unless excessive cost proves an over-riding factor. If the job has to be done it is then more important to know how long the maintenance activity is likely to take, both for planning purposes and for costing, than to put any time limit on the activity. The $Z$ measure would seem more relevant in these circumstances.

So far as specification is concerned, it may be desirable to specify $M$ if there is an operational requirement, but it is clear that the undue length of time involved in many maintenance activities is of as much if not more practical concern. It seems essential therefore that a $Z$ value is set as a specific requirement in most specifications.

*Exercise 9.9*

In order to calculate the number of failures to expect with replacement, it is necessary to express the failure pattern as a Weibull distribution. Since the failures are near normally distributed the shaping parameter, $\beta$, = 3.44.

For $\beta = 3.44$, $\dfrac{\mu}{\eta} = 0.899 \approx 0.9$ from equation (6.36) or from figure 6.5

$$\text{Locating constant } t_0 = 30$$
$$\mu = 150 - 30 = 120$$
$$\eta = 133$$

Therefore the Weibull distribution is given by $t_0 = 30$, $\eta = 133$, $\beta = 3.44$.

We can calculate the number of tiles which fail and have to be replaced in each ten year interval as described in section 9.2, taking the time interval equal to 10. The table below gives the number which fail in each interval, the cumulative failures and the cumulative cost of replacement.

| *Age* | *Failure in previous 10 years* | *Cumulative failures* | *Cumulative replacement cost (unit = initial cost of single tile)* |
|---|---|---|---|
| 10 | 0 | 0 | 0 |
| 20 | 0 | 0 | 0 |
| 30 | 0 | 0 | 0 |
| 40 | 0 | 0 | 0 |
| 50 | 6 | 6 | 12 |
| 60 | 22 | 28 | 56 |
| 70 | 49 | 77 | 144 |
| 80 | 90 | 167 | 334 |

| | | | |
|---|---|---|---|
| 90 | 145 | 312 | 624 |
| 100 | 211 | 523 | 1046 |
| 110 | 286 | 809 | 1618 |
| 120 | 364 | 1173 | 2346 |
| 130 | 436 | 1609 | 3218 |
| 140 | 494 | 2103 | 4206 |
| 150 | 529 | 2632 | 5264 |
| 160 | 534 | 3166 | 6332 |

The cumulative cost has been plotted against age in figure 15.21. It should be noted that the initial cumulative replacement cost is usually assessed as the acquisition cost less the resale value. In this case the resale value is negative –

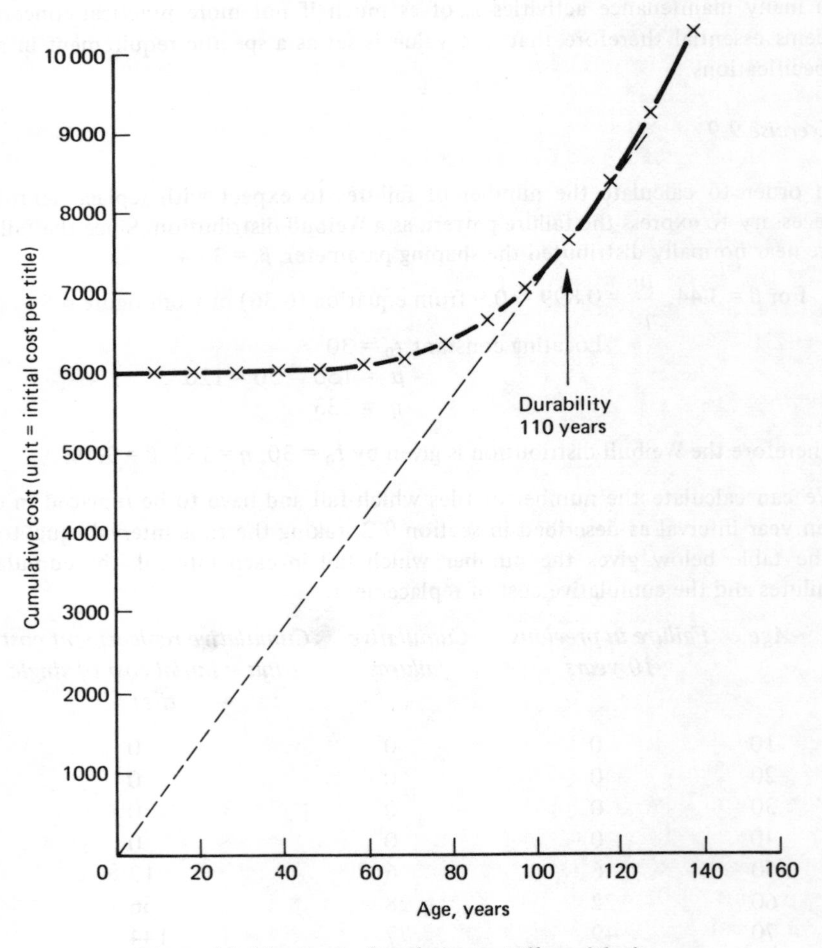


**Figure 15.21** Cost of replacement tiles with time

you've got to pay to get them taken away! The minimum cost per year is then deduced graphically as the tangent to the curve passing through the origin. (The slope of any line through the origin to the cumulative cost at any year measures the average cost per year at that time, and has a minimum when the line is a tangent to the curve.) From this the durability can be assessed as 110 years. At that time it would pay to replace the whole roof tiling. (It should be noted that, for simplicity, the cost of replacement tiles has been expressed as an average value using the average cost of the original tiling as a unit. Real costs of repair are more likely to consist of a constant for access plus a cost dependent on the number of tiles. While this may give a slightly higher value for the durability, the principle of the calculation remains the same.)

The above example refers to a multi-component item in which all components are identical. Once again the same principle would apply to a multi-component item of different components. The relationship between the durable life and the mean life would be different in each instance, and depend on the components' failure-age distributions.

As a separate but important side issue it will be observed that the number of failures in 60 years is given in the above table as 28, whereas the value calculated in exercise 5.2 is only 7. Since no second generation failures are involved, this is a significant difference. It stems from the representation of the failure pattern by the Weibull and the normal distributions respectively. It should be emphasised that there is no fundamental reason why either distribution is more valid. It illustrates the importance of correct representation, particularly of the tails, of the actual failure pattern before any real life calculation can be undertaken.

*Exercise 9.10*

The change of slope in the cumulative cost curve is due to the introduction of second generation components. It is a reflection of the changes in slope of the cumulative failures seen at about 60 per cent of the life for the near normally distributed failure pattern illustrated on figure 9.5. It may more readily be illustrated by extending the curve of cumulative costs of roof tiles (exercise 9.9) as in figure 15.22. This shows that in addition to the original minimum cost per unit time which defines durability, there is a second minimum at 275 years whose value is however greater than the first minimum, that is, the cost per unit time is greater. It is to be expected that the $k$ value for the second branch of the curve will be less than that for the first, for as the age increases and more later generation components are introduced so the failure pattern approaches the pseudo-random form. Hence $k$ approaches unity. In the simple case the second minimum must have a higher value because the cost of the replacements are necessarily higher than the cost of the original component – if only because the failed component must be removed. However, this may not necessarily be true if many different components of differing failure-age patterns are involved.

It should also be noted that there is a maximum cost, and if the durability

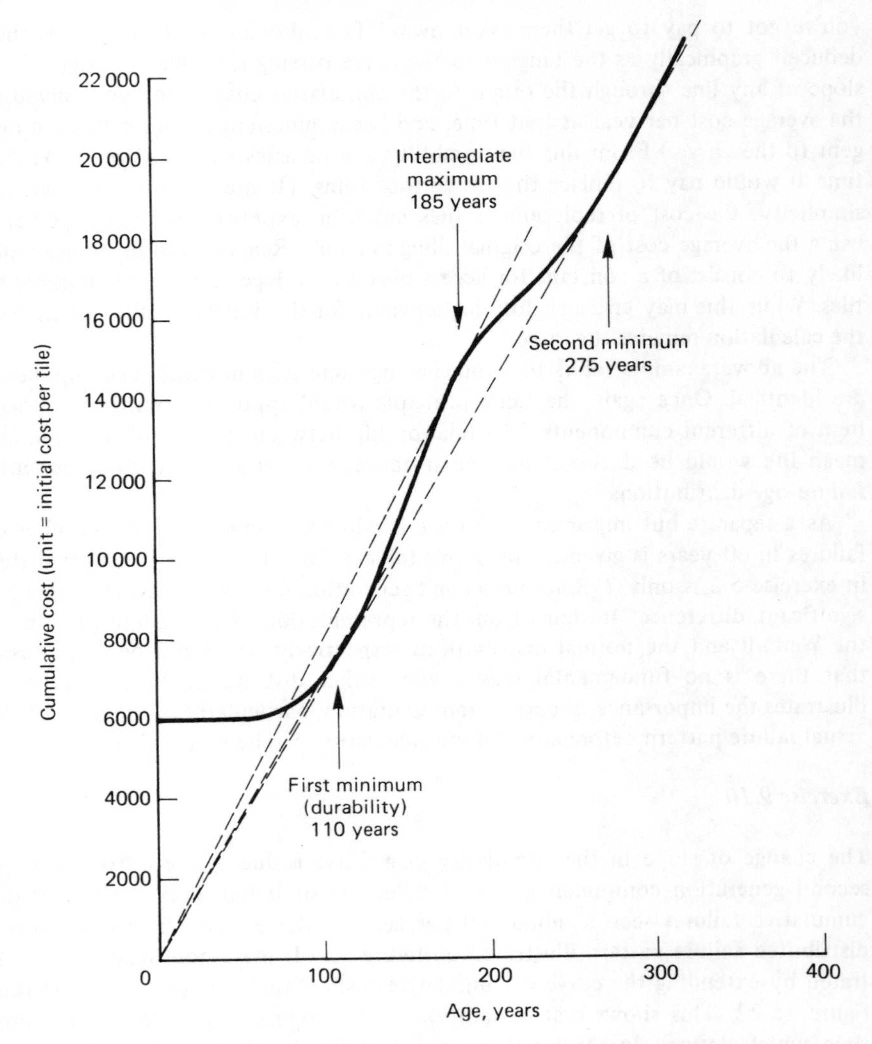


**Figure 15.22** Cost of replacement tiles over an extended period

(that is, the initial minimum) is exceeded the cost per unit age will increase until this maximum is reached. Thereafter it decreases to the second minimum. It would pay to continue using the item once complete replacement had been delayed beyond the age where the average cost equalled the second minimum (130 years in this particular case) until the second minimum is reached: in colloquial terms, 'getting one's money back' after a period of heavy repair bills. The optimum procedure in each case would, of course, depend on precise failure pattern and repair costs applicable to the situation, and in real life the cost of money.

*Exercise 9.11*

The cumulative maintenance cost has been plotted against age on log–log paper in figure 15.23. These costs can be represented reasonably by the straight line drawn on the graph, from which it follows that the equation

$$C_{\mathrm{m}} = Bt^k$$

is a valid representation with $B = 40$ and $t = 1.3$. Substituting into equation (9.35) with $A = 550$ gives $r(t) = 141.5$ when $t = 8$. The estimated cost of repair is therefore greater than the repair limit $r(t)$ and the boiler should be scrapped and replaced by a new one.

In this example the estimated repair cost and the repair limit cost are reasonably close. The decision to scrap is based largely on previous experience which may, or may not, be representative and which may not be valid for the future. (Unless he has a number of colleagues with boilers older than his.) Decisions of this nature do depend on having substantial evidence available if they are to be made with any degree of confidence.

The durability of this boiler calculated from the evidence presented, using equation (9.33) is just under 20 years, so that the owner may be considered unlucky to have to scrap it after 8, but he would be incurring extra expense by repairing it.

*Exercise 10.1*

1. We can identify the 'mission endurance' with the period of a college lecture. One may suppose that a student is not prepared to be handicapped by an inability to take notes as a result of a pen failure for more than one lecture in his whole course (you may specify another figure to satisfy your own temperament and purpose). Assuming 20 contact h/week for 33 weeks per year for 3 years the reliability required during the mission is

$$100 - \frac{1 \times 100}{20 \times 33 \times 3} = 99.95 \text{ per cent}$$

   Assuming failures are of a random nature you could express this as the mean time between failures (see *Note a*).
2. Ten minutes are allowed between lectures, of which one may assume that about seven are taken up in moving to the appropriate lecture room and preparing for the lecture. Hence minor servicing or adjustments which can be carried out with a penknife in less than 3 minutes are permissible every hour (see *Note b*).
3. Assuming 20 per cent utilisation, the pen must be capable of writing for one month (4 weeks) without replacing the ink cartridge. Means must be provided to give at least one day's indication (5 h) that the ink in the cartridge is about to be expended (see *Note c*).

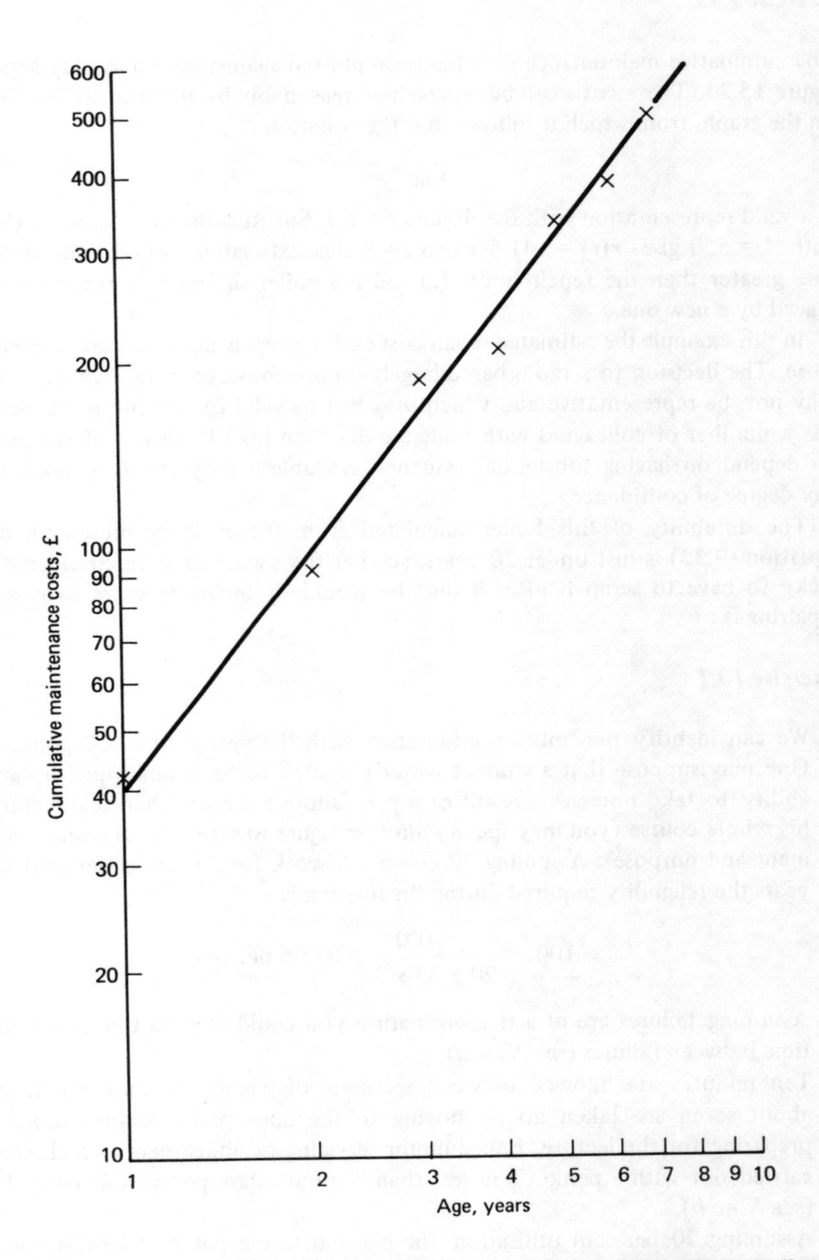


**Figure 15.23** Cumulative maintenance costs of central heating boiler (log–log scales)

4. Any retracting mechanism which is provided may be serviced/replaced once a term (see *Note d*).
5. It should be assumed that the pen will be carried loose in a jacket pocket with other articles. In use it may be knocked from a laboratory bench and sustain a free fall on to a concrete floor (see *Note e*).
6. For the purpose of this specification a failure is defined as the inability to produce a line between 0.2 mm and 0.8 mm thick on demand when used in the normal way (see *Note f*).

*Note a:* This pre-supposes the nature of the failures. It is better practice to quote the required reliability (in this case 99.95 per cent) over the time period for which it is required (in this case 3 years).
*Note b:* The constraint of 3 minutes placed on the maintenance activity is the maintenance time constraint adopted in chapter 9.
*Note c:* The figure of one month is a performance feature, but the period of 5 hours corresponds to that of a series of missions quoted in section 10.2. A higher level of maintenance – in this case changing the cartridge – is possible at this time. Steps must be taken in the specification to ensure the appropriate level of reliability and maintenance over this period.
*Note d:* This specifies an even longer term level of maintenance.
*Note e:* This paragraph specifies the environment. Your specification may differ.
*Note f:* As we have seen, it is necessary to specify what constitutes a failure.

The above specification is, of course, over-elaborate for the article in question but illustrates a fairly rigorous approach. You should note the type of thing that has been specified rather than the actual quantities and consider if the specification is adequate.

*Supplementary exercise 10.1A*

After producing the specification of exercise 10.1 you find that the best reliability being quoted by the various manufacturers is only 95 per cent. What action would you take?
(The solution to this exercise is given at the end of this chapter.)

*Exercise 10.2*

1. Failure will be deemed to occur when either (see *Note a*)
   i. the charge detonates in storage, transit, or on site when the appropriate command has not been given
   ii. the charge does not detonate when the appropriate command has been given
   iii. the charge is physically incompatible with associated equipment or is rejected on simple visual inspection.

2. The charge must achieve the following failure rates
   fail safe for failure modes covered by 1 i (see *Note b*)
   1 in $10^3$ for failure modes covered by 1 ii (see *Note c*)
   1 in $10^2$ for failure modes covered by 1 iii (see *Note d*).
3. The charge should be designed for storage, transportation and use anywhere in Europe.

   When packaged for transportation the charge should be immersion proof, air portable (including drops by parachute) and compatible with regulations for storage and safety during transportation below the water line in ships.

   The design should be suitable for use on all building site and demolition areas under all weather conditions in the specified area of operation (see *Note e*).
4. No maintenance is permitted (see *Note f*).
5. The charge can only be subject to a simple on-site visual check before use.
6. The charge should have a shelf life of 10 years. (see *Note g*).

*Note a:* In some cases, such as the one under discussion, different reliabilities are acceptable or required for different failure modes. This is particularly true when safety is involved. In such instances the various failure modes must be defined.
*Note b:* The writer prefers to call for an absolute limit via a 'fail safe' requirement when safety is involved. Others claim that a truly absolute limit is impossible and should not be quoted in a legal requirement. It is usually stated that the 'person on the Clapham omnibus' is prepared to accept a 1 in $10^3$ chance of a serious/fatal accident in situations under his own control, but only a 1 in $10^6$ chance situations not under his control, rising to 1 in $10^8$ when psychological factors, such as aircraft landings systems, nuclear installations etc., are involved. It is often argued that the appropriate figure should be quoted. However, I believe such figures are meaningless in that it is categorically impossible to design for such accuracy, and it would be prohibitively expensive to establish the figure at any reasonable level of confidence during development. It is possible, of course, that quoting such figures can be good window dressing.
*Note c:* A lower level of reliability may be acceptable if safety is not involved. The figure quoted must be deduced on economic grounds or other considerations. It has been set here at the lowest of the safety levels listed at b since the situation is fully under operator control.
*Note d:* An even lower level may be acceptable in a trivial failure mode that is not likely to cause other than inconvenience.
*Note e:* It is important that the environmental conditions of use be fully specified. A more elaborate specification than the one given here may well be called up.
*Note f:* Maintenance procedures should always be specified even when, as in this case, none is allowed. A very slight relaxation is given at 5. It is doubtful if it means very much, but the designer should be allowed every dispensation possible. By so doing the designer is given a freedom to employ techniques which may not be envisaged at the specification stage. Generally, the object should be to give the

designer as much freedom as possible while defining the essential requirements in detail.

*Note g:* It is clearly necessary to specify a time-scale over which the specification must apply.

*Exercise 10.3*

From figures 3.5, 3.6 and the data provided we can deduce the following regarding the loading conditions

$$\text{Maximum torque} = 2.5 \times \text{rated maximum torque}$$
$$\text{Mean torque} = 0.5 \times \text{rated maximum torque}$$

Although reasonable data was recorded, it cannot be regarded as comprehensive and hence it would be advisable to treat the extreme as the oft quoted $3\sigma$ value. Hence

$$3\sigma_L = (2.5 - 0.5) \text{ maximum rated torque}$$
$$\sigma_L = 2/3 \text{ maximum rated torque}$$
$$\bar{L} = 0.5 \text{ maximum rated torque}$$

Therefore

$$\sigma_L/\bar{L} = \frac{2}{3} \times \frac{1}{0.5} = 1.3$$

A value of $\gamma = 0.1$ would be appropriate to this application to account for variation in basic material properties, differences in fabrication, etc. Hence from figure 10.2

$$\text{Factor of safety} = 10$$

But note that this factor of safety is based on the mean conditions (that is, $= \bar{S}/\bar{L}$) whereas it is usual to design engines for the maximum rated conditions. The factor of safety referred to in these conditions is then 5. This value is higher than would normally be adopted in design, and it would result in a heavy engine. It would probably be more advantageous to limit the peak load, which is associated directly with a gear change, than to design the engine to take that excessive load. The above calculations do bring out the penalty of an excessively wide distribution of load (that is, $\sigma_L/\bar{L}$), and reference to figure 10.2 shows how the factor of safety can be reduced significantly, with considerable improvements in the design, if the load distribution is restricted.

*Exercise 10.4*

Using the data established as part of exercise 4.7 we can deduce the following parameters

$$\gamma = \frac{\sigma_S}{\bar{S}} = \frac{15}{80} = 0.188$$

$$\frac{\sigma_L}{\bar{L}} = \frac{185.2}{705.6} = 0.262$$

Hence from graph on figure 10.2, factor of safety = 5.1, whence

$$\text{design load} = 5.1 \times 705.6 = 3598$$

As is to be expected, this is close to the value of 3614 deduced by the more laborious method of successive approximations in exercise 4.7. The small difference probably arises from the linearisation of the safety margin–loading roughness curve and inaccuracies in reading figure 10.2, but is really insignificant. The remainder of the solution follows that of exercise 4.7 to give a rung diameter of 1 inch, as before.

Current practice is best represented by British Standard Specification *BS 1129: 1966 (amended) Timber Ladders, Steps, Trestles, and Lightweight Stagings for Industrial Use*, which states in para. 13 b (p. 15) regarding circular rungs: "A cylindrical end shall have a diameter at the point of entry into the stile of between 1 and $1\frac{1}{8}$ in . . . ." The distance between the stiles of 240 mm (9.5 inches) taken in the solution to this and exercise 4.7 is close to the minimum allowed in the above specification for short ladders. The maximum distance between stiles allowed in the British Standard Specification is 1.5 times the minimum distance, and theoretically would require a rung diameter of $1 \times \sqrt[3]{1.5} = 1.14$ inches for intrinsic reliability.

Agreement between the above calculated and specified values is very good. However, the British Standard Specification calls for a greater distance between stiles for longer ladders. Thus a 15 foot ladder would require a calculated rung diameter of 1.15 inches for minimum distance between stiles, increasing to 1.32 inches for maximum distance, to achieve intrinsic reliability. This is greater than the diameter required in the specification. It would be as well to note, though, that the specification lays down stringent conditions for the selection of timber. If the selection rejects 5 per cent of the total population, which we could well expect, the factor of safety on figure 10.4 would be more appropriate. This gives a factor of safety of 3.3 for the values of the parameters quoted above. The rung diameter for intrinsic reliability then varies from 0.99 inch to 1.14 inches for minimum to maximum stile distance compared with 1 to $1\frac{1}{8}$ inches called for in the British Standard Specification. Thus very close agreement is obtained between the values calculated to give intrinsic reliability and those required by the British Standard Specification.

*Exercise 10.5*

The design criterion

$$\bar{S} - 3k\sigma_S \geqslant \bar{L} + 3k\sigma_L$$

can be written as

$$\bar{S} - \bar{L} \geqslant 3k(\sigma_S + \sigma_L)$$

But loading roughness (LR) = $\dfrac{\sigma_L}{\sqrt{(\sigma_S^2 + \sigma_L^2)}}$

giving

$$\sigma_S = \sigma_L \sqrt{\left\{\frac{1}{(LR)^2} - 1\right\}}$$

Hence substituting for $\sigma_S$ in the design criterion

$$\bar{S} - \bar{L} = 3k\sigma_L \left\{\sqrt{\left[\frac{1}{(LR)^2} - 1\right]} + 1\right\}$$

or

$$\frac{\bar{S} - \bar{L}}{\sqrt{(\sigma_S^2 + \sigma_L^2)}} = 3k \frac{\sigma_L}{\sqrt{(\sigma_S^2 + \sigma_L^2)}} \left\{\sqrt{\left[\frac{1}{(LR)^2} - 1\right]} + 1\right\}$$

which can be written as

$$SM = 3k\,\{(LR) + \sqrt{[1 - (LR)^2]}\}$$

Thus this design rule expresses the safety margin (SM) in terms of the loading roughness in the same way as the rule for intrinsic reliability given in figures 4.18 and 10.1. A direct comparison is therefore possible and is made in figure 15.24 using values of $k = 1.0$, 1.2 and 1.4.

The agreement between the curve based on intrinsic reliability and that based on a fixed number of standard deviations is only superficially adequate, though for some particular ranges it is quite good. Close numerical agreement is obtained at low loading roughness. As the loading roughness increases the shapes of the curves strongly resemble one another and, although there are numerical differences, reasonable agreement is obtained for values of $k$ lying between 1 and 1.2 up to a loading roughness of 0.7. For loading roughness above 0.7, agreement is not obtained either numerically or even in the shape of the curve.

Some workers in the U.S. go further and use a cut-off in the load/strength distribution, that is, in the distribution of $(S$–$L)$, for design purposes. This leads to a unique value of the safety margin for all values of the loading roughness, as a design criterion. It also leads to a reliability which is independent of the number of load applications, giving a failure rate which is zero except at the application

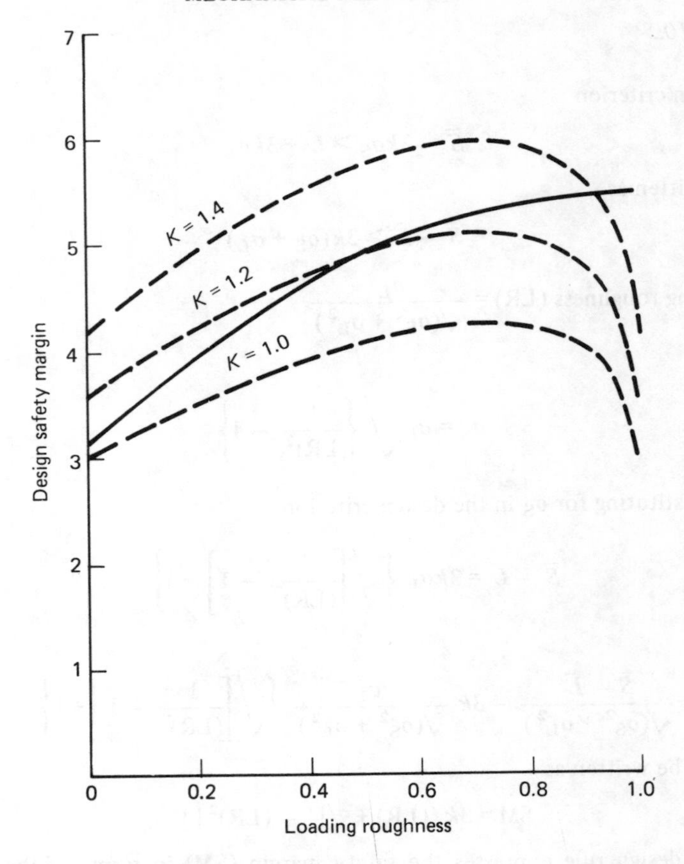


**Figure 15.24** Comparison of design safety criteria. ——— Intrinsic reliability. – – – – Means separated by constant number of standard deviations ($K$ values marked on curves)

of the first load. It thus applies only to ideally smooth loading, and should not be used for mechanical items where the loading roughness may be high. Both the design rules referred to in this exercise were in fact developed in the electronic industry, for which they may be quite valid.

*Exercise 10.6*

Let $\overline{S}_b$ and $\sigma_b$ be the mean and standard deviation of the basic material.
Let $\overline{F}$ and $\sigma_f$ be the mean and standard deviation of the improvement factor of the finishing process.

Then the final mean strength $\overline{S} = \overline{S}_b \times \overline{F} = 1.2\ \overline{S}_b$

having a standard deviation $\sigma = \sqrt{(\bar{S}_b{}^2\sigma_f^2 + \bar{F}^2\sigma_b{}^2 + \sigma_f^2\sigma_b{}^2)}$

$$= \bar{F}\bar{S}\sqrt{(0.07^2 + 0.07^2 + 0.07^4)}$$

$$= 0.11\bar{S}$$

that is, the mean finished strength is 1.2 × the basic material strength, but the standard deviation of the finished material has increased from 7 to 11 per cent of the mean.

It should be noted that this applies to all processes, which must themselves be distributed, so that the standard deviation of the products' strength is increased by a greater factor than the mean, that is, the coefficient of variation increases at each operation. It is thus seen how multi-process production techniques can lead to a wide distribution of strength unless they are closely controlled.

*Exercise 10.7*

Let the cost per unit cross-section be $c$.
Also the relation

$$\text{LR} = \frac{\sigma_L}{\sqrt{(\sigma_S{}^2 + \sigma_L{}^2)}} \quad \text{gives} \quad \sqrt{(\sigma_S{}^2 + \sigma_L{}^2)} = \frac{\sigma_L}{\text{LR}}$$

(a) For intrinsic reliability the safety margin, SM, equals 5.5. Therefore

$$\frac{\bar{S} - \bar{L}}{\sqrt{(\sigma_S{}^2 + \sigma_L{}^2)}} = 5.5$$

or

$$\bar{S} = 5.5\sqrt{(\sigma_S{}^2 + \sigma_L{}^2)} + \bar{L}$$

$$= 5.5\,(\sigma_L/\text{LR}) + \bar{L}$$

Substituting for $\sigma_L = \bar{L}/3$ and LR = 0.9

$$\bar{S} = 3.037\,\bar{L}$$

$$\text{Cross-section} = \bar{S}/\bar{E}$$

$$\text{Initial cost} = c\bar{S}/\bar{E}$$

$$= 3.037\,(c\bar{L}/\bar{E})$$

Maintenance cost = 0 since intrinsic reliability is retained

$$\text{Total cost} = 3.037\,(c\bar{L}/\bar{E})$$

(b) Fatigue failure-free life = rated life.
From figure 5.14, SM = 4.92, hence proceeding as before

$$\text{Initial cost} = 2.822\,(c\bar{L}/\bar{E})$$

Maintenance cost = 0 since failure-free

Total cost

(c) For a safety margin of 4.5, figure 5.12 gives cumulative failures of 2 per cent at the rated life. Hence

$$\begin{aligned}\text{Initial cost} &= 2.666\,(c\bar{L}/\bar{E})\\ \text{Maintenance cost} &= 0.02 \times 2.5 \times 2.666\,(c\bar{L}/\bar{E})\\ &= 0.133\,(c\bar{L}/\bar{E})\\ \text{Total cost} &= 2.799\,(c\bar{L}/\bar{E})\end{aligned}$$

(d) For a safety margin of 3.5, figure 5.11 gives cumulative failures of 70 per cent at the rated life. Hence

$$\begin{aligned}\text{Initial cost} &= 2.296\,(c\bar{L}/\bar{E})\\ \text{Maintenance cost} &= 4.018\,(c\bar{L}/\bar{E})\\ \text{Total cost} &= 6.314\,(c\bar{L}/\bar{E})\end{aligned}$$

The minimum cost occurs at (c), which is the design criterion which should be adopted in this case. A more precise value could be obtained by plotting and interpolation, but it is doubtful if the accuracy of the calculations warrants such refinement. What is important to note is that the minimum cost arises at a safety margin well below that for intrinsic reliability or indeed for the fatigue failure-free life. In this particular example the rated life has been chosen to lie on the graphs already presented to save excessive calculation. Had a lower rated life been chosen, the minimum cost would have occurred at an even lower safety margin. Hence fatigue design for minimum cost implies safety margins well below those that ensure a failure-free operation. This follows from the shape of the fatigue failure pattern. On the other hand, increasing the cost of maintenance would put a higher premium on avoiding failures and hence demand a higher safety margin.

*Exercise 10.8*

The various examples have been chosen to illustrate common faults leading to unreliability under the headings of attachments, stress raisers, maintenance, environment and materials.

*Figure 10.5a.* The major defect of this pendant switch is the absence of any cord grips. One could expect the wires to pull out from the simple screw pressure fastening under any significant load. (This fastening provides electrical contact, not mechanical strength.) In addition to being unreliable, this switch is not safe. This is an obvious example of bad design in an article of everyday use. Care must be taken with all kinds of attachments to ensure that they are suitable for their intended application. Attachments and connectors are a notorious cause of unreliability.

*Figure 10.5b.* The holes for the end attachment to the horizontal control rod have been drilled vertically. With the load as shown this removes metal from the highly stressed regions of the rod and attachment. If the holes were drilled horizontally, metal would only be removed at or near the neutral bending axis, that is, where the stresses are low. Hence a much stronger structure is obtained at no weight cost or performance penalty. Additionally, you will note that the end attachment has been secured by two nuts and bolts bearing down on curved surfaces. The tightening load must be restricted to unacceptably low values or distortion will occur. It is suggested that riveting would be a better practice.

*Figure 10.5c.* The bolts holding the bracket will be subject to alternating shear and tension. As drawn, the threads on the bolt extend across the shear plane and would act as stress raisers, leading to fatigue failure and to failure by crushing of the bolt threads. Walker gives a good description of the fatigue of nuts and bolts. Additionally, in this case no means of locking the nuts has been provided. Even with ordinary locking techniques it would be advisable to insert the bolts from the top so that if the nuts did wear off, the bolts would not fall out under gravity and cause an immediate failure.

*Figure 10.5d.* It is virtually impossible to get a grease gun on the nipples, which are all shielded by adjacent parts. Inadequate lubrication would then lead to failure. This is a glaring example, but instances where maintenance is difficult though still possible are quite common. With normal standards of labour available for maintenance the job will not be done properly, leading to failures. It is important to consider all aspects of maintenance at the design stage.

*Figure 10.5e.* This example contains the same feature as figure 10.5c, in that no means have been provided for locking the large nut holding the bearing on the hook. In the typical environment associated with a crane this nut would soon work off. Full consideration must be paid to the environment and this exercise draws attention to random vibration and shock loading.

*Figure 10.5f.* This is another example where environment has been neglected. The suspension arrangement shown contains a member which has a concave surface pointing upwards. This will collect dirt and in particular moisture. One can, therefore, expect corrosion and subsequent failure. This again is a common source of failure and care should be taken to avoid designs which encourage corrosion.

*Figure 10.5g.* This is a simple illustration of incorrect material. The web of the swivel pulley bracket is stressed in tension, but the drawing calls for cast iron which is weak in tension. Either another material must be used or the bracket redesigned to avoid tension in cast iron parts. The incorrect specification of material is also a common source of failure. Reference 4 is an excellent reference on selection of materials.

*General notes*

This exercise illustrates only a few examples of frequently occurring types of design faults. You should look up ref. 64 to see illustrations of very many more common design faults. Some of these exercises should also show that it is often quite difficult to detect simple faults by straightforward inspection of drawings. It is, therefore, highly desirable that a systematic approach be adopted in any design review.

*Exercise 10.10*

(a) Significant modes of failure arising from overall consideration

1. Failure to shut off as a result of failure of washer and/or valve seat
2. Leakage past packing and gland nut.

(b) Detail modes of failure of individual components

Main body of stopcock

1. Integrity of casting
2. Alignment of valve seat and head nut thread
3. Surface finish of valve seat
4. Accuracy and finish of inlet/outlet unions
5. Accuracy of head gear/body mating surfaces
6. Compatibility with soil content if used underground
7. Compatibility with domestic water.

Washer

8. Retention of adequate surface contour and softness (wear)
9. Retention of adequate strength
10. Compatibility with domestic water (hot and cold)
11. Lack of adhesion to valve seat.

Jumper

12. Alignment of washer face and stem
13. Accuracy of stem diameter
14. Surface finish of stem
15. Compatibility with domestic water.

Spindle

16. Concentricity of stem and jumper guide
17. Surface finish of jumper guide
18. Accuracy of jumper guide
19. Wear between head nut and spindle
20. Wear between gland and spindle
21. Compatibility with soil content if used underground.

Head nut

22. Alignment of thread and spindle bearing
23. Accuracy of body/head gear mating surface
24. Compatibility with soil content if used underground.

Packing

25. Resistance to rotting
26. Adequate impermeability
27. Lack of adhesion to spindle (low, but adequate, friction)
28. Adequate softness and flexibility.

Gland nut

29. General dimensional accuracy
30. Compatibility with soil content if used underground.

Crutch handle

31. Adequate strength at fastening to spindle
32. Compatibility with soil content if used underground.

Water

33. Dissolved matter
34. Excessive pressure
35. Expansion on freezing
36. Solid particles (after repair work, etc.)
37. Compatibility with all parts.

Secondary modes

Lack of alignment as at modes 2 and 16 will be taken up by the washer and result in excessive wear. Hence we should include this aspect under mode 8 which should be redesignated as

8. Retention of adequate surface contour and softness (wear) when subject to both axial and slightly non-axial loads.

*Exercise 10.11*

The suggested frequency that each of the possible failure modes recorded in exercise 10.7 would be experienced in a 10 year period is listed below. The frequency is expressed as $\log_{10}$ and recorded against the serial number of the failure mode used in exercise 10.7.

| | | |
|---|---|---|
| 1, − 4.5 | 13, − 4.0 | 25, − 3.0 |
| 2, − 3.0 | 14, − 3.0 | 26, − 1.7 |
| 3, − 2.7 | 15, − 6.0 | 27, − 3.5 |
| 4, − 4.0 | 16, − 4.0 | 28, − 1.7 |
| 5, − 4.0 | 17, − 3.0 | 29, − 4.0 |
| 6, − 3.7 | 18, − 4.0 | 30, − 3.7 |
| 7, − 6.0 | 19, − 3.2 | 31, − 4.5 |
| 8, − 2.2 | 20, − 3.2 | 32, − 3.7 |
| 9, − 2.2 | 21, − 3.7 | 33, − 3.3 |
| 10, − 3.2 | 22, − 3.0 | 34, − 3.7 |
| 11, − 3.0 | 23, − 4.0 | 35, − 4.0 |
| 12, − 3.0 | 24, − 3.7 | 36, − 5.5 |
| | | 37, − 6.0 |

In estimating the above values a number of assumptions were made. For example, in assessing the frequency of washer failures a mean life of 15 years was assumed, together with a normally distributed failure-age pattern (associated with wear) and a standard deviation of 2 years. This was taken to include non-axial loading. The failures to be expected in 10 years were calculated from this distribution. As regards the gland packing, it was believed that no failures would occur if the gland nut were tightened when required by inspection, that is, if an on-condition maintenance policy were adopted. This assumption is open to some doubt, and an over-riding assumption is made that it would not be followed by 1 in 50 of the householders. This gives rise to certain failures in the time period under consideration for that proportion of values. Other failure frequencies were either estimated directly or ratioed from the more positively calculated ones. Very little field data is available and it is well known that subjective assessments can vary widely, so no great accuracy is claimed and your estimates may differ.

A Pareto plot of the frequency of each failure mode is presented in figure 15.25. The vital few are clearly modes 26, 28, 8, 9, and possibly 3. The first two both refer to gland-packing failures, and the second two both refer to washer-failure modes. The detail analysis thus gives much the same result as the more

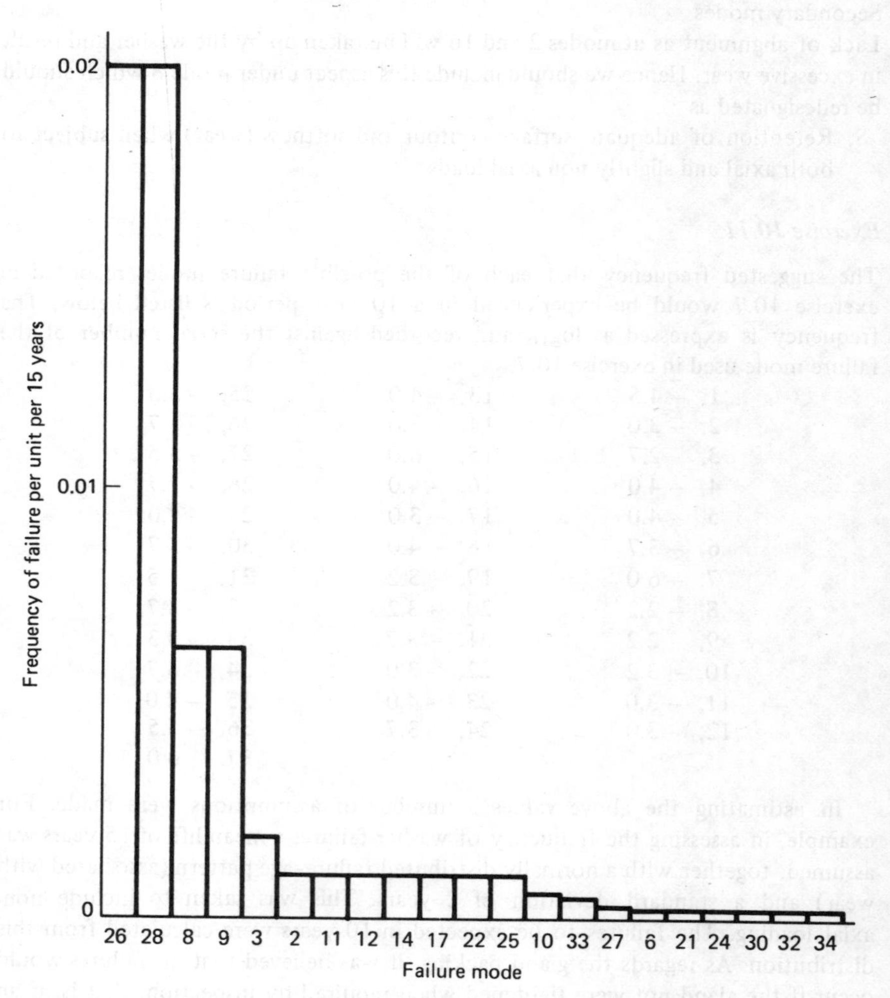


**Figure 15.25** Pareto diagram of failure modes in domestic stopcock. Key to failure modes as in exercise 10.7

brief and cursory overall appraisal, except that the detail assessment places the first two items in a reversed 'pecking order'. The detailed analysis does draw separate attention to failure mode 3, the valve seat, which was absorbed into the washer failures in the general overall survey. To this extent the detailed analysis can be more precise.

*Exercise 10.12*

(A) The effect of each failure on the performance of the stopcock itself (that is, the next higher level) is listed below

1. Mild leaking at cock
2. Wear on washer, leading to incomplete cut-off
3. Incomplete cut-off
4. Leaking at stopcock unions
5. Mild leaking at cock
6. Corrosion of cock body
7. Contamination of water supply
8. Inadequate seal and incomplete cut-off
9. Breakdown of washer, leading to no cut-off
10. Contamination of water supply
11. Sustained seal or breakdown of washer, leading to no cut-off
12. Inadequate seal and incomplete cut-off
13. Sticking of jumper and loss of non-return feature
14. Sticking of jumper and loss of non-return feature
15. Contamination of water supply
16. Wear on washer, leading to incomplete cut-off
17. Sticking of jumper and loss of non-return feature
18. Sticking of jumper and loss of non-return feature
19. Spindle not turning true, leading to leakage at gland
20. Spindle not turning true, leading to leakage at gland
21. Corrosion of spindle – could be significant at crutch handle joint – leading to no cut-off
22. Wear on spindle, leading to leakage at gland
23. Mild leakage at cock
24. Corrosion – could make maintenance difficult
25. Leakage at gland developing and increasing with age
26. Leakage at gland
27. Disturbance of package, leading to leakage. Cut-off difficult
28. Leakage at gland developing and increasing with age
29. Wear on spindle – leakage or difficult maintenance
30. Corrosion leading to difficult maintenance
31. Fracture and inability to operate – no cut-off
32. Corrosion and fracture at spindle joint – no cut-off

33. Corrosive deposits at any slight leakage, possibly leading to operational or maintenance difficulties
34. Deformed washer, leading to incomplete cut-off
35. Fracture of stopcock
36. Prevents adequate seal, leading to incomplete cut-off
37. Contamination of water supply and corrosion.

While considering the effects of first level failure on the second level, we must also take into account failures arising at this level. These can be listed as

38. Fitted so that stopcock handle is inaccessible, leading to no cut-off while remedial action is taken
39. Fitted so that flow is in wrong direction so stopcock is inoperable, but should be immediately apparent
40. Leakage at the gland can result in solid corrosive deposits, making operation difficult.

As regards frequency of above additional failure modes, it is considered that about 1 in 100 are fitted where access is difficult or where the occupation or use of building etc. will subsequently make access difficult. Failure mode 39 is considered to be so apparent as to have only a 1 in 10 000 chance of affecting operation. The frequency of failure mode 40 is considered to be 1/5 of total gland leakage frequency.

(B) It is possible to trace the effect of each failure listed at (A) on the next higher level (that is, the water system) in turn, but to save repetition we identify below failures of the cock itself which are common to a number of primary failures, and examine the effect of these on the system. The reader should find no difficulty in following each of the primary failure modes listed in the solution to exercise 10.7 through the effects on the next level listed under (A) above to the ultimate effect at (B).

1. The stopcock may fail to shut off the water at all.
   i. This could delay and increase the cost of any maintenance action being taken on the water system since it would require prior action on the stopcock itself.
   ii. In an emergency, for example in the case of a burst pipe, it could lead to an extensive discharge of water
2. The stopcock may not completely shut off the supply. The conditions i and ii of 1 above would apply but the unintentional discharge from the system would be less and could possibly be contained, by use of buckets, opening of normal outlet taps, etc.
3. Shutting off the water supply is difficult.
   i. This could delay maintenance as 1 i above, though not so seriously
   ii. In an emergency it could have the same effect as 1 ii above
4. Contamination of the water supply. This is unlikely to have any effect on the operation of the system itself, but would result in a contaminated supply
5. The non-return effect does not function. This is unlikely to affect the opera-

tion of the system itself but could lead to possible contamination of the mains supply if mains pressure were lost
6. Leakage past the gland. This would have no effect on the system itself unless it involved corrosion and seizure of the valve when circumstance 1 ii would apply. Otherwise mild leakage
7. Maintenance is made difficult. This would prolong maintenance action, but all good plumbers should anticipate these circumstances and be prepared and capable of dealing with them.

(C) We can now identify the effects on the household as follows, using the same reference as in (B) above.

1. Failure to close in an emergency could lead to flooding and extensive damage to house and household effects
2. Similar effects to 1 above, though not on such a serious scale
3. Could lead to similar effects as 1
4. Could constitute a health hazard
5. No effect on household. Possible health hazard to the general supply in the area, but remote
6. (a) General small leakage could lead to dampness and its consequences in vicinity of stopcock. Could be more serious if leakage dribbled on to wooden floor, leading to wood rot etc.
   (b) Leakage involving corrosion and operational difficulty could lead to similar effects as 1
7. Inconvenience during maintenance.

*Exercise 10.13*

The following effects rating was allocated to each of the complete system failures identified in section (C) of exercise 10.9.

(C) 1. Considered a serious failure and given a rating of 10.
(C) 2. Since not so serious as (C) 1, was given a rating of 6, but some other grading between 6 and 10 may be justified in particular circumstances.
(C) 3. Rating of 10 since similar to (C) 1.
(C) 4. Since health is involved, given highest rating of 10.
(C) 5. Given rating of 3. Should be higher, but is not likely to register high on householder's (purchaser's) list of priorities. May rate much higher if Public Utilities approval is necessary.
(C) 6. (a) Standard dribble was not considered important and given a value of 0.5.
(b) Was given a derived rating of

$$(4/5 \times 0.5 + 1/5 \times 10) = 2.4$$

(C) 7. Since no loss of revenue is involved, given low rating of 0.1.

Using these factors and equation (10.9) the criticality of each failure mode listed in exercise 10.7 can be calculated and plotted as a Pareto diagram as in figure 15.26. It is then apparent that the most critical failure mode arises from the accessibility of the installation. This did not appear at all in figure 15.25 because it is a higher level failure mode which would not appear at first level analysis. The remaining vital few are the same in both figures 15.25 and 15.26, but the order of importance has been changed. Break-up of the washer due to inadequate strength is revealed to be more critical than the quality of the gland

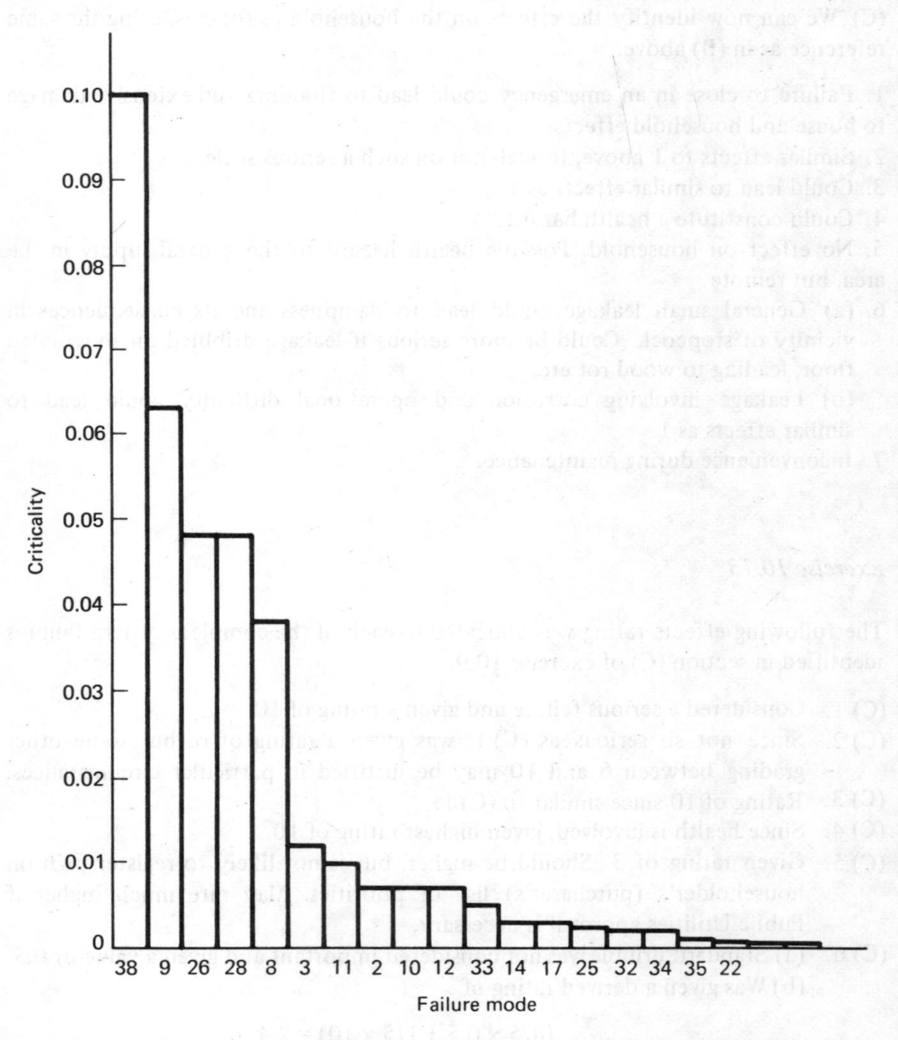


**Figure 15.26** Pareto diagram of criticality in domestic stopcock. Key to failure modes as in exercise 10.7

packing, with the softness and surface contour following closely behind. The remaining items can be considered insignificant.

The object of the FMEA should now be kept firmly in mind, and the quantitative result not be over-emphasised. Instead we specify the vital few as areas needing particular attention during the design of the complete system. They may be designated as

1. The accessibility of the stopcock during system design
2. The quality of the washer
3. The quality of the gland packing.

In comparing the conclusions of the detailed FMEA with those arrived at subjectively, no significance should be attached to the order in which the vital few appear since that should have no influence on the consequent action. We then see that the subjective approach revealed two of the three items isolated in the detailed FMEA. (It is possible, of course, that you identified the third as well: that would be unusual, but not exceptional.)

*Exercise 10.14*

The initial reaction of the reader (particularly if he has taken the trouble to work out exercises 10.10 to 10.13 for himself!) may justifiably be, what a lot of work to produce an obvious result. Of course, we have taken a very common item on which to carry out the exercises, so that experience could well lead to the final conclusion. It would be fair to ask the reader, though, whether he did list 'installation accessibility' in his first answer to exercise 10.10. In the author's experience it is often overlooked, and is only brought out by the rigour of the FMEA. Nevertheless, we must recognise that a great deal of work has to be carried out in a full FMEA, as these exercises have illustrated what is really a very simple item indeed. Furthermore, these exercises should have brought out the fact that the basic data on which to estimate the frequency of failure is often not available. Since the final Pareto diagram depends fairly heavily on the assumptions (change the mean life of the washer from 15 to 12.5 years and increase the standard deviation to 2.5 years, for example, modifies figure 15.26 considerably though in this particular case the overall conclusion – the set of vital few modes – is unchanged), the absence of this data is important. It has to be multiplied by an 'effects rating' which is also often largely subjective. It would be surprising if your rating agreed with the author's in exercise 10.13. The final numbers must therefore be taken with 'a very large pinch of salt'. It is open to question, therefore, if all the numerical work is worth while. While having considerable sympathy with this viewpoint, the author is firmly convinced that the rigorous examination of all components for all possible forms of failure, together with *some* assessment of the consequences, is essential – both in the early feasibility stages and also when the detail design has been finalised, if we are to avoid all those stupid errors which lead to an unreliable product.

It is a matter of judgement in any particular case which route should be followed. It is clear from the above exercises that the more experience is at hand on the item under review the less valuable a numerical analysis becomes. It is in the newer areas that a quantitative FMEA can pay off. It is ironic that these are the very circumstances in which greatest doubt attaches to estimated failure frequencies. However, forcing designers to estimate the life of intrinsic reliability and the subsequent failure–age relationship is of itself a salutary experience which is essential if a planned level of reliability using a known maintenance policy is to be achieved. But whether the quantitative or qualitative route is followed, the first essential is that every failure mode should be identified. No calculations can be carried out on an unidentified failure mode. It thus seems to the author that the first priority must rest with the method of recognising the failure modes that may arise. The prime tool for that job is experience, and in the absence of experience, which may well have departed for 'greener pastures', it depends on the experience retention system available to the designer and the design auditor, FMEA analyst or whatever other title he is given.

Finally, you may think we have spent too much time on an irrelevant item like the domestic stopcock; but it has only been chosen as the most simple version that the author could imagine of an item which has probably received more attention by reliability workers than any other because of its ubiquitous application. I refer to the industrial control valve. Space completely ruled out the treatment of such an item in this book, but the principles are identical. If you wish to pursue the matter, a good starting point is ref. 80 which also gives many other references: and if you are a glutton for hard work try to develop a FMEA for an industrial pneumatically operated fluid control valve. If you do not think valves are important enough to warrant the attention they have received, study accounts of the Three Mile Island nuclear generating plant failure.

*Exercise 10.15*

The principal top events and their immediate causes have been listed in the table on pages 452 and 453 (adapted from ref. 66).

*Exercise 10.16*

It is considered that the Design Review should ensure that the following items have been resolved, and endorse the conclusions reached. The items have been listed in associated groups and the particular Design Review at which they would be taken is indicated using the notation P = Preliminary, EI = one of the Early Intermediate, LI = one of the Later Intermediate, and F = Final Reviews. The symbol ? denotes optional if necessary. Each item has been given a star priority rating in which the larger number of stars indicates a higher priority.

1. The interpretation of the requirement in respect of safety and relative importance of reliability, maintainability and performance **** P F

| | | |
|---|---|---|
| 2. The cost, including whole life cost, and the time-scale of the project | *** | P |
| 3. The design philosophy and design concepts to be adopted. The proposed solution and alternatives | *** | P |
| 4. The environmental conditions that the equipment will meet and any special environmental conditions imposed by the equipment on standard components | **** | P, EI, LI? |
| 5. Any operating constraints and methods of use entailed in the proposed design | *** | P, EI, F? |
| 6. An appreciation of any stress or other engineering analysis techniques, defining difficult areas | ** | EI, LI |
| 7. An appreciation of any reliability or failure rate predictions | ** | EI, LI |
| 8. Consideration of failure modes and effects analyses | **** | P F |
| 9. An appraisal of maintainability | **** | P?, LI, F |
| 10. Assessment of the maintenance policy, including manpower, tools and special equipment | *** | LI |
| 11. An appraisal of any research programmes to support any of the above aspects | ** | P?, LI |
| 12. Any standardisation and interchangeability involved in production or maintenance | * | P?, LI |
| 13. The distribution of work within a contractor's organisation where it is applicable, and where appropriate, between sub-contractors | * | LI |
| 14. Prototype test plan including data collection, interpretation, analysis techniques, communication of results and implementation | **** | EI, LI |
| 15. Any demonstration trials or evaluation trials | ** | LI |
| 16. An appraisal of the production processes and identification of critical areas | ** | LI |
| 17. Quality control methods | ** | LI |
| 18. Inspection techniques required | * | LI |
| 19. The effect of any value engineering on reliability | * | LI |
| 20. An assessment of any post design services that may be required | ** | LI |

*Exercise 10.17*

Since each filter is only inspected at 1000 mile intervals, the exact failure age is unknown, but as an unbiassed estimate can be assumed to be mid-way between successive inspections, that is, the failure noted at 23 000 miles is assumed to have occurred at 22 500 miles. On this basis, failures can be ordered and the mean rank calculated. When plotted on Weibull paper a slight curve is obtained which can be

Causes of primary failure of solenoid valve (from ref. 81)

| | PARTIAL FAILURES | COMPLETE FAILURES | *ROW A* PRINCIPAL FACTORS INFLUENCING FAILURE | *ROW B* PRINCIPAL FACTORS LEADING TO MIS-USE FAILURES |
|---|---|---|---|---|
| SUDDEN FAILURES | | STRUCTURAL COLLAPSE | Material Strength<br>Fluid Pressure<br>Fluid Flow Rate<br>Frequency of Operation<br>Quality of Brazing<br>Fatigue Strength<br>Component Misalignment | Excessive Fluid Pressure<br>Fluid Attack<br>Atmospheric Attack<br>Excessively Low Temperature |
| | | BROKEN SPRING | Mains Voltage<br>Component Misalignment<br>Fluid Pressure<br>Frequency of Operation<br>Stroke<br>Mass of Disc/Core/Spring Assembly<br>Fluid Flow Rate<br>Fatigue Strength of Spring<br>Core/Core Tube Clearance | Fluid Attack of Spring<br>Excessively Low Temperature |
| | | 'STICKING' | Residual Level in Core<br>Residual Level in Plugnut<br>Mains Voltage<br>Frequency of Operation<br>Spring Force<br>Surface of Plugnut<br>Surface of Core<br>Component Misalignment<br>Core/Core Tube Clearance | |
| | | COIL BURN-OUT | Mains Voltage<br>Mains Frequency<br>Coil Construction<br>Component Misalignment<br>Spring Force<br>Frequency of Operation<br>Orifice Size<br>Stroke<br>Fluid Pressure<br>Fluid Flow Rate<br>Fluid Temperature<br>Mass of Disc/Core/Spring Assembly<br>Magnetic Circuit Construction | Excessive Voltage<br>Insufficient Frequency<br>Excessive Ambient Temperature<br>Excessive Fluid Temperature<br>Mechanical Damage<br>Dirt<br>Excessive Fluid Pressure<br>Interference with Magnetic Circuit<br>Atmospheric Attack of Coil |

| | | | | |
|---|---|---|---|---|
| GRADUAL FAILURES | LEAKAGE | To atmosphere | Fluid Flow Rate<br>Force on 'O' Ring<br>Porosity of Structure<br>Quality of Brazing | Fluid Attack of Structure<br>Atmospheric Attack of Structure<br>Mechanical Damage<br>Fluid attack of 'O' Ring |
| | | Across Seat | Fluid Pressure<br>Fluid Flow Rate<br>Fluid Temperature<br>Hardness of Disc<br>Surface of Seat<br>Orifice Size<br>Spring Force<br>Frequency of Operation<br>Component Misalignment | Fluid Pressure too high or too low<br>Fluid Temperature too high or too low<br>Fluid Attack on Disc<br>Fluid Attack on Seat<br>Mechanical Damage<br>Dirt<br>Fluid Attack of Spring |
| | NOISE | | Mains Voltage<br>Mains Frequency<br>Fluid Pressure<br>Fluid Flow Rate<br>Spring Force<br>Frequency of Operation<br>Magnetic Circuit Construction<br>Surface of Plugnut<br>Surface of Core<br>Component Misalignment<br>Core/Core Tube Clearance | Low Mains Voltage<br>Low or High Mains Frequency<br>Excess Fluid Pressure<br>Dirt<br>Fluid Attack of 'Shading Coil'<br>Fluid Attack of Plugnut or Core Face<br>Mechanical Damage<br>Insufficient Fluid Flow |
| | LOSS OF SPEED | | Mains Voltage<br>Mains Frequency<br>Fluid Pressure<br>Fluid Flow Rate<br>Spring Force<br>Frequency of Operation<br>Stroke<br>Coil Construction<br>Mass of Disc/Core/Spring Assembly<br>Orifice Size<br>Magnetic Circuit Construction<br>Core/Core Tube Clearance | Insufficient Voltage<br>Excess Frequency<br>Excess Fluid Pressure<br>Excess Fluid Viscosity<br>Excess Fluid Temperature<br>Excess Ambient Temperature<br>Mechanical Damage<br>Insufficient Fluid Flow |
| | LOSS OF PRESSURE RATING | | Mains Voltage<br>Mains Frequency<br>Fluid Pressure<br>Fluid Flow Rate<br>Spring Force<br>Frequency of Operation<br>Stroke<br>Coil Construction<br>Component Misalignment<br>Mass of Disc/Core/Spring Assembly<br>Orifice Size<br>Magnetic Circuit Construction<br>Core/Core Tube Clearance | Low Mains Voltage<br>High Mains Frequency<br>Excess Fluid Pressure<br>Excess Fluid Viscosity<br>Excess Fluid Temperature<br>Excess Ambient Temperature<br>Mechanical Damage |

straightened, so that the data can be represented by a Weibull distribution having the following parameters

$$t_o = 10\,500 \text{ miles}$$
$$\eta = 24\,150 \text{ miles}$$
$$\beta = 4.3$$

The first conclusion to be drawn is that the filters are intrinsically reliable (just) for the existing scheduled replacement interval of 10 000 miles, but would not be so for the proposed interval of 15 000 miles. It would therefore be inadvisable to recommend an increase of the scheduled servicing interval.

An unusual feature of this distribution is the high value of $\beta$, indicating a skewed distribution having an extended tail on the low life side. For $\beta = 4.3$ we can read off Weibull paper that $\mu/\eta = 0.91$ and $\sigma/\eta = 0.24$. It follows that

$$\sigma = 5796 \qquad \begin{aligned} \mu &= 21\,976 \\ t_o &= 10\,500 \\ \bar{t} &= 32\,476 \end{aligned}$$

It may well be thought that the standard deviation is on the high side and, taken in conjunction with the skewing, suggests that the quality control is not good enough.

Retaining the same mean life of 32 476 miles, but reducing the standard deviation to 5000 miles with a more or less normal distribution ($\beta = 3.44$), would increase $t_o$ to 16 686 miles which is well in excess of the required service interval.

It is therefore recommended that the service interval should not be increased to 15 000 miles with the existing filters, but that the quality control be carefully examined for possible improvements. It is reasonable to expect that with improved quality assurance the 15 000 mile service interval target could be reached.

*Exercise 10.18*

The data is plotted on Weibull paper using standard techniques. From the plot we deduce

a. Failures after 1 month = 2.5 per cent  Reliability = 97.5 per cent
b. Failures after 2 months = 6 per cent  Reliability = 94 per cent
c. $t_o = 0$ since raw data plots as straight line
d. $\beta = 1.3$, that is, roughly random failures throughout

Clearly the failures are too high but, of more importance, they are occurring in a random manner whereas we should expect failures to arise through wear. It follows that there is something radically wrong with the construction or the material. This is all that a statistical analysis of the overall data can tell us, but does indicate that a specialised physical investigation is urgently required.

*Exercise 10.19*

The data given in this exercise is typical of the mass of information received from prototype trials. Some of it must be relevant, but we should expect quite a proportion to be misleading. The first step is to separate the wheat from the chaff. Recurring failures in any specific mode must first be identified and grouped together for assessment. In some cases a Pareto plot may be necessary, but in most development work it is reasonably obvious without resorting to such techniques. In this particular case we can list the following areas as worthy of attention

1. Throttle clamping bolt (reported loose)
2. Throttle cable (reported frayed or broken)
3. Distributor (a number of miscellaneous faults reported)
4. Plugs and ignition leads (a number of failures reported, but it is claimed that these are covered by a modification. This claim still needs to be verified).

Before attempting any detailed appreciation of these failures, it should be recognised that difficulties can arise from the lack of any firm definition of failure. Prototypes in the hands of a trial unit will receive close attention, so that an incident which may have passed unnoticed in the hands of a representative user could be classified as a failure in their assessment. Thus an examination of the reports will show that many 'failures' were noticed during a routine inspection before the actual failure had developed sufficiently to be reported by the user. Recurring failures are known and sought out. We should not therefore over-emphasise the factual data and should expect a great deal of scatter in the information. Recognising this, each of the major trouble areas is examined separately below.

1. Throttle clamping bolts. Some 14 cases of loose throttle clamping bolts were recorded. Although many of the cases were clearly of an incipient nature, more serious failures could be expected in service when operation would usually be continued to ultimate failure with sudden loss of engine power. This could be dangerous. Possible remedies are redesign or more frequent checks. In order to obtain more quantitative information on which to base a decision, the data has been plotted on Weibull paper on figure 15.27. In so doing the distance between successive failures on the same vehicle has been included. This is not good practice: ideally the distance to first failure only should be used, but more often than not the lack of data from prototype trials makes this impossible. The failures on each prototype vehicle are shown in brackets against the prototype number in the following list. A dash indicates no failure had been recorded on that prototype and its maximum mileage was less than that of the next recorded failure, that is, it was a drop-out. The data in ascending life are

1(478), 15(–), 9(583), 18(–), 1(759), 24(801), 9(834), 1(944), 1(949), 10(–), 7(–), 21(–), 3(–), 1(1534), 20(–), 16(–), 22(–), 14(–), 5(–), 1(3904), 6(–), 11(5328), 9(5562), 23(6531), 13(–), 17(–), 12(–), 2(7773), 1(11 503)

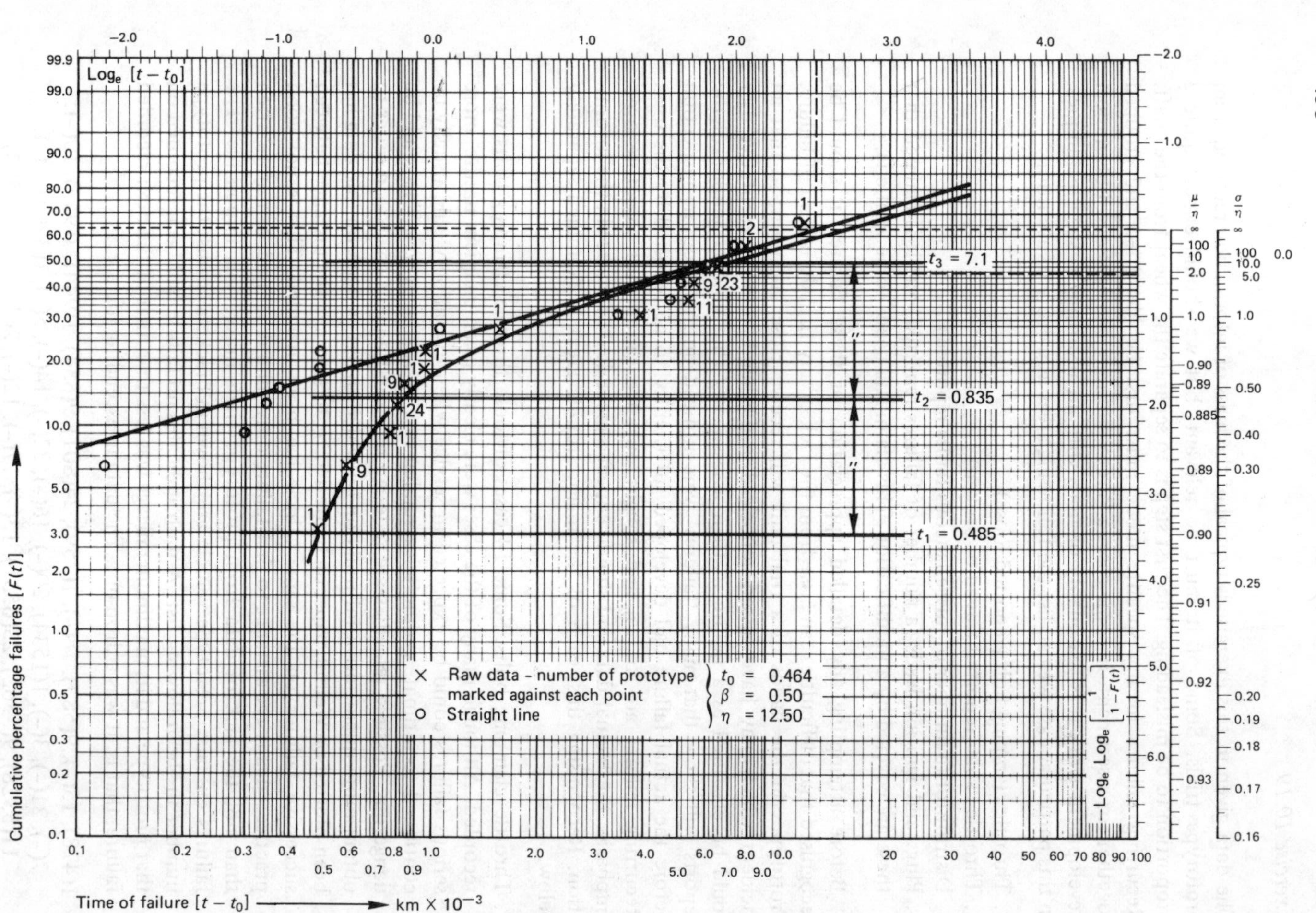


Figure 15.27 Clamping bolt failures in prototype vehicle

Different failures from the same vehicle seem randomly dispersed in the above data, so it would seem justified to use repeated failures from the same vehicle, and we proceed with the analysis by constructing the usual table.

| *Order number* | *km to failure* | *Mean rank* |
|---|---|---|
| 1.000 | 478 | 3.03 |
| 2.03226 | 583 | 6.16 |
| 3.10011 | 759 | 9.39 |
| 4.16797 | 801 | 12.63 |
| 5.23582 | 834 | 15.87 |
| 6.30367 | 944 | 19.10 |
| 7.37152 | 949 | 22.34 |
| 8.69295 | 1534 | 26.34 |
| 10.39202 | 3904 | 31.49 |
| 12.27602 | 5328 | 37.20 |
| 14.16002 | 5562 | 42.91 |
| 16.01402 | 6531 | 48.62 |
| 18.47001 | 7773 | 57.18 |
| 22.40251 | 11503 | 67.89 |

The above data is plotted on figure 15.27 where it gives a curve which is straightened by the usual techniques to obtain the Weibull distribution representing the failure pattern. The analysis has been given in detail to illustrate the procedure with multiple drop-outs.

Turning to the interpretation of the data, the points on figure 15.27 define a recognisable curve with surprisingly little scatter, and no one prototype yielding a significantly different pattern. The form of the Weibull distribution is one that could be associated with the loosening of a clamping bolt under successive application of load or vibratory conditions. The life of intrinsic reliability (464 km) is far too low to permit a solution of the problem by maintenance action, even allowing for the fact that the values plotted may be pessimistic compared with service use because of the rigour of prototype inspection. At 6000 km, nearly 50 per cent of the clamps could be expected to fail. The very low value of $\beta$ (0.5) suggests that the clamping action is not nearly strong enough in relation to the load (actual, not design) being applied. Redesign is clearly called for. You may be of the opinion that a similar conclusion would have been reached on a simple failure count basis. This may be so, but the conclusion would not have been nearly so justified in that the magnitude of the problem is not revealed by that approach.

2. Throttle cable. Five cases of throttle cable breaking or fraying were reported, together with four cases of trouble with the linkage which may or may not be associated with cable disintegration of some kind. The five cases of true cable fracture can be analysed by Weibull or other techniques, but with 11 drop-outs before the first failure, little credence can be placed on any such analysis. All

that can be deduced, by means of a simple head count, is that something is wrong with the cables. Physical examination is essential. Appraisal of the data can only draw attention to the problem, and you should have requested further information in your solution. This in fact was done in the real instance. Physical examination established that the failures could be attributed to corrosion due to lack of lubrication of the inner cable. It was subsequently established that the cables had not been packed with graphite during manufacture as required.

3. Distributor. It is impossible to analyse the distributor failures. There appears to be no systematic failure mode. Failure of the points on three occasions is noteworthy but more data is really needed. It is for consideration whether the distributor is waterproof, even if all the failures in this mode occurred under exceptional conditions. The distributor is a typical area of doubt. A physical examination is clearly called for and any action must be decided on the basis of that examination and past experience. It is a matter for engineering judgement. For the record, no action was taken on these failures which were all judged to be one-off unrepresentative failures. The author would have demanded a closer look at the waterproofing.
4. Plugs and leads. Although a number of failures were recorded on these items, no failures were recorded after modified plugs and leads were fitted. It would appear that the manufacturer's modification is satisfactory.
5. Two cases of an oil filter cap seal being ineffective have been reported. This could be a serious fault and hence warrants further physical investigation.
6. There are several instances of one-off failures which may or may not be significant. The reader should look at each of these and decide for himself if he would pursue the matter. The author has considered them all insignificant. In practice it *may* be possible to follow up one-off failures by an engineering appraisal, but in many cases the evidence will have been destroyed. This is not unusual in these circumstances since it is generally more important to keep the prototype running for other reasons. Unless the one-off failure results in highly critical effects, the tests tend to go on and the evidence is lost. The real life development engineer is thus often in little better position than you are in this exercise. It is thus of some importance to ensure a well-informed decision can be taken *at the time*.

There are some general and important lessons we can learn from this exercise, which can be summarised as follows

1. It is often difficult to decide whether a single failure is significant or not. Remember you are looking at only a small fraction (less than 5 per cent) of the total data reported on these trials, so you are giving them a more concentrated attention than could be expected in practice.
2. Even recurring failures can be classified as 'trivial' and possibly discarded from further consideration. Note how many throttle clamping bolt failures were often so described, even though this is the most often recurring failure mode.
3. Virtually all the evidence came from the three vehicles operated on an acceler-

ated test programme by the trials unit. Although these vehicles probably received rough treatment, there was no overload testing involved. The practical problem is to accumulate sufficient running time. The second three prototypes (Nos 4, 5 and 6) provided relatively little data, even though operated by the trials unit itself because insufficient testing had been accumulated by the end of the trials because of lack of resources. The necessity and difficulty of adequate testing is well brought out by this data. Simply producing prototypes is not enough: the support resources must also be available.

4. We may well query whether value for money was obtained from the prototypes in the hands of actual users. Although real conditions were undoubtedly achieved, the low mileage recorded detracts substantially from the potential value of such tests. However, they do, in this instance, confirm that the major failure modes experienced by the trials unit were also present in the users' hands, and hence that they are significant and detailed analysis is justified. To use 19 of the 25 prototypes in this way does, however, seem a gross extravagance. It should be noted that, on the other hand, the trials unit did not have the resources to properly test the vehicles allocated to it. So retaining more prototypes in the trials unit may not lead to more testing, and in these circumstances it was probably the right thing to ensure that what testing was possible was representative. The main lesson here is that extensive representative testing is essential for successful prototype trials, and the resources must be made available: otherwise the duration of the test must be increased.
5. These trials bring out the latter point: the time-scale for purposeful trials is a long one. These trials took one year to complete. It is clear that only the minimum information applicable to a fraction of the total life was obtained. To be sure that no major failure is likely to arise in the major portion of the equipment's planned life would require very prolonged testing: at least three times that actually undertaken.
6. Finally, we come to the crunch point: what effective use can be made of the conclusions? The reader can now be told that these prototypes led to the production vehicles on whose subsequent service experience figure 10.5 was produced. The major source of service failures as recorded in that figure is excessive play in throttle linkage, which turns out to be clamping bolts loosening! The very fault which was identified in the prototypes as a major source of failures becomes the major source in service. Historically, minor modifications were made after the trials to rectify the fault instead of the proper redesign surely demanded by the low value of $\beta$, indicating a totally inadequate initial design. This assumes $\beta$ was evaluated at the time: most likely not! The modifications were put straight into the service vehicle without retest – for the usual reason that production could not be held up. All so typical! Even more appalling, the seventh most common cause of in-service failure was broken or seized accelerator cables, which had been identified as the second most frequent cause of failure in the area covered by this exercise during the prototype trials. Again no action was taken in spite of positive evidence being available. (The other

sources of service failure lying between these two were not covered by the area of the exercise.) If these two areas alone had been properly dealt with, some 20 per cent of the in-service failures would have been eliminated. It does not seem to the writer to be beyond the wit of man to design a clamping bolt that clamps or to ensure a control cable is properly lubricated during manufacture. This exercise is typical of so much experience. The most recurring failures are trivial ones, and the lessons of prototype test data are often ignored – or not properly acted upon for what, at the time, are 'good and valid' reasons.

7. There are less decisive areas. We were suspicious of the distributor, and in-service experience confirms that it is still giving trouble – see items 15, 17 and 19 on figure 10.5, accounting for 3.7 per cent of all failures. It should be noted though that the failure modes encountered in service differed from those experienced during the prototype trials. Although cleared in the prototype trials, plug leads (item 11) still account for 2 per cent of all failures. Cracking of the oil cooler pipes, accounting for 1.5 per cent of all service failures, was not reported at all the prototype trials. One wonders why?
8. There seems to be some anomaly in that prototype No. 4 covered a reasonable distance but recorded few failures. This would warrant further attention, because it could imply complacency at the time when the tests were assessed. For example, how was it used? Loading conditions are only recorded when failure occurs, which is surely inadequate. We need to know the whole load cycle for proper assessment.

It has only been possible to cover the major aspects in the suggested solution to this exercise and the additional comments. The reader is strongly advised to have another look at the exercise. There is a lot to be learned from this minor, but typical, trial.

*Exercise 10.20*

A development programme for the bicycle specified in the exercise is given below, broken down into the various steps essential in planning any development programme.

*1. Failure modes and effects analysis*

The basis of any development programme is an FMEA, and if this has not already been carried out as part of the design, it is the first step to be taken. It will in turn be based on checklists applicable to the item. At the preliminary stage with which we are here concerned, a detailed design is not available and so the FMEA will be limited in scope and need not be based on numerical techniques. It is very important, however, that the FEMA recognises integral sub-systems which must be treated as entities in the development programme – see choosing the starting level of FMEAs in sub-section 10.3.5. Applying these concepts to the bicycle gives the following FMEA which is appropriate for this stage in the planning.

| *Component* | *Failure mode* | *Effect* |
|---|---|---|
| Frame | Stress-rupture fracture | Catastrophic – involves safety |
| | Fatigue failure (particularly at the joints) | Catastrophic – involves safety |
| | Bending or buckling | Detracts from performance. May lead to failures in other parts such as gears, brakes, etc. because of misalignment of parts. Could involve safety if excessive |
| | Any of above | Redesign of the frame to avoid any of the above failure modes must hold up development. Any major fundamental redesign requires a review of the whole concept |
| | Corrosion | Detracts from appearance; possible failure if excessive |
| Handle bars and front forks | Loosening of adjustments (for example, bars to forks) | Catastrophic – involves safety |
| | Stress-rupture failure | Catastrophic – involves safety |
| | Fatigue failure (particularly at joints) | Catastrophic – involves safety |
| | Seizure of bearings | Catastrophic – involves safety. Could affect frame design |
| | Any of above | Redesign of the forks could have major effects on the development programme and possibly reflect on the whole project |
| | Wear of bearings | Gradual loss of steering and stability |
| | Corrosion | Detracts from appearance; possible loss of adjustment, possible failure if excessive |
| | Loosening of grips | Loss of ride comfort |
| Brakes | Cable fracture | Catastrophic – involves safety |
| | Yielding of cable | Loss of braking power |
| | Wear of block | Gradual loss of braking power |
| | Malalignment | Loss of braking power |
| | Block/rim incompatibility | Excessive wear of block, loss of braking power |

| | | |
|---|---|---|
| | Excessive friction | Loss of braking power |
| | Loosening of fasteners and adjusters | Loss of braking power at first, but catastrophic if allowed to develop |
| Wheels | Buckling | Unsatisfactory ride; loss of braking power; requires extra pedal effort |
| | Loss of spoke tension | Leads to buckling |
| | Loss of spoke anchorage | Leads to buckling |
| | Brake block/rim incompatibility | Excessive wear of block; loss of braking power |
| | Wear of bearings | Requires extra pedal effort |
| | Loosening of wheel nuts | Could be catastrophic – involves safety |
| | Fatigue of hub | Could be catastrophic – involves safety. Could affect frame design if hub dimensions are changed |
| Saddle | Loosening of adjustments | Could be catastrophic – involves safety |
| | Creep of springs | Loss of comfort |
| | Wear | Detracts from appearance; possible loss of comfort |
| Drive line (to include bottom bracket, its bearing, and rear hub bearing) | Wear (of all parts) | Increased friction and hence extra pedal effort; if excessive could lead to failure of drive line and immobility |
| | Loss of adjustment | Increased friction and hence extra pedal effort |
| | Malalignment | Increased friction and hence pedal effort |
| | Fatigue failure (of any part) | Vehicle immobile |
| | Stretch of chain | Increased friction and hence pedal effort; possible inability to change gears |
| | Failure of change mechanism | Loss of traction, but could be catastrophic if gears jam – involves safety |
| | Inadequate lubrication | Increased friction |
| | Cable failure | Inability to change gear |

| | | |
|---|---|---|
| | Any of above | Could have serious consequences on development costs and time, but not thought to jeopardise project |
| Fasteners | Work loose | Varies from minor inconvenience to catastrophic. Remedial action not difficult |

It should be noted that in the above list considerable integration of the components and of the failure mechanisms has been adopted. For example, the failures of the gear-change mechanism could have been listed in greater detail. However no advantage is seen in doing this at the preliminary design stage – the object being to identify the failure modes which are likely to endanger the whole project and which will need special development attention. The FMEA would, of course, be updated as the design progressed, so some modes would be given greater importance and others less.

*2. General test requirements*

It is suggested that the average use of such a bicycle would be half an hour per day – a greater use would justify a motorised version – giving about 125 hours usage per year. This could be accommodated in one week of accelerated rig testing. A five-year planned life would then require 5 weeks of such testing: this is thought to be a reasonable target for development programming.

In order to obtain the load cycle for any test, an existing bicycle should be instrumented and ridden over typical routes. Defining what is 'typical' is likely to prove the most difficult part of this preliminary work.

*3. Development programme*

(i) Even before design begins, it is necessary that tests be carried out on any new material to be used in the new model and for which adequate data is not already available. Although material scientists are producing many new materials, they often fail to produce the data for engineering design. It is necessary that the data refer to the material in its finished form; for example, as tube for the frame, to quote one example – not as test pieces.

(ii) The FEMA shows that the frame is the critical item. An experimental unit should be tested as early as possible in the programme. It is for consideration how many frames are necessary. At least six would be necessary to establish strength variation. Unless multiple rigs were available, the author would suggest that the time scale for testing a representative sample is going to exclude this approach. It would be preferable at the early stages of the programme to test just one sample well beyond the planned life to establish the strength reserve, leaving any consideration of strength variation to be verified on complete prototypes. This may not be an ideal course of action when the ability of the production process to hold specified levels is one of the major unknowns. On the other hand, it is extremely doubtful if early samples will conform to the production variations. Indeed some care will have to be taken to ensure that test units are manufactured to standards which are as close as possible to the final production version. This is seen as the major source of 'unrepresentative' development findings. The author would therefore prefer the thorough testing of one specimen. This would first be loaded under static conditions. The stress levels and deflections at critical points would then be measured and checked against the design values to ensure that the design was basically sound. If all was well, the frame should then be loaded dynamically (with computer-controlled hydraulic jacks), using the load cycle determined at 1 above, and testing continued to failure. The life to failure must be well in excess of the planned life (a factor of 5 or even 10 is not unreasonable), and testing must continue, if necessary in parallel with other work, to determine this life and the nature of the failure. Failure is here defined as any fracture or any deformation outside the design limits. If the testing should show the frame to be unsatisfactory the design will have to be revised, of course, and retested. Should a series of tests prove unsatisfactory the whole project may have to be reconsidered, but the minimum financial investment will have been made.

(iii) Similar tests to prove structural integrity should be carried out on the front fork. The proving of bearing reliability must be left to the prototype stage, since it involves integration with the frame.

(iv) A drive line, which includes the bottom bracket and its bearing, chain wheel, chain, the Deraileur gears, and rear wheel and bearing, should be set up in a small rig using a second test frame and tested as a whole. The interaction between the parts of the drive line demands an integral test – tests on the individual components would have little meaning. The rig need consist only of two jacks driving the pedals, with the rear wheel loaded by a dynamometer programmed to the load/speed cycle determined at 1 above – the so-called 'rolling road'. Means must also be provided to change the gears in a typical sequence. This is straightforward for Deraileur gears but more complicated for hub gears. Besides noting any obvious failures, measurements would be taken at short intervals on all mating parts to determine the wear rate. As for the frame and front forks, testing would be carried out to failure. Repair of failures and continued testing to promote other likely failure modes are desirable in this case.

(v) The final critical sub-assembly is the wheels. Since novel features would be common to the front and rear wheels, and since the rear wheel would be included in the drive line, no further test is called for. Special attention must be given to this feature in its own right, as well as part of the drive line, in evaluating test data.

(vi) Assuming that the rig tests at 3(ii), 3(iii), and 3(iv) are satisfactory, 20 complete prototype bicycles would be assembled. All 20 prototypes would be subject to full inspection and quality evaluation before being freed for test purposes.

(vii) Six prototypes would be subject to normal use, though on an accelerated basis. At the equivalent of six month's use, each of these would be completely stripped and all parts subjected to a full inspection. Departures from the 'as new' condition would be recorded. If any significant departures were recorded the time between inspections would have to be reduced, and if serious failures were encountered the whole programme would have to be revised. Any redesign action demanded by the results would, of course, be taken as quickly as possible and incorporated in all the prototypes under test.

(viii) Another six prototypes would be subject to normal use though on an accelerated basis, but these would be subject only to a cursory inspection to allow testing to go on faster. The results of these tests and those under 3(vii) would be combined to give a sample of 12 for appraisal of variations in strength attributes, and hence life arising from the basic material variations or more likely from variations in the production process.

(ix) Six prototypes would be subject to rough usage and moderate abuse. They need receive only cursory inspection unless significant defects were revealed. Rough usage should include extreme climatic conditions.

(x) The two remaining prototypes would be subject to artificially imposed vibration and shock. This could be on a continual basis and is intended to reveal inadequate fasteners etc. ahead of tests 3(vii), 3(viii) and 3(ix). Any defects would be rectified on the prototypes used for evaluation under real conditions.

(xi) After vibration and shock tests, the above two prototypes would be used for very extreme environmental or climatic tests – that is, extremes of heat, humidity and cold – to establish both corrosion resistance and performance under extreme conditions. It is here assumed that those prototypes tested under 3(vii) and 3(viii) will be subject to normal environmental variations, and those under 3(iv) to harsh conditions, but since environmental conditions are a notorious source of trouble, which should never be allowed to occur, additional tests under the most extreme conditions are considered necessary. Those proposed here are, in fact, crude overload tests, which could be extended if time and effort were available.

There are some further factors which need to be considered. Manufacturers usually produce more than one size of frame. It is thought that any scale effect will be negligible in this case and only one size of prototype is necessary. However, if more than one frame style is considered, a similar series on each style is required. Further complications arise if a multiplicity of saddles and handlebars and so on is to be produced. Strictly all combinations should be tested, but this is a counsel of perfection, and some limit must be put on the scope of the test work. Although not all combinations may be tested, it is essential that every variation of a component be tested and proved in one version of the bicycle.

Finally it should be noted that space limitation has prevented a detailed specification of the test programme in this book. Thus the detailed measurements to be taken during the drive line tests have not been specified, for example. Even so this outline specification does bring out the magnitude of a thorough development programme, and hence its high costs, in time as well as money. Yet without such a programme the manufacturer would have no idea of the reliability to be expected. In the real life, short cuts have often to be taken. The reader might well consider in this case whether the drive line tests are essential, the conditions under which they might be omitted, and the consequences of so doing.

*Exercise 10.21*

a. It is essential to bring the specimen into the correct condition for the test path A in figure 10.18. It is readily seen that even if one follows a parallel path to this from zero time, the fatigue zone would hardly be encountered in any practical sequence.
b. One would replace the component with a new one and continue with the test to locate any other failures that lie on the test path.

*Supplementary exercise 10.21A*

Accepting the above, how would you then proceed for a very-long-life component (take 10 years, for example) for which the pre-test sequence is not possible? (The solution to this exercise is given at the end of this chapter.)

*Exercise 10.22*

Let us assume that the overload test is a valid one. The test results can be expressed as a stress-life, represented by the total number of rounds fired, for each failure. It will be convenient to express this as a percentage of the rated life as in the table below

| *Failure* | *Stress-life (per cent rated)* | *Mean rank* |
|---|---|---|
| 1 | 138.1 | 7.7 |
| 2 | 151.3 | 15.4 |
| 3 | 162.9 | 23.1 |
| 4 | 174.4 | 30.8 |
| 5 | 183.9 | 38.5 |
| 6 | 189.8 | 46.2 |
| 7 | 198.8 | 53.9 |
| 8 | 200.5 | 61.6 |
| 9 | 210.1 | 69.3 |
| 10 | 215.2 | 77.0 |
| 11 | 220.2 | 84.7 |
| 12 | 235.6 | 92.5 |

These results can be plotted on Weibull paper to give a straight line which can be extrapolated backwards to obtain a failure rate of 1 per cent at 100 per cent stress-life, that is a reliability of 99 per cent. This would apparently meet the specification.

One must, however, consider the validity of the test and subsequent calculation. A failure mode can obviously be associated with overheating of the barrel – the test schedule allowed the barrel to cool between each burst. In these circumstances ambient temperature must be an important parameter controlling the failure process. If this is so, it should also have been increased during the overload testing. It may have a critical influence. It is not possible to say from the given data what the effect of temperature would be, but one would expect it to be an adverse effect and so would treat the results of the above test with a good deal of suspicion. It is thus not possible to say that the rifle meets the specification.

You are asked to note that this is an artificial example in which the data have been arranged to give a straight line on the Weibull plot. In practice a curved line would probably be obtained. Do remember that there is no fundamental reason whatever why test data should correspond to the Weibull distribution. The techniques discussed in section 2.2 would have to be employed to obtain a straight line.

Neglecting any doubts of the validity of the test, we can assume that the life at 100 per cent rated stress time is the true life. With no overload testing, the failure points would be expected to lie on a straight line passing through this point and having a shaping parameter of 3.44 (normal distribution). The expected total life of the longest surviving rifle is obtained as 4900 rounds, that is more than double that required in an overload test technique. This illustrates the practical value of overload testing.

*Exercise 10.23*

The data of figure 10.20 has been plotted on figure 15.28 as a Duane diagram on log-log scales. The first point has been omitted on the grounds that failure patterns had not developed in the short distance to which it refers. The points can be represented reasonably by the straight line shown on the diagram, even though there are signs of a sequence of rapid growth followed by slow growth (more noticeable if the first point were included). The slope of the line, that is, the value of $\alpha$, is about 0.6 – which suggests a fair effort was being put into the achievement of the target reliability in this case, if the conventional interpretation of the $\alpha$ values is accepted. The specified value of the MBTF for this vehicle is 6000 miles, which is, of course, the instantaneous value. The instantaneous MTBF line has also been plotted on figure 15.28. Extrapolation suggests the target will be achieved by some 520 thousand miles cumulative testing, that is, after about three times as much development effort as has already been undertaken.

How far is this conclusion justified? The straight line drawn has ignored the first point on the basis that the Duane curve cannot represent early data, as noted. But how far we can ignore early data is unresolved, and it is clear from figures 10.20 and 15.28 that any allowance for that data will reduce the slope of the line

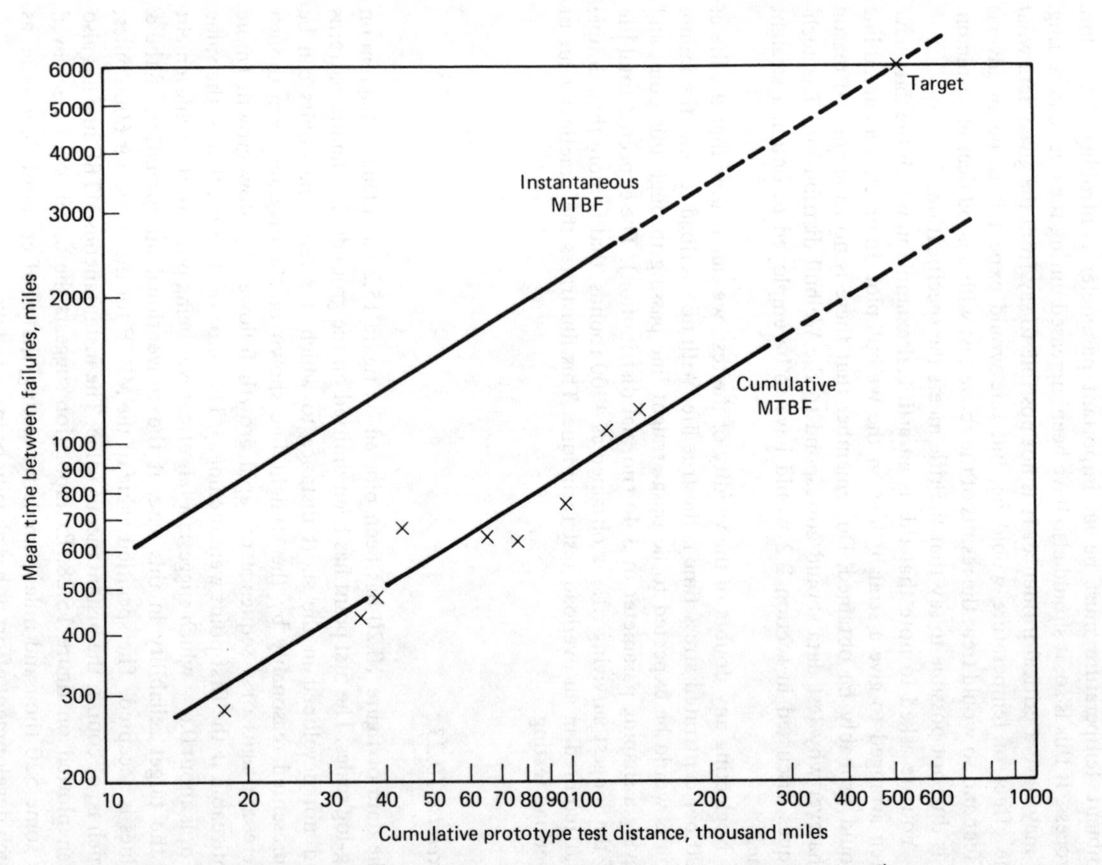


Figure 15.28 Duane plot of data on figure 10.20 (first point omitted)

and increase the time necessary to achieve the target MTBF. Considerable errors are possible. More seriously, the initial MTBF is well below a quarter of the target value one might reasonably expect from a new design. Two explanations are possible: either the specification is reasonable and the design is poor, or the design is adequate but the specified target has been set unrealistically high. In the former case we may wonder if the high level of achievement ($\alpha = 0.6$) is merely a reflection of an easy task in the initial stages and will not be maintained as the going gets harder, or in the latter case will it ever be possible to achieve the target – a lot of extrapolation is required. The author would need much more engineering data before reaching any conclusion, and believes it would need a very brave man to predict the successful outcome of this development programme solely on the basis of a Duane analysis. Nevertheless we see that the Duane representation is plausible. The problem is extrapolation. For example, on receipt of the eighth reported data (including the first) the best straight line would have a lower slope and therefore suggest much more development time would be necessary to achieve the target. We are now led to believe that is untrue. Or will subsequent points follow the earlier data? On the other hand, after receipt of the fifth reported data we could have been much more optimistic: we now know it would have been unjustified.

*Exercise 10.24*

The total number of failures recorded in the prototype trials was 50, excluding plug and lead changes, which are supposedly rectified. The total distance covered in the trials was 211 332 km. Now it seems reasonable from an engineering point of view that the failures arising from the faulty design of the throttle clamping bolt, the lack of lubrication in the throttle cable, and the leaks in the filler cap seal could all be cured. The remaining failures are all isolated ones, and it will be assumed that these will also be experienced in service. Assuming also that they are random, the failure rate will remain constant. Note that any marked wear-out arising at a later life could invalidate this assumption and this underlines the necessity of covering the specified service life in the development trials. On the basis of the above assumptions the number of failures in the same distance would be reduced from 50 to $(50 - 21) = 29$. The total distance covered in the tests was 211 332 km, so that the failure rate is $50/211\,332 = 0.000\,24$ per kilometre, and when the above failure modes are removed it becomes $29/211\,332 = 0.000\,14$ per kilometre.

The author believes that an assessment following along the above lines is in every way as simple as a Duane analysis. It is more rational in that an engineering judgement is brought to bear on each failure mode and the time necessary for rectification can be assessed in the light of the resources available and the workload on hand. Furthermore, an allowance can be made for the anticipated technical difficulty of each problem. It does, of course, demand technical experience. Such decisions are usually made by reliability 'sentencing' or 'scoring' committees, comprising project managers, design engineers, test personnel and perhaps the user. The difficulty is to obtain agreement when non-technical matters intrude, such as a manufacturer trying to have a single incident (though possibly a recurring fault) declared not a failure so as to meet a specification and hence obtain an order. (By the time it is proved to be a recurring fault, the order will have been placed!). Factors such as these have brought disrepute on this essentially simple and basically sound technique. Sound engineering judgement is required to make it work, and must be the sole arbiter.

*Exercise 10.25*

Of course, because of some design features not achieving their objective, some random failure modes may be encountered in prototypes. Clearly, standard growth models represent these modes. However, in the majority of cases where the design does not achieve its objective the result is premature wear-out rather than random mode failures – that is, the mode mechanism is still wear-out, not stress rupture.

The author must confess to having some inability to understand what constant failure rate growth models are measuring in the presence of wear-out. If failures only up to the $B_5$ or $B_{10}$ age are included, then the models may just represent the onset of wear times (that is, intrinsically reliable lives). These are most likely to be randomly distributed in a multi-component equipment. But surely a development programme is trying to establish the whole wear-out pattern of all (ideally) components as a basis for further action, and the extent to which this is achieved nullifies the conditions outlined above in which a constant failure rate model is applicable. Alternatively, the conditions may be satisfied by only evaluating the onset of each wear-out mode. Crow seems to be following this approach in ref. 70 when he uses equation (10.22) to represent the "mean number of *new* Type B modes" (my italics) – Type B modes in ref. 70 are those on which development action is to be taken. If that is so then the growth model is evaluating the rate of onset of modes, not failures. It would seem to be a perfectly valid approach, but the service failure rate must be evaluated from a knowledge of the failures rates that the specific modes will entail. The expected service failure rate will not be equal to the intensity function at the end of development. However, the failure rate will continue to increase with time as new modes are encountered at a rate given by the intensity function at the end of development. This interpretation would be in accordance with the Clapham/Hastings model for repair costs, but it would seem to the author that great care is necessary to ensure that the raw data is processed into a format yielding random phenomena before such models can be used, and to interpret the findings of the model in a clearly defined manner which reflects the nature of the model, and which is fully understood by all concerned.

If wear phenomena have developed significantly and the data are not selectively analysed as above, then the assumption of constant failure rate must break down. Even if the pseudo-random state is reached (unlikely in a development programme) it is the occurrence of failures which is constant, not the failure rate. It is fundamentally bad statistics to confuse these quantities even if in some cases the numerical error is small. If pseudo-randomness had not actually been reached so that transients in the occurrence of failures with time have not disappeared, errors in projections to greater ages by means of constant rate growth models could be very large, of course – and in either direction. Furthermore, if a system having a large number of components is repaired, the condition after repair (involving no development action) depends on the failure modes of all the other components. Thus if all the components have true random failure modes (for example, most electronic equipment) then the state after repair is 'good as new'. This is what most models model. But if the other components have wear modes (for example, most mechanical equipment) then the condition after repair is 'bad as old'. Consequently the former system would show zero reliability growth if the design and manufacturing process remain unchanged (that is, no development) but the latter would show a negative growth in the same circumstances – that is, there can be a variation of reliability with time in the absence of any development action at all. To assess the real growth during the development of mechanical equipment experiencing wear, allowance must be made for this.

It is also important to note that the relevant age to which the deduced failure rates refer is not the cumulative test time but the actual age of the individual prototypes, that is, roughly the cumulative test time divided by the number of prototypes. This distinction was made clear by using $T$ and $t$ for the distinctive distributions in the earlier analysis. Thus $t \approx T/N$. Wear-out may occur subsequently, that is, in service which extends beyond the development test duration, in entirely new modes. The Comet aircraft may be taken as a classic lesson.

*Exercise 12.1*

The first step in planning any reporting system is to define its objective. Realistically, in this case, many repairs will be undertaken by those not participating in the reporting system, including many individual owners. The best one could hope to achieve in circumstances such as these is a rudimentary Pareto analysis. This is not to be despised. In the author's opinion a properly designed and executed Pareto analysis is nearly always the most powerful initial technique, even for the most sophisticated equipment. In this case, then, the simple proposed reporting code is considered quite adequate. More elaborate reporting codes giving the detailed nature of the failures etc. could not be analysed with any degree of confidence on account of the missing data. The inability of the average repairer to assess failure mechanisms, and the unnecessary form-filling probably leading to even more corrupt data, are additional reasons for not extending the coding. A major positive feature of the proposed coding is that the main defect of nearly all failure-reporting systems – no information on the loading – has to some extent been avoided by the subdivision of the primary faults. Although the information will be highly subjective, it does give a first-order guide to the treatment that the bicycle has received. However, one doubts if the average repairer could differentiate between manufacturing and design faults, under codes 08 and 09. Combining these under one code may be more advisable. It is also worth noting that in this code the cause of failure has to be indicated from a given set of possibilities. It assumes that the person reporting the failures can ascertain the source of failure. This may not be possible, but in spite of this a code number has to be quoted. It could lead to guess-work, and spurious statistics. It may be more satisfactory to have a separate code labelled 'undetermined', to be used in all cases of doubt. It should be appreciated, however, that this could be used on an excessive number of occasions in order to save examination of the parts and subsequent thought. On balance, the author would accept this disadvantage for the sake of greater accuracy in the positively identified failures.

Although the nature of the use of the bicycle has been at least subjectively recorded, the amount of use (that is, miles) cannot be given (unless the machine is fitted with an odometer as standard, in which case its reading should be called for). This information is not required for the simple Pareto analysis envisaged earlier, but would be useful for assessing wear. On the other hand, the frame number called for in the report does allow failures to be related to the quality control documentation of the manufacture.

It is concluded that, maybe with some minor modifications, the proposed coding meets the requirements of the simplified analysis envisaged, and should allow the manufacturer to 'home-in' on any sub-standard part for a more detailed engineering analysis.

The reader is asked to consider if the data so collected would be any better than that obtained from the numbers of spare parts supplied to dealers. The latter data is often readily available and certainly cheaper to obtain. Analysis can then proceed on the basis that spares supplied equal failures, once a steady state has been reached. For what it is worth, the author has found that such analyses can be misleading, but it may provide a rough first approximation when no other data is available.

*Exercise 13.1*

This is, or should be, an irrelevant question. It is not the job of the quality manager to get involved in the detail or even schematic design of any project. He has to ensure that the organisation is present in the firm or institute to achieve the quality that is sought, to ensure that that organisation is adequately staffed, and that all sections are functioning properly. The guidelines are his responsibility, but these should not inhibit design. The quality manager may demand formal Design Reviews as part of his quality plan, and it is probably here that friction is most likely to arise. It must be remembered that design is a creative act. It is not the result of any mechanical or systematic process. Consequently designers tend to regard criticism of the design as criticism of the designer. On the other hand, while designers can be quite sensitive about their creations, they are usually only too happy to hold an informed discussion about their designs. The sting lies in the word 'informed'. The creative mentality is apt to be very intolerant of the uninformed. It is a great pity that so many in responsible positions have little idea what design is about and how it is carried out. If necessary management must educate. However, it would be idle to pretend that differences of opinion do not exist, or that the resolution of these difficulties is easy when there is a large number of equally acceptable (or unacceptable!) solutions. We then return to the prime function of management, which is to understand what all members of its staff are doing and how they go about it, and to create those conditions in which all staff work at their best – that is, create the team spirit which is essential for success in all but the smallest project.

*Exercise 3.1A*

One might expect the strength of the basic material to be more or less normally distributed. As a rough guide, one could anticipate a hot finished carbon steel of 600 $MN/m^2$ ultimate strength would have a standard deviation of strength of about 20 $MN/m^2$, that is, $\sigma_s$ is about three or four per cent of the mean strength. If the nominal strength is quoted as 600 $MN/m^2$ it would be reasonable to expect that this is some two to three standard deviations greater than the mean, giving a mean strength of about 550 $MN/m^2$. Hence one might expect a distribution of strength which follows the normal distribution with a mean of 550 $MN/m^2$ and a standard deviation of 20 $MN/m^2$. However, these conclusions need to be treated with some reservation for two main reasons

1. An examination of actual distributions shows that there are often significant departures from the normal. Reference 6 contains about the best information on the distributions to be expected from various materials and should be consulted for further data.
2. There is no agreed margin between the nominal quoted strength and the mean strength, and it is very often nowhere near three standard deviations. Indeed the nominal could be quoted as the mean. Some care is necessary in this connection.

The strength of the finished product will vary partly because of the variations in the basic strength of the material, which may be further modified by the manufacturing process, and partly because of dimensional variations. In general, one would expect the standard deviation to be higher than for the basic material. Again one would be reasonable in assuming a basically normal distribution, but the material factors already mentioned and inspection processes on dimensional accuracy could introduce significant variations. In particular, the inspection process is likely to cut off the tail of the distribution, which has a marked effect on reliability. The reader is referred to Conway (reference 83) for some interesting examples of the latter effect, and he could well consider how the distribution shown in figure 3.3 came about.

*Exercise 5.1A*

It is generally accepted that a fatigue crack starts from an intrusion, many of which are scattered at random throughout an average machine component. As it travels down the slip band into the material, the incipient fatigue crack travels more slowly until when it is about 0.01 to 0.1 mm in length it becomes virtually dormant. Well over three-quarters of the fatigue life is spent in this dormant stage, and in certain circumstances the crack may become entirely non-propagating. Exactly what is going on in the material during the substantially dormant phase is not understood at present, certainly changes are only taking place on a microscopic or sub-microscopic scale. There is no physical change in the condition of the item so far as its functioning as a machine part is concerned, and clearly failures do not occur during this portion of the fatigue life. So while the fatigue process starts when the part is first used there is, up to this point, no outward indication of any wear-out process. The final phase, occupying only about 10 per cent of the total life, starts when the crack turns out of the slip plane into a path perpendicular to the tensile axis, producing the characteristic ripple-surfaced crack. This crack is, of course, visible and propagates rapidly until fracture occurs. The period of wear-out is essentially confined to the 10 per cent of fatigue life remaining at completion of the dormant phase. We can thus conclude that, although fatigue processes are active over the whole life, there is no outward manifestation until some time after the item is first used. For practical purposes, therefore, fatigue belongs to category 2.

*Exercise 6.11A*

Plot Weibull line as already described in section 6.4.2 but using 50 per cent median ranks instead of the mean rank. Also calculate appropriate upper and lower median ranks, for example, for a normal 90 per cent confidence limit calculate 95 per cent and 5 per cent ranks. Then for each cumulative failure value, plot the two points as shown on figure 15.29. Repeat for all the cumulative failure values and draw curves through the upper and lower limit points as shown in the figure 15.30.

Some analysts draw common tangents to the upper and lower limit curves as shown by dotted lines in figure 15.30 to obtain the extreme two parameter Weibull distributions that fit the data. However, this does depend on the *ab initio* assumption that the two parameter distribution is the relevant one. As we have seen, most mechanical failure modes have to be represented by a three parameter distribution, of which the neglected locating constant is often the most important one. Such procedures are, therefore, not to be encouraged. Once a three parameter distribution is accepted, the widely different values of the upper and lower limits for the locating constant have to be accepted. These may lie between zero and the first statistic for any reasonable level of confidence, for example. The significance

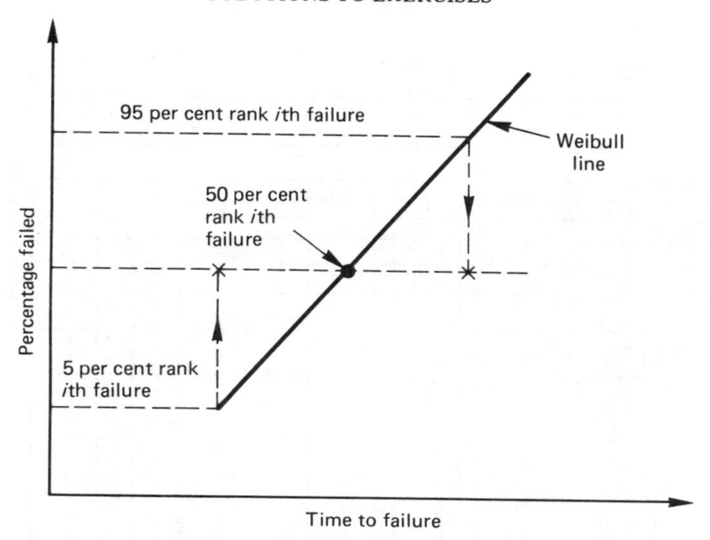


**Figure 15.29** Method of plotting points to give confidence interval

of confidence limits for three parameter Weibull distrbutions is therefore very doubtful.

It is also important to note that the method of estimating the confidence limits is invalid if any of the data is censored or contains drop-outs. When we re-ordered the failure data to allow for drop-outs, as in section 6.4.2, the revised order values are the best expected estimates – that is, the 'mean' values. It is not possible to calculate the limits from the expected order numbers. Hence, since most mechanical field failure data contain a large number of drop-outs and should be assumed to require a three parameter function unless proved otherwise, it is virtually impossible to quote confidence limits. It follows that, since the only virtue in using median ranks is the ability to calculate confidence limits, there is little point in using them for analysis of most mechanical failure data. On the contrary, there is good reason for basing the analysis of such data on the mean rank, as has been done throughout this book. The lower *B*-life values estimated from a median rank analysis are always higher than those estimated from a mean rank analysis. Hence a median rank analysis would indicate a higher scheduled maintenance interval than a mean rank analysis. If there is any doubt, the lower value is safer in that it is likely to incur the least cost error, and so this is the preferred method from that standpoint. Both methods give practically the same mean life estimate, so there is no reason to adopt one rather than the other on that score. The higher estimate of high *B*-life values given by a mean rank analysis is not thought to have any significance for reliability evaluation. Thus, in general, a mean rank analysis is to be preferred to median rank for mechanical reliability analysis.

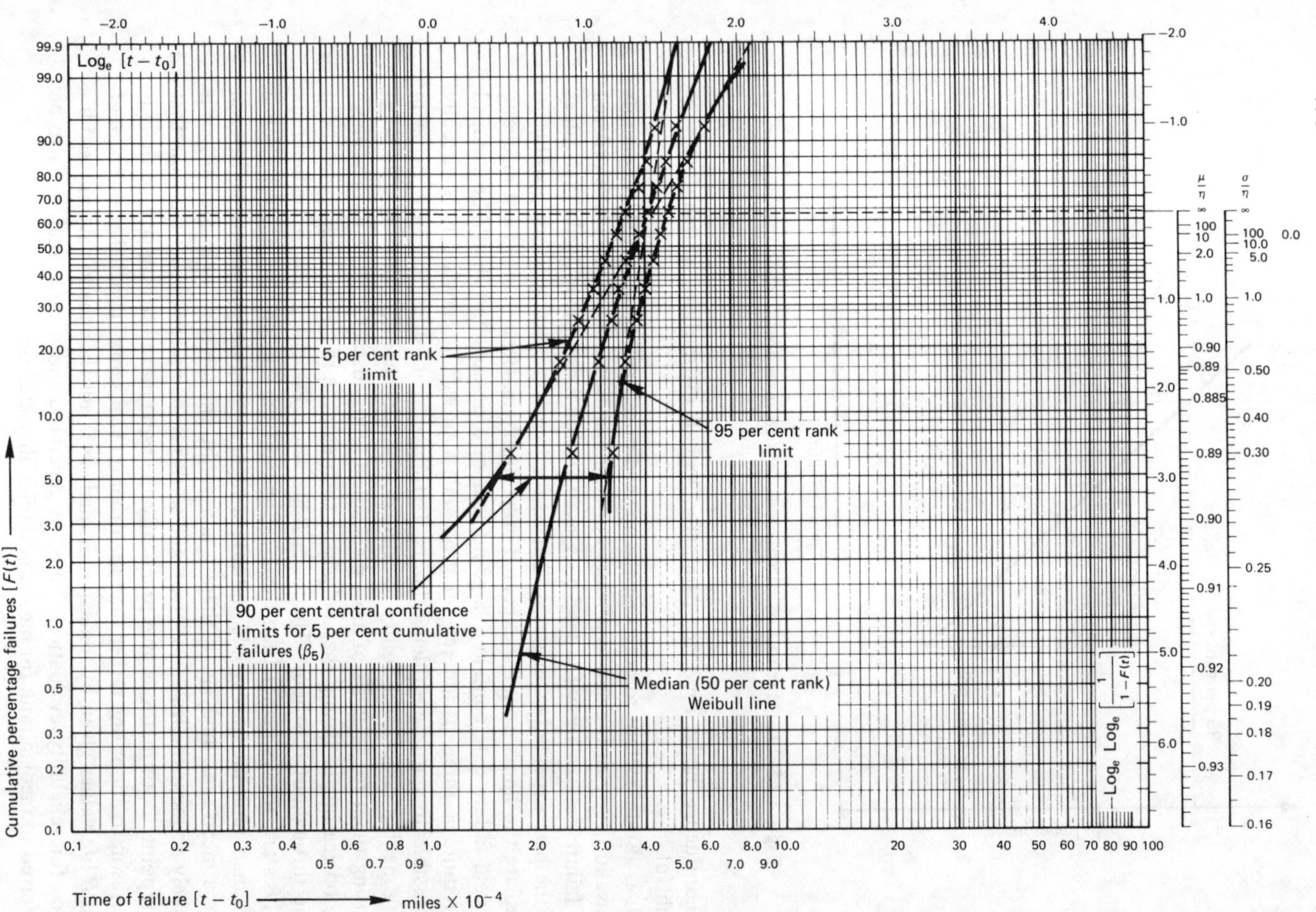


**Figure 15.30** Plotting of 5 per cent, 50 per cent (median) and 95 per cent ranks for exhaust system data given in chapter 6 (sub-

*Exercise 7.1A*

Failure – the system is not functioning as designed. Note how environment determines how a system can function and that success or failure is not an absolute property of the equipment.

*Exercise 10.1A*

You should do what every good secretary does and carry a spare pen. In this particular case $\bar{R}$ = 95 per cent and $R$ has been specified as 99.5 per cent, so that substituting in equation (3.6) we obtain

$$(1 - 0.9995) = (1 - 0.95)^n$$
$$\text{giving } n = 2.32$$
$$= 3 \text{ expressed as whole pen}$$

It follows that it would be necessary to carry two spare pens to achieve the required reliability.

The use of spare or standby units is a well-known and standard technique of achieving very high reliability from systems which may be built up of components of only moderately high or even fairly low individual reliability. For example, multi-engined aircraft are always considered more reliable than single engined aircraft. Full power is only required at take off under full load and hence, for most of its operating life, spare engine capacity is available and can be called for in the event of failure. A more direct illustration is the spare wheel carried by all automotives. Other examples may spring to mind. As an extension, a whole unit may be duplicated; for example, standby generators are always installed in hospitals. The technique can be very powerful, as simple substitution of numbers in equation (3.6) will illustrate. Thus two components each of 90 per cent reliability can give a combined reliability of 99 per cent, and two of 80 per cent reliability can still achieve a combined reliability of 96 per cent. Using more standbys can yield very high reliabilities indeed. Technically it is referred to as parallel redundancy. It is discussed more fully in section 11.5.

*Exercise 10.19A*

In this case the pre-test sequence is physically just not possible and one is forced to accept a reduced pre-test. In these circumstances a single overload sequence of tests along the direction A in figure 15.31 would, if continued long enough, reveal all types of failures. Such a test would not specify the likelihood of these failures being encountered in service. At this stage, common sense and experience can be called upon to decide if the failures warrant attention. Alternatively, additional pre-test sequences can be carried out followed by overload tests along the paths indicated by arrows B and C.

Plotting the failure density distribution as illustrated would give a good idea of

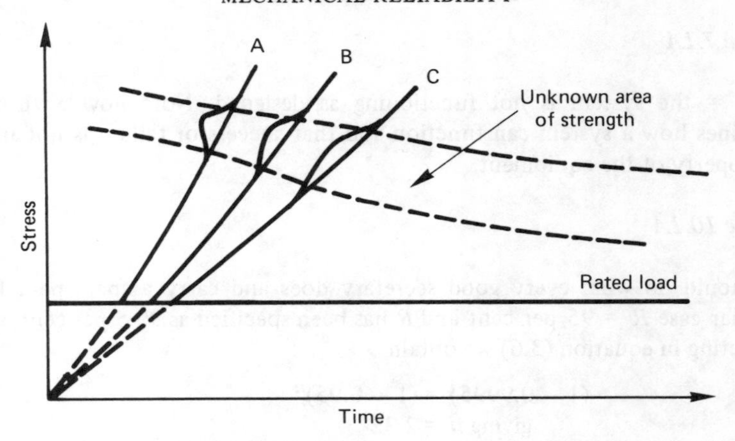


**Figure 15.31** Overload test sequence for a very-long-life item

the critical strength area and, taken in conjunction with a critical appraisal of the failure, it should enable a possible failure to be anticipated to some extent. A more refined technique would be to determine the failure rate at 100 per cent rated load for each of tests A, B and C, using the technique illustrated by exercise 3.6. This can be plotted against the time of pre-test, and extrapolated. There is clearly an element of doubt over this procedure, but it is better than no evidence at all. Obviously, the doubt increases with the amount of extrapolation.

It may be possible at a later stage to withdraw some items from service and subject them to further overload testing. In this case, actual service forms the pre-test sequence and the variation of strength with time can be determined up to the time of test. Once again extrapolation is necessary. The main difficulty in this situation, however is that the type of equipment which has a very long life is usually fairly expensive. Items cannot be withdrawn purely for overload testing.

# Reading list


I. Bazovsky, *Reliability: Theory and Practice*, Prentice-Hall, New Jersey, 1962

S. R. Calabro, *Reliability Principles and Practices*, McGraw-Hill, New York, 1962

E. Pieruschka, *Principles of Reliability*, Prentice-Hall, New Jersey, 1963

A. S. Goldman and T. B. Slattery, *Maintainability*, Wiley and Sons, New York, 1964

N. H. Roberts, *Mathematical Methods in Reliability Engineering*, McGraw-Hill, New York, 1964

R. E. Barlow and F. Proschane, *Mathematical Theory of Reliability*, Wiley and Sons, New York, 1965

M. J. Moroney, *Facts from Figures*, Penguin, London, 1965

W. G. Ireson, *Reliability Handbook*, McGraw-Hill, New York, 1966

F. Nixon, *Managing to Achieve Quality and Reliability*, McGraw-Hill, London, 1967

A. Green and A. J. Bourne, *Reliability Technology*, Wiley and Sons, New York, 1972

J. H. Bompass-Smith, *Mechanical Survival – the use of reliability data*, edited by R. N. W. Brook, McGraw-Hill, London, 1973

J. M. Juran, *Quality Control Handbook*, McGraw-Hill, New York, 1974

N. Mann, R. E. Schafer and N. D. Singpurwaller, *Methods for Statistical Analysis of Reliability Data*, Wiley and Sons, New York, 1974

R. E. Barlow and F. Proschane, *Statistical Theory of Reliability and Life Testing*, Holt, Rinehart and Winston, New York, 1975

K. Lloyd and M. Lipow, *Reliability: Management, Methods and Mathematics*, 1st edn, Prentice-Hall, New Jersey, 1962, 2nd edn, published by authors, 1977

J. Shigley, *Mechanical Engineering Design*, McGraw-Hill, New York, 1977

C. Chatfield, *Statistics for Technology*, 2nd edn, Chapman and Hall, London, 1978

U.S. Army Communications R & D Command, Fort Monmouth, New Jersey, *MIL–HDBK–XXX: Reliability Growth Management*, 1978

Ministry of Defence, *Def. Stan. 00–5 Design criteria for reliability, maintainability and maintenance of land service material, Part 1 General requirements, Part 2 Mechanical aspects, Part 3 Electrical and electronic aspects, Part 4 Optical aspects*, 1979

Ministry of Defence, *Def. Stan. 00–40 Achievement of reliability and maintainability, Part 1 Management responsibilities and requirements for R & M programmes and plans,* 1981

R. F. de la Mare, *Manufacturing Systems Economics*, Holt, Rinehart and Winston, New York, 1982

USAF Rome Air Development Center, *MIL–HDBK–217D: Reliability Prediction for Electronic Systems*, available from National Technical Information Service, Springfield, Virginia, 1982

British Standards Institution, *BS 3527 Glossary of terms used in data processing, Part 14 Reliability, maintainability and availability*, 1983

British Standards Institution, *BS 5760 Reliability of systems, equipments and components, Part 1 Guide to reliability programme management, Part 2 Guide to the assessment of reliability, Part 3 Guide to reliability practices: examples*

Ministry of Defence, *Def. Stan. 00–5 Requirements for achieving reliability and maintainability of MGO procured material, Part 0 Management requirements*, 1983

Ministry of Defence, *Def. Stan. 00–41 MoD practices and procedures for Reliability and Maintainability, Part 1 Reliability design philosophy, Part 2 Reliability apportionment modelling and calculation, Part 3 Reliability prediction, Part 4 Reliability engineering, Part 5 Reliability testing and screening*, 1983

P. D. T. O'Connor, *Practical Reliability Engineering,* 1st edn, Heyden and Son, London, 1981, 2nd edn, Wiley and Sons, 1985

D. J. Smith, *Reliability and Maintainability in Perspective – Practical, Commercial and Software Aspects*, 2nd edn, Macmillan, London, 1985

# References

1. R. R. Carhart, 'A survey of the current status of the electronic problem'. *RAND Corp. Res. Memo. RM 1131* (1953)
2. R. Lusser, ***Reliability of Guided Missiles***. US Army Ordnance Missile Command, Redstone Arsenal (Sept. 1954)
3. R. Lusser, ***Reliability through Safety Margins***. US Army Ordnance Missile Command, Redstone Arsenal (1958)
4. F. Nixon, 'Testing for satisfactory life'. ***Symposium on Relation of Testing and Service Performance***. American Society for Testing and Materials, Annual Meeting (1966)
5. F. Nixon, 'Quality in engineering manufacture'. ***Colloquium on Reliability, I. Mech. E., London*** (1958)
6. American Society of Metals, ***Metals Handbook Vol I. Properties and Selection***, 8th edn (1966)
7. D. E. Powell and D. B. Wedmore, 'Establishing Diesel engine reliability'. ***Conference on Reliability of Diesel Engines and its Impact on Cost. Inst. Mech. Engrs., London*** (1973)
8. N. Natarajan, 'Presentation and comparison of duty cycles'. MSc Thesis, RMCS (1974)
9. R. G. Watkins, 'Analysis of duty cycles for tracked military vehicles'. MSc Thesis, RMCS (1975)
10. W. J. Howard, 'Chain reliability – a simple failure model for complex mechanisms'. *RAND Corp. Res. Memo. RM 1058* (1958)
11. P. Hearne, 'Automatic Control – a new dimension in aircraft design'. *R. Aeronaut. Soc.* (1974)
12. J. E. Gordon, ***Structures – or why things don't fall down***. Pelican Books (1978)
13. A. D. S. Carter, 'Reliability Reviewed'. *Proc. Inst. Mech. Engrs.*, Vol. 193, No. 4 (1979)
14. F. Jensen and N. E. Petersen, ***Burn-in; an engineering approach to the design and analysis of burn-in procedures***. Wiley (1982)
15. M. R. Allen, 'The mechanism governing fatigue lives'. MSc Thesis, RMCS (1981)

16. Z. W. Birnbaum and S. C. Saunders, 'A probablistic interpretation of Miner's rule'. *SIAM Journal of Applied Mathematics*, Vol. 16, p. 637 (1968)
17. Z. W. Birnbaum and S. C. Saunders, 'A new family of life distributions'. *Journal of Applied Probability*, Vol. 6, p. 319 (1969)
18. Z. W. Birnbaum and S. C. Saunders, 'Estimations for a family of life distributions with applications to fatigue'. *Journal of Applied Probability*, Vol. 6, p. 328 (1969)
19. J. Leiblein and M. Zelen, 'Statistical investigation of the fatigue life of deep grooved ball bearings'. *Journal of Research, National Bureau of Standards*, Vol. 57, p. 273 (1956)
20. A. M. Freudenthal and M. Shinozuka, 'Structural safety under conditions of ultimate load failure and fatigue'. *WADD Technical Report 61-77* (1961)
21. N. R. Mann, R. E. Schafer and N. D. Singpurwaller, *Methods for Statistical Analysis of Reliability and Life Data*. Wiley (1974)
22. D. Kececioglu, R. E. Smith and E. A. Felsted, 'Distributions of cycles to failure in simple fatigue and the associated reliabilities'. *Proceedings – Annals of Assurance Science*, pp. 357-74 (1969)
23. D. Kececioglu, 'Reliability analysis of mechanical components and systems'. *Nuclear Engineering and Design*, Vol. 19, pp. 259-90 (1972)
24. E. D. Gardner, 'Reliability of components subject to cumulative fatigue'. University of Arizona, Aerospace & Mechanical Eng. Dept., Master's Degree Report (1971)
25. A. N. Kinkead and P. Martin, 'An approach to the fatigue reliability of sequentially loaded demountable joints'. *8th Advances in Reliability Technology Symposium, National Centre for Systems Reliability and University of Bradford* (1984)
26. Engineering Science Data Unit, 'Fatigue life estimation under variable amplitude loading'. *Item 77004* (1977) – see also *items 76014* and *76016*
27. W. J. Evans, 'Fatigue limitations in the design of gas turbine discs'. *Conference on Environmental Engineering Today (SEECO 79). Soc. Environmental Engrs.* (1979)
28. H. J. K. Kao, 'A summary of some new techniques on failure analysis'. *Proc. 6th Nat. Symp. on Reliability and Control in Electronics, American Society for Quality Control, Washington DC* (1960)
29. R. A. Mitchell, 'An introduction to Weibull analysis'. *Pratt & Whitney Aircraft Rep. PWA 3001* (1967)
30. E. A. Lehtinen, 'An estimation for three-parameter Weibull distributions'. *Second National Reliability Conference, Birmingham* (1979)
31. R. F. C. Hill, 'Obtaining hazard functions for Weibull plots'. *J. Naval Science*, Vol. 3, No. 3 (1976)
32. A. D. S. Carter, 'Summary of papers to Conference'. *Conference on the Reliability of Diesel Engines and its Impact on Cost. Inst. Mech. Engrs., London* (1972)
33. T. A. Harris, 'Predicting bearing reliability'. *Machine Design*, Vol. 35, No. 1, p. 129 (1969)

34. British Standards Institution, *Guide on the assessment of reliability of engineering equipments and parts. DD13* (1974)
35. E. T. Allen, 'Reliability engineering applied to the production of a quality motor car'. *Conference on Improvement of Reliability in Engineering. Inst. Mech. Engrs., Loughborough* (1973)
36. A. K. S. Jardine, 'Component and system replacement decisions'. *Electronic Systems Effectiveness and Life Cycle Costing*. Ed. J. K. Skwirzynski. NATO ASI Series, Springer-Verlag (1983)
37. A. E. Green, 'Reliability prediction'. *Conference on Safety and Failure of Components. Inst. Mech. Engrs., Sussex* (1969)
38. J. H. Bompass-Smith, 'The determination of distributions that describe the failure of mechanical components'. *Eighth Annals of Reliability and Maintainability*, p. 343 (1967)
39. C. J. Harris, M. Webster, R. S. Sayles and P. B. Macpherson, 'Bi-modal failure mechanism in tribological components'. *Third National Reliability Conference – Reliability 81. National Centre for Systems Reliability and Inst. Quality Assurance, Birmingham* (1981)
40. R. J. Parker, *Rolling element fatigue lives of AISI 52100 steel balls with several synthetic lubricants*. Heyden & Son on behalf of Inst. Petroleum, London (1976)
41. M. F. Griffey and B. H. Harvey, 'The management and assessment of availability of Diesel engines in the Royal Navy'. *Conference on the Reliability of Diesel Engines and its Impact on Cost. Inst. Mech. Engrs., London* (1972)
42. N. A. J. Hastings, 'A stochastic model for Weibull failures'. *Proc. Inst. Mech. Engrs.*, Vol. 182, Part 1, No. 34, pp. 715–23
43. C. Lipson, 'Wear considerations in design'. Prentice-Hall, New Jersey (1967)
44. *Conference on Lubrication and Wear, Inst. Mech. Engs.* (1967)
45. D. W. Botstiber, 'Equipment reliability through oil monitoring'. *Tedco Bulletin A65B*
46. N. G. Yates, 'Vibration diagnosis in marine geared turbines'. *Trans. NE Coast Engrs. and Shipbldrs.*, Vol. 656, p. 225 (1949)
47. T. C. Rathbone, 'Vibration tolerance'. *Pwr. Pl. Engng.* (1939)
48. Verein Deutscher Ingenieure, *VDI Vibration Chart* (1960)
49. *Conference on Acoustics as a Diagnostic Tool, Inst. Mech. Engrs., London* (1970)
50. J. R. Nutter and D. E. Kirby, 'Trends in power plant maintenance'. *Orient Lines Research Bureau Safety Seminar, Orient Airlines, Tokyo* (1969)
51. F. H. Thomas and M. I. Littlewood, 'Logical aero engine inspection, rejection, and component replacement/failure algorithms'. *8th Advances in Reliability Technology Symposium, National Centre for Systems Reliability and University of Bradford* (1984)
52. A. S. Goldman and T. B. Slattery, *Maintainability*. Wiley (1964)
53. M. B. Kline and N. F. Schneidewind, 'Life cycle comparisons of hardware and software reliability'. *Second National Reliability Conference – Reliability 81. National Centre for Systems Reliability and Inst. Quality Assurance* (1981)

54. F. A. Hodden, R. R. Howard and W. J. Howard, 'Study of downtime in military equipment'. *Fifth Annual Symposium on Reliability and Quality Control. IRE–EIA–ASQC*, p. 402 (1959)
55. A. Z. Keller and N. A. Stipho, 'Reliability analysis of chlorine production plant'. *First National Reliability Conference. National Centre for Systems Reliability and Inst. Quality Assurance* (1979)
56. D. E. Mathews, 'Reliability of vehicles for the distribution of petroleum products'. *Conference on the Reliability of Diesel Engines and its Impact on Cost. Inst. Mech. Engrs., London* (1972)
57. J. C. R. Clapham, 'Economic life of equipment'. *Operational Research Quarterly*, Vol. 8, p. 181 (1957)
58. R. W. Drinkwater and N. A. J. Hastings, 'An economic replacement model'. *Opl. Res.*, Vol. 18, No. 2, p. 121 (1967)
59. V. Metcalf, Contribution on discussion to ref. 57
60. N. A. J. Hastings and D. W. Thomas, 'Overhaul policies for mechanical equipment'. *Proc. Inst. Mech. Engrs.*, Vol. 184, p. 565 (1970–1)
61. Commission of the European Communities, *Report No. COM(83), 350FINAL* (1983)
62. J. Green, 'Systematic design and fault review'. *Conference on Safety and Failure of Components. Inst. Mech. Engrs., Sussex* (1978)
63. P. H. Reed, 'The nature of aircraft and complex system reliability and maintainability characteristics'. *Conference on Reliability of Aircraft Mechanical Systems and Equipment. Inst. Mech. Engrs., MoD, and R. Aeronaut Soc., London* (1978)
64. *Conference on Reliability of Aircraft Mechanical Systems and Equipment, Inst. Mech. Engrs., MoD, and R. Aeronaut. Soc., London* (1978)
65. S. N. Wearne, 'A review of reports of failures'. *Proc. Inst. Mech. Engrs.*, Vol. 193, p. 125 (1979)
66. Joseph H. Levine, 'NASA approaches to Space Shuttle reliability'. *Electronic Systems Effectiveness and Life Cycle Costing*. Ed. J. K. Skwirzynski. NATO ASI Series, Springer-Verlag (1983)
67. H. W. Cerda and D. H. Napier, 'Fault tree analysis and the optimised allocation of funding for safety and reliability'. *Second National Reliability Conference. National Centre for Systems Reliability and Inst. Quality Assurance, Birmingham* (1979)
68. A. C. Lovesey, 'The art of developing aero engines'. *J. R. Aeronaut. Soc.*, Vol. 63, p. 429 (1979)
69. D. Shainin, 'A statistical approach to manufacturing and engineering management'. *Seminar at Rolls Royce Ltd* (1969) and other private seminars
70. J. T. Duane, 'Learning curve approach to reliability monitoring'. *IEEE Trans. on Aerospace*, Vol. 2, p. 563 (1964)
71. L. R. Crow, 'On methods of reliability growth assessment during development'. *Electronic Systems Effectiveness and Life Cycle Costing*. Ed. J. K. Skwirzynski. NATO ASI Series, Springer-Verlag (1983)

72. P. V. White, 'A standardised approach to the quantification of common mode failure probabilities'. *8th Advances in Reliability Technology Symposium, National Centre for Systems Reliability and University of Bradford* (1984)
73. USNRC, 'Reactor safety study'. *WASH 1400* (1975)
74. W. E. J. Haywood and V. Metcalf, 'Army reliability requirements'. *Conference on Reliability of Service Equipment. Inst. Mech. Engrs., London* (1968)
75. G. Miaoulis and V. L. O'Brian, 'The international automobile: the impact of quality and reliability perceptions upon competition. *7th Advances in Technology Symposium, Bradford* (1982)
76. F. Nixon, 'Quality management in Europe's expanding market'. *Internal Management Congress, Combined Intelligence Subjective Sub-committee, New York* (1963)
77. W. A. Shewhart, *'Economic control of quality and manufactured product'*. Van Nostrand, New York (1931)
78. R. V. Austin, 'A corporate framework for quality'. *20th International QA Conference. Economic and Legislative Pressures on Industry – Quality Assurance Involvement. Inst. Quality Assurance, London* (1981)
79. Ministry of Defence, 'Quality control system requirements for industry'. *Defence Standard 05-21 Issue 1* (1973)
80. Ministry of Defence, 'Inspection system requirements for industry', *Defence Standard 05-24 Issue 1* (1973)
81. British Standards Institution, 'Reliability of systems equipment and components'. Parts 1, 2, and 3. *BS 5760* (1979)
82. T. R. Moss, 'The reliability characteristics of process control equipment'. M.Phil. Thesis, University of Bradford (1982)
83. J. R. Mercer, 'Reliability of solenoid valves'. *Conference on Safety and Reliability of Components. Inst. Mech. Engrs., Sussex* (1969)
84. H. G. Conway, *Engineering Tolerances*, 2nd edn. Pitman (1962)

# Index